TROPICAL FORESTS IN PREHISTORY, HISTORY, AND MODERNITY

Tropical Forests in Prehistory, History, and Modernity

PATRICK ROBERTS

OXFORD
UNIVERSITY PRESS

OXFORD
UNIVERSITY PRESS

Great Clarendon Street, Oxford, OX2 6DP,
United Kingdom

Oxford University Press is a department of the University of Oxford.
It furthers the University's objective of excellence in research, scholarship,
and education by publishing worldwide. Oxford is a registered trade mark of
Oxford University Press in the UK and in certain other countries

First Edition published in 2019

Impression: 1

Published in the United States of America by Oxford University Press
198 Madison Avenue, New York, NY 10016, United States of America

British Library Cataloguing in Publication Data
Data available

Library of Congress Control Number: 2018950701

ISBN 978–0–19–881849–6

Printed and bound by
CPI Group (UK) Ltd, Croydon, CR0 4YY

Mynd allan
Für Jana, Livia, Ida, Julia, Neil, Tom (Bob), and Ali—proof that humans can always make new homes.

Acknowledgements

Although by now bored of me constantly talking about forests, it is Michael Petraglia that must take the credit, whether he would like to or not, for getting me involved in multidisciplinary research into tropical forests. Working with him in Sri Lanka, I was struck by the fact that the arrival of humans on the island was apparently associated with rapid, specialized adaptation to tropical evergreen rainforests, environments that many palaeoanthropologists and archaeologists had considered 'barriers' to human dispersal and long-term occupation. I have also been lucky enough to interact with a variety of enthusiastic scholars demonstrating significant hunter-gatherer manipulation, the development of agricultural practice, and even the rise of urban forms in these settings. The growing temporal and geographic diversity of this evidence, alongside its general neglect in palaeoanthropology, archaeology, conservation science, ecology, anthropology, and amongst the wider public, have led me to the writing of this book.

This book is the product of the work of the growing number of teams and multidisciplinary specialists working in tropical forest settings over the last two decades. Moreover, it stands as a testament to the generosity and openness of the people inhabiting the nascent sub-field of 'tropical archaeology'. A huge number of people have contributed to the final form of this manuscript, reading chapters, discussing ideas, and providing images. Special thanks go to Peter Mitchell for sending me down the perilous path of book writing in the first place (it must also be said that he did read and comment on the whole thing). I am also immensely grateful to Julia Lee-Thorp, Jay Stock, Matt Sponheimer, Nicholas Taylor, Michael Petraglia, Robin Dennell, Christopher Hunt, Jaboury Ghazoul, Carlos Sierra, Glenn Summerhayes, Alison Crowther, Robert Spengler, Anabel Ford, Roland Fletcher, Lisa Lucero, Nuria Sanz, Etienne Patin, Scott Fitzpatrick, Douglas Sheil, and Kathy Morrison for reading and providing comment on the various chapters.

A number of these colleagues, as well as a huge variety of others, also provided pictures for use in the figures of this book (and I have highlighted their contribution in this regard throughout). In particular, thanks go to my brother Thomas Roberts for beautifying the cover photograph, as well as to Michelle O'Reilly, Hans Sell, Nils Vanwezer, and Ulrike Thuering for their creation of a variety of schematics and maps used in this book. Thanks also to the Max Planck Society for their funding of this book; it feels fitting that I have finished it less than a kilometre away from where Alexander von Humboldt discussed a number of his theories about the natural world based on exploration of the tropics. I am also immensely grateful to Nicole Boivin for giving me the opportunity to write this book. Without her friendship, patience, as well as the willingness to fund the last thing she thought she would probably ever end up funding, this would not have been possible. Furthermore, without the gracious support of Siran Deraniyagala and Oshan Wedage, as well as the people of Sri Lanka more widely, my long journey towards the completion of this book would never have begun. It truly is the 'Resplendent Isle'.

I am indebted to my parents for continuing to support my academic endeavours since the moment I first expressed an interest in visiting an archaeological monument at the age of 5. Even the often-reluctant trudging steps of my brother towards that monument have been crucial. Although life has its ups and downs, they have all contributed to my career more than they can ever know. I also know that Kay is up there looking on in amazement that 'trees' and 'soil' were what 'I wanted to do with my life' but that she supports it nonetheless. Finally, the biggest thank you goes to Jana, Livia, and Ida for showing me that although academia is one of the most fascinating and privileged careers someone can ever have, it is not everything. I apologize for not being able to be a horse or princess more often in the past year, but I promise to make up for that in the future. I hope that the awe-inspiring habitats and human sites discussed throughout this book are still there for them to see in the years to come.

Contents

List of Figures

1

Introducing Tropical Forests in Prehistory, History, and Modernity

> It gives me the impression of a primeval forest full of the most remarkable things, a monstrous and boundless thicket, with no way of escape, into which one may well dread to enter.
>
> —Friedrich Wöhler
>
> Letter to J. J. Berzelius (28 Jan. 1835). In *Bulletin of the Atomic Scientists* (Nov. 1949), 310. Date of letter as given in Mackay (1991: 267).

Friedrich Wöhler was referring to the field of organic chemistry during the early 1800s when he wrote the above but his comments would not be out of place in the context of embarking upon a global study of past and present human relationships with tropical forests. Dense vegetation, difficulty of navigation, issues of preservation, political and health concerns, poisonous plants, animals, and insects, and the prospect of carrying out sampling or excavation in high humidity have all meant that our knowledge of human history and prehistory in these environments is under-developed relative to temperate, arid, or even polar habitats. There have been theoretical questions as to what kind of human activity one would even expect to find in tropical forest environments, which seem hostile to human foraging (Hart and Hart, 1986; Bailey et al., 1989) let alone thriving agricultural or urban settlements (Meggers, 1971, 1977, 1987). This has, until relatively recently, left the state of archaeological tropical forest research in a similar position to popular conceptions of these environments—untouched, primeval wilderness.

Public ideas of an archaeologist investigating a tropical forest are probably synonymous with someone in a shabby-looking leather hat being chased, if not by a large stone boulder then by a group of Indigenous people with blowpipes, as they wade through dense undergrowth and vines while clutching a golden discovery that has been lost to the western world for thousands of years (Spielberg, 1981). The more recent development of the best-selling *Uncharted* video game series has done little to change these ideas amongst the next generation of media consumers, with players taking on the role of Francis Drake's mythical ancestor in search of long lost treasure, frequently hidden within caves and ruins surrounded by vines and dense canopies (Naughty Dog et al., 2016). The idea of treasure hidden within tropical forest is also not a modern conception. The long-term myth of *El Dorado*, a city covered in gold, fuelled exploration of the tropical forests of South America by renowned individuals, including Sir Walter Raleigh, from the sixteenth to the nineteenth centuries (Nicholl, 1995).

Figure 1.1. The Classic Maya site of Tikal (*Tik'al*), Guatemala emerging from dry tropical forest. Taken by, and used with permission from, Lisa Lucero.

Albeit with considerably less gold at stake, these depictions are not so far beyond reality in terms of the common archaeological focus in tropical forest regions. For a long period, tropical forest archaeological research centred on the description and evaluation of 'lost' cities (Figure 1.1). In the case of the particularly evocative example of the Maya, following intense monumental construction during the 'Classic' (AD 250–900) and early 'Post-Classic' (AD 900–1539) periods, urban sites were abandoned at various points in time, both pre- and post-Spanish arrival (Demarest, 2004). In the absence of human clearance and maintenance, this left these areas of land to be reclaimed by tropical forest and allowed a spate of archaeological 'rediscovery' over the course of the nineteenth century. Even today, the 2015 finding of similarly complex urban networks in the Honduran rainforest, including the 'legendary City of the Monkey God', was attributed to 'explorers' rather than 'archaeologists' by the media (Preston, 2015) (Figures 1.2 and 1.3).

This idea of 'discovery' is also associated with people who still dwell as foragers within tropical forests. Despite the clear presence of ancient cities in some tropical forests, Native peoples living in these environments have often been publicly presented as 'pristine' until the unrelenting machine of modern urbanism, industrialization, and global trade honed into view (Stiles, 1992; Rowland, 2004). Environmental activists have drawn on these perceptions to portray these groups as a reiteration of the 'noble savage', caretakers of environments that are now under threat by capitalism, agriculture, and increasingly dense human populations. Television documentaries such as Bruce Parry's *Amazon* (BBC, 2007) and Sting's Amazon publicity drive in the late 1990s (Syler, 2011) have only reinforced

Figure 1.2. Archaeological expedition transferring equipment by boat along the Kapuas River in the interior of equatorial Kalimantan, Borneo. Taken by, and used with permission from, Vida Kusmartono.

these ideas, while local communities have also occasionally encouraged these perspectives as part of political strategies to maintain their own land rights. In 2007, the Brazilian government officially recognized a total of sixty-seven 'uncontacted tribal groups', highlighting the idea that many populations still remain to be discovered in the tropics (Ghazoul and Shiel, 2010).

Together with their inhabitants and ruins, tropical forests have themselves come to be seen as pristine, unchanging 'wildernesses'. In literature, these environments frequently provide the stifling backgrounds to individual turmoil. In perhaps the most vivid of Victorian colonial dramas, *Heart of Darkness*, it is the Congolese tropical rainforest that forms the oppressive, dark, maddening background to one man's grappling with the concept of 'civilization' (Conrad, 1999). The term 'jungle', often used in common reference to tropical forests, comes from the Hindi word *jangal* which has long been used in reference to untouched, impenetrable forests surrounding human settlements (Whitmore, 1998; Ghazoul, 2015). Such perspectives have even been adopted in archaeological and palaeoenvironmental reviews, with equatorial regions, and their resident tropical forests, being considered as relatively static compared to other global environments (e.g., Basell, 2008). Indeed, despite the romantic conceptions of 'discovery' noted above, these environments, and their prehistoric inhabitants, have remained somewhat impenetrable, until recently, to archaeological and historical research.

Certainly, we should not underestimate the huge, novel impact the unrelenting industrial and urban powers of the modern world are having on tropical forest

Figure 1.3. Archaeological expedition entering evergreen rainforest in the interior of equatorial Kalimantan, Borneo for reconnaissance survey. Taken by, and used with permission from, Vida Kusmartono.

environments relative to previous human activities. Every day *c.* 320 km^2 of tropical rainforest are destroyed, and 135 plant, animal, and insect species along with it (Scientific American, 2009), as a result of expanding human populations, monoculture plantations and livestock systems, increasing demand for wood and mineral resources, and human induced climate change (Shukla et al., 1990; Bruenig, 1996; Goldsmith, 1998; Ghazoul and Shiel, 2010). These alterations threaten ecosystems that are central to the stability of Earth's atmosphere and climate (Houghton et al., 2015). However, despite popular belief, significant human manipulation and alteration of tropical forests are not recent phenomena (Willis et al., 2004; Boivin et al., 2016; Roberts et al., 2017a), and it is increasingly recognized that each forest is, in fact, a dynamic 'artefact' of millennia of human interactions (Barton et al., 2012).

This book is an attempt to highlight tropical forests as an *ongoing* part of human evolutionary, cultural, demographic, and economic prehistory and history. From being blanks on the archaeological map, tropical forests across the globe are now being increasingly investigated by archaeologists, anthropologists, conservationists, climate scientists, and ecologists in more systematic ways than previous searches for a single, monumental discovery. This work is revealing the sheer time-depth of our species', and its ancestors', interaction with these environments. Moreover, it is highlighting the geographical and temporal diversity of tropical forests, which has been previously neglected in archaeology, as well as human responses to, and impacts on, this diversity. Combining data from state of the art, developing methods with traditional excavation and survey, I will enter this 'monstrous and boundless thicket' with a multidisciplinary toolkit in the hope of escaping with a comprehensive, structured analysis of tropical forests as a key part of the human story and evaluation as to how we can continue to live sustainably within them.

FORESTS OF PLENTY OR 'GREEN DESERTS'?

The characterization of tropical forests in archaeological and anthropological discourse has been intimately tied up with ideas regarding the nature of their ethnographic inhabitants. In 1971 Manuel Elizalde—the head of a Philippine government agency tasked with protecting the interests of cultural minorities—reported the discovery of a 'tribe' living completely separated from the rest of Philippine society. Isolated in a tropical forest environment this tribe was known as the Tasaday. Living in caves, with no words for 'war', the Tasaday people were characterized as the epitome of the 'noble savage'. Furthermore, on the cover of *National Geographic*, the subject of popular films, and the focus of the book, *The Gentle Tasaday: A Stone Age People in the Philippine Rainforest* (Nance, 1975), the Tasaday were described as an unchanged relic of past human society. Yet this idyllic picture was soon disrupted by claims that Elizalde, and members of the Tasaday, had created a hoax in order to obtain political and economic capital (Iten, 1986; Headland, 1993; Hemley, 2003) (Figure 1.4). Today, while the Tasaday are considered to legitimately represent long-term tropical forest hunter-gatherers, the concept of them having had no contact with neighbouring agricultural groups has been discredited (Reid, 1992; Headland, 1993).

The Tasaday debate was the focus of my first ever undergraduate Archaeology and Anthropology essay, acting as a classic warning that anthropological beginners should forget any western preconceptions that social and cultural anthropological research deals only with pristine, primitive, uncontacted societies. The Tasaday were not the last example of high profile debate of this kind. In 2008, *Survival International* was at the centre of controversy when it portrayed painted

Figure 1.4. Tropical forest on the island of Mindanao, the Philippines where the supposedly 'isolated' Tasaday people were 'discovered'. Such environments have traditionally provided a backdrop for claims about uncontacted populations around the world. Picture taken by Keystone Pictures USA/Alamy Stock Photo. Reproduced with permission from Alamy.com.

people in the Amazon as part of an 'uncontacted tribe'. Consequently, a barrage of media describing this group as 'lost' or 'ancient' appeared, forcing *Survival International* to quickly counter that 'uncontacted' by the western media was not necessarily the same as a population having no social connections with the outside world. In both these instances, while undergraduate anthropology essays invariably focus on the questionable characterization of the people, it is interesting for our purposes to note the way in which tropical forests act as an unquestioned setting for the isolation of timeless culture. In these cases the potential of these environments to host 'pristine' groups was not in doubt.

The reaction to such ideas of forest 'tribes' as 'pristine primitives' also directly informed anthropological and ecological characterization of the viability of tropical forests for long-term human occupation. In the 1980s and early 1990s Thomas Headland, Robert Bailey, and Lawrence Reid demonstrated that many, if not all, supposedly isolated tropical forest foragers had recorded evidence of historical or ethnographic contact with local agricultural societies (Bailey et al., 1989; Headland and Reid, 1989). They argued that in the absence of this trade, for agricultural produce and metal materials, long-term human forager habitation of tropical forests was impossible (Bailey et al., 1989; Bailey and Headland, 1991). Subsistence based purely on foraging was considered impractical due to the scarcity, and seasonal availability, of energy-rich wild foods, such as fat-rich animals, oil-rich seeds, and carbohydrate-rich tubers such as those of wild yams (Hart and Hart, 1986). When combined with the perceived barrier of dense tropical forest vegetation to human movement (Hutterer, 1983), the likelihood of any population existing in these 'green deserts' without external contact or farming based on introduced resources appeared slim.

In archaeology, the viability of tropical forest soils to support dense agricultural populations had already been questioned (Meggers, 1971, 1977, 1987). The eventual collapse of the urban settlements of the Maya in Guatemala, Belize, Honduras, and Mexico was argued to be a result of the inherent vulnerability of tropical forest environments (Harrison and Turner, 1978; Rice and Rice, 1984) and the inevitably unsustainable soil erosion resulting from dense agricultural occupation (Deevey et al., 1979). These ideas were also related to the fact that groups living in tropical forests described by colonial powers in the twentieth century were inevitably small-scale hunter-gatherers or cultivators (Roosevelt, 1999). However, the anthropological debate surrounding 'pristine' hunter-gatherer societies even led to the doubt in archaeology as to whether small-scale foragers, including early members of our species and its hominin ancestors, migrating beyond Africa, could live in tropical forest habitats (Gamble, 1993). Combined with taphonomic research highlighting the unlikelihood of organic remains being preserved in these environments (Tappen, 1994), archaeological research in tropical forests, beyond the search for abandoned temples and monumental structures, was somewhat put on hold.

Responding to Bailey and Headland (1991), in particular, anthropologists and ecologists rapidly refuted the impossibility of tropical forest foraging. Bahuchet et al. (1991) used direct assessments of the ecological and nutritional availability of wild yams in the western Congo Basin to demonstrate, among others (Brosius, 1991; Dwyer and Minnegal, 1991; Endicott and Bellwood, 1991), that it *is* nutritionally possible for ethnographic foragers to live off wild resources alone in a rainforest. Moreover, many other carbohydrate-rich plant resources exist in tropical forests, including honey and palm starch, while tropical game can be both fat- and protein-rich (Brosius, 1991; Dentan, 1991). It was also demonstrated that historical and ethnographic tropical forest foragers are not simply passive caretakers of a 'natural' forest but often actively alter their environments, burning forest to encourage an abundance of sub-canopy plant growth and its consumers (Posey, 1984; Hill and Baird, 2003), relocating important plant resources (Ichikawa, 1999; Yasuoka, 2009), and even corralling high-protein freshwater resources such as fish and turtles (Colinvaux and Bush, 1991; Hemming, 2009) (see Chapter 2). Specialized tropical forest foraging completely independent of agriculture has now been definitively demonstrated (Hewlett, 2014).

Nevertheless, the viability of tropical forest occupation, farming, and extensive settlement has taken some time to filter through into archaeology. Despite the long-held knowledge of Maya cities, and suggestions of early tropical forest plant domestication and land modification in New Guinea (Golson, 1989), it is only in the last two decades that the scale of prehistoric modification of tropical forests has begun to be understood. The pre-Columbian Amazon has now emerged as an important centre of plant domestication (Clement et al., 2010, 2015) and urban development (Heckenberger et al., 2008; Heckenberger and Neves, 2009), and deforestation is continuously revealing large-scale prehistoric landscape modification and management (Rostain, 2014). Indeed, it is now evident that the only reason small-scale foraging communities remain the sole Indigenous occupants of the Amazon Basin, for example, is due to disease, genocide, and slavery commencing with the arrival of European colonial powers in the Americas (Roosevelt, 1999; Lewis and Maslin, 2015). Nevertheless, the long-term sustainability of prehistoric agricultural and urban existence in tropical forests remains a source of persistent debate (Beach et al., 2006; Penny et al., 2014; Grollemund et al., 2015).

While hominin occupation of African rainforests was hinted at decades earlier (Clark, 1963, 1971, 1988), the work of Larry Barham (2001, 2002) and Julio Mercader (2002a, 2002b) has also recently stimulated interest in Middle and Late Pleistocene *Homo sapiens*' 'rainforest prehistory' within Africa. Evidence discovered in the last decade has, furthermore, extended the human use and occupation of tropical forests back to at least 45,000 years ago in Southeast Asia and Melanesia, and 38,000 years ago in South Asia (Barker et al., 2007; Summerhayes et al., 2010; Perera et al., 2011; Roberts and Petraglia, 2015). Despite this, many still see tropical forests as marginal, particularly in the context of *H. sapiens*' hominin ancestors (Dennell and Roebroeks, 2005). In pre-Holocene archaeology and palaeoanthropology tropical forests are frequently perceived as barriers to hominin subsistence, occupation, and migration. Studies cite difficulties of navigation, thermoregulation, limited large game opportunities, and a lack of easily-processed foods, in labelling these habitats as unattractive, at least without significant advances in technology (Bird et al., 2005; Dennell and Roebroeks, 2005; Cosgrove et al., 2007; Boivin et al., 2013). This is despite the fact that the earliest hominin species in Africa and our closest relatives, the non-human great apes, evolved within a predominantly tropical forest setting.

ANCESTRAL FORESTS

Extant non-human great apes, as our closest living relatives, have frequently been used as analogues for 'ancestral' locomotion, subsistence, and cultural behaviours in early hominins and, in particular, the Last Common Ancestor (LCA) of the non-human great apes and *H. sapiens* (Tuttle, 2014). Association of all three modern non-human great ape genera—orangutans (*Pongo* spp.), gorillas (*Gorilla* spp.), chimpanzees, and bonobos (*Pan* spp.)—with primary and secondary tropical forests has, in turn, led to these environments being considered a primeval part of hominin evolution (Figure 1.5). Theories concerning early forms of bipedal locomotion, forest subsistence, reliance on plant foods, and small, isolated hominin populations have been largely based upon models of extant non-human great apes (Tuttle, 2014), as well as their relationship to evergreen and semi-evergreen forests of Africa and Southeast Asia that have been conceived of as broadly homogeneous. Furthermore, the expansion, and then diversification, of Miocene (23.1–5.3 Ma) ape populations has been linked to the expansion of tropical forest biomes across vast swathes of Africa and Eurasia as a result of warmer temperatures and an increased partial pressure of carbon dioxide in the atmosphere (Hamon et al., 2012).

It has been difficult to determine the locomotion and subsistence of the LCA due to the variety of ecological adaptations, independent evolution of knuckle walking, different forms of bipedalism, and cultural variability seen among the modern non-human great apes (e.g., Kivell and Schmitt, 2009). Nevertheless, it appears that the earliest members of the hominin clade evolved habitual bipedalism while retaining forms of locomotion that were relevant to forest ecosystems. For example, the *Orrorin tugenensis* (6.1–5.7 Ma) femur from Kenya has a long,

Figure 1.5. Eastern chimpanzee (*Pan troglodytes schweinfurthii*) in the Kibale Forest National Park, Uganda. Picture taken by Bernard Dupont and used with permission from www.flickr.com.

narrow neck and a wide proximal diaphysis, features shared with later bipedal Australopithecines (Richmond and Jungers, 2008), yet the structure of the humerus and the shape of the phalanges have been linked to tree climbing (Senut et al., 2001). Similarly, the lower limb proportions and the hand and foot morphology of *Ardipithecus ramidus* (*c.* 4.4 Ma) fossils from Ethiopia have been associated with tree climbing, while its pelvic morphology is linked to ground-based bipedal walking (White et al., 2009; Kimbel et al., 2014).

Yet, as will be seen in Chapter 3, despite these early origins in tropical forest settings, the emergence of the hominin clade and the genus *Homo* are most prominently linked to more open environments. The increasingly specialized adaptation to terrestrial bipedal movement (Harcourt-Smith and Aiello, 2004; Ruff, 2009), the development of stone tool technology (Harmand et al., 2015), and an apparent growing reliance on the meat of large game (Bunn, 2001) appear to occur during the period 4–2 Ma in Africa when tropical forests retreated and fragmented in the face of expanding 'savanna' environments (Reed, 1997; Uno et al., 2011, 2016). From this point onward, tropical forests tend to be neglected, if not completely ignored, in the evolution and expansion of our genus within and beyond Africa during the Early-Middle Pleistocene. This is in spite of the fact that some of the earliest *Homo* fossils beyond the African continent occur in Southeast Asia in a region that is argued to have been dominated by tropical forest during this period, just as it is today (Sémah et al., 2002; Sémah and Sémah, 2012).

A similar picture exists for *H. sapiens*. Although the earliest fossils of our own species date to *c.* 300,000 years ago in North Africa (Hublin et al., 2017) and *c.* 200,000 years ago in East Africa (White et al., 2003; McDougall et al., 2005) there has, until recently, been little consideration of pre-Holocene human interaction with tropical forests in Africa. Technological and cultural florescence in Africa from *c.* 100,000 years ago, including evidence for personal ornamentation and symbolism, has been associated with reliable protein opportunities available at coastal sites (Henshilwood et al., 2002; Bouzouggar et al., 2007; Marean et al., 2007; Henshilwood et al., 2011) or large game hunting facilitated by new microlithic technologies that emerged in savanna environments (Ambrose, 1993). Since Julio Mercader's (2002a, 2002b) discussion of us as 'Forest People', work in South Asia, Southeast Asia, and Melanesia has revealed that our species rapidly adapted to rainforest environments in all of the regions it encountered them on its expansion beyond Africa during the Late Pleistocene. However, this has taken time to filter into dispersal models (see, for example, Bird et al., 2005; Boivin et al., 2013).

It is clear that the earliest hominins emerged from tropical forests in the Miocene and Early Pliocene. Nevertheless, there still remains a large gap in our knowledge of the importance of tropical forest environments to the adaptations of Early and Middle Pleistocene *Homo* in Africa and Southeast Asia. There has been little attempt to interrogate or even investigate the evidence for tropical forest use by hominins during this period. There is also currently no direct evidence for tropical forest use by *H. sapiens* from its evolution in Africa *c.* 300 ka until 45,000 years ago onwards in Southeast Asia, South Asia, and Melanesia (Barker et al., 2007; Summerhayes et al., 2010; Roberts et al., 2016, 2017b). Are these gaps real, representing significant adaptive shifts in our species' and its ancestors' abilities to colonize tropical forest habitats, or are they simply the product of preservation biases against earlier remains in these challenging environments? Where there is clear evidence for tropical forest use by our species in the Late Pleistocene, in various parts of the world, there has thus far been little attempt to compile and evaluate the diversity of these early human adaptive strategies and how these may have been influenced by, and in turn influenced, the diversity that exists within the category of tropical forests.

TROPICAL FOREST DIVERSITY, PAST AND PRESENT

Tropical forests are some of the oldest land-based ecosystems on the planet, present as far back as 100 million years ago in the mid-Cretaceous period (Couvreur et al., 2011). The term 'tropical forest' is often associated with tropical rainforests, a phrase coined by the botanist A. F. W. Schimper in 1898 (*tropische Regenwald*) (Allaby, 2010) (Figures 1.6 and 1.7). However, while tropical rainforests are defined on the basis of high precipitation (Grainger, 1996a, 1996b), high temperature (White, 1983), and low seasonality in temperature and rainfall (Ghazoul and Sheil, 2010), where significant dry periods exist in the tropics monsoonal forests, deciduous forests, open forest structures, and mosaics prevail (Grainger, 1996a, 1996b). As will be seen in Chapter 2, there is also considerable diversity within the category of 'tropical rainforest' itself, and geological substrate,

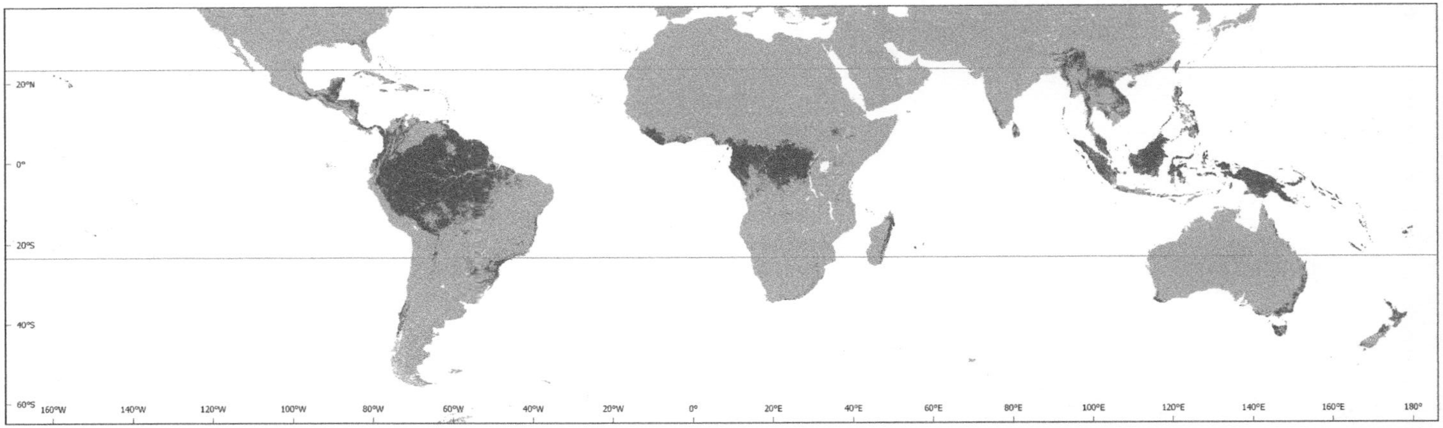

Figure 1.6. Map of the extent of the world's tropical forests determined by the MODIS (Moderate Resolution Imaging Spectroradiometer) Land Cover MCD12Q1 majority land cover type 1, class 2 for 2012 (spatial resolution of 500 m). Downloaded from the US Geological Survey Earth Resources Observation System (EROS) Data Center (EDC). Temperate evergreen rainforest is also visible in Tasmania, New Zealand, and Chile, and dry subtropical forest can be seen on the coasts of Australia and Argentina. With thanks to Ulrike Thuering.

Figure 1.7. Evergreen lowland rainforest in the vicinity of the site of Batadomba-lena in the Wet Zone of Sri Lanka. Taken by the author.

altitude, and precipitation dynamics all lead to different tropical forest ecosystems with different challenges and potentials for hominin populations. They include semi-evergreen rainforest, montane rainforest, heath forest, peat swamp forest, and freshwater swamp forest, habitats found in varying frequency across the different tropical continents.

There is also underappreciated diversity *within* a given tropical forest formation in the context of archaeological studies of these habitats. In biology, forests are schematically separated into vertically stratified layers, including the forest floor, sub-canopy layer, canopy layer, and emergent canopy layer. Each of these layers sustains different plant and animal resources (Denslow, 1987). In addition, horizontal variation exists as gaps in the canopy form as a result of natural tree death and disturbance as a result of storms, cyclones, natural burning, and anthropogenic deforestation. Gaps at ground level will be rapidly colonized by light-seeking seedlings and often contain the greatest species-richness. Larger-bodied fauna also often gather at these locations (Dentan, 1991; Roberts et al., 2016). Furthermore, while tropical forests are richer than most environments in terms of access to freshwater from well-fed streams and rivers, waterways vary with topography and influence the surrounding vegetation they cut through (Whitmore, 1998). Moreover, high volumes of rainfall can turn life-sustaining channels into life-threatening floods and mudslides (Larsen, 2017).

Alongside this geographical diversity, the temporal diversity of tropical forests has frequently been ignored in tropical forest archaeological research. The equatorial regions and their vegetation compositions have been considered relatively static through time (e.g., Basell, 2008). As we will see in Chapter 2, however, this has been refuted by increasingly detailed palaeoclimatic and palaeoenvironmental research that has demonstrated considerable change in the extent, structure, and composition of tropical forests in Africa, South Asia, Southeast Asia, South America, Melanesia, and Australia on a number of different temporal and spatial scales. Low percentage CO_2 concentrations in the atmosphere during glacials reduce the competitive advantage of plants that photosynthesize through the so-called 'C_3' pathway, including many shrubs, bushes, and trees found in forests, relative to 'C_4' grasses—leading to forest retreat (Jolly and Haxeltine, 1997). Similarly, decreased temperature and precipitation can cause forest fragmentation, particularly at high altitudes (Jolly and Haxeltine, 1997; Mayle et al., 2004). Although more difficult to track into the past, changes in rainfall and temperature seasonality will also have had considerable regional impacts on tropical forest composition and extent.

Therefore, while tropical forests are some of the most species-rich terrestrial environments, containing over half of the world's existing plant and animal species (Wilson, 1988; Whitmore, 1998), they are also some of the most complex. As will be seen throughout this book, this complexity has had a major impact on human foragers, farmers, and urban dwellers across space and time. Early-Middle Pleistocene hominin dispersal into tropical Southeast Asia appears to have been linked to a reliance on access to freshwater and mosaic tropical forest, swamp, and grassland environments (Chapter 3). Our own species demonstrates radically different, tool-supported adaptations to the montane rainforests of Melanesia relative to the peat swamp forests and semi-evergreen rainforest of Southeast Asia and South Asia, respectively, during the Late Pleistocene (Chapter 4). Tropical forest diversity also influenced the nature of early hominin impacts on different tropical forest environments. Early-Middle Pleistocene *Homo* apparently exploited large mammalian fauna across varied tropical forest and grassland mosaics in Southeast Asia (Chapter 3), while a lack of protein on the tropical forest islands of Melanesia led to deliberate human translocation of small mammals by *H. sapiens* when it reached this part of the world (Chapter 4).

Tropical forest diversity also had a clear impact on different prehistoric agricultural regimes (Chapter 5). A long history of plant manipulation and rich carbohydrate resources in Melanesia provided the backdrop to early agricultural experimentation in the Highlands of New Guinea (Golson, 1989; Denham et al., 2003). By contrast, the acidic soils and abundant peat swamps in the tropical forests of Southeast Asia have been argued to lead to maintenance of hunter-gatherer strategies until the onset of forest clearance and rice agriculture in the region (Bellwood, 1993, though see Denham, 2011). There are also differences in the challenges faced by prehistoric farmers expanding agricultural systems into tropical forest regions (e.g., Bantu-linked agriculture in Central and West Africa), who may have relied on waterways and tropical forest retreat to expand (Grollemund et al., 2015), and those developing indigenous horticultural regimes within tropical forest ecosystems themselves (Iriarte et al., 2007). Both of these systems, in turn, have different impacts on tropical forest environments relative to the expansion of monoculture palm, rubber, tea and fruit plantations that followed colonial arrival in these habitats and now pose one of the greatest threats to tropical forest ecosystems in the modern world.

Prehistoric urban populations in tropical forests appear to have developed a diversity of strategies to their surroundings. State, and even empire, level societies in the seasonal tropical forests of Mesoamerica, South Asia, and Southeast Asia managed to support large populations across extensive areas, combining agrarian land-use with administrative centres and monumental construction (Higham, 2001; Coningham et al., 2007; Fletcher, 2009, 2012; Lucero et al., 2015). The 'Garden City' networks of the Amazon similarly saw a spread of hunting, garden agriculture, and deforestation across large areas (Heckenberger et al., 2008). These latter groups focused occupation around the Amazon's tributaries, with annual flooding ensuring access to fertile soils and the calculated 'farming' of freshwater resources (Chapter 6). The sustainability of prehistoric urban systems in tropical forests remains a hot topic of discussion, both academically and in the media. Ruins in tropical forests provide evocative pictures of dramatic abandonment linked variously to untenable human alterations to soils and vegetation. Given that by 2050 over half of the world's population will live in the tropics, understanding the challenges faced by prehistoric people in this regard is key to modern outlooks (The State of the Tropics Project, 2016).

NEW METHODS OF 'DISCOVERY'

The emergence of this geographical and temporal diversity of human adaptations to tropical forests is, in part, a result of the development and application of new methodological approaches in tropical forest archaeology. Detailed preservation and site formation studies have revealed that rock shelters and cave sites can preserve long records of human occupation at least as far back as the Middle Pleistocene (Mercader et al., 2003; Barker et al., 2007; Westaway et al., 2009; Perera et al., 2011; Barker, 2013; Barker and Farr, 2016). There has also been a realization that the diversity of tropical forest formations means that there is varying potential for organic preservation in these environments rather than the wholly negative picture previously painted by limited research into skeletal degradation in hydrologically active evergreen rainforest (Tappen, 1994). For example, open-air tropical sites dating to the Late Pleistocene have been found in the cool montane rainforests of the Ivane Valley in the Central Highlands of New Guinea (Summerhayes et al., 2010). Even within evergreen tropical rainforest it has been shown that local site conditions can lead to the preservation of bone collagen in human and animal remains for radiocarbon dating and dietary analysis through stable isotope approaches (Hermenegildo et al., 2017).

Acknowledgment of the potential for good preservation in tropical forest environments has encouraged the increased multidisciplinary analysis of human occupation sequences in these habitats. While these are routine in archaeology elsewhere, a long period of neglect of tropical forests in archaeological research means that even basic zooarchaeological, archaeobotanical, and technological research has only recently begun to produce revolutionary results. At the Niah Caves, Sarawak, fragments of fruits, nuts, and plant tissue, and starch grains have indicated that the collection of rainforest tuberous plants such as taro, yam, and sago palm occurred as early as 45 ka (Barker, 2005; Barton, 2005; Paz, 2005; Barker et al., 2007; Barker, 2013; Barker and Farr, 2016). Many of these tubers require extensive processing to make them digestible, suggesting that the early human occupants were experienced in rainforest manipulation (Barton, 2005; Barker et al., 2007). Detailed use-wear and residue analysis of tools is also showing some potential, indicating how bone points and stingray spines in Southeast Asia were hafted and used as projectile points (Barton et al., 2009; O'Connor et al., 2014).

Nevertheless, while these methods can provide episodic insight into human strategies in tropical forest environments they are not necessarily indicative of overall subsistence reliance. Furthermore, in earlier time periods, issues of preservation mean that crude inferences from tool morphology are relied upon to suggest tropical forest exploitation (Groube, 1989; Barham, 2001; Mercader, 2002a, 2002b; Leavesley, 2006). Stable isotope analysis of human tooth enamel has recently emerged as a means of directly assessing human tropical forest resource reliance (Krigbaum, 2003, 2005; Roberts et al., 2015a, 2016, 2017b). While this methodology has been used for a long time in the study of the relative reliance of African hominins on forest or savanna environments (Lee-Thorp et al., 1989a, 1989b; Sponheimer and Lee-Thorp, 1999; White et al., 2009; WoldeGabriel et al., 2009; Sponheimer et al., 2013), it is only more recently that it has been applied to questions of human tropical forest adaptations. Future application to *Homo erectus*, and the small-bodied *Homo floresiensis* (the 'hobbit') discovered on the tropical island of Flores, Indonesia (Brown et al., 2004; Morwood et al., 2005), in Southeast Asia holds much potential for determining the tropical forest interactions of Early-Middle Pleistocene hominins in this region (Chapter 3).

As we saw above, the archaeological study and understanding of tropical forests has been intimately connected to our understanding of foraging populations living within them today. Moving beyond the debate as to whether these habitats are viable for human foraging, researchers are beginning to systematically study, and communicate with, extant tropical forest hunter-gatherers and farmers in order to better understand the potential resources, impacts, and challenges faced by the prehistoric human inhabitants of these environments (Xhauflair et al., 2017). This work is becoming increasingly prominent in the Amazon Basin as archaeologists seek to work directly with local communities to understand past occupation and adaptation, but also inform the future conservation of endangered tropical forest environments (e.g., Heckenberger et al., 2003; Heckenberger and Neves, 2009). Work amongst current Maya communities in Mesoamerica is beginning to reveal the complex approaches they take to tropical forest 'gardening' and how this might have been relevant for the Classic and Post-Classic urban approaches to the maintenance of dense populations in these habitats (Ford and Nigh, 2015).

Genetic studies are also being increasing utilized within archaeological frameworks in order to understand human demographic trends and adaptations through time. Working on modern populations, the last decade has seen a large amount of research focusing on the adaptive, genetic basis of the 'pygmy' phenotype that is associated with extant tropical forest foragers in Africa, Southeast Asia, and South America (Verdu et al., 2009; Perry et al., 2014; Perry and Verdu, 2017). Moreover, divergence dates obtained from a comparison of 'pygmy' populations in Africa with neighbouring populations is providing tantalizing hints at the time-depth of the tropical forest influence on human demography (Perry and Verdu, 2017). While obtaining ancient DNA in tropical forest regions has long been thought impossible, recent work in the tropical Pacific suggests there may be hope (Skoglund et al., 2016; Posth et al., 2018). Linguistic analysis of modern populations is also providing a means of understanding modern cultural perceptions of tropical forest environments, as well as the spread of external farming populations into tropical forests and the nature of their contact with Indigenous hunting and gathering communities (e.g., Bellwood, 1993; de Maret, 2013; De Luna, 2017).

Huge temporal variability in tropical forest ecosystems has led to the need for greater methodological refinement in the study of these environments in the past. Study of leaf wax compounds (n-alkanes) is being shown to track the proportions of different types of forest at different elevations (Feakins et al., 2016), while the stable isotope analysis of these compounds in sediments has been used to reconstruct the regional prevalence of tropical forests in Africa throughout the Pleistocene (Uno et al., 2016). When combined with existing regional records from pollen, phytoliths, and palaeontological sequences, these novel methodologies are providing increasing resolution of tropical forest extent and structure (Rabett et al., 2017; Garcin et al., 2018). In addition to regional records, the importance of 'on-site' palaeoenvironmental records is beginning to be understood and faunal profiles, stable isotope records of sediments and fossil teeth, pollen, and phytolith records are all now being used to build up multi-proxy, local tropical forest records. The computer modelling of human and climatic impacts on forests is becoming increasingly powerful and offers a further potential avenue for our understanding of tropical forest distributions in the past, particularly through the comparison of local records to regional and global predictions (Roberts et al., 2018a).

The progressively more enthusiastic application of old methods, and the development of the above-mentioned new methods, in the study of human tropical forest prehistory is beginning to take archaeology well beyond opportunistic exploration and the search for monumental ruins. However, one of the most publicized recent developments in archaeological tropical forest research has been the form of remote sensing known as LiDAR (Light Detection and Ranging) scanning (Figure 1.8). The transmission of a highly dense network of infrared lasers from an aerial vehicle enables the development of a 3D model of the ground surface. Algorithms can be used to strip away forest canopies and reveal landscape features on the forest floor. This methodology has revealed the vast urban networks of the Khmer Empire around Angkor Wat and beyond in Southeast Asia (Evans, 2016) and has also revealed previously unknown urban settlements under the canopy of the Honduran rainforest in Central America (Fisher et al., 2017) (Chapter 6). These images provide archaeologists with high-resolution maps of where to look in the search for large-scale, anthropogenic landforms, and attempt direct 'boots on the ground' verification. While exponentially more hi-tech, these novel methods of 'discovery' mean that the archaeological 'exploration' of tropical forest civilizations is not yet completely gone.

It is now slowly being appreciated that humans, from their earliest evolution, could and did occupy and modify tropical forest environments. Furthermore, humans developed agricultural systems and dense urban networks in tropical forests across the globe prior to the appearance of colonial powers and modern global trade networks and capitalism. This has resulted in increased willingness to apply for permits and funding to investigate tropical forests and to develop new methods for their study. However, as of yet there has been no attempt to bring our understanding of the temporal and geographical diversity of human, and ancestral

Figure 1.8. Oblique perspective of the centre of Angkor Thom, Cambodia with layers created from LiDAR (light detection and ranging) data. The top layer reproduces a conventional view including dense vegetation; the bottom shows the bare-earth terrain model with the vegetation digitally removed, revealing traces of intensive occupation around the monuments. With thanks to Damian Evans.

hominin, interactions with tropical forests together. Methodological specialists, and archaeologists, anthropologists, historians, and conservationists, have tended to write separate narratives in discussions of different aspects of human occupation and modification of these environments. This book is the first attempt to put together a truly global human tropical forest prehistory. Moreover, it is the first attempt to take a multidisciplinary, multi-method approach to understanding what the past can teach us regarding the sustainability and ongoing significance of human occupation in some of the most threatened ecosystems on the planet.

A PATH THROUGH THE JUNGLE

So how should one approach the 'boundless thicket' of past and present human relationships with tropical forests? Two clear avenues have already been highlighted. Firstly, it is essential to understand the geographical and temporal variability of these environments and how this might have influenced human occupation and adaptive strategies. How have the tropical forests facing different hominin species, different human populations, different agricultural communities, and different urban systems varied, and how did *H. sapiens* and its ancestors respond to these different challenges? Secondly, it is important not to retain a reliance on one method or discipline, but rather to use multiple approaches. It is clear that to study the variability of human responses to, and modifications of, different tropical forest habitats we need an understanding of changing tropical forest structure and extent, the subsistence strategies employed by human communities, the resources being utilized, and the scale of landscape modification and settlements. Moreover, we need to understand the reciprocal effects tropical forests have had on human and hominin biology, demography, and culture. I have outlined above how various archaeological methodological approaches are being developed and applied to do just this, and these methods provide a key source of data for this book.

I have so far only touched upon the combined role of archaeology, anthropology, conservation science, and history in the multidisciplinary approach to be undertaken here. A huge corpus of anthropological and historical descriptions of tropical forest foragers, agriculturalists, and urban dwellers exists in indigenous texts such as those of the Classic Maya, as well as colonial sources written in English, Spanish, Portuguese, French, and Dutch from the sixteenth century onwards. Reviews of this literature reveal specific details regarding human hunting practices, prey sources, the variety of uses and categories for tropical forest plants, the scale of pre-Columbian agricultural landscapes, and the wealth of prehistoric cities (e.g., Dentan, 1991; Junker, 1999; Morrison, 2002a, 2002b, 2014; Hemming, 2009). It is also possible to discern temporal changes from these sources, for example, the differences between the small-scale foragers studied in the Amazon Basin, Southeast Asia, and Melanesia today, relative to those encountered by European colonial powers. Modern conservation science literature also offers insights into the climatic and human challenges posed to tropical forests today, the nature of human impacts on tropical forest flora and fauna, and the relative sustainability of different agricultural systems and urban policies in these environments (e.g., Corlett and Primack, 2011; Larsen, 2017).

It may seem that the relatively rare traces of material information from the archaeology of tropical forests pale in comparison to the detail available in the conservation, ethnographic, and historical literature, and this may be why few tropical forest archaeology-focused books to date have gone far beyond the realm of recorded history. However, none of these records can document the skeletal, environmental, and material culture traces of hominin tropical forest interactions from the Miocene, Pliocene, Pleistocene, or even the Early and Middle Holocene. It is ignorance of archaeological work in tropical forests that has led to the popular perception that millennia of human adaptations to, and modifications of, tropical forests simply did not occur. Moreover, even over historical timescales, history, and to some extent anthropology, can only reveal what the writer thought, or thought they knew, of tropical forest adaptations and indigenous cultures. This leads to partial histories and documentation, and here material remains—artefacts, rock paintings, settlement distributions—produced by ethnographic and historical communities can contribute to reports of tropical forest lifestyles that are often dominated by European perspectives (cf. Mitchell, 2015). Even modern conservation science is often beleaguered by political bias and western priorities.

Any attempt to produce a comprehensive picture of human relationships with tropical forests must weave archaeological, anthropological, historical, and conservation science datasets together. This facilitates multi-layered understandings of what is possible for humans in a tropical forest, what challenges are faced by humans in a tropical forest, what factors influence human sustainability in tropical forests, and how these parameters have changed across space and time. With this, and alongside the ever-increasing toolkit of tropical forest archaeology, we can begin to outline some key themes for 'Tropical Forests in Prehistory, History, and Modernity'. Firstly, at the furthest temporal edge of this study, how have different hominin species adapted to tropical forests and how have their adaptive capabilities in this regard changed through time? The early hominins of the Miocene and Early Pliocene are often thought of as analogous to great ape populations, but the appearance of the genus *Homo*, tool-use, and greater ecosystem impacts are generally thought to have led to fundamentally different relationships with the world's environments, including tropical forests. With the emergence of our species, 300–200,000 years ago, this relationship gained a whole new level of intensity, leading to the appearance of agriculture, cities, and present-day ecosystem impacts.

With this in mind, how much did foraging *H. sapiens* modify tropical forests and their biota on their expansion within and beyond Africa? The 'megafaunal extinctions' of the Late Pleistocene have been popularly documented in Eurasia, Australia, and the Americas. Each region saw a massive decline in megafaunal (any animal over 44 kg) diversity between the period 100,000 and 12,000 years ago. These extinctions have been variously attributed to human hunting pressures (Carrasco et al., 2009), climate change (Cooper et al., 2015), disease, and even the impact of an asteroid (Firestone et al., 2007), with different explanations taking precedence in different regions. Yet, few of these discussions have focused on the potential impacts of humans on tropical forest megafauna (though see Malhi et al., 2014; Doughty et al., 2015). This is despite the fact that humans were physically moving, (albeit small) mammals, between tropical forests as far back as 20,000 years ago (Gosden and Robertson, 1991). Furthermore, Late Pleistocene human impacts on tropical forest flora are only beginning to be studied. Extensive anthropogenic fires have been suggested in Melanesia, Southeast Asia, and Australia as far back as 45,000 years ago (Hunt et al., 2012). To what extent did such interactions permanently modify these forests?

From early experimentations with tropical forest horticulture across the Terminal Pleistocene/Holocene boundary, the Holocene epoch was to see an increase in human tropical forest impacts in the form of large-scale agriculture, deforestation, and even urban settlement. Tropical forests have driven nuanced discussions of what the term 'agriculture' actually means and even questioned the viability of the dichotomy between 'hunter-gatherers' and 'farmers'. To what extent do different types of 'agriculture', including indigenous horticulture, the external introduction of cereal crops, and modern monoculture and plantation, introduce different challenges? Similarly, what is prehistoric tropical forest 'urbanism' and is it always the same? How does it differ from modern urban settlements in tropical forest settings? In both cases, this book will attempt to draw out the unique nature of different tropical forest agricultural and urban strategies and make some assessment regarding the sustainability of different practices. Furthermore, it will review the challenges faced by these types of communities in the past and how these might inform modern farming and urban practices in tropical forest settings.

Together, reviewing prehistoric foragers, farmers, and urban dwellers in tropical forests will also get to the heart of one of the core themes of this book, as to whether tropical forests are fundamentally attractive and resilient environments for human occupation and adaptation relative to other environments. Are complex, specialized technological devices and economic strategies necessary in order to occupy tropical forests, or are they simply just one other ecosystem that our species has inserted itself into since its evolution and expansion beyond Africa? Furthermore, do environmental constraints in different tropical forest types, variable spatially and geographically, place certain limitations on human societies living in these habitats? These questions lead into broader considerations of not only how humans have altered these environments but how, in turn, these environments have made unique contributions to human ecology, biology, economic networks, and political systems. Furthermore, how has research in tropical forests influenced definitions and approaches used in western archaeology and anthropology? I do not want readers to see this evaluation as environmentally deterministic—rather I hope to highlight tropical forests as a dynamic, reactive stage for human cultural and adaptive agency, with broader relevance for the human past, present, and future than they are often given credit for.

More widely, this book sits within current, popular debates regarding the 'Anthropocene' and when our species began to have significant impacts on the world's environments and earth systems (see Ellis et al., 2016, Malhi, 2017). Geologists have currently placed the origin of this epoch at the start of the nuclear bomb-testing curve. Yet archaeologists, in tropical forests and beyond, are increasingly demonstrating that humans have had significant continental and global environmental and geological impacts for millennia prior. What is different is that most of the prehistoric human societies discussed in this book had long-term practical experience with tropical forest environments. Multi-national corporations, governments, and urban planners threatening these habitats today do not have such experience. This brings me to my final theme, namely—what is the relationship between ethnographically documented Indigenous communities, prehistoric tropical forest foragers, farmers, and urban dwellers, and modern forces threatening to destroy these habitats today? What can archaeology, history, and palaeoenvironmental research in tropical forests do to inform more productive conservation initiatives and human adaptations in the modern world?

APPROACHING TROPICAL FORESTS IN PREHISTORY, HISTORY, AND MODERNITY: STRUCTURE AND ORGANIZATION

With these themes and the general rationale of the book in mind it is now important to elaborate on the structure of how these will be addressed. As I have stated, tropical forests have immense geographical and temporal diversity, both within and between formations, which is often ignored archaeologically and anthropologically. Chapter 2 will therefore use a plethora of ecological, fossil, and palaeoenvironmental literature (e.g. Whitmore, 1998; Ghazoul and Shiel, 2010; Ghazoul, 2016) in order to outline the evolutionary history and current ecological situation of tropical forests across the world, pointing out differences in structure, climate, species diversity, and the potential for different types of human habitation through time. Palaeoclimatic, palaeoenvironmental, and modern genetic methods are compared with regards to their ability to provide us with tropical forest natural history information of relevance to archaeological records. Ethnographic evidence for different hunting, foraging, and farming strategies is also evaluated in order to provide insight into how different tropical forest formations may have been navigated by our species.

From this point, the structure of the book is largely chronological and Chapter 3 will take us back to the very beginning, the emergence of the hominin evolutionary clade in tropical forest settings. It will look at the diversity of ecological adaptations, locomotor practices, and even cultural behaviours of the non-human great apes and how they have been used as analogues for the LCA between humans and apes. From here, Chapter 3 will review the morphological and environmental information for the changing importance of tropical forests to hominin evolution from the Late Miocene into the Early Pleistocene. In particular, it will question the 'savanna hypothesis' that emphasizes the unilinear, growing importance of grasslands to hominin locomotion, technology, and subsistence over this time period. Chapter 3 will also review our knowledge of potential tropical forest adaptations amongst Early and Middle Pleistocene hominins dispersing from Africa to Southeast Asia. In doing so, Chapter 3 will review whether there is a real adaptive shift in the 'reoccupation' of tropical forests by our own species, *H. sapiens*, during the Late Pleistocene or whether this is a product of preservation bias.

Chapter 4 looks at the beginning of a truly 'global' tropical forest prehistory with the expansion of our species within and beyond Africa during the Late Pleistocene. Until very recently, it was generally considered that *H. sapiens* only successfully colonized the world's tropical forests during the Holocene period (Gamble, 1993, 2013; Bird et al., 2005). However, the last two decades have yielded evidence that indicates the occupation and exploitation of tropical forests by our species from at least *c.* 45 ka and, more controversially, perhaps *c.* 200 ka (Roberts and Petraglia, 2015). This part of the book approaches Africa, South Asia, Southeast Asia, Melanesia, Australia, and South America in turn to evaluate the variety of tropical forests occupied by early members of our species, the different strategies employed by foragers in these habitats, and the varying human impacts on tropical forest plants and animals.

Chapter 5 moves us to one of the major ecological thresholds of our species in tropical forests—the experimentation with the domestication of different plant and animal species. This chapter explores the distinctive nature of early 'agricultures', and their predecessors, that emerged in tropical forest environments. This includes early human influence through forest burning and the deliberate movement of plants and animals between tropical forest environments. Chapter 5 will explore how the indigenous strategies involved differ globally, ranging from long-term forest interaction and tree-cropping in Melanesia (Denham et al., 2003), to the maintenance of diverse hunting, fishing, and cultivation in the Amazon Basin (Roosevelt, 2000). It also looks at the differences between these indigenous experiments with 'agriculture' and the arrival of exogenous agriculture strategies including livestock farming, cereal systems, and modern monoculture plantations into the tropics of Africa, Southeast Asia, the Caribbean, and the Pacific, in particular.

The exploration of tropical forest monumental 'ruins', some more visible than others, will be undertaken in Chapter 6. Contrary to popular stereotypes of tropical forests only being able to house small groups of pristine hunter-gatherers, this chapter builds on the volume of Bacus and Lucero (1999) in outlining the growing evidence for the emergence and sustainability of large, complex social forms in the tropical forests of the Americas, Asia, and Oceania. Here, debates regarding definitions of urbanism and the state in archaeology, as well as the traditional focus on a dramatic 'urban revolution' following an 'agricultural revolution', are explored. Tropical forests challenge Eurocentric obsessions with dense, monumental cities, hereditary hierarchies, and an agricultural focus on a few staple resources in discussions of social complexity. Comparison of tropical forest urbanism with that in neighbouring ecologies highlights the unique challenges presented by these environments, as well as the various adaptations developed by prehistoric societies to densely occupy them sustainably for hundreds if not thousands of years.

Chapter 7 uses ethnographic, historical, and archaeological datasets to evaluate how, despite the plethora of evidence for diverse prehistoric tropical forest use, these habitats have become blank points on the archaeological map. While much ethnographic research has been undertaken on Indigenous groups living in tropical forest ecologies, there has often been a limited attempt to study the dynamic histories of these groups, particularly beyond a purely western perspective. This chapter will also look at the impacts of local states and European colonial powers on tropical forests and their inhabitants around the world, in the form of political marginalization, economic exploitation, and movement away from traditional subsistence strategies and settlement patterns (e.g., Junker, 1999; Hemming, 2009). Chapter 7 will highlight how Indigenous foragers and small-scale farmers were not just passively influenced during this process, but also actively manipulated new economic, political, and cultural frameworks within their own intimate local knowledge of tropical forest environments.

In Chapter 8 I try to make a case for the inclusion of tropical forest archaeology in modern conservation studies, policy making, and protection of the rights of Indigenous communities. The crucial position of tropical forests in earth systems is highlighted prior to a discussion as to when human modifications to tropical forests may have begun to have regional, continental, or global repercussions in the form of the 'Anthropocene'. I seek to evaluate to what extent the 'Anthropocene' concept can be used as a framework to bring the archaeological evidence for human tropical forest modifications to bear on present-day interests, threats, and conservation solutions. This is not to suggest that the advance of urbanism, industrialization, targeted exploitation of mineral, flora, and faunal resources by local groups and multi-national corporations, and the position of tropical forests in discussions of anthropogenically induced climate change, is not leading to a break from the past in terms of sustainable environmental use (Malhi et al., 2014). Rather, that archaeological and historical datasets can help to inform and monitor these changes on a quantitatively rather than qualitatively different scale.

Chapter 9 returns to the main themes of 'Tropical Forests in Prehistory, History, and Modernity' with a discussion as to how the extent and intensity of human occupation and alteration of tropical forests has varied through space and time. I then evaluate to what extent human adaptations to tropical forests are unique rather than just one part of a broader human ecological niche. Are there constraints in tropical forests, relative to other environments, and does this lead to frequent limitations on resilience and sustainable use? Throughout this book it should become evident that our species has influenced and manipulated tropical forests, but to what extent have these environments shaped the biological, cultural, economic, and political characteristics of hominins? Finally, are tropical forests the ultimate environmental marker for changes in human–environment relationships as we push the Earth to the very limit of sustainability? Chapter 9 closes the book with an analysis of the potential demographic, climatic, economic, and social thresholds for human tropical forest occupation, and whether the crossing of these in the future risks not only particular social forms, but entire environments that have stood on Earth for 100 million years.

A NOTE ON NOMENCLATURE

As will be seen in Chapter 2, throughout this book I define 'tropical forests' as those forest formations that lie between the Tropic of Cancer (23°26′14.0″N) and the Tropic of Capricorn (23°26′14.0″S) (Chapter 2). While this encompasses tropical rainforests (Allaby, 2010), it does neglect temperate rainforests in places such as Chile. Although these forest formations will be touched upon, particularly with discussion of the early human occupation of South America (Chapter 4), they are not the main subject of this book. The word 'human' I reserve solely for our own species, *H. sapiens*, not other members of the genus *Homo*. 'Hominin' is used to refer to those hominoid species, from *Orrorin tugenensis* onwards, that indicate some form of adaptation to terrestrial bipedalism. This category includes the genus *Homo* and our own species. The term 'hominoid', on the other hand, includes both hominin and non-human ape taxa, while 'hominid' refers specifically to hominin and non-human great ape taxa (Chapter 3).

The geological epochs of the Miocene (23.030 to 5.332 million years ago), Early Pliocene (5.333 to 3.600 million years ago), Late Pliocene (3.600 to 2.588 million years ago), Early Pleistocene (2.588 to 0.781 million years ago), Middle Pleistocene (781,000 to 126,000 years ago), Late Pleistocene (126,000 to 11,700 years ago), and Holocene (11,700 years ago to present) are defined following geological convention (Gradstein et al., 2004). The 'Anthropocene' remains a source of debate and has been defined by the International Anthropocene Working Group (Zalasiewicz et al., 2015) as beginning with the Trinity nuclear test on 16 July 1945. Others suggest it could begin as early as 1610 (Lewis and Maslin, 2015). Where this epoch is discussed I make sure to highlight this ambiguity as well as the potential role of data in this book in evaluating its definition.

Throughout the book, where a particular radiocarbon chronology for a site, artefact, or finding is referred to I use the convention of calibrated years before present (cal. years BP) (Bronk Ramsey et al., 2010). This will be based on the publication referenced at the point in the text and the reader should be aware that different publications may use different calibration curves depending on the region and point in time at which the study was published. Otherwise I use available chronological information from the published and cited literature of chronometric ages and historical texts to define species, sites, and cultural periods in terms of millions of years ago (Ma) and thousands of years ago (ka). Where convention in the literature describing a given period or region uses years before (BC) and after (AD) the Christian era for periods within the Middle and Late Holocene, I adopt this format.

As per Mitchell (2015), I distinguish people from plants and animals when discussing indigenous occupants of tropical forests by capitalizing Indigenous and Native whenever they are employed for the former, though I recognize that this is potentially problematic (e.g., Kuper, 2003). Where familiar English names exist I have chosen to employ them, rather than communities' self-designations (which in some cases cannot be extrapolated back beyond the twentieth century). Where no name is likely to be recognized by English readers I follow common practice in the historical and ethnographic literature. I apologize in advance for any inadvertent offence caused in this regard.

2

Tropical Forests

Natural History, Diversity, and Potentiality as Theatres of Human Adaptation and Negotiation

> Trees are sanctuaries. Whoever knows how to speak to them, whoever knows how to listen to them, can learn the truth. They do not preach learning and precepts, they preach, undeterred by particulars, the ancient law of life.
>
> —Herman Hesse (1984)

The above quote by the German poet, novelist, and painter Herman Hesse highlights the cultural significance of forests in nineteenth- and twentieth-century western culture as the 'natural' contrast to growing urban populations and industrial expansion. Hesse's focus on the 'ancient' element of these environments is certainly valid in a tropical context, given that tropical forests are some of the oldest land-based environments on the planet, existing for over one thousand times longer than *Homo sapiens* (Upchurch and Wolf, 1987; Davis et al., 2005; Ghazoul and Shiel, 2010; Couvreur et al., 2011). This antiquity also makes them one of the richest and most diverse terrestrial ecosystems on the planet (Whitmore, 1998; Ghazoul and Shiel, 2010). Tropical rainforests, for example, contain over half of the world's existing plant, animal, and insect species (Wilson, 1988). A significant portion of the developed world's diet today originated in tropical forests—including staples such as squash and yams, spices such as black pepper, cinnamon, cloves, and sugar cane, and fruits including bananas, coconuts, avocados, mangoes, and tomatoes (Iriarte et al., 2007; Roberts et al., 2017a). Tropical forests also often provide ample freshwater for their inhabitants. However, despite popular perceptions of forests, and specifically tropical forests, as uniform, they are, in fact, highly variable across space and time. In tropical evergreen rainforests productivity is often primarily allocated to wood products, meaning that edible plants and animals for human subsistence have been considered lacking, or at least more difficult to extract, relative to more open tropical forest formations (Whitmore, 1998; Ghazoul and Shiel, 2010). Similarly, while evergreen tropical rainforests generally receive significant precipitation and freshwater, seasonally dry tropical forests are subject to sub-annual periods of aridity.

Therefore, while archaeologists and anthropologists have tended to see 'tropical forest' as a uniform environmental block, it is important to explore the diversity within this category. Although tropical forest is usually equated with dense, dark, and humid 'tropical rainforest' in the public eye, this is not the case ecologically and, in regions with prolonged periods of drought, tropical forests present very different prospects to the engulfing canopies of evergreen rainforest. Furthermore, even amongst 'tropical rainforests' there are a number of different vegetation formations that present different spatial patchworks of plant and animal resources (Ghazoul and Shiel, 2010; Corlett and Primack, 2011). In addition, despite ongoing discussion of equatorial regions as climatically stable relative to other parts of the world (e.g., Wallace, 1878; Basell, 2008), growing palaeoenvironmental datasets are demonstrating that tropical forest extent and structure has fluctuated, often dramatically, in different regions (Hope et al., 2004; Bonaccorso et al., 2006; Hunt et al., 2012). This geographical and temporal variability has had significant impacts on the challenges posed to human and hominin hunter-gatherers, farmers, urban settlements, and present-day communities (e.g., Roberts and Petraglia, 2015; Roberts et al., 2017a). Yet, tropical forest diversity has thus far received little systematic review in an archaeological and historical context, despite popular interest in them in the face of their potential disappearance (Moritz et al., 2014).

WHAT ARE TROPICAL FORESTS?

The term 'tropical forest' is popularly synonymous with 'tropical rainforests'. These forest formations have been variously defined on the basis of their composition (warmth-loving botanical formations dominated by broad-leaf angiosperms; White, 1983), precipitation (an annual precipitation in excess of 1,400 mm, and often around 2,000 mm or more; Grainger, 1996a, 1996b), and even temperature (mean temperature of coldest month >18°C, annual temperature >24°C; Holdridge, 1947). However, it is primarily low seasonality, in both temperature and rainfall, that is the key climatic parameter behind the persistence of tropical rainforests (Whitmore, 1998; Sternberg, 2001; Ghazoul and Shiel, 2010). Rainforests generally develop, and persist, where every month is wet (with around 100 mm rainfall or more—though see Ghazoul and Shiel, 2010), or where there are short dry spells lasting only a few days or weeks (Whitmore, 1998; Sternberg, 2001). This characteristic separates the 'permanently' wet rainforests from monsoon forests, savanna forests, and thorn tropical forests, and their corresponding deciduous species, open forest structures, and ecological mosaics (Grainger, 1996a, 1996b), although closed forests can form in the lowland tropics with as little as 800 mm of annual rainfall (Ghazoul and Shiel, 2010).

Tropical forests, as well as tropical rainforests more specifically, occur in all five of the tropical continents (Sayer et al., 1992; Blasco et al., 1996; Harcourt and Sayer, 1996; Malhi and Wright, 2004). Most extensive are the Neotropical rainforests that occur in the Amazon and Orinoco Basins, on the eastern slopes of the Andes, along the Pacific coasts of Ecuador and Colombia, in North and Central America, and on the Atlantic coast of Brazil (Figure 2.1). The second largest tropical forest region is in Southeast Asia, centred on the Malay Archipelago and extending up the mainland into Burma, Thailand, and Indo-China (Figure 2.2). Smaller again is Africa's tropical forest zone, focused on the Congo Basin, with an 'outlier' in Madagascar (Figure 2.3). Tropical forests also extend into Melanesia, the Pacific, and form a narrow coastal strip in Queensland, northeast Australia (Figure 2.2). Furthermore, small blocks of tropical forest, including rainforest proper, exist

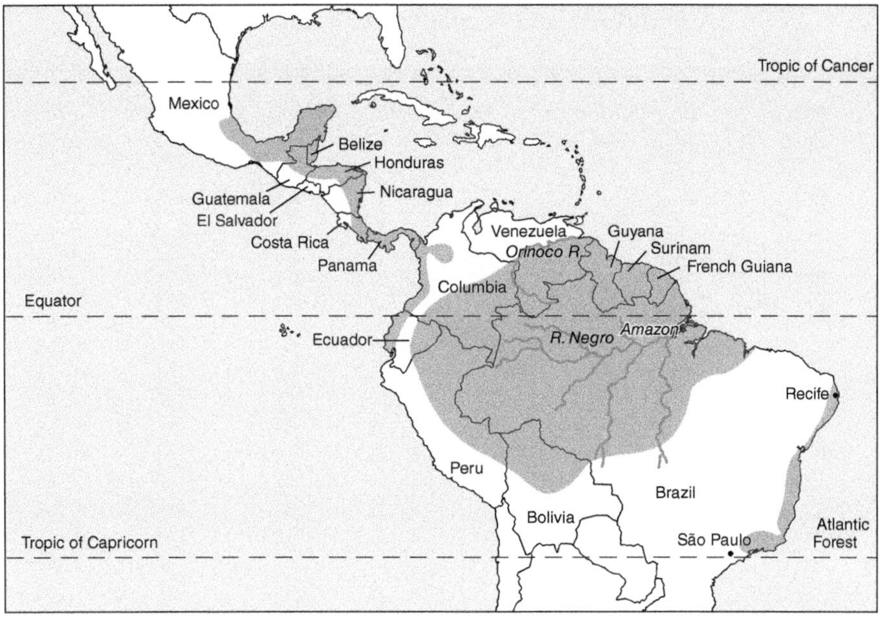

Figure 2.1. Extent of Neotropical forest based on the MODIS (Moderate Resolution Imaging Spectroradiometer) Land Cover MCD12Q1 majority land cover type 1, class 2 for 2012 (spatial resolution of 500 m). Downloaded from the US Geological Survey Earth Resources Observation System (EROS) Data Center (EDC). With thanks to Hans Sell.

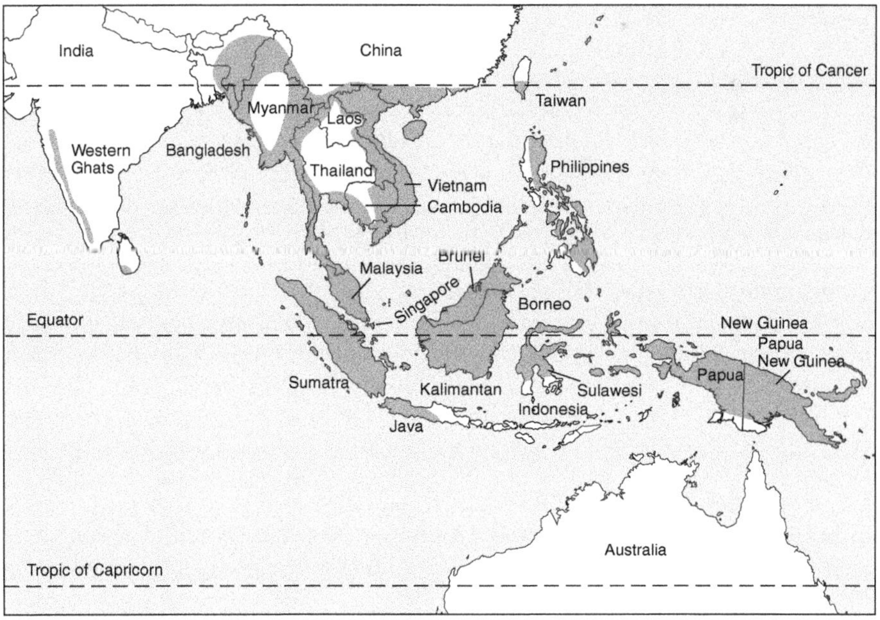

Figure 2.2. Extent of tropical forest in South Asia, Southeast Asia, and Melanesia based on the MODIS (Moderate Resolution Imaging Spectroradiometer) Land Cover MCD12Q1 majority land cover type 1, class 2 for 2012 (spatial resolution of 500 m). Downloaded from the US Geological Survey Earth Resources Observation System (EROS) Data Center (EDC). With thanks to Hans Sell.

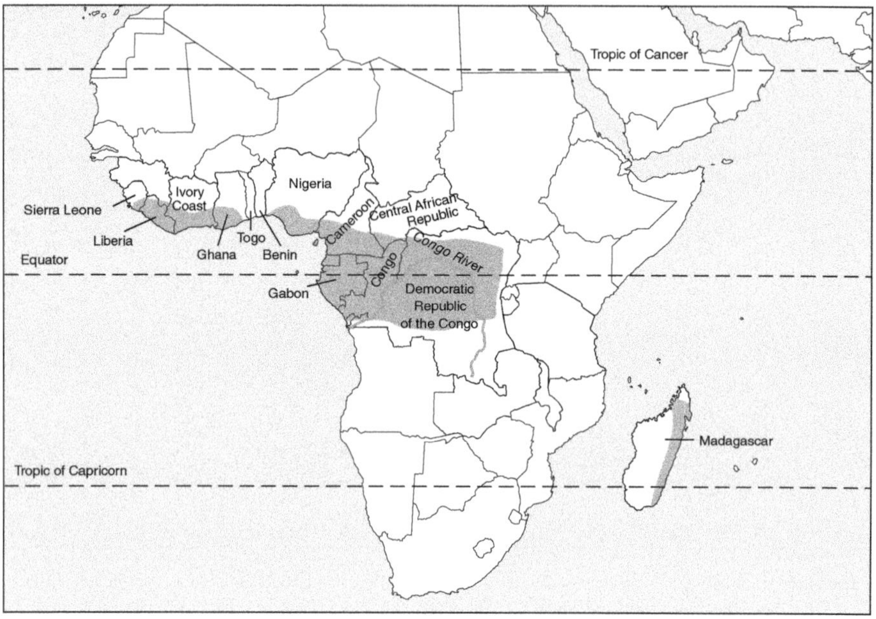

Figure 2.3. Extent of tropical forest in Africa based on the MODIS (Moderate Resolution Imaging Spectroradiometer) Land Cover MCD12Q1 majority land cover type 1, class 2 for 2012 (spatial resolution of 500 m). Downloaded from the US Geological Survey Earth Resources Observation System (EROS) Data Center (EDC). With thanks to Hans Sell.

in Sri Lanka and along the Western Ghats of India (Figure 2.2). Importantly for our purposes here none of these 'blocks' are neat. In South America and Southeast Asia there is often an ill-defined boundary with subtropical forests as one moves away from the Equator, while in South Asia various layers of tropical forest persist through strong monsoonal seasonality in precipitation (Whitmore, 1998; Barboni et al., 2003; Ghazoul and Shiel, 2010). In Australia, where tropical forests reach their temperate limit, they become increasingly restricted to the wettest areas and the deepest, most fertile soils (Webb, 1968).

While tropical forests, and particularly tropical rainforests, are recognized as being of considerable antiquity, the location and timing of their origin remain uncertain (Corlett and Primack, 2011). This is primarily a result of the fact that the fossil plant record for tropical regions is sparse and fragmentary (Jaramillo et al., 2006). Direct evidence from fossil leaves and wood place the origins of swamp forests in North America 385 million years ago (Stein et al., 2007), becoming widespread, broad-leaved, multi-storied forests that reached 30 metres in height by 360 million years ago (Algeo and Schekler, 1998). Gymnosperms (seed-producing plants) became increasingly dominant during the Carboniferous period and fossil evidence for root systems characteristic of some modern rainforests date to 305 million years ago (Ghazoul, 2016). The major evolution of flowering plants is thought to have occurred before Gondwanaland began to break up 120 million years ago and 59 families of flowering plants are pantropical (Whitmore, 1998) (Figure 2.4). Nevertheless it has been argued that the first palaeoflora truly representative of tropical forest formations only appear during

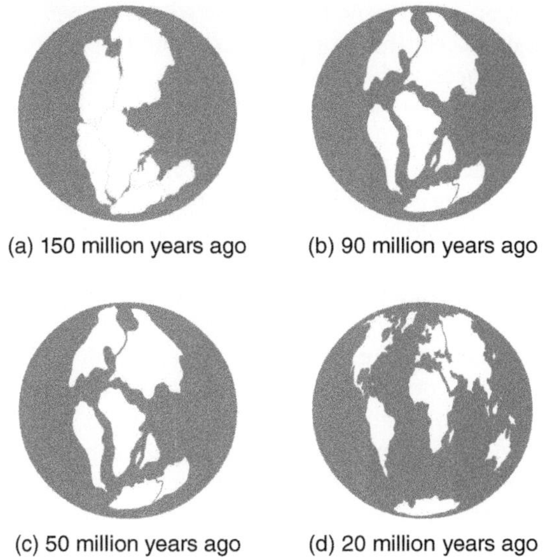

Figure 2.4. Schematic of the breaking up of Gondwanaland and the formation of current continental distributions. With thanks to Hans Sell.

the Early Palaeocene (66.0 to 61.6 million years ago) in North America (Johnson and Ellis, 2002) and the Late Palaeocene (59.2 to 56.0 million years ago) in Africa (Jacobs, 2004) and South America (Wing et al., 2009). More recent phylogeny studies of the palm family (Arecaceae/Palmae), major components of tropical forests worldwide, appear to confirm an earlier date of *c.* 100 million years ago, however (Couvreur et al., 2011) (Figure 2.5).

Difficulties in determining the exact origins of tropical forests are inevitably linked to issues of definition. Equating tropical forests with tropical rainforests is problematic given that the latter are often considered to have a very specific composition and climate today (Ghazoul and Shiel, 2010; Ghazoul, 2016). Furthermore, it is open for debate as to whether the presence of a plant that is important across present-day tropical forests in the fossil record is enough to signal the origin of a tropical forest formation. Whereas previous ecological research has suggested that tropical forests have been ecologically stable (Wallace, 1878; Basell, 2008), more recent data indicate highly dynamic forest formations that fluctuate both in extent (Morley, 2000) and in the diversity of the plants that they sustain (Jaramillo et al., 2006). As reviewed by Couvreur et al. (2011) this has in turn led to broad, differing evolutionary hypotheses to explain high levels of plant and animal species diversity in tropical forests today: (i) early, rapid speciation in response to favourable climatic conditions followed by a deceleration of diversification due to global cooling and forest contraction ('ancient cradle model' (e.g., Morley, 2000)); (ii) constant diversification rates and low extinction rates leading to gradual accumulation and long-lasting ecosystem stability ('museum model' (e.g., Wallace, 1878)); and (iii) an increase in diversification rates towards the present in response to climatic, tectonic, or biotic changes (the 'recent cradle model' (e.g., Haffer, 1969; Gathorne-Hardy et al., 2002)).

With the exception of the now largely discredited 'museum model', the other hypotheses generally accept that tropical forests, and tropical rainforests, will have had different appearances, with regards to extent, structure, and species, from their first appearance through to the present day. Indeed, plate tectonics have clearly played a key role in shaping the current distribution and diversity of tropical forest genera across the planet. The Dipterocarpaceae family, that represents some of the main economically important timber trees in tropical forests today, is playing an increasingly prominent role in studies of tropical forest origins and spread. While one hypothesis of origins suggests Dipterocarpaceae arose in Southeast Asia in the late Mesozoic before spreading into India and then Africa (Shukla et al., 2013), fossil, molecular, and morphological evidence support a Gondwanan origin for this family (Ashton and Gunatilleke, 1987; Gregory et al., 2009; Ghazoul, 2016). The Dipterocarpaceae family originated in Gondwanaland pre-120 million years ago, and migrated towards the Laurasian Plate following the break-up of this supercontinent. Fossil evidence demonstrates that after the separation of Gondwana in the early Cretaceous, this family drifted north as part of the Deccan Plate, holding the Indian subcontinent and Sri Lanka, until it collided with Laurasia 45 million years ago and allowed the spread and diversification of this family across Southeast Asia (Morley, 2003; Ghazoul, 2016) (Figure 2.5).

Figure 2.5. Photographs of fossil Dipterocarpaceae leaves from the Cerrejon Formation dated to the Late Palaeocene (60–58 Ma). a) CJ6 leaf with water, leaf of *Menispermites cerrejoneses* (Menispermacea), b) CJ26 leaf, leaf of *Malvaciphyllum macondicus* (Malvaceae). Taken by, and reproduced with permission from, Carlos Jaramillo.

Regardless of their exact origins, it seems likely that the world's existing tropical forests owe much to the high, annually-stable temperatures and precipitation, coupled with periods of aridity, temperature decline, and seasonality, that have characterized the climate of much of the tropics since at least the onset of the Quaternary period (2.6 million years ago) (Ghazoul and Shiel, 2010). It is, in part, for this reason, that this book characterizes tropical forests as those forest formations that exist within the tropics today. However, perhaps more importantly for our purposes, while this latitudinal definition faces the problem that palaeolatitudes have varied in the past (van Hinsbergen et al., 2015), it facilitates the discussion and comparison of diverse tropical forests across space and time. This is particularly important from the start of the Quaternary period when, as will be seen, more resolved palaeoclimatic and palaeoenvironmental records demonstrate significant changes in tropical forest structure, species composition, and extent through time. Moreover, perhaps from *c.* 7–5 Ma, and certainly from the evolution of our species *c.* 300 ka, hominins began to have significant impacts on the structure, plant species, and animal species present in different tropical forest formations around the world. So much so that the idea of a 'virgin' or natural tropical forest existing anywhere today is looking increasingly unlikely (Willis et al., 2004; Barton et al., 2012; Bayon et al., 2012). In order to be able to compare tropical forests across different spatial and temporal scales, this broader definition of tropical forests is key.

GEOGRAPHICAL DIVERSITY

The diversity within the categories of 'tropical forest' and 'tropical rainforest' have been reviewed comprehensively elsewhere (Ghazoul and Shiel, 2010; Ghazoul, 2016). However, here I wish to highlight this variability in the context of human adaptations within these habitats. In forest ecology, the forest canopy refers to the habitat zone formed by mature tree crowns, and canopy structure (its organization or spatial arrangement) is often a key parameter in the definition of forest types (Parker, 1995). In tropical rainforests low climatic seasonality encourages the formation of closed canopies, whereby the crowns of individual trees overlap to form a virtually continuous layer. This characteristic canopy layer intercepts around 95 per cent of sunlight (Whitmore, 1998) and has resulted in the classification of rainforests into different abstract layers, with vegetation organized into a vertical pattern from the top of the soil to the canopy (Bourgeron, 1983; Denslow, 1987). Each stratum is home to a unique biotic community adapted to life within a particular microclimate and of different relevance to human activity (Figure 2.6). For example, the forest floor is home to a few plants adapted to low light conditions and is relatively clear, facilitating the movement of large mammals, reptiles, and amphibians (Bourgeron, 1983). The understory and canopy contain a larger variety of usable plants and plant parts (including herbs, climbers, mosses, and tree fruits) and are home to diverse collections of semi-arboreal and arboreal mammals, reptiles, birds, and insects. The highest, emergent layer is home to smaller numbers of very large trees and specialized birds, primates, and other mammals (Whitmore, 1998).

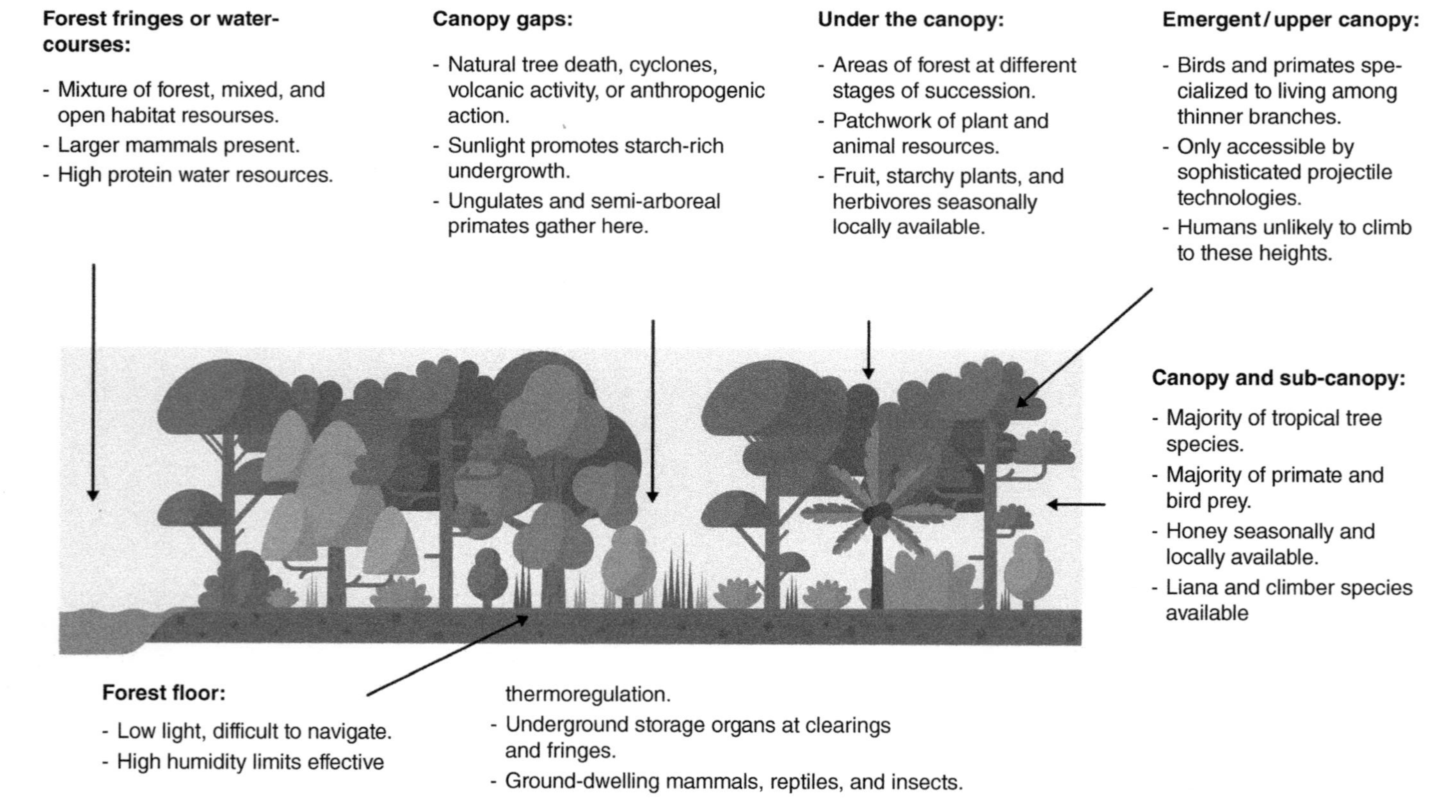

Figure 2.6. Schematic of the distribution of plant and animal resources, as well as conditions relevant to human subsistence, in a tropical forest. Adapted from Roberts et al. (2016). With thanks to Hans Sell.

Yet, even in rainforests, schematics of forest strata belie considerable dynamism within these ecosystems. Vegetation lifecycles mean that gaps constantly occur in the canopy, making this stratification a simplification, and ensuring that various patches exist at all stages of the growth cycle (Colinvaux and Bush, 1991) (Figure 2.7). These gaps are rapidly colonized by seedlings and often facilitate increased densities of vegetation growth and competition and is where most carbohydrates in the form of yams, aroids, and plantains will be found (Whitmore, 1977; Denslow, 1987). The nature and frequency of gap formation varies over space and time, and is influenced by anthropogenic and climatic factors including disease, fire, and extreme weather events (Ghazoul, 2015). For example, Southeast Asia and Papua New Guinea are prone to cyclones, earthquakes, and volcanic eruptions, all of which can clear vast swathes of forest (Johns, 1986; Sémah and Sémah, 2012). Given their reliable access to water, greater ease of navigation, and relative faunal and floral richness, tropical forest canopy gaps around rivers likely provided crucial corridors for the expansion of hunter-gatherers (Bird et al., 2005; Venkataraman et al., 2017) and farming communities (Grollemund et al., 2015), and populations in the Amazon Basin often relied upon tributaries for access to high protein freshwater resources such as fish and turtles (Heckenberger and Neves, 2009). The unique volume of water that cuts through tropical forests, which challenges agricultural and urban settlements today (Larsen, 2017), and the riverine and stream paths this can create, is something that is often ignored in palaeoanthropological and archaeological discussions.

Moreover, there is not a single 'tropical forest' formation (Figures 2.8 and 2.9). Climatic factors, geology, and topography lead to division by ecologists and botanists into a series of vegetation 'formations' (Whitmore, 1998). Rainforest formations are defined by forest structure (i.e., canopy coverage, tree height, the presence of tree crowns or layers, and the presence of climbers) and by their physiognomy (whether trees are buttressed, i.e., extensive horizontal root span—Tang et al., 2010; crown shape; the nature of leaves; to what extent the forest is evergreen or deciduous; and where flowers and fruits form on trees) (Ghazoul and Shiel, 2010). These formations can, in turn, be grouped according to the main physical characteristics of their habitat (Holdridge, 1967; Holdridge et al., 1971; Webb and Tracey, 1981; Brünig, 1983; Whitmore, 1998; Ghazoul, 2016). Other tropical forest formations, including dry tropical forests, monsoon forests, thorn forests, and mosaic forest–woodland–savanna habitats, are also defined by structure, species composition, and physiognomy. However, here it is often climate, and in particular the extent of seasonality in temperature and rainfall, that dictate the presence of these formations. The frequency of tropical rainforest, and other tropical forest, formations, as well as their corresponding resource distributions, differs across the tropics, and has varied in the past—particularly during the Late Pleistocene. Human hunter-gatherers, farmers, and urban dwellers will each have been faced with different possibilities and constraints within each of these forest types.

Evergreen Tropical Rainforest

This dense, tall ($\geq$45 m) forest type is the most species-rich of all plant communities. Evergreen tropical rainforest is the model of vertical forest stratification

Figure 2.7. Schematic of gap formation and forest succession in an evergreen tropical rainforest. With thanks to Hans Sell.

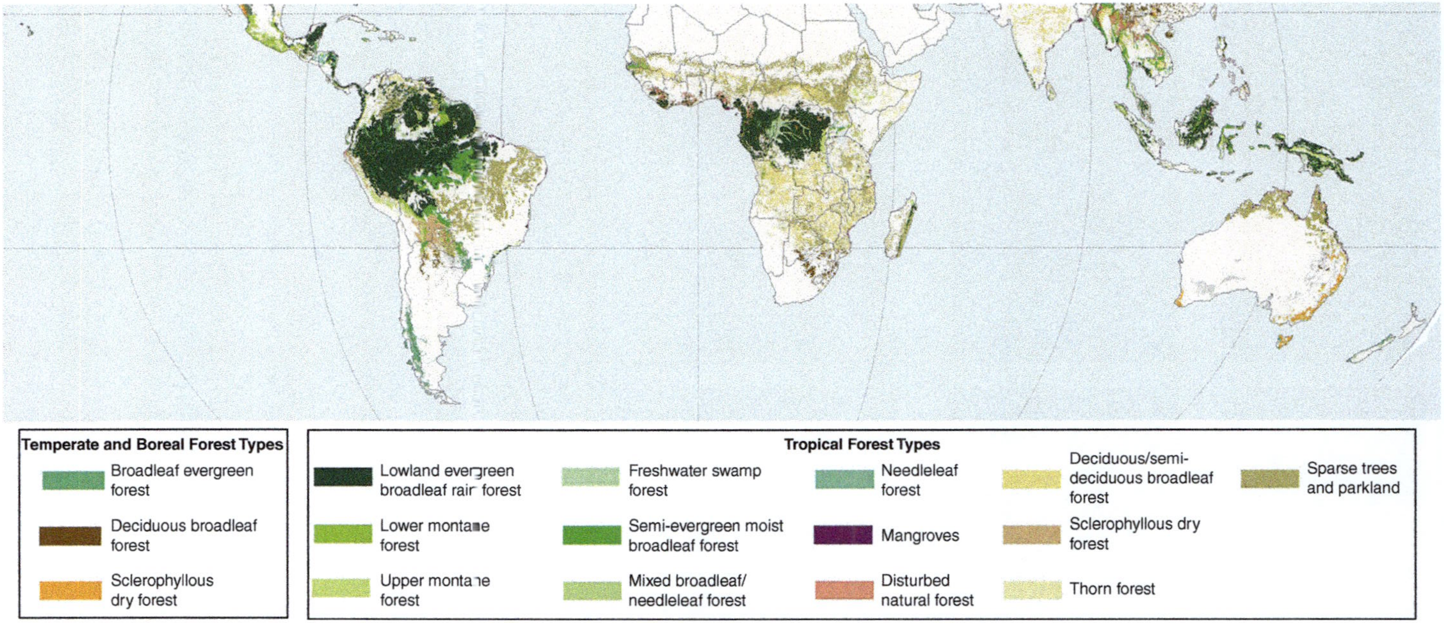

Figure 2.8. Map of different tropical forest distributions adapted from Ghazoul and Sheil (2010). Temperate broadleaf evergreen forest, deciduous broadleaf forest, and sclerophyllous dry forest are also shown. Map courtesy of WCMC-UNEP. <http://www.unep-wcmc.og/forest/global_map.htm>. With thanks to Nils Vanwezer.

Figure 2.9. Tropical forest formations discussed in the text (see also Whitmore, 1998): Tropical lowland evergreen rain forest in southeastern Cameroon (Photograph taken by, and used with permission from, Fernando Ramirez Rozzi), Tropical semi-evergreen forest in southwestern Sri Lanka (Photograph taken by the author), Montane rainforest in the Ivane Valley of Papua New Guinea (Photograph taken by, and used with permission from, Glenn Summerhayes), Peat swamp forest in southern Thailand (Photograph taken by Thipwan and reproduced with permission from shutterstock.com), Freshwater swamp forest in southeastern Cameroon (Photograph taken by, and used with permission from, Fernando Ramirez Rozzi), and Dry tropical forest in northeastern Thailand (Photograph taken by Neenawat Khenyothaa and reproduced with permission from shutterstock.com). With thanks to Jana Zech.

mentioned above. The absorbance of light by the canopy means that the forest floor is often relatively bare, leading to an anthropological focus on a lack of carbohydrate-rich ground plants for human exploitation (Hart and Hart, 1986; Bailey et al., 1989), as well as a paucity of large game beyond the canopy (Bailey et al., 1989). Nevertheless, canopy fruits can offer crucial resources, for example the masting Dipterocarp trees of Southeast Asia (Knott, 1998; Ghazoul, 2016) and the fruit and fibre of palms in evergreen moist forests more widely (Whitmore, 1998). The formation of canopy gaps through tree death, natural disaster, or human action can also foster increased plant productivity and animal presence on the forest floor. Empty forest floors also facilitate the movement of hominins, though without rivers or canopy gaps reference points for navigation can be lacking. The density of the canopy also makes hominin thermoregulation difficult under highly humid conditions (Perry and Dominy, 2009; Roberts et al., 2016), while constant moisture makes trees relatively difficult to burn effectively. This forest type is the main lowland tropical forest formation of Southeast Asia and Melanesia. It also occupies western Amazonia and the Pacific coast of South and Central America. In Africa it is today restricted to the western coast, from Guinea-Bissau down to the Democratic Republic of Congo. All regions with this formation have no regular annual dry season.

Semi-Evergreen Tropical Rainforest

This is closed, tall forest that includes significant proportions of evergreen and, in the top of the canopy, deciduous trees. The number of species remains high in this formation but less so than in evergreen rainforest. The stature of semi-evergreen rainforest is also usually less than its evergreen counterpart, with local canopy stratification and emergents occurring as scattered individuals or groups (Ghazoul and Shiel, 2010). Large woody climbers and bamboos are abundant in these forests, and can represent important resources for tool and structure manufacture (e.g., Xhauflair et al., 2017). Semi-evergreen rainforest occurs instead of evergreen tropical rainforest where there is a strong, regular annual period of moisture stress. This seasonality leads to the opening of the canopy that can lead to greater plant diversity accessible to human foragers on the forest floor and within the canopy itself, as well as slightly less humid conditions (at least during parts of the year). Semi-evergreen forests often form a transitional formation between evergreen rainforest and more open, mixed forest environments. They occupy most of the African rainforest block and occur around the main Southeast Asian rainforest zone. Tropical semi-evergreen rainforest also forms the majority of the South Asian rainforest, as a result of seasonal precipitation changes dictated by the monsoon, and most of the Australian tropical rainforest (Whitmore, 1998; Ghazoul and Sheil, 2010).

Montane Rainforest

In contrast to evergreen and semi-evergreen rainforest formations, which are dominated by trees with multi-veined mesophyll leaves, montane rainforests have lower, more even, single-vein microphyll canopies dominated by slender trees with dense subcrowns (Whitmore, 1989). Upper montane rainforest is often only *c.* 10 m tall, with trees heavily covered in ferns. On small mountains, upper montane rainforest directly borders lowland rainforest. However, on larger mountains, lower montane rainforest occurs as an intermediate formation. On higher peaks still, upper montane rainforest is itself replaced by shorter, small-leaved formations called subalpine forests (Ghazoul and Shiel, 2010). Montane rainforests have open canopies, though while precipitation can be low plants specialized at the capture of moisture can still lead to high humidity. Their high altitudinal position, however, does mean that human thermoregulation must increasingly focus on the need to keep warm, particularly during periods of climate change (Gosden, 2010). Montane rainforest formations have plentiful access to numerous small streams but not large rivers and their corresponding resources. Increased light penetration to the forest floor leads to increased plant and animal resources where temperatures allow, and these forests are relatively easy to burn. Extensive montane forests exist in Southeast Asia and Melanesia (Denham and Haberle, 2008; Haberle et al., 2012). They are less extensive in Africa, but do occur in Cameroon and in the Congo Basin along the Albertine Rift (Whitmore, 1998; Plumptre et al., 2007).

Heath Forest

As with montane rainforests, this tropical formation has a low, but orderly, canopy formed of small, young, thin trees (Whitmore, 1989; Ghazoul and Shiel, 2010). The canopy is closed and dense with no trace of layering. Like montane forests, this formation has a predominance of microphyll trees, while long slender climbers and epiphytes dominate non-tree vegetation. Occurring on water-limited, podsolized, siliceous, acidic soils, heath forest is extensive in South America, including the Orinoco Basin, Brazil, and Guyana. In Sundaland (including mainland and Island Southeast Asia) these forests are known as *Kerangas*, a term the Dayak Iban of Borneo use to refer to soils that cannot grow rice even following deforestation (Whitmore, 1984; Riswan and Kartawinata, 1991). Very low levels of phosphorus have been observed in Amazonian podsols (Quesada et al., 2011) and the sandy soils of heath forests are also often lacking in nitrogen (Proctor, 1999). As a result, many plant species utilize symbiotic relationships with rhizobia (nitrogen-fixing bacteria) and insects in order to survive (Proctor, 1999). These soils are poor for traditional crop agriculture. Given that burning or clearance is unlikely to improve this situation, human foragers and agriculturalists are more likely to focus on tree-cropping, including for starch-rich sago palm species (Ruddle et al., 1978; MacKinnon et al., 1997). While insects, reptiles, and amphibians may provide resources, these forests are poor in reliable animal protein as a result of the toxicity and low nutritional value of much of the flora. Minor heath forests also occur in Gabon, Cameroon, and Ivory Coast in Africa.

Peat Swamp Forest

This formation is defined primarily by its special habitat. Peat swamp forests are tropical forests where waterlogging prevents the full decomposition of organic matter. Over time this creates thick layers of peat (Page et al., 2006). Tropical peat swamp formations occur between better-drained semi-evergreen forest and mangroves on the coast. Parts of Southeast Asia and Melanesia physiographically favour the formation of peat. Since sea-level rose at the end of the Last Glacial Maximum, rivers have deposited silt on flood plains as levees, behind which swamps have developed, becoming less saline as their soil level was raised by continuous organic matter deposition (Whitmore, 1998). Tropical peat swamps can be home to varied forest types, ranging from open, stunted forest at the centre to >50 m tall forests on the outer edges. Tropical peat swamp forests, like heath forests, have poor soils, making agriculture impossible in the absence of clearance and drainage. However, their rich organic content means these habitats represent carbon sinks (Sorensen, 1993) that can support rich tree and animal resources, including our great ape relatives, the orangutans, as they are pushed from their preferred evergreen and semi-evergreen forest habitats. Mass conversion of peat swamps to rice padis and agricultural lands in Southeast Asia today represents a major conservation threat to tropical forest plant and animal taxa, as well as the supply of water that many communities extract from peat swamps (Yule, 2010). Peat swamp forests also occur in the Amazon Basin, as well as in Uganda and Congo-Kinshasa.

Freshwater Swamp Forest

Again defined by habitat, this group includes a diverse assemblage of forests flooded by river water that leads to soils richer in nutrients for plants, and of a much higher pH, than those of peat swamp forests (Whitmore, 1998; Ghazoul and Shiel, 2010). Where flooding is episodic the periodic swamp forest group encapsulates additional variability. In the Amazon Basin of Brazil, a seasonally flooded forest is known as a *várzea* or *igapó*. These forests consist of a mosaic of habitats, as a result of topographical variation and the instability of river courses, including sequences of successional vegetation, forest mosaics, aquatic vegetation and ecosystems, and permanent swamp forest. This provides hominins with ample access to a diversity of forest plant, animal, reptile, and insect resources and, perhaps significantly in the context of protein access, reliable access to freshwater fish and fauna. Of significance to farmers in these habitats is the fact that freshwater swamp forests are locked in a rapid cycle of nutrient capture and recycling (Wittman et al., 2004). Moreover, regular or periodic flooding enriches soils on the banks for crop growth, providing the associated hazards can be mitigated. The most famous, extensive, and diverse permanent and periodic freshwater swamp forests occur in the Amazon and the Congo Basins. The alluvial plains of Southeast Asia were once home to extensive swamp forests, but little now remains as a result of clearance for wetland rice cultivation (Whitmore, 1998; Ghazoul and Shiel, 2010).

Dry Tropical Forest

Where regions of the tropics have a long dry season, often lasting several months, tropical dry forest formations occur (Murphy and Lugo, 1986; Bullock et al., 1995; Dirzo et al., 2011). Deciduous trees predominate in these forest formations and during the dry season these trees shed their leaves, facilitating water conservation as well as the seasonal growth of plants below the canopy to form a thick understory. Within these formations, succulent and evergreen species also occur where they have access to plentiful groundwater. Although these forests are less biologically diverse than rainforests, they are home to a large variety of wildlife, including monkeys, deer, large cats, parrots, various rodents, and ground-dwelling birds (Dirzo et al., 2011). The mammal biomass tends to be much higher in dry forests than in rainforests, although large predators also exist, such as the tiger, which present different complications for hominin foragers. Plant and animal species within dry tropical forests tend to have larger ranges than moist forest species and connectivity is often essential for the maintenance of viable populations (Bullock et al., 1995). Dry forests are highly sensitive to excessive burning, overgrazing, and the expansion of exotic species. This makes initial colonization and deforestation easier for human foragers and farmers. However, the long-term balance of these ecosystems is more challenging to maintain, especially for sedentary societies. Indeed, recent research in the Caribbean and Latin America has highlighted these forests as the most threatened tropical ecosystems in the world (DRYFLOR et al., 2016).

Monsoon Tropical Forest

The nature and extent of this type of dry tropical forest is defined by a climatic system, the monsoon. The monsoon is a seasonal change in wind direction (McKnight and Hess, 2000). In Asia, during the summer there is an onshore flow of air due to areas of low pressure forming over rapidly heating landmasses. In winter, an offshore air-flow is prevalent as the land cools down faster than the oceans. These pressure patterns affect the amount and seasonality of precipitation. For example, in South Asia the majority of annual precipitation is delivered by the Southwest monsoon while less and more spatially limited precipitation results from the winter Northeast monsoon (Clemens et al., 1991; Enzel et al., 1999; Wang et al., 2005). This leads to a period of drought in many parts of the region, with corresponding impacts on vegetation and forests. Monsoon tropical forests are, as a result, highly sensitive to any changes in climate, and in particular, the position of the Inter-Tropical Convergence Zone (ITCZ). As with other dry or seasonal tropical forests, large bodied mammals, including deer and primates, are prevalent, although large fauna is limited to large birds and reptiles in the monsoonal forests of Australia (Bowman and Woinarski, 1996). Honey and a variety of other plant resources provide ample carbohydrates (e.g., Deraniyagala, 1992). However, farming and urban communities in particular would need to develop ways of dealing with unpredictability of water in these habitats, with floods and droughts occurring in some regions over the course of a few years (e.g., Gilliland et al., 2013).

Thorn Tropical Forest

Thorn forests consist of dense, scrub vegetation, with seasonal rainfall averaging 250–500 mm, and exist only right at the edge of the tropics (Wikramanayake et al., 2002). These forests exist at the lower end of the tropical rainfall threshold in South America, South Asia, Australia, and southwestern Africa. In South America, thorn tropical forest is also known as *Caatinga*. Thorn tropical forests consist of open woodland with thorny trees typically less than 10 m tall, alongside xeric shrubland and patches of grassland. In the *Caatinga* forests of South

America the small thorny trees shed their leaves seasonally while cacti, thick-stemmed plants, and arid-adapted grasses make up the understory (Klinge et al., 1977). A large diversity of annual plants grows, flowers, and dies during the brief rainy season, providing a highly-seasonal abundance of resources for hominin populations. The available faunal resources can vary significantly within this category of forest. While *Caatinga* forests are generally limited to rodents, birds, bat, and primate species (Klinge et al., 1977), the thorn tropical forests of South Asia support large mammals, including the Indian elephant (*Elephas maximus indicus*) as well as the nilgai antelope (*Boselaphus tragocamelus*). Long dry seasons mean that any settled agricultural or urban communities must develop sophisticated irrigation technologies to sustain any semblance of a large, sedentary population (Meggers et al., 1973). These forests are highly vulnerable to both natural and anthropogenic burning (Rodríguez-Trejo et al., 2011).

Tropical Forest-Savanna Mosaics

At the fringes of the tropics, tropical forests grade into grassland and savanna ecosystems in boundaries that are often difficult to demarcate (Dexter et al., 2015) (Chapter 3). These environments are potentially highly productive for hominins offering access to resources from both forest and savanna ecosystems (Blome et al., 2012). They are anthropogenically created in drier regions of the tropics where urban and agricultural settlement disrupts tropical forest formations and will become an increasing focus of conservation as climate change alters the boundary of tropical forests around the world. Large predators and large mammals, such as elephants, frequently make use of semi-evergreen and dry tropical forest resources on the one hand, and savanna opportunities on the other (Wikramanayake et al., 2002; Fernando et al., 2008). This has also been suggested for chimpanzee populations in Fongoli, Senegal (Pruetz and Bertolani, 2009; Pruetz et al., 2015). The same was likely true of hominin (Figure 2.9), including human, populations in the past as these environments would have allowed access to the huge biodiversity of tropical forests without the same challenges of navigation, thermoregulation, and a lack of large game that might be experienced as part of a tropical forest specialization. Moreover, while climate change might have influenced the position of the boundary between forest and savanna, this environment would have persisted. Forest-savanna mosaic environments will also have been important for ancient settled agricultural and urban populations, with the persistence of forest maintaining soil balance and stability and savannas providing land for grazing and crop farming.

ETHNOGRAPHIC PERSPECTIVES ON HUMAN OPPORTUNITIES IN TROPICAL FORESTS

From the above it can be seen that while some anthropologists and archaeologists have characterized tropical forests as uniformly inhospitable from the perspective of human foraging and farming (Chapter 1), no one human population will have the same experience of a tropical forest as another. This is clear from the rich record of ethnographic information of hunter-gatherers, horticulturalists, and farmers in tropical forest environments, and highlights how universal ideas of plant and animal availability, and soil fertility, will miss local nuance and agency. As seen in Chapter 1, ethnographic exploitation of botanical resources has been at the heart of the anthropological debate regarding the viability of tropical forests for long-term human occupation without access to agricultural crops. It has been argued that in tropical forests, and particularly evergreen lowland tropical rainforests, most of the energy stored in plants is in trunks or leaves too high in the canopy to be accessible to human gathering, even if edible (McElroy and Townsend, 1985; Bailey et al., 1991; Ghazoul and Shiel, 2010). Furthermore, the depressed seasonality of evergreen tropical rainforests makes storage of energy and moisture in tubers less adaptive for plants than in zones with marked winters or dry seasons. Where such resources are present, they have been argued to be highly scattered (Kikkawa and Dwyer, 1992) and to not provide enough reliable starch to support foragers without agricultural connections (Hart and Hart, 1986; Bailey et al., 1989).

Nonetheless, as we have seen, seasonality can differ between tropical forest formations, with some having short dry seasons or longer, less pronounced dry seasons. Furthermore, to focus on 'yams' and starch as the only sources of carbohydrate or botanical benefit to hunter-gatherer subsistence is to greatly understate the diversity of rainforest microhabitats and resources available (Dentan, 1991). In the Amazon Basin, for example, the Chácobo of Bolivia employ 82 per cent of the rainforest plants available, while the Quijos Quichua of Ecuador use more than 90 per cent of identified tree species (Bennett, 1992). Tropical forest fruits represent a major source of carbohydrate and sugar. For example, in Malaysia, cycad seed flour can provide a useful stored carbohydrate (Thieret, 1958), while the grains of Job's tears (*Coix lachrymae jobi*) are about 60 per cent soluble carbohydrates, mostly starch (Burkhill, 1966). Malaysian ethnographic groups have been shown to tend wild trees, deliberately defecate seeds near campsites, and organize fruit tree collection in order to ensure stable availability (Ellen, 1988). The lack of seasonality cited as causing reduction in underground storage organ availability is, by contrast, likely to make many tropical forest fruits a reliable year-round source of carbohydrate. In addition, the biggest starch storage organs in the plant kingdom are not underground but in the trunks of the hapaxanthic palms, especially those of the genus *Corypha*, that grow in clumps across Southeast Asian, Melanesian, and Sri Lankan rainforests (Corner, 1966).

The starch from palm trees is available throughout the year and, once extracted and dried, is light and easy to carry (Strickland, 1986). In the case of the Penan of Malaysia it has been argued that a combination of wild yams and sago palm (*Eugeisonna utilis*) can comfortably provide enough starch to maintain independent lowland evergreen rainforest foraging (Hong, 1987; Brosius, 1990, 1991; Dentan, 1991). The Penan cut a small number of mature trunks from a single clump of roots, never felling the entire root-stock and allowing for sustainable regeneration (Brosius, 1986). In addition to intervening in the reproduction of a single species, human groups will also foster entire microhabitats. Hill and Baird

(2003) demonstrate that the Kuku-Yalanji of Australia will burn vegetation to preserve different 'types' of tropical forest. On the one hand, the most important wet season staple, *Beilschmiedia bancroftii* yams, are preserved by the protective burning of rainforest margins in order to ensure that no high-temperature fires develop. By contrast, in the dry season, fires are lit to improve access to the palm-like *Cycas media* growing in open forest areas. This also enhances the seed productivity of this crop, an essential source of carbohydrate during the dry season. Diversity in these open forest areas is maintained by fire strategies that prevent these habitats reverting to rainforest (Hill et al., 2001), yet permanent conversion to grassland is also prevented.

Beyond active modification, foraging and farming groups frequently focus on particular microhabitats. Ichikawa (1999) has argued that most of the food plants important to the Mbuti communities (Aka and Baka groups) of north-eastern Congo and southeast Cameroon are light-demanding trees, including the nut-bearing *Antrocaryon nannanii* and *Ricinodendron heudelotii*, and are often exploited around previous clearings. The fruit-bearing vines of *Landolphia* spp. are also said to grow best in disturbed areas (Ichikawa, 1999). Similarly, while some tropical forest agriculturalists disturb whole areas to grow a new crop, such as rice in Southeast Asia, in the Amazon Basin many of the important foodstuffs exploited today are South American natives, including manioc (*Manihot esculenta*), papaya (*Carica papaya*), sweet potato (*Ipomoea batatas*), and pineapple (*Ananas comosus*), that can be managed by foragers and farmers alike (Dufour, 1990; Bennett, 1992) (Chapter 5). Particular plants can also be moved into particular tropical forest habitats to boost subsistence. This can be clearly seen in the growth of manioc use in Central Africa to supplement yam yields (Ghazoul and Shiel, 2010), as well as in the historic tending of Brazil nut (*Bertholletia excelsa*) groves that has left its mark on the modern distribution of this tree across the Amazon Basin (Clement et al., 2015).

The dismissal of tropical forest soils as nutrient-poor, waterlogged soils that are unfavourable for crop growth (Meggers, 1971, 1977, 1987; Grollemund et al., 2015) also clearly neglects the variability discussed above, as well as the complex strategies exploited by ethnographic agriculturalists. Although the introduction of light-loving crops with low humidity tolerances into evergreen rainforests has been seen as necessitating permanent, large-scale clearance (Bellwood, 1993; Ghazoul and Shiel, 2010; Bayon et al., 2012) (Chapter 5), crops such as rice and maize can also be incorporated into small-scale and shifting management practices (Hecht et al., 1988; Toledo et al., 2003; Rerkasem et al., 2009). Indigenous communities inhabiting environments ranging from evergreen rainforest to dry tropical forest frequently combine the clearing of patches of forest, artificial fertilization of soils through burning, and the growth of reliable carbohydrate crops within broader systems of forest regrowth and regeneration (Posey, 1984; Hecht, 2003; Freire, 2007; Ford and Nigh, 2015). These systems contrast with western ideas of monoculture farming and the wholesale economic conversion of landscapes (Chapter 5). Indeed, the deliberate cultivation or propagation of yams in forest clearing microhabitats and the tree-cropping systems discussed above blur the lines of farming and foraging in tropical forest settings (Denham, 2009,

2011), and emphasize the variety of strategies available to even larger, sedentary populations in different forest formations (Chapter 6).

The fact that many tropical evergreen rainforest plants cover their leaves and fruits with toxic substances (McKey, 1979), and the rapid decomposition of leaf-litter and fallen fruits (Anderson and Swift, 1983), has also been traditionally argued to lead to a relative paucity of faunal opportunities, and fat-rich protein sources, for tropical forest human populations (Odum and Pigeon, 1970; Medway, 1972; Kikkawa and Dwyer, 1992). On the other hand, many animals play an important role in tropical forest ecosystems in the form of forest disturbance, pollination, seed predation, seed dispersal, folivory, frugivory, and as top predators (Terborgh, 1988, 1992; Cuarón, 2000). Species densities can vary significantly from place to place in relation to soil fertility, forest structure, and inter-specific competition (Emmons, 1984; Robinson and Redford, 1986). However, while the carrying capacity of tropical forests for hunter-gatherers can be low relative to grasslands and drier woodlands (<1 person km^2; Robinson and Bennett, 2000), the ethnographic literature lists a diversity of hunting techniques and strategies practised by human tropical forest hunters to capture faunal prey that varies greatly between different tropical forest contexts. One particularly interesting relationship in this regard is that between humans and closely-related monkey taxa that are (a) frequently among the larger sources of protein (Mittermeier, 1987) and (b) often conspicuous due to their social nature, diurnal feeding strategies, and loud calls (Cowlishaw and Dunbar, 2000).

The reliance on primates is particularly prominent in dense evergreen or semi-evergreen tropical forests where they form the largest protein package (Mittermeier, 1987). By contrast, in drier, semi-evergreen and deciduous settings, the choices of prey and hunting strategies become more varied, as can be seen in the prey choice of the Wanniya-laeto forest foragers of Sri Lanka which includes large-bodied monitor lizards (*Varanus bengalensis*), macaques (*Macaca sinica*), langurs (*Semnopithecus priam thersites*), pigs (*Sus* sp.), mouse-deer (*Moschiola meminna*), barking deer (*Muntiacus muntjak*), spotted deer (*Axis axis*), and sambhur (*Rusa unicolor*) (Chapter 8). Dwyer (1983) has shown similar variability on an altitudinal gradient of montane tropical forest in Papua New Guinea. In the lowland rainforest areas the Gadio Enga focus their hunting on wild pigs that provide an efficient and plentiful source of protein. Further up, the Etolo use axes and dogs to corner, isolate, and then cut down small-medium sized monotremes, marsupials, and rodents. In the Highlands, the Rofaifo people gain little return from hunting limited resources and sporadically ambush individual animals at night (Dwyer, 1983). The more diverse, and often more mobile, use of montane gradients or evergreen to forest edge gradients is likely to have played a key role in the past, as well as the present.

Of the various hunting techniques available to tropical forest foragers, it is perhaps the blowpipe, or the poison-tipped arrow, that the Hollywood portrayals of *Indiana Jones*, *Apocalypto*, and the *Lost City of Z* have ensured are most linked to rainforest hunting in the public domain. Frequently made from a kind of bamboo, blowpipes can range up to 2 m in length and are used to deliver small, thin darts, frequently tipped with poison, into prey. Kuchikura (1988), studying blowpipe use among the Semaq Beri in Peninsular Malaysia, argued that this technique provides an extremely reliable source of protein, accounting for

41.5 per cent of animal foods consumed. Other records from Southeast Asia have noted the use of blowpipes and poisoned darts to also target pigs and deer (Bellwood, 1999; Endicott, 1999). Furthermore, the Waorani of Ecuador, with larger, higher velocity blowpipes, shoot at large, ground-dwelling birds, such as currasows and tinamous, as well as peccaries and deer (Yost and Kelly, 1983). That said, research in both regions has maintained that projectile shot accuracy and strike-distances are optimized in the shooting of arboreal, rather than ground-dwelling, fauna (Yost and Kelly, 1983; Kuchikura, 1988). Other forms of direct hunting include the bow and arrow, that can even be employed alongside stealth in the hunting of both large suids and small, semi-arboreal and arboreal monkeys (Deraniyagala, 1992), and more recently, the use of firearms where they are available.

A more indirect hunting technique employed by tropical forest hunter-gatherers is trapping. Noss (1998) has argued that cable snare hunting is becoming the most widespread hunting method in Central African tropical forests today as a result of the low cost of wire and ability to leave traps while engaging in other activities, offsetting potentially wide foraging radii required in the search for reliable plant foods. Noss (1998) notes that all mammals identified were captured in snares, including semi-arboreal primates (where traps were set both on the floor and in low tree platforms), although the majority of captures were composed of three small antelope (duiker) species (*Cephalophus callipygus*, *C. dorsallis*, and *C. monticola*) and the brush-tailed porcupine (*Atherurus africanus*). The latter demonstrates a further use of traps and snares, the capture of nocturnal prey that is more difficult to stalk. Another interesting tropical forest hunting technique comes in the form of communal net hunting practised by communities such as the Aka of the Central African Republic (Noss, 1997). Male and female hunters form a circle or semicircle with the nets, hanging the nets on small trees or vines and pegging the bottom to the ground. Some, armed with spears, search the interior to drive out animals towards the nets, seizing them and killing them with spears or machetes. Prey taken in this manner includes large rodents, small to medium-sized bovids (Noss, 1997) and, in the case of the Raute of Nepal, semi-arboreal monkeys (Fortier, 2009).

In the context of farming, a paucity of domesticable animals (Smith, 1999; Diamond, 2005), and the barriers of climate and disease presented by tropical forests, have tended to make them seem hostile to herding (Gifford-Gonzales, 2000, 2015; Grollemund et al., 2015) and even the use of dogs as hunting aids (Mitchell, 2017). As with crop growth, the effective feeding of cattle, goats, and sheep in tropical forests is often associated with permanent deforestation (Ockwell and Lovett, 2005). This and the effects of indiscriminate mass hunting for 'bush-meat' to meet a growing commercial demand further the popular perception that it is difficult to sustain large human populations in tropical forests (Harrison, 2011) (Chapter 8). Yet, as with plants, ethnographic foraging and farming communities will often exploit all animal protein sources available to them. Many communities prize the honey of various bee species (Dentan, 1991). Furthermore, opportunistic hunting of lizards, snakes, frogs, tortoises, terrapins, and various other small vertebrates is commonplace (Bennett, 1992; Luiselli, 2003). The consumption of insects, high in protein, essential minerals, and vitamins, is also widespread (Posey, 1978; Dufour, 1987; Bennett, 1992; DeFoliart, 1999). Finally, fishing, through fish poisons, temporary damming, or harpooning, can form a major part of tropical forager diets, and in many Amazonian communities is the predominant source of protein (Bennett, 1992). For example, Tukanoan-speaking groups actively protect flooded forests, an important feeding ground for fish, to ensure high yields (Chernela, 1985).

PERCEPTIONS OF PRESERVATION

It is not just generalized ideas of nutritional capacities that have inhibited the study of tropical forests in archaeology and anthropology. Based on anecdotal evidence and a few token studies it has been perceived that hydrological activity, acidic soils, and active decompositional environments would mean that, even if archaeological remains are identified in tropical forests, the reconstruction of structures, subsistence activities, human remains, and the accurate dating of sites would be difficult at best. Martha Tappen (1994), for example, demonstrated that the decomposition of bones in the Ituri evergreen tropical rainforest in the Democratic Republic of Congo is much more rapid than in open savanna settings as a result of a greater intensity of microbe, plant, and animal taphonomic processes. Similarly, greater than 95 per cent of mass loss of decomposing leaf-litter in tropical forests around the globe has been shown to occur within a year (Powers et al., 2009). There is also the potential problem that the living spaces, serving vessels, and technology of sedentary agricultural, or even foraging, communities are likely to be made from wood and other plant products (Bennett, 1992; Friesem and Lavi, 2017; Xhauflair et al., 2017). This may mean that entire urban hinterlands, farming communities, and foraging camps are invisible to a basic archaeological survey (Friesem and Lavi, 2017). It has been similarly argued that Early to Middle Pleistocene hominins could have based their technology on bamboo and other forest products (Bar-Yosef et al., 2012), leading to a paucity of stone tool assemblages (Chapter 4).

Nevertheless, tropical forests and the activities of their human inhabitants are not taphonomically uniform and more comprehensive taphonomic studies of archaeological materials across different tropical forest formations and locales are greatly needed. Researchers have already demonstrated that organic decomposition varies with annual precipitation (Powers et al., 2009), local soil nutrients, and local microbiotic communities (Szefer et al., 2017). Decomposition rates have even been shown to decrease from early successional to primary forest locales associated with increased humidity in the latter (Szefer et al., 2017). Hättenschwiler et al. (2011) have argued that leaves forming in nutrient-poor swamp forest settings actually decompose more slowly than in a temperate forest when they fall to the ground. The preservation potential of archaeological remains accordingly varies. For example, Peterhans et al. (1993) have noted that well-drained neutral soils in the Kibale evergreen, semi-evergreen, montane, and deciduous forests of western Uganda can be conducive to bone accumulation. Similarly, Mercader and colleagues demonstrated that nut use and stone modifications by chimpanzees are documented in open-air tropical evergreen rainforest sites in the Tai forest of the Côte d'Ivoire back to 4,300 years ago (Mercader et al., 2002, 2007).

Perhaps most impressively, although the preservation of biomolecules was thought improbable in the tropics, growing datasets have already found the preservation of usable bone collagen in Amazonia (Hermenegildo et al., 2017) and ancient DNA in Pacific island contexts (Skoglund et al., 2016; Posth et al., 2018).

A growing awareness of this variability, increasingly enthusiastic survey and excavation, as well as the diverse application of new methodologies, is thus now resulting in the rapid growth of tropical forest archaeology datasets and an increase in our knowledge of the human past in these environments. In a Pleistocene context, cave and rock shelter sites have demonstrated stratigraphic integrity and long-term organic preservation of organic tools and faunal and hominin remains in a number of tropical forest settings around the world (Mercader et al., 2003; Barker et al., 2007; Westaway et al., 2009; Perera et al., 2011; Demeter et al., 2012; Barker, 2013; Barker and Farr, 2016) (Chapters 3 and 4). In a Holocene setting, the introduction of the LiDAR scanning form of remote sensing discussed in Chapter 1 has allowed large-scale mapping of urban complexes through dense evergreen rainforest canopies in Honduras (Fisher et al., 2017) and dry tropical forests in Cambodia, Southeast Asia (Evans, 2016). While stone monuments are more conducive to such studies, there is also the potential of mapping ditches, post-holes, and the traces of wooden structures (Fisher et al., 2017). Furthermore, detailed geomorphological analyses of soil layers and soil features (such as drainage ditches) have also been shown to yield previously unforeseen details relating to plant cultivation, domestic spaces, and human land-use (Denham et al., 2003; Arroyo-Kalin, 2012, 2014; Friesem et al., 2017), even in the absence of living spaces characteristic of more temperate or arid settings.

TEMPORAL DIVERSITY

The potential for human occupation and archaeological preservation in a given tropical forest does not just vary geographically, but also through time. Debate still surrounds the relative impacts of temperature change, atmospheric CO_2 concentration, precipitation change, and differences in seasonality on past tropical forest structure and extent (Malhi and Wright, 2004; Mayle et al., 2004; Feeley et al., 2011). Over glacial-interglacial cycles that characterize the Quaternary, and for which we have the most palaeoenvironmental information, tropical regions are characterized by distinctly low seasonality in temperature, with the annual variation in daily mean temperature as low as 1–4°C in lowland regions straddling the Equator (Osborne, 2000). In contrast to temperate regions, therefore, temperature has been considered to be a climatic parameter of limited importance in the tropics in the past. That said, estimates of temperature differences between glacial minima and maxima of *c.* 4–8°C have been recorded for equatorial Africa, Sahul, and Amazonia (Bonnefille et al., 1990; Thompson et al., 1995; van der Kaars, 1998; Kull et al., 2008; Hope, 2009; Hunt et al., 2012). Although these changes may not have been as extreme as elsewhere (Mercader, 2002a; Boivin et al., 2013), such decreases in temperature in glacial periods would have had a significant impact on regional humidity and growing season temperatures, particularly affecting the distribution of montane, evergreen, and seasonal tropical forests.

The influence of precipitation changes on tropical forests in the past is a particular source of contention, even on scales associated with the periodic expansion and contraction of ice sheets in the northern hemisphere (glacial/interglacial) that have been linked to changes in aridity at subtropical and tropical latitudes. For much of the tropics, precipitation is a significant parameter today and, as seen above, often dictates the extent and structure of different tropical forest formations. However, the relevant palaeoenvironmental proxies are less predictable and more regionally and temporally variable than records of temperature (Hooghiemstra and Van der Hammen, 1998; Mayle et al., 2004). It has been argued, particularly for the Pleistocene, that biogeographical patterns of endemic distribution in the Amazon and Congo Basins are linked to increased past aridity (Haffer, 1969; Hamilton, 1972). In this 'refuge hypothesis' scenario, increased aridity, for example during glacial periods, is argued to have caused the dramatic removal of rainforest except for a few patches, with high levels of groundwater, where high species diversity, recognized in modern studies of floral and faunal endemism (Haffer, 1969), was fostered and preserved until later rainforest expansion. Others have argued, by contrast, that the wet tropics remained consistently hydrologically hospitable to tropical forests, even during glacial periods, with relatively minor reductions of precipitation and forest extent (between 20 and 40 per cent—Maley and Brenac, 1998; Thomas, 2000).

Particularly over longer timescales, atmospheric CO_2 concentration has also been seen as a major driver of tropical forest extent and structure. Dynamic vegetation models, ecophysiological experiments, and stable carbon isotope records have shown that carbon-depleted periods, including glacials, would have reduced the water-use efficiency of plants photosynthesizing following the C_3 pathway, which dominate tropical forest ecosystems (Polley et al., 1993; Street-Perrot et al., 1997; Cowling et al., 2001). In some places, including the montane areas of Africa, CO_2 reduction has, in itself, been argued to lead to complete forest replacement (Jolly and Haxeltine, 1997). Forest plants have also been argued to adapt their physiology to changing CO_2 conditions by producing smaller leaf sizes during CO_2-starved conditions. As a result, tropical forest structures likely changed in many regions during glacial periods, leading to more open forest and increased frequencies of herbaceous taxa on the forest floor, and the expansion of plants photosynthesizing using the C_4 pathway, notably in grassland mosaics, that are more adapted to lower CO_2 concentrations (Mayle et al., 2004). The local impacts of changing atmospheric CO_2 concentration are currently much harder to discern in palaeoenvironmental proxies, relative to the impacts of precipitation and temperature. However, broad computer circulation models and datasets from ice and marine cores have provided insight into potential changes in this parameter in the past (e.g., Hamon et al., 2012; Hare et al., 2018).

It is evidently difficult to disentangle the relative impacts of precipitation, CO_2, and temperature on past tropical forest environments. Pollen analyses, commonly used to look at past forest type and presence, can sometimes struggle to determine the relative impacts of temperature and precipitation, while global changes in CO_2

are difficult to correlate chronologically with regional floristic changes. It is particularly challenging to document temperature or precipitation seasonality that, as noted above, is the greatest influence on tropical forest persistence. The future documentation of seasonality is reliant on effective models built from modern datasets (Malhi and Wright, 2004; Mayle et al., 2004; Gond et al., 2013) or the future potential of high-resolution speleothem analysis in tropical regions (Bertaux et al., 2002; Tagami, 2013). The stable carbon and oxygen isotope analysis of animal tooth enamel (Bernard et al., 2009; Birch et al., 2016) or shell (Stephens et al., 2008) associated with hominin populations offer further potential to track seasonal changes in vegetation and precipitation of immediate relevance to local tropical forest environments and resources utilized by our species and its ancestors. Moreover, growing understanding of how different species reflect environmental change in the stable oxygen isotope values of their tooth enamel may provide a higher-resolution picture of local water dynamics (Levin et al., 2006; Krigbaum et al., 2013; Blumenthal et al., 2017).

There is also a current lack of detailed knowledge of the impacts of atmospheric, marine, and terrestrial climatic systems on local and regional tropical forest distribution. For example, the El Niño-Southern Oscillation (ENSO) is the dominant mode of variability in tropical precipitation on interannual timescales today (Dai and Wigley, 2000; Ghazoul and Shiel, 2010), altering cloud cover, rainfall, and temperature in ways that directly limit plant function, and the resources available for reproduction, in tropical rainforests in Amazonia, Africa, and Asia (Laurance et al., 2001; Wright, 2014) (Figure 2.10). However, Malhi and Wright (2004) modelled regionally specific impacts of temperature, precipitation, oscillations of the ENSO, and seasonality on the world's rainforests since the 1970s. They demonstrate that while ENSO is known to impact temperature variations and precipitation fluctuations in the tropics of the Americas and Southeast Asia, it is in the rainforests of Africa and India where clear increases in dry-season intensity are documented. Furthermore, only African rainforests appear to demonstrate a significant overall 'drying' trend (Malhi and Wright, 2004). The potential rapid, sub-millennial environmental effects caused by systems such as ENSO are increasingly being picked up in palaeoenvironmental records over the course of the Holocene (Moy et al., 2002; Turney et al., 2004), though they will be more difficult to observe back into the Pleistocene given the error margins that currently exist with all existing chronological methods.

The above demonstrates that researchers must understand all of the different potential climatic drivers in order to model past fluctuations in tropical forest extent and structure. Moreover, regionally and temporally specific approaches to tropical forest reconstruction are essential (Ghazoul and Shiel, 2010; Corlett and Primack, 2011). Simplistic ecological overlaps of species endemism with areas of glacial remnant rainforest are proving dangerous as phylogenetic estimates suggest that high levels of genetic divergence preceded the Pleistocene, and likely extend back to the early Quaternary and the Late Cenozoic period (da Silva and Patton, 1993; Bermingham and Dick, 2014; Patton and da Silva, 2014). On a more refined level, in equatorial Africa it has been shown that while palaeoenvironmental sequences from forest borders or altitudinal and vegetation transitions may show significant changes in forest extent during glacial periods, and even within the Holocene (Van Noten, 1977), lowland palynological records

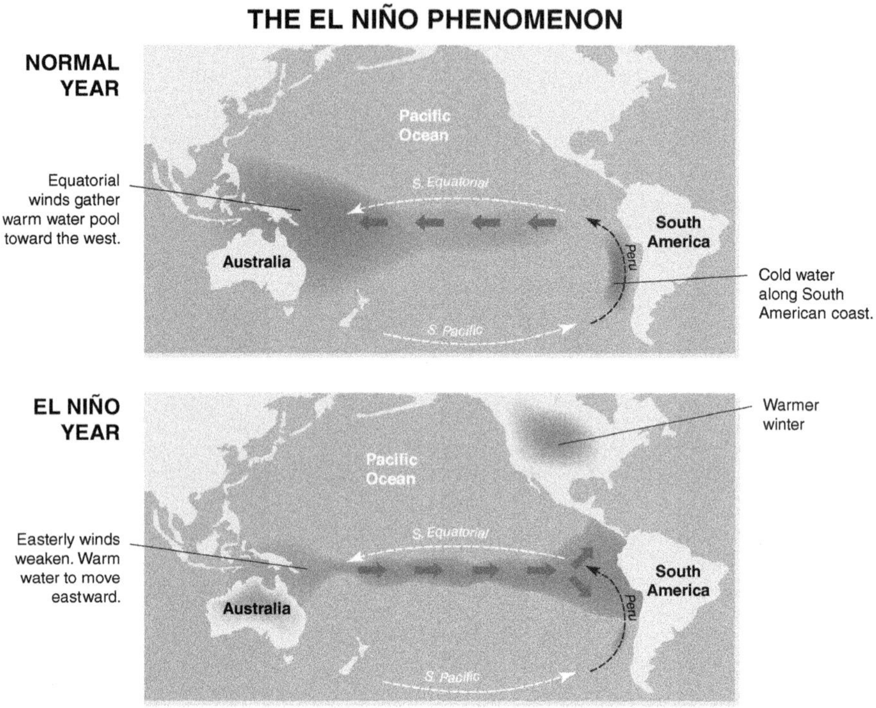

Figure 2.10. Map of El Niño regional impacts. During an El Niño year easterly winds across the Pacific Ocean weaken causing warm water to move eastward. This results in warmer and drier conditions across tropical South and Central America. Created by Designua and reproduced with permission from shutterstock.com.

demonstrate forest persistence over the last 30,000 years (Maley and Brenac, 1998; Colinvaux, 2014). Similarly, marine cores indicative of savanna encroachment cannot be correlated directly with lowland rainforest ecologies (Mercader et al., 2000). While 'off-site' palynological and phytolith sequences are often the only ones to survive in the vicinity of tropical rainforests, one must carefully assess to what extent the 'catchment' of these records is relevant to human behaviour (see Roberts and Petraglia, 2015).

It is also important to note that, at least since the arrival of our species on the planet, hominins have modified tropical forest environments, maximizing resource potential through hunting, gathering, plant and animal translocation, forest burning and deforestation, water channeling and other forms of landscape manipulation, and soil modification (Hunt et al., 2012; Roberts et al., 2017a). These behaviours will, at times, have exacerbated climatic changes and led to the production of different tropical forest formations. For example, in Amazonia, it has been shown that the ecological impacts of the ENSO were amplified by the intensification of agricultural activity *c.* 2,200 cal. years BP and faded when agricultural activity declined *c.* 800 cal. years BP (Bush et al., 2017). Historical ecology approaches are also highlighting the influence of past human populations on current plant species distributions in various tropical forests (e.g., ter Steege et al., 2013; Maezumi et al., 2018). The varying human impacts on tropical forest formations are a crucial component of the variability in tropical forest formations geographically and temporally, varying with population size, technological innovations, and subsistence strategies. Indeed, all tropical forests encountered by our species, and perhaps also its hominin ancestors, are, to some extent, 'novel ecosystems' (Hobbs et al., 2009; Boivin et al., 2016; Roberts et al., 2017a). This should be borne in mind when reconstructing past tropical forests and, indeed, tracking and comparing human impacts on these environments is a key component of this book.

A BROAD PICTURE OF CHANGE

Despite these caveats of complexity, growing global, regional, and local palaeoenvironmental proxy datasets enable a broad picture of global tropical forest dynamics, in relation to climatic changes, to be constructed for the periods of relevance to *H. sapiens* and its hominin ancestors. Throughout this book we will see how the actual impacts of these changes on local vegetation and faunal resources, and hominin populations, may have differed between regions and between the different forest types discussed above. However, Figures 2.11–2.13 and the following description provide a working guide of gathered global knowledge within which to place hominin tropical forest adaptations in context.

Miocene (23–5.3 Million Years Ago)

Warm global temperatures during the Miocene saw forests persist even on Greenland throughout this period and well into the Pliocene. Between 21 and 14 Ma tropical-adapted forest biomes expanded across much of Africa and Eurasia (Hamon et al., 2012). This drove the expansion of forest-adapted ape populations across much of the Old World (Hamon et al., 2012), with at least 100 species of ape estimated to have existed on the planet at this time. Profound changes in vegetation apparently began *c.* 10 million years ago, on the basis of faunal abundance records and dietary evidence gained from stable isotope analysis (Reed, 1997; Uno et al., 2011) (Figure 2.11). In particular, the evolution and expansion of C_4-photosynthesizing grasses, which had evolved a form of photosynthesis that was more competitive under low carbon dioxide concentrations (Cerling et al., 1997), occurred at the expense of forests in eastern and southern Africa (*c.* 9–8 Ma), as well as much of Asia (*c.* 7 Ma) (Quade and Cerling, 1995). The fragmentation of global forests from this point onwards seemingly led to the separation and diversification of ape populations, in part dictating the distribution of our closest living relatives today and, potentially, the emergence of the hominin lineage. Molecular evidence implies that the Last Common Ancestor between modern non-human great apes and humans also existed at some point between 8 and 7 million years ago (Langergraber et al., 2012). The first hominins appeared in Africa at the very end of the Miocene in the form of *Sahelanthropus tchadensiis*, *Orrorin tugenensis*, and *Ardipithecus* spp. (Chapter 3) (Figure 2.11).

Pliocene (5.3–2.6 Million Years Ago)

During the Pliocene, global temperatures generally remained 2–3°C higher than today (Robinson et al., 2008), though the gradual global cooling that was to eventually lead towards Pleistocene glacial cycles continued. Shifts towards cooler and drier conditions led to further reduction in tropical forests until these environments, as well as their associated fauna, became limited to a thin equatorial region (Whitmore, 1998; Ghazoul and Shiel, 2010) (Figure 2.12). It is during this time, as deciduous forests, savannas, and even deserts, appeared in Africa, Asia, and South America, that the hominin lineage witnessed major diversification and expansion in Africa. As will be seen in Chapter 3, a number of Australopithecine fossils and taxa, including the iconic 'Lucy' (*Australopithecus afarensis*), retained adaptation to arboreal locomotion while also demonstrating various evolutionary adaptations to bipedal locomotion. Moreover, recent discoveries in Ethiopia have shown that our own genus, *Homo*, emerged, during the Pliocene *c.* 2.8 Ma (Villmoare et al., 2015). The development of stone tool technology also occurs towards the end of the Pliocene *c.* 3.3 Ma (Harmand et al., 2015) (Figure 2.12). It currently appears that these trends towards specialized bipedal locomotion, and the emergence of stone tool technology, are, in some way, adaptively associated with the continued expansion of C_4 grasslands and savannas (Sponheimer et al., 2013), although, as Chapter 3 will demonstrate, the picture is perhaps more complicated than has often been assumed.

Early Pleistocene (2.6–0.8 Million Years Ago)

The long-term cooling of the Pliocene continued into the Pleistocene. However, the latter period is particularly characterized by the intensification of climatic cycling between the expansion of continental glaciers, reductions in sea-level, and the decline of tropical forests, and glacial contraction, sea-level rise, and increased temperature and humidity in the tropics (Ghazoul and Shiel, 2010). Milankovitch cycles, global changes in solar radiation linked to patterns in the Earth's movement and orientation, are thought to be the main factor behind these changes, although regional environments and climate systems complicate the local expression of these trends. Glacial cycles were frequent, but less extreme, during the Early Pleistocene. At this time hominin diversification continued in Africa with the evolution of Paranthropocine hominins and their heavy reliance on C_4 resources. Our own genus continued to evolve until *Homo erectus*, arguably the first truly specialized bipedal hominin, emerged. During the Early Pleistocene this species, or at least our genus, gave rise to the first hominin populations to expand beyond Africa into central Asia (Gabunia et al., 2000), southwest Europe (Falguères et al., 1999), and the tropics of Southeast Asia (Zaim et al., 2011) (Figure 2.12). It has been argued that these expansions were linked to savanna corridors, opened up as a result of glacial cycles (Dennell and

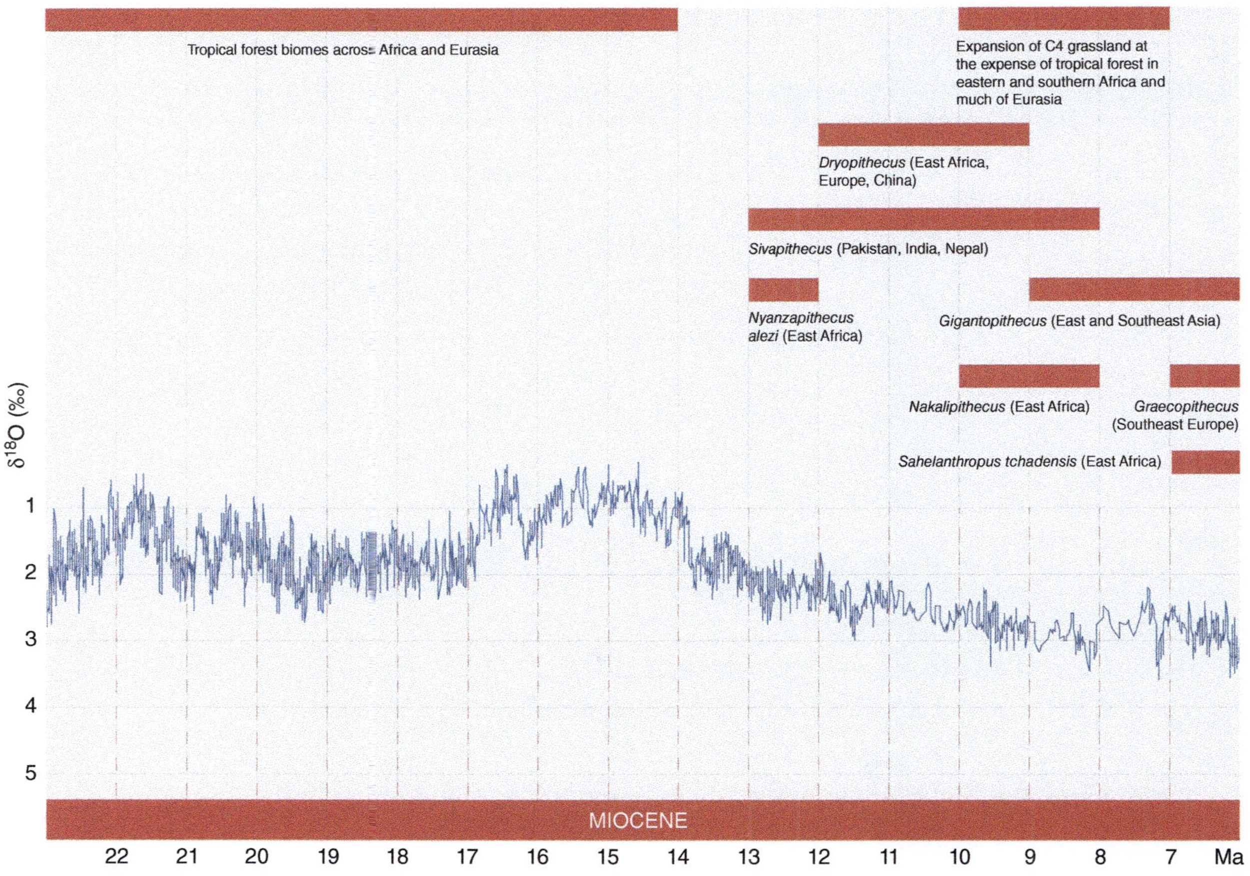

Figure 2.11. Major events discussed in the text during the Miocene between 23–6 Ma plotted against a stable oxygen isotope curve from benthic foraminifera from Zachos et al. (2001). Higher $\delta^{18}O$ indicates warmer climates with fewer ice sheets and lower $\delta^{18}O$ indicates cooler climate with more ice sheets. With thanks to Hans Sell.

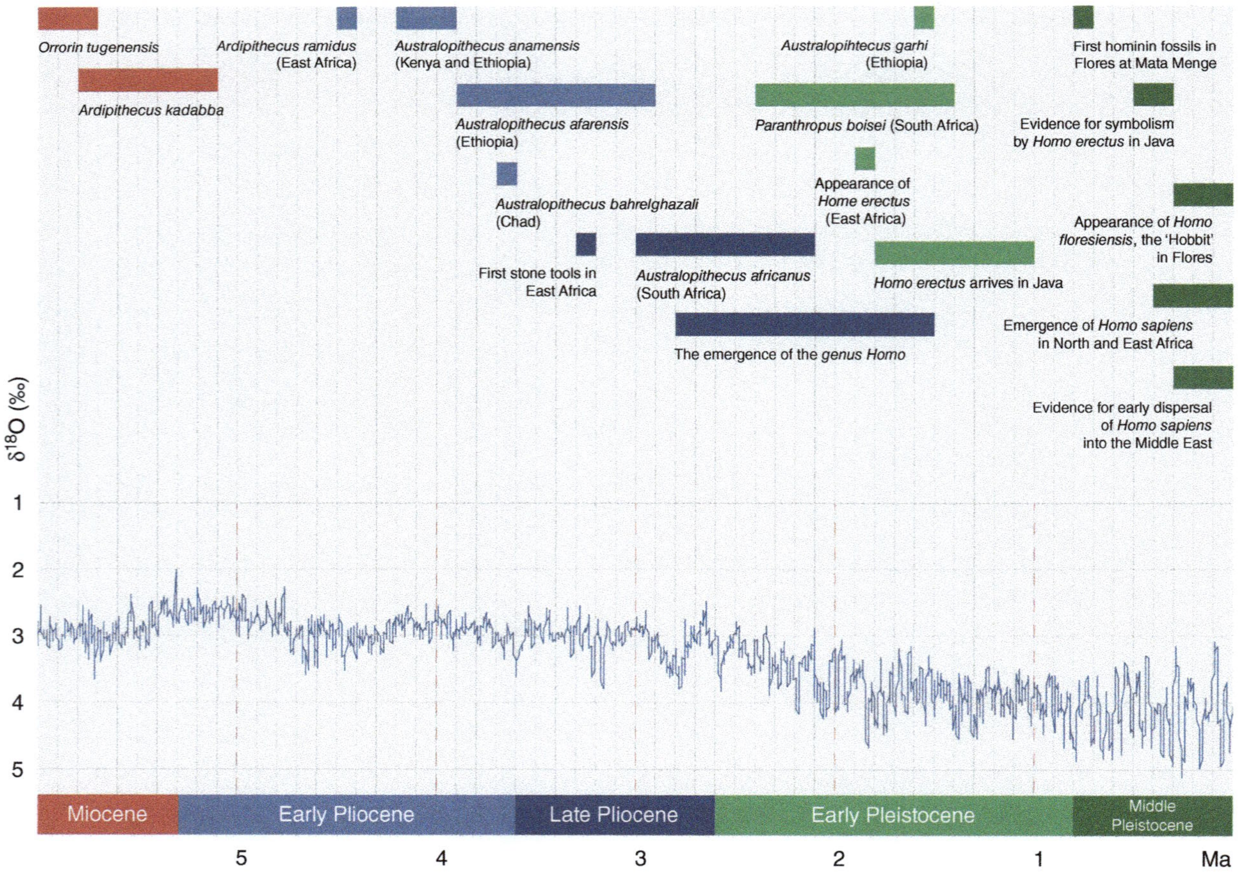

Figure 2.12. Major events discussed in the text during the Miocene-Middle Pleistocene (6–0.12 Ma) plotted against a stable oxygen isotope curve from benthic foraminifera from Zachos et al. (2001). Higher $\delta^{18}O$ indicates warmer climates with fewer ice sheets and lower $\delta^{18}O$ indicates cooler climate with more ice sheets. With thanks to Hans Sell.

Roebroeks, 2005). However, Chapter 3 will explore the proposition that tropical forests were plentiful during the arrival of *Homo erectus* in Southeast Asia *c.* 1.5 Ma (Sémah et al., 2002; Sémah and Sémah, 2012).

Middle Pleistocene (0.8–0.1 Million Years Ago)

Glacial cycles became less frequent during the Middle Pleistocene, but also more intense. Tropical forests faced more extreme, rapid periods of contraction, with consequent effects on faunal communities. Hominin species during this period showed increased innovation and diversity in the context of local geographical and environmental challenges, with hominins making it across a significant body of water to the island of Flores, eastern Indonesia, *c.* 1–0.7 Ma (Van den Bergh et al., 1996a, 2016; Brumm et al., 2010) (Figure 2.12). On this island, *Homo floresiensis*, or 'the Hobbit', appears *c.* 0.2 Ma, perhaps as a result of island dwarfing of the initial occupants of the region (Sutikna et al., 2016). The non-*H. sapiens* hominin taxa of tropical Southeast Asia, including *Homo erectus*, persisted through Middle Pleistocene glacial and interglacial periods until at least 0.2–0.1 Ma (Ciochon, 2009; Marwick, 2009). It has been argued that a change to closed, evergreen rainforest environments, and the appearance of modern rainforest fauna in this region, as well as increased island isolation with sea-level fluctuations, may have led to the demise of these archaic hominin species (Marwick, 2009). In addition, a significant biogeographical event occurred with the appearance of our own species in Africa *c.* 0.3 Ma. It has been suggested that *H. sapiens* may have instantly occupied the tropical forests of Africa following its evolution (Mercader, 2002a, 2002b) (Figure 2.12). While Chapter 4 will explore whether this is the case, Late Pleistocene tropical forest human occupation was to have clear ramifications for existing flora and fauna, including other hominin taxa, in these environments (Figure 2.13).

Late Pleistocene (0.1–0.01 Million Years Ago)

The Last Glacial Maximum (*c.* 26–20 ka) (LGM) (Clark et al., 2009) represents the last, great extension of ice sheets across much of North America, northern Europe, and Asia. Evidence worldwide now demonstrates that the LGM did affect tropical forests, causing significant contraction and structural change, particularly at low latitudes (Jolly and Haxeltine, 1997). Reduction in tropical forest extent appears to have occurred in the subtropics and tropics of Africa (Basell, 2008; Blome et al., 2012), Southeast Asia (as summarized by Rabett, 2012), South Asia (Roberts et al., 2017b), and Melanesia (Haberle, 2005; Summerhayes et al., 2017). However, it was during the Late Pleistocene that our species expanded into every tropical forest region on the planet. While early evidence for human tropical forest use in Africa and China remains debated (see Roberts et al., 2016), as will be seen in Chapter 4, the first definitive human fossils in Southeast Asia are apparently associated with tropical forest presence (Demeter et al., 2012; Westaway et al., 2017), if not clear exploitation (Barker et al., 2007) (Figure 2.13). Montane rainforest exploitation occurred in New Guinea as early as 45 ka (Summerhayes et al., 2010), while the earliest date

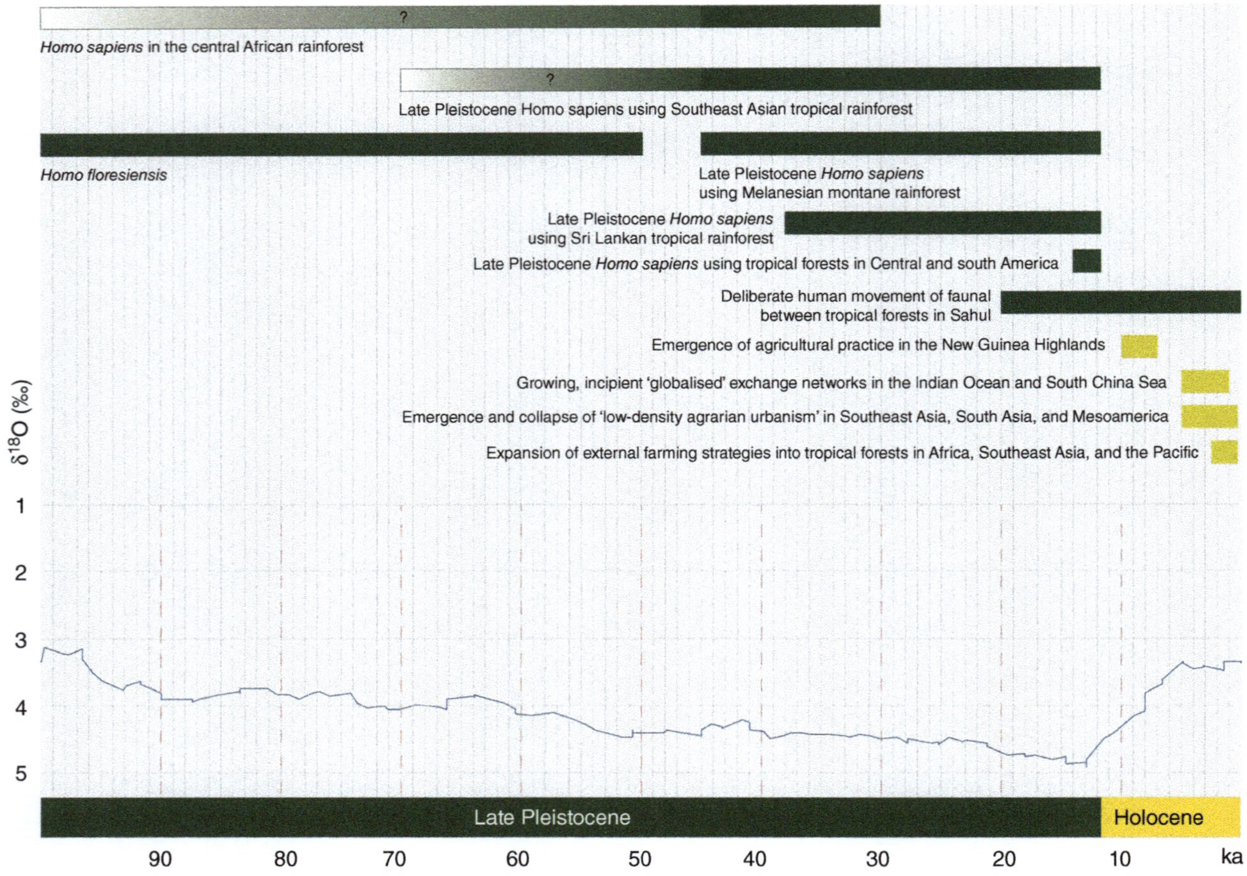

Figure 2.13. Major events discussed in the text during the Late Pleistocene-Holocene (100–0 ka) plotted against a stable oxygen isotope curve from benthic foraminifera from Zachos et al. (2001). Higher $\delta^{18}O$ indicates warmer climates with fewer ice sheets and lower $\delta^{18}O$ indicates cooler climate with more ice sheets. With thanks to Hans Sell.

for *H. sapiens* fossils in South Asia is directly associated with tropical rainforest resource reliance (Roberts et al., 2015a, 2017b). Some of these occupations are also associated with the earliest anthropogenic alteration of tropical forest environments (Chapter 4). By the end of this period, our species had also entered the tropical forests of the Neotropics *c.* 13–12 ka (Piperno, 2011a), where significant human-related changes were to be seen for much of the Holocene.

Holocene (0.01 Million Years Ago–Present)

The Holocene is synonymous with the generally warmer temperatures that have characterized the Earth's climate since the LGM, leading to re-expansion of tropical forests worldwide (Figure 2.13). While this appears to have broadly led to an increase in the area covered by tropical forest, more refined chronologies and palaeoenvironmental records for the Holocene enable us to document the impact of millennial and sub-millennial events on these environments. These include the rapid cold–warm, dry–wet fluctuations that characterize the onset of this period (Flenley and Bush, 2007), changes in the ENSO (Koutavas et al., 2002) mentioned above, and changes in tropical monsoon climate systems (Kumaran et al., 2014). These rapid climate shifts likely had impacts on increasingly sedentary agricultural and urban populations in tropical forests (e.g., Fagan, 2009; Masson, 2012), and more resolved Holocene palaeoenvironmental and archaeological records enable studies of short-term human adaptivity, behavioural shifts, or population decline (Figure 2.13). Shorter-term climate impacts on tropical forests such as these have almost certainly been a constant fixture of hominin experience in these environments. Yet, as discussed above, beyond the Holocene, and certainly beyond the Late Pleistocene, we have no good way of identifying these climatic changes or human responses to them. As will be seen in Chapters 5 through 8 the Holocene is also the period when human impacts on tropical forests intensify, involving modifications of these environments in terms of species composition, structure, and extent. This makes the relative impacts of climate versus humans on tropical forests difficult to determine during this period, and has even contributed to the argument that a new geological period, the 'Anthropocene', is warranted on pre-industrial timeframes (Lewis and Maslin, 2015).

In 2002 Julio Mercader (2002a, 2002b) encouraged what he termed the development of a 'Rainforest Prehistory' for our species. He argued that tropical forests were valid environments within which to search for traces of our ancestors. A decade and a half on, with the spate of discoveries discussed in this book since then, he has been vindicated in this position (see reviews in Roberts and Petraglia, 2015; Roberts et al., 2017a). However, with the increased archaeological datasets that have developed since 2002, and the clear observation that no single tropical forest is the same, both across space and time, it should now be the goal of tropical forest archaeology to formulate 'tropical forest *prehistories*'. This is a key aim of this book and throughout I will seek to present the available local and regional palaeoenvironmental information of tropical forest presence, type, and change, alongside information from archaeological sequences of human behaviour and tropical forest use. Over the course of this book we will see how exploited tropical forest plants and animals vary with hominin species, subsistence approaches, tropical forest formation, and climatic changes. We will also see how different human groups have modified tropical forest formations to meet their own subsistence needs, preferences, or simply enhance the existing natural longevity of certain plant and animal resource availability. However, we first turn to the disjoint between the fact that tropical forests represent the evolutionary cradle of our ape and earliest hominin ancestors and the commonly held perception that these habitats are marginal environments for the genus *Homo*, including *H. sapiens*.

3

Cradle Under the Canopy

The Forest Origins of our Ape and Hominin Ancestors and the Tropical Forest Forays of the Genus *Homo*

> Man's ability to exploit the planet, to take of its resources as he needs, and to usurp entire forests and all living creatures therein, rests upon the unwritten assumption that the chasm between himself and all other creatures is impassable.
>
> —Sue Savage-Rumbaugh and Roger Lewin (1996)

The evolutionary proximity of the non-human great apes to us is often stressed in studies of animals, such as Kanzi, a bonobo (*Pan paniscus*) bred in captivity, that demonstrate their capacity to undertake tool-use and even utilize and comprehend language (Toth et al., 1993; Savage-Rumbaugh and Lewin, 1996; Schick et al., 1999). Likewise, studies of chimpanzees (*Pan* spp.) have highlighted the similarity of their emotional and empathetic capacities to those of humans (Parr et al., 2005; Campbell and de Waal, 2014). However, as noted by Savage-Rumbaugh and Lewin (1996), in palaeoanthropology and archaeology more broadly, the emergence of the hominin clade and, later, our species, is referenced in terms of the 'chasm' between ourselves and other extant great apes. Indeed, despite our genetic and behavioural proximity, extant non-human great ape taxa are often popularly characterized as living fossils of how we *used* to be. They are used as analogues for the subsistence and behaviour of the Last Common Ancestor (LCA) of humans and non-human great apes (Clutton-Brock and Harvey, 1977; Goodall, 1986; Foley and Lewin, 2004) and it is almost as if the fact that they still occupy the tropical environments in which these hominoids likely evolved (though see Elton, 2008) allows them to be treated as static comparisons (Figure 3.1). Since Darwin wrote the *Descent of Man* in 1871, the forests of the tropics, and their modern non-human great ape inhabitants, have tended to be perceived as being left behind as the hominin clade gained increasingly 'human' traits of tool-use, medium to large game hunting, and upright locomotion on open 'savanna' landscapes (Dart, 1925; Potts, 1998; Klein, 1999).

From this perspective it is perhaps unsurprising that tropical forests are seen as alien to the genus *Homo* and its closest hominin ancestors. However, while a review of Miocene to Pleistocene palaeoenvironmental and palaeodietary evidence

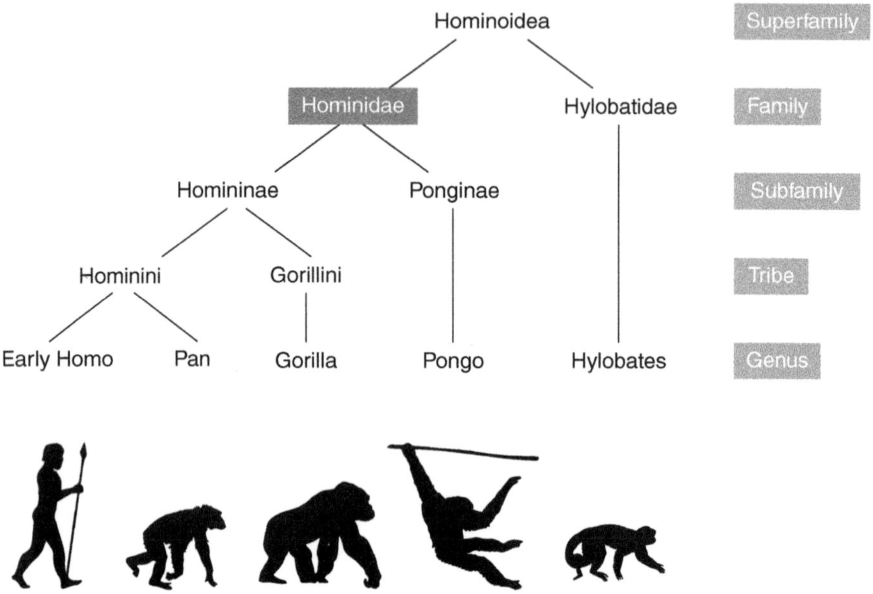

Figure 3.1. Schematic of the relationship between hominoids, hominids, and hominins. With thanks to Hans Sell.

suggests a general temporal shift in hominin adaptations towards 'savanna' and C_4 plant-based habitats (Dart, 1925; Domínguez-Rodrigo, 2014), there is growing evidence for complexity in this trend and understandings as to what this actually means in the context of ecological and dietary adaptations. In this chapter I briefly review what researchers mean when they talk about 'savanna' as a springboard for re-evaluating the ongoing role of tropical forest habitats in the evolution and diversification of the hominin clade, from the Miocene through to the appearance of the first members of our genus in the Late Pliocene/Early Pleistocene fossil record (Figure 3.1). From this point onwards there is clear evidence for the expansion of our genus into tropical Southeast Asia. However, there has so far been little systematic review of the adaptive implications of hominin finds in this region and their associated palaeoenvironmental and palaeodietary data. Here, I examine to what extent pre-*Homo sapiens* Pleistocene hominins occupied and utilized tropical forests in Africa and Southeast Asia, and to what extent this represented a marked 'return' to the tropical forest evolutionary cradle rather than a continuity of adaptation.

THE PERSISTENCE OF THE 'SAVANNA HYPOTHESIS'

In 1924 Raymond Dart proclaimed that a fossil uncovered in the lime quarries of Taung, South Africa should be classified as the first 'hominin' (Dart, 1925, 1926). The position of the foramen magnum, the hole through which the spinal cord joins the brain, was considered characteristic of bipedal locomotion. Although the 'human-like' hands, and in particular the thumbs, of what became known as

Australopithecus africanus were only discovered later, Dart argued that this hominin was capable of making and using tools to hunt medium to large game (Dart, 1925, 1926). Most prescient here, however, was the fact that these traits were located in arid, open Kalahari environments rather than the dense tropical forests of extant non-human great apes. With additional findings of hominin fossils in the xeric Rift Valley of East Africa (Leakey, 1959) this lineage became increasingly perceived in terms of decreased arboreality and increased terrestrial focus. Baboons were used as a useful referent for australopith foraging strategies in 'savanna' grasslands (DeVore and Washburn, 1963; Jolly, 1970) and ecological adaptations became an increasingly important focus for palaeoanthropology (Howell and Bourlière, 1963). As Domínguez-Rodrigo (2014) notes, perhaps the most blatant exposition of this was Coppens' argument (1982, 1991, 1994) that hominins evolved in Rift Valley 'savanna' ecosystems while non-human great apes remained to the west of the Rift in tropical forests.

The continued discovery of Pliocene (*c.* 5.3 to 2.6 Ma) and Early Pleistocene (*c.* 2.6 to 0.8 Ma) hominin remains in eastern and southern Africa only furthered this ecological association. *Australopithecus anamensis* (*c.* 4.2 to 3.9 Ma) is known from roughly one hundred fossils, making up twenty individuals, in Kenya and Ethiopia (Leakey et al., 1995a, 1998; White et al., 2006). Although limited, the existing postcranial morphology data from *A. anamensis* suggest that it may have been habitually bipedal, walking terrestrially in an upright fashion (Ward et al., 2001). The infamous 'Lucy' skeleton of *Australopithecus afarensis*, found in the Afar region of Ethiopia, was argued to provide the clearest example of the emergence of terrestrial bipedalism *c.* 3.9–2.9 Ma (Johanson and Maitland, 1981). The pelvic and femoral morphology, as well as the orientation of the big toes, implied that this hominin was almost exclusively in an upright bipedal position, and lacked feet with any capacity for effective grasping (Susman and Susman, 1983; Latimer and Lovejoy, 1989). Finds of the gracile *Australopithecus bahrelghazali* in Chad (3.6 Ma) (Brunet et al., 1995), *A. africanus* (3.0–2.1 Ma) (Herries et al., 2010) in South Africa, and *Paranthropus boisei c.* 2.4–1.4 Ma all point towards a radiation of habitual bipedalism. This culminates in evidence for more energetically-efficient obligate bipedalism in *Homo habilis* (2.1–1.5 Ma) (Bramble and Lieberman, 2004; Harcourt-Smith and Aiello, 2004). Among other things, upright bipedalism was argued to provide the most efficient form of locomotion in open settings (Rodman and McHenry, 1980) and expose a smaller surface area to the hot sun (Wheeler, 1984; Jablonski, 2004).

Although Dart dubiously associated the use of animal bones as tools with his 'man-like apes' in the Kalahari of southern Africa (1925, 1926), the earliest evidence for actual stone tool technologies also comes from xeric, grassland regions in Africa. Stone tools dating to as early as *c.* 2.5 Ma were found at Gona in Ethiopia just 100 kilometres away from finds of *Australopithecus garhi* (Asfaw et al., 1999; de Heinzelin et al., 1999). More recently, stone tools dated to *c.* 3.3 Ma have been found in the open, arid setting of Lake Turkana in Kenya (Harmand et al., 2015). Due to their apparent ecological contexts, the use of stone tools has been hypothesized to reflect hominin adaptations to the butchery, or even hunting, of medium to large 'savanna'-dwelling game (Potts, 1994; Plummer et al., 2009; Domínguez-Rodrigo, 2014). Although this evolutionary leap has traditionally been associated with our genus, particularly in line with cranial capacity increases in *Homo habilis* (2.1–1.5 Ma) (Bramble and Lieberman, 2004; Harcourt-Smith and Aiello, 2004), our genus remains poorly characterized and *Paranthropus boisei* and *Australopithecus garhi* could potentially also be culpable. Regardless, critical mass of evidence, and the subsequent developments of Oldowan and Acheulean lithic complexes in the Late Pliocene and Early Pleistocene, seem to link the formalization of stone tool technologies with the appearance of our own genus, *Homo*, now dated to 2.8 Ma within the arid grasslands of eastern Africa (DiMaggio et al., 2015; Villmoare et al., 2015), and, alongside specialized bipedalism, they were certainly part of the initial hominin movements Out of Africa from this point onwards (Gamble, 1993).

Nevertheless, assumptions of a linkage between hominin fossils, stone tool technologies, and 'savanna' habitats cannot be dependent on the modern environmental context in which these remains have been found. Importantly, fossil fauna found in association with Pliocene and Pleistocene hominins have also tended to be primarily indicative of open 'savanna' biomass (L'environnement des hominidés au Plio-Pléistocène, 1985). In particular, faunal assemblage evidence indicates a gradual, and in some places a more punctuated, shift towards more open and varied habitats between 2.9 and 2.4 Ma (Reed, 1997; de Menocal, 2004), including the appearance of arid-adapted bovid species *c.* 2.7 Ma (Vrba, 1995). Rodent assemblages from the Shungura Formation in Ethiopia also imply increased aridification from 3 Ma to 2.3 Ma (Wesselman, 1985). Large-scale studies of habitat-specific taxa, including antelopes, monkeys, and pigs in the Omo sequence of southern Ethiopia demonstrate a decline in closed woodland and forest species and an increase in grassland species between 3.6 and 2.4 Ma (Bobe et al., 2002). This has been used to further associate the growth in hominin fossil finds, and taxonomic diversity, in the Late Pliocene and Early Pleistocene with the development of more open and variable habits (de Menocal, 2004). Nevertheless, different methodologies in fossil faunal analysis, difficulty of associating hominin finds with faunal assemblages, and variability in chronologies of change (Reed, 1997) have made it difficult to directly link these changes to hominin evolution.

As will be seen throughout this chapter, and the next, one of the main tools that has emerged in assessing hominoid reliance on forest versus grassland environments is the stable isotope analysis of their tooth enamel, along with that of associated fauna. Given the frequency with which I refer to this evidence it is worth giving a brief outline here. Stable isotopes of an element differ slightly in their nuclear mass as a result of differences in the number of neutrons. For a certain chemical element (E) the ratios of the abundances of the heavier (i) versus lighter (j) isotope are measured in a sample and in a standard reference material with an internationally accepted δ value (Brand et al., 2014). The delta notation ($\delta^{i}E_{sample}$), defined through Equation 1, is commonly employed in reporting stable isotope results:

$$\delta^{i}E_{sample} = \frac{\left(\frac{^{i}E}{^{j}E}\right)_{sample} - \left(\frac{^{i}E}{^{j}E}\right)_{reference}}{\left(\frac{^{i}E}{^{j}E}\right)_{reference}} \qquad \text{(Equation 1)}$$

The mass difference in isotopes of the same element, caused by differences in the number of neutrons, results in small but significant variation in their thermodynamic and kinetic properties (Sharp, 2006). Kinetic effects involve lighter isotopes being more 'mobile' and reacting faster than heavier ones. In enzyme-driven biochemical reactions, the enzyme discriminates against the heavier isotope in the system (Hoefs, 1997; Sharp, 2006). These fundamental fractionation phenomena cause subtle, but measurable, differences in isotope abundance (δ) in different biogenic materials that can be related to biological processes such as photosynthesis and food metabolism, and environmental factors such as temperature (Koch, 1998; Sharp, 2006).

In land plants, the two dominant photosynthetic pathways, C_3 and C_4 (so-called because of the number of carbon atoms fixed in the first organic product), differ in their net discrimination against the isotope ^{13}C during fixation of CO_2 (Smith and Epstein, 1971; Farquhar et al., 1989). Most terrestrial plants, including trees, herbs, shrubs, and shade-loving, temperate, or high-altitude grasses follow the C_3 photosynthetic pathway (Calvin and Benson, 1948). C_3 plants discriminate against ^{13}C by a small amount during diffusion and dissolution of CO_2 into their chloroplasts. However, during carboxylation, the enzyme ribulose bisphosphate carboxylase/oxygenase (RuBisCO) strongly discriminates against ^{13}C, leading to an overall average discrimination of –19‰ relative to atmospheric $\delta^{13}C$. In C_4 or Hatch-Slack photosynthesis, followed by most tropical grasses and sedges with warm growing seasons, plants that have often been taken as synonymous with 'savanna', CO_2 is fixed by phosphoenolpyruvate carboxylase (PEP) before it enters the RuBisCO cycle (Hatch et al., 1967). This enables increased efficiency in CO_2 assimilation and a reduced discrimination of RuBisCO against ^{13}C in C_4 plants. C_4 $\delta^{13}C$ values are consequently closer to atmospheric CO_2 compared to C_3 plants (Ehleringer and Osmond, 1991). C_3 $\delta^{13}C$ values vary from about –24 to –36‰ (global mean –26.5‰), while C_4 values range from approximately -9 to –17‰ (global mean –12.0‰; Smith and Epstein, 1971). C_3 and C_4 plants therefore have distinct and non-overlapping $\delta^{13}C$ values (Tieszen, 1991) (Figure 3.2).

In a tropical context, the isotopic distinction between C_3 and C_4 biomass is useful for studying the relative proportion of C_4 grassland and C_3 woodland or forest in faunal and hominin diets and, indirectly, their associated environments (Lee-Thorp et al., 1994; Sponheimer et al., 2006a). Within tropical forests themselves, vegetation growing under a closed forest canopy is strongly depleted in ^{13}C, due to low light (Farquhar et al., 1989) and large amounts of respired CO_2, especially near the forest floor (van der Merwe and Medina, 1991) (Figure 3.3). The 'canopy effect' causes CO_2, soils, leaves, and fruits within a closed canopy to have low $\delta^{13}C$ values that are also reflected in the tissues of animals feeding in the same environments (van der Merwe and Medina, 1991; Cerling et al., 2004) (Figure 3.3). There is a ^{13}C enrichment factor between plant resources and herbivore tooth enamel. For the purposes of this book I use an enrichment $\delta^{13}C$ value of 13‰ between vegetation and tooth enamel as the midpoint between estimates of *c.* 12‰ (Lee-Thorp and van der Merwe, 1987) and *c.* 14 (Cerling and Harris, 1999) for mammals, though much more work is needed to refine our understanding of this process in primates. In faunal, or hominin, diets, prior to the impact of significant fossil fuel emission from the eighteenth and nineteenth centuries on atmospheric $\delta^{13}C$, this results in tooth enamel with $\delta^{13}C$ *c.* –14‰ representing reliance on dense or closed canopy forest, while average herbivore values for C_3 and C_4 reliance are *c.* –12‰ and *c.* 0‰, respectively (Lee-Thorp et al., 1989a, 1989b; Levin et al., 2008).

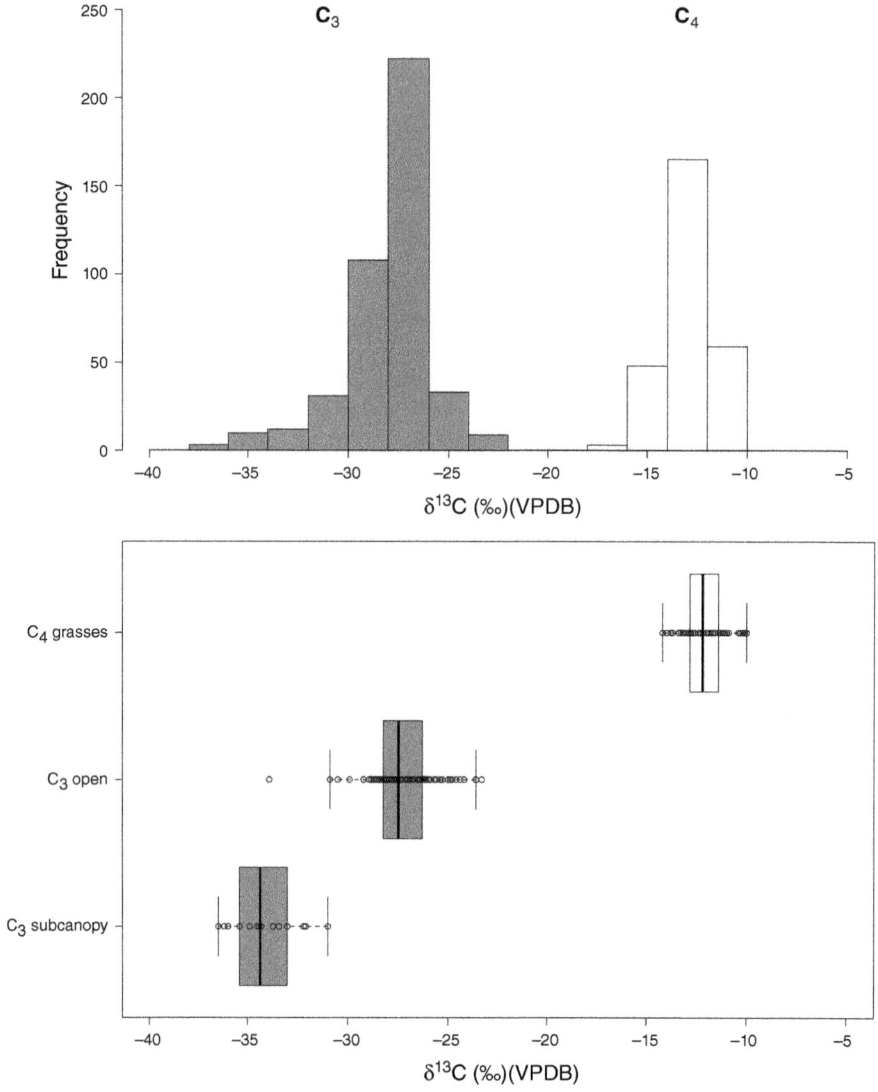

Figure 3.2. Top: Histogram of stable carbon isotope ($\delta^{13}C$) values of East and Central African plants collected between 1997 and 2006 (data from Cerling et al., 2010). Bottom: $\delta^{13}C$ values of plants from Mongolia, Argentina, Utah, Zaire, and Kenya collected between 1962 and 1997 (data from Cerling and Harris, 1999).

As the environmental end-points can vary between systems it is important to have a local faunal baseline, often with known or hypothesized feeding behaviours, against which the $\delta^{13}C$ of the hominin population can be measured. The oxygen isotope composition of animal tissues can provide additional palaeoecological information about water and food. In tropical ecosystems it has been shown that vegetation $\delta^{18}O$ reflects either evaporative potential or the source-effect of

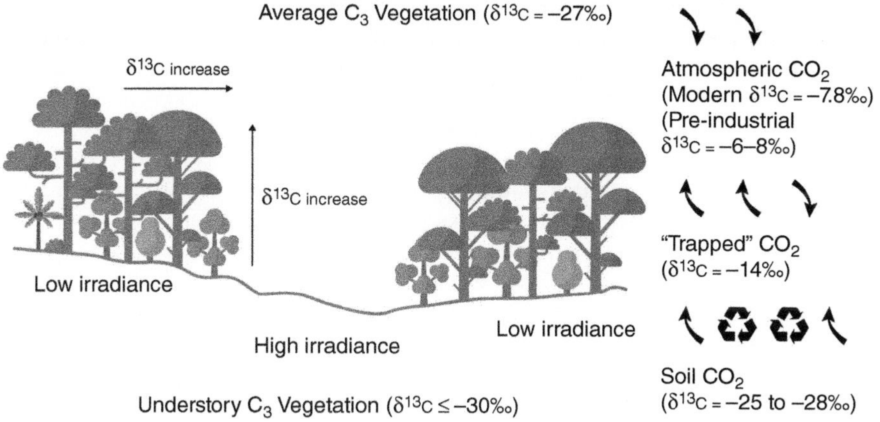

Figure 3.3. Diagram representing C_3 plant stable carbon isotope ($\delta^{13}C$) variation in a tropical forest context. Low irradiance and recycled CO_2 result in lower $\delta^{13}C$ values in air, soil, and vegetation growing under a closed canopy relative to more open areas. In closed areas, understory vegetation is significantly depleted in ^{13}C, and vegetation has lower $\delta^{13}C$ values, compared to that growing higher in the canopy. Adapted from Jackson et al. (1993), Quade et al. (1995), and Krigbaum (2003). With thanks to Hans Sell.

rainfall (Buchmann et al., 1997; Buchmann and Ehleringer, 1998; McCarroll and Loader, 2006). The relationship between plant $\delta^{18}O$ and evaporative potential may be used to infer canopy density in a forest environment. CO_2 and vegetation $\delta^{18}O$ have also been shown to demonstrate differences depending on height within the canopy, and between different plant parts, given changing evaporative potentials (Sternberg et al., 1989; Buchmann et al., 1997; Buchmann and Ehleringer, 1998). Those animals, such as folivorous primates, which obtain the majority of their water requirements from plants should strongly reflect environmental factors that influence plant transpiration and plant $\delta^{18}O$ (Kohn et al., 1996; Levin et al., 2006; Carter and Bradbury, 2016). However, interpretation remains problematic and subtle $\delta^{18}O$ differences are not realistically interpretable in the hominin record (Roberts et al., 2017b, 2017c). That said, obligate drinkers obtaining their water from open, highly irradiated water sources can be expected to show higher $\delta^{18}O$ relative to those drinking shaded standing water. Similarly, non-obligate drinkers feeding on plants living in open regions will have more positive $\delta^{18}O$ than those in shaded areas.

Stable carbon isotope evidence from the sites of Makapansgat, Sterkfontein, Kromdraai, and Swartkrans, spanning a period of about 3–1.5 Ma all demonstrated a significant contribution of C_4 ecosystems to the diets of *Paranthropus robustus* and *Australopithecus africanus* in southern Africa (Lee-Thorp et al., 1994, 2003, 2010). Unfortunately stable isotope analysis of tooth enamel cannot distinguish between the consumption of animals feeding on C_4 plants or the direct consumption of C_4 grasses and sedges. Regardless, this data directly implicated 'savanna' ecosystems in hominin diets (Lee-Thorp et al., 2003, 2010; Sponheimer et al., 2005). $\delta^{13}C$ data from *Paranthropus boisei* in Oludvai, Tanzania revealed a strong dependence of this hominin on C_4 resources (van der Merwe et al., 2008).

Based on a dramatic expansion of hominin stable isotope datasets, it is also now evident that there is a temporal progression towards increasing inputs of C_4 resources into hominin diets from 4 Ma to 2 Ma, culminating in the C_4 resource specialization of *P. boisei* and significant C_4 contributions to the diet of the earliest members of our genus (Sponheimer et al., 2013). There is also no evidence for hominin reliance on closed forest environments from *A. anamensis* and *A. afarensis* onwards. This work has invigorated the 'savanna hypothesis' and for the last decade or so palaeoanthropological focus has switched towards determining which C_4 resources Pliocene and Pleistocene hominins used, rather than whether they occupied 'savanna' settings at all (Lee-Thorp et al., 2010; Sponheimer et al., 2013).

Nevertheless, while the exploitation of open 'savanna' environments broadly remains associated with the evolution of the hominin lineage (though see White et al., 2009; White, 2014), this does not preclude the continued and complex use of tropical forests. Evolution acts under extremes and even occasional part-time occupation of open settings could drive hominin evolution towards the usage of these environments. Indeed, the reader may also have noticed that 'savanna' has so far been used persistently in inverted commas within this chapter. This is because the term 'savanna' has been equated with various environmental entities in palaeoanthropological and archaeological discussions of hominin evolution (Domínguez-Rodrigo, 2014). In popular imagination 'savanna' conjures images of sweeping African grasslands and migrating large mammals, and indeed some palaeoanthropologists see 'savanna' as being primarily composed of grassland (Lovejoy, 1981; White et al., 2009; White, 2014). Yet, when other palaeoanthropologists and archaeologists now use 'savanna' it can incorporate significant elements of woodland, forest, lacustrine, and riverine settings (Cerling et al., 2010, 2011). There also remains a clear debate of timing in the emergence of modern 'savanna' and whether it can be associated with the expansion of C_4 grasses. Much attention has been focused on this process beginning *c.* 4 Ma but lake, marine, faunal stable isotope, and soil palaeoenvironmental datasets now highlight the emergence of C_4 resources from 10–7 Ma onwards (Cerling et al., 2010, 2011; Uno et al., 2011, 2016; Feakins et al., 2013). In either case, local, selective utilization of forested settings utilized by hominins apparently persisted until 2 Ma and beyond (Barboni et al., 2010; Ashley et al., 2014).

In many ways, the 'savanna hypothesis' no longer provides added-value, at least as a scientific hypothesis, or perhaps even as a 'suggested explanation for a phenomenon' (contra Domínguez-Rodrigo, 2014). At the broad level it is clear that growing C_4, 'savanna' landscapes characterized East, and South, Africa from at least 4 Ma, if not earlier, and that these landscapes increasingly contributed to hominin diets through the Pliocene to the Early Pleistocene. Yet, if we actually want to look at the context of hominin evolution we now need to take a locale- and temporal-specific approach. Here, it is the different habitat elements and food sources, and climatically-driven fluctuations in their availability, that are most important. Utilizing broad definitions of 'savanna' (Domínguez-Rodrigo, 2014), or opposing the dense evergreen rainforest habitats of extant non-human great apes with open grasslands, will no longer provide meaningful advances in studies of hominin adaptations, the emergence of bipedalism, or stone tool usage and development. Similarly, the evolutionary history of extant non-human great apes is not necessarily one of uniform tropical rainforest (Elton, 2008). In the following,

I hope to demonstrate that it is environmental variability and ecological flexibility that has driven hominid and hominin evolution and its earliest forays beyond the African continent (Potts, 1998, 2013; Grove, 2011), and that tropical forest formations have played a persistent role as part of that variability.

ECOLOGY OF THE APES: TROPICAL FORESTS AS DYNAMIC THEATRES OF GREAT APE CULTURAL, DEMOGRAPHIC, AND MORPHOLOGICAL VARIABILITY FROM THE MIOCENE TO THE PRESENT DAY

While the LCA between extant non-human great apes and *Homo sapiens* is often taken as the starting point for discussions of human evolution it is important to note that all apes and humans, collectively known as Hominoidea, share a long evolutionary history prior to this divergence (Foley and Lewin, 2004). The first hominoid fossils are grouped collectively within the genus *Proconsul* that existed *c.* 25–35 Ma, though dates back to 31 Ma in the Early Miocene, and even the early Oligocene, have also been suggested (Fleagle, 1983; Leakey et al., 1995b; Tuttle, 2006) (Figure 3.4). Some researchers argue whether this genus should be placed within the hominoid superfamily at all given a mixture of Old World monkey traits (including horizontal postures, curved fingers, and morphological adaptations for quadrupedal branch walking) and ape traits (the lack of a tail, stabilized elbow, and facial structure). Nevertheless, its discoverers, Louis and Mary Leakey, placed it at the base of the hominoid clade (Andrews and Martin, 1987). The majority of fossils that make up the four species of *Pronconsul* (with differing body sizes) come from Kenya and Uganda, where Early Miocene environments appear to have ranged between tropical forests and more open, arid grasslands. Alongside other 'primitive apes', for example *Afropithecus turkanensis* (Leakey et al., 1988), these specimens show the facial and dental anatomy, and manipulative dexterity, that was to define the apes as they diverged from Old World Monkeys (McNulty et al., 2015). The recently discovered fossil ape cranium of *Nyanzapithecus alezi* in Kenya has been argued to indicate that the full divergence of hominids (extant *great* apes, including hominins) from this group occurred in Africa by *c.* 13 Ma (Nengo et al., 2017).

The first hominoid for which stable isotope data exists is *Sivapithecus* (Figure 3.4). This genus was part of a hominoid expansion beyond Africa and into Eurasia *c.* 17 Ma at a time of expanding tropical forests facilitated by warmer global conditions and increased atmospheric partial pressure of CO_2. *Sivapithecus* fossils have been found in Pakistan, India, Nepal, and possibly also the Middle East, dating from 13–8 Ma (Kelley, 2002) (Figure 3.4). This large-bodied ape, with large face and thick enamel, has been described as ancestral to the modern orangutan, although this remains debated (Pilbeam and Young, 2001). Of particular interest here, however, is the fact that stable carbon isotope analysis of *Sivapithecus* tooth enamel in Pakistan revealed that it occupied dense, closed tropical forest similar to that inhabited by orangutans, gorillas, bonobos, and chimpanzees today (Nelson, 2003) (Figure 3.5). Moreover, comparison of isotope data

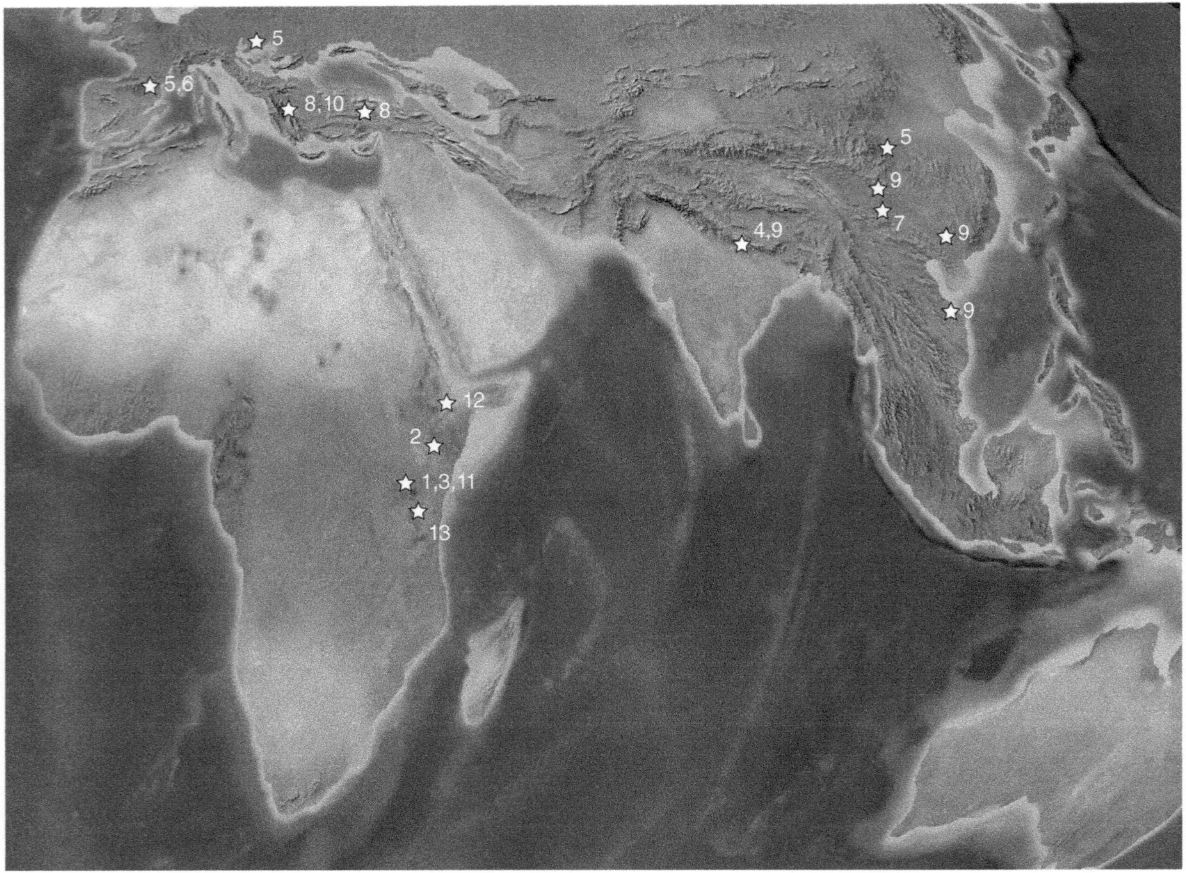

Figure 3.4. Map of Miocene Africa, the Middle East, and Eurasia from Deep Time Maps with key hominoid sites mentioned in the text plotted. With thanks to Ron Blakey and Nils Vanwezer. 1. *Proconsul*, 2. *Afropithecus turkanensis*, 3. *Nyanzapithecus alezi*, 4. *Sivapithecus*, 5. *Dryopithecus*, 6. *Anoiapithecus brevirostris*, 7. *Lufengpithecus*, 8. *Ouranopithecus*, 9. *Gigantopithecus*, 10. *Graecopithecus*, 11. *Nakalipithecus*, 12. *Chororapithecus abyssinicus*, 13. *Pan*.

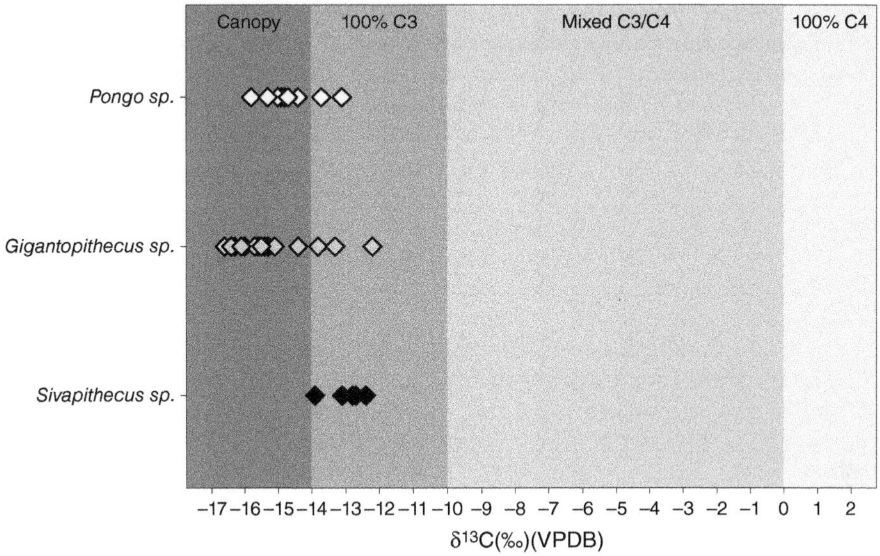

Figure 3.5. Stable carbon isotope ($\delta^{13}C$) data from the tooth enamel of *Sivapithecus* from Siwalik deposits in Pakistan (Nelson, 2003) and *Gigantopithecus blackii* and fossil orangutan (genus *Pongo*) from Plio-Pleistocene sites in southern China and Thailand (Bocherens et al., 2017). The data is shown against estimates of 100% canopy resource consumption, 100% C_3 resource consumption, a mixed C_3/C_4 resource diet, and 100% C_4 resource diets highlighted in the text.

from *Sivapithecus*-bearing and post-*Sivapithecus* layers, suggests that its local extirpation may have been driven by forest fragmentation (Nelson, 2003). Although no isotopic data exists for this group, the genus *Dryopithecus* (*c.* 12–9 Ma), represented by fossils from East Africa, Europe, and China, further demonstrates the expansion of apes made possible by widespread tropical forests and a land bridge across the Mediterranean during the Middle Miocene (Begun, 2004, 2010). This genus has been argued to have been adapted to tree dwelling and the consumption of soft fruit (Xue and Delson, 1989). Other hominoids representative of this radiation include *N. alezi* in Kenya (Nengo et al., 2017), *Anoiapithecus brevirostris* in Spain (Moyà-Solà et al., 2009), *Lufengpithecus* in China (Harrison et al., 2002; Chaimanee et al., 2004) and *Ouranopithecus* in Greece (de Bonis et al., 1990) (Figure 3.4).

The Middle Miocene was the most diverse period in hominoid evolutionary history and also witnessed the splitting of hominids (great apes, including hominins and humans) from the broader hominoid tree (Moyà-Solà et al., 2009; Nengo et al., 2017). From 10 Ma cooler temperatures and reduced rainfall led to the gradual break-up of tropical forests and the expansion of drier woodland and grassland. These environmental changes played a major role in shaping the geographical distribution of non-human great apes today. The majority of non-human great ape taxa, including those mentioned above, had disappeared outside of Africa after 8 Ma. By 2 Ma orangutans (*Pongo* spp.) were one of only two groups of non-human great apes left in Asia. The other, *Gigantopithecus*, was the largest hominoid genus ever to exist (Figure 3.4). Popularized in a recent portrayal of King Louis from the *Jungle Book* (Favreau, 2016), *Gigantopithecus* could weigh in at a maximum of 270 kg (Simons and Ettel, 1970). This hominid had thick tooth enamel and its dietary preferences have been argued to variously include carnivory, bamboo, and fruits (Ciochon et al., 1990; Daegling and Grine, 1994). *Gigantopithecus* has been identified in Pakistan, China, and Thailand and lived from 8 Ma to 0.1 Ma (Tougard et al., 1996; Louys et al., 2007). Stable carbon isotope evidence from *Gigantopithecus* tooth enamel suggests that, like *Sivapithecus*, this ape was focused on dense tropical forest environments (Loudon et al., 2016), though it was also able to occupy forest–woodland–'savanna' mosaics until *c.* 0.1 Ma (Bocherens et al., 2017) (Figure 3.5).

Another intriguing player in Middle-Late Miocene hominoid diversity is *Graecopithecus*. Tooth morphology of members of this taxon discovered in Greece and Bulgaria has been used to argue that the origins of the hominin clade were in Europe *c.* 7.2 Ma at a time of waning tropical forest presence in this region (Böhme et al., 2017; Fuss et al., 2017). While this remains hotly disputed (Benoit and Thackeray, 2017) it highlights this period as key to the divergence of hominins from extant non-human great apes. Beyond the Middle Miocene, the only other non-hominin great apes found in the fossil record were apparently the direct relatives of those extant today. Modern genetic analysis of chimpanzees (*Pan troglodytes*), bonobos (*Pan paniscus*), and gorillas (*Gorilla* spp.) indicates the gradual divergence of their lineages in Africa between 10–7 Ma (Patterson et al., 2006; Prado-Martinez et al., 2013). Nevertheless, divergence times estimated from modern genetic information have broad error ranges (Figure 3.6). For example, hominins and chimpanzees have been argued to diverge anywhere between 13 Ma (Arnason et al., 1998) and 4 Ma (Patterson et al., 2006; Prado-Martinez et al., 2013). The sparse fossil evidence has not helped to clarify the timing of this

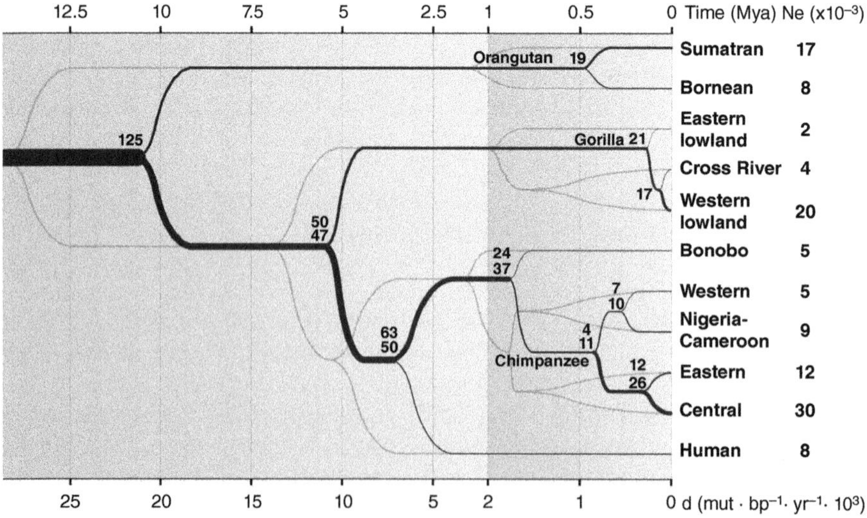

Figure 3.6. Schematic of evolutionary divergence dates based on modern great ape genetics. Adapted from Prado-Martinez et al. (2013). With thanks to Tomas Marques-Bonet.

divergence. A maxilla dated to *c.* 10–8 Ma in Samburu, Kenya (Ward and Duren, 2002) and specimens of *Nakalipithecus* found in Kenya's Rift Valley (Kunimatsu et al., 2007) have been used to suggest that hominins, gorillas, and chimpanzees diverged after 8 Ma (Kunimatsu et al., 2007). Nevertheless, the new find of *N. alezi* in Kenya suggests that the ancestor of extant great apes was alive *c.* 13 Ma (Nengo et al., 2017), and the earliest fossil member of gorillas, *Chororapithecus abyssinicus*, in Ethiopia *c.* 8 Ma has been used to argue that humans diverged from gorillas by at least 10 Ma (Katoh et al., 2016).

From this point there is a gaping hole in the non-hominin great ape fossil record until the earliest fossil of *Pan* from the East African Rift Valley dated to the Middle Pleistocene (*c.* 550–300 ka) (McBrearty and Jablonski, 2005). The palaeo-environmental associations of the sparse non-hominin African great ape fossil record remain largely unknown. However, the earliest fossil of *Pan* is best associated with tropical forest–woodland mosaic habitats rather than dense evergreen rainforest (McBrearty and Jablonski, 2005; Elton, 2008). Moreover, changes in tropical forest environments appear to have played a major role in the genetic make-up of extant African great ape populations. The genetic divergence of bonobos from chimpanzees *c.* 2 Ma has been argued to be a result of the dissection of the West African rainforest by the Congo River (Prüfer et al., 2012), and the dynamic variation in the flow of this river through time (Takemoto et al., 2015). Meanwhile, a dearth of genetic diversity in eastern gorillas in Africa has been linked to a consistent contraction of their forest habitat (Prado-Martinez et al., 2013). Indeed, it is highly likely that flux in tropical forest environments has contributed significantly to modern genetic diversity observable in great ape populations today, with Guillén et al. (2005) arguing that climate change during the Pleistocene led to cyclical periods of forest isolation and expansion resulting in the current distribution, phylogeny, and genetic diversity of African non-human great ape populations. This evidence reveals that tropical forests were locations of incredible dynamism in great ape evolution, rather than sites of stasis.

It would therefore be a mistake to consider extant non-human great apes, and their tropical forest homes, as static relics of the time when the LCA walked the planet. Variability in tropical forest environments across space and time has not only shaped the genetic make-up of extant populations, but also their cultural and adaptive niches. For example, in the dense evergreen rainforests of Southeast Asia, that have been argued to have been relatively stable in the face of Miocene climate change (Whitmore, 1998), orangutans gorge themselves on seasonally available fruit from dense stands of masting dipterocarp trees in order to obtain enough carbohydrate to form fat stores for the remainder of the year (Knott, 1998). By contrast, in the tropical evergreen and deciduous forests of Central and West Africa chimpanzee groups show varied subsistence practices. These range from the 'savanna' populations that have been documented utilizing the tool-assisted hunting of small mammals in the mosaic tropical forests of Fongoli, Senegal (Pruetz and Bertolani, 2007) to groups in the dense evergreen rainforests of Cameroon that fragment and coalesce their social groups in line with seasonal variation in the abundance of resources (Macho and Lee-Thorp, 2014). Krützen et al. (2011) have also argued that subtle variations within and between tropical forest environments have heavily influenced the observation of different great ape tool-use (including hunting with spears and termite-fishing with long sticks), grooming and courtship behaviours, and social learning (e.g., Whiten et al., 1999; Van Schaik et al., 2003) on an intra- and inter-specific basis.

Finally, while increasingly specialized bipedal locomotion characteristic of the hominin clade has tended to be associated with 'savanna' environments, increasing morphological and observational analysis of extant non-human great ape locomotion has highlighted the variety of locomotion strategies practised by our closest living relatives within tropical forest environments (Crompton et al., 2010). While all apes can and do walk bipedally, most do so in a forest context (Thorpe and Crompton, 2005; Crompton et al., 2010). Indeed, it is the most arboreal, the orangutan, rather than chimpanzee groups known to inhabit open 'savanna' regions, that uses bipedal locomotion most often (Thorpe and Crompton, 2005). Terrestrial knuckle walking, quadrupedalism, vertical climbing, and forelimb suspension make up *c.* 67 per cent of bonobo locomotion, 93 per cent of chimpanzee locomotion, and 97 per cent of mountain gorilla locomotion, but only 39 per cent of orangutan locomotion (Crompton et al., 2010). Thorpe et al. (2007) have argued that straight-backed, straight-limbed bipedalism evolved in orangutans as an adaptive strategy to the navigation of thin, flexible branches when gathering fruit high in the canopy of the masting dipterocarps discussed above. It has also been suggested that this form of locomotion between tree gaps would combat the threat of ground-dwelling predators (Crompton et al., 2010). This research questions the significance of solely focusing on grassland and 'savanna' environments in the evolution of bipedal locomotion and, perhaps, encourages re-evaluation of the importance of tropical forests in the early stages of our own evolution.

A SHADED CRADLE FOR LATE MIOCENE HOMININS IN AFRICA?

The environmental context of the origin of our own ancestral lineage remains murky, primarily as a result of the extreme rarity of fossil material of the right age, though there is now a growing fossil record that extends beyond the Pliocene. The environmental associations of some of the first members of the hominin clade have tended to be with more complex tropical forest–woodland–grassland mosaics and have even been used to refute the 'savanna hypothesis' as a meaningful theory of hominin origins (White et al., 2009; White 2014). The earliest suggested hominin in Africa is *Sahelanthropus tchadensis* (Figure 3.7). A team led by Michel Brunet found six fossils of this species, including several jaw fragments, teeth, and a cranium. These remains have been dated to *c.* 7 Ma (Lebatard et al., 2008), although this is based on cosmogenic nuclide dating of sand that may have been subject to taphonomic disturbance (Beauvilain, 2008). The research team argued that the fossils demonstrated a series of derived 'hominin' features including a tooth enamel thickness intermediate between living non-human great apes and australopiths, cheek teeth that are similar in size to Australopithecines, a thickened brow ridge similar to that of later fossil *Homo*, and a location of the foramen magnum

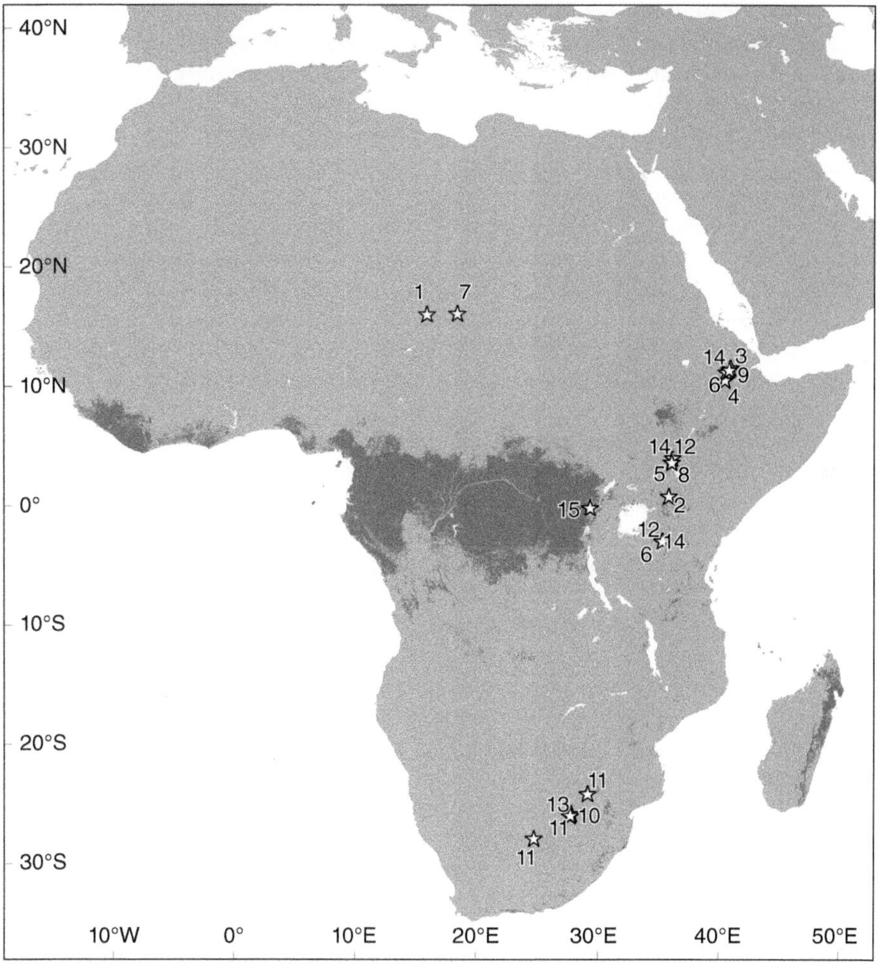

Figure 3.7. Map of major Pliocene and Pleistocene hominin sites in Africa discussed in the text plotted against tropical forest distributions estimated on the basis of the MODIS (Moderate Resolution Imaging Spectroradiometer) Land Cover MCD12Q1 majority land cover type 1, class 2 for 2012 (spatial resolution of 500 m). Downloaded from the US Geological Survey Earth Resources Observation System (EROS) Data Center (EDC). 1. *Sahelanthropus tchadensis*, 2. *Orrorin tugenensis*, 3. *Ardipithecus kadabba*, 4. *Ardipithecus ramidus*, 5. *Australopithecus anamensis*, 6. *Australopithecus afarensis*, 7. *Australopithecus bahrelghazali*, 8. *Kenyanthropus platyops*, 9. *Australopithecus garhi*, 10. *Australopithecus sediba*, 11. *Australopithecus africanus*, 12. *Paranthropus boisei*, 13. *Paranthropus robustus*, 14. *Homo*, 15. Ishango hominin. With thanks to Ulrike Thuering.

that suggests *S. tchadensis* may have been bipedal (Brunet et al., 2002). Environmental information, based primarily on the ecological preferences of associated fauna, was used to suggest that these traits developed in a mixed lacustrine, forest, woodland, and 'savanna' landscape (Brunet et al., 2002, 2005).

A number of more 'ape-like' traits, including the shape of the rear of the skull, protruding face, large canine fossa, and wide eye sockets, have also been used to suggest that *S. tchadensis* was a common ancestor of both humans and chimpanzees, a separate hominoid lineage, or an ancestor of gorillas (Wolpoff et al., 2002, 2006). Moreover, the lack of postcranial fossils, with the possible exception of a femoral midshaft (Hawks, 2009), makes it difficult to firmly assert that this species was bipedal (Wolpoff et al., 2002, 2006). Nevertheless, CT scans and a morphological comparison of over thirty skull features between *S. tchadensis*, chimpanzees, gorillas, and other hominins placed the cranium of the former firmly in the hominin range (Guy et al., 2005). Stable isotope analysis of carbonates and leaf waxes from East African terrestrial sequences (Cerling et al., 2011; Uno et al., 2016) and marine cores (Feakins et al., 2013) suggest that there was an early expansion of C_4 grassland in this part of the world between 10–7 Ma, potentially hinting at the earlier importance of open environments to hominin evolution. A similar environmental context of expanding woodland and grassland at the expense of dense forest canopies has been noted for *Graecopithecus* in Europe *c.* 7 Ma (Böhme et al., 2017; Fuss et al., 2017), though this remains to be confirmed as a hominin.

The picture is somewhat different for *Orrorin tugenensis* (Figure 3.7), the next proposed member of the hominin clade. Discovered by a team led by Brigitte Senut and Martin Pickford in the Tugen Hills region of Kenya, twenty fossils of this hominin have been dated to between 6.1 and 5.7 Ma (Senut et al., 2001; Sawada et al., 2002). While the small teeth and thick enamel of these fossils have been used to argue its affinity to the genus *Homo*, it is an upper femur that has taken on the greatest importance in placing this species within the hominin evolutionary tree. This femur has a long, antero-posteriorly narrow neck and wide proximal diaphysis that are typical of later Australopithecines and *Homo*, suggesting that *O. tugenensis* practised obligate bipedalism (Senut et al., 2001; Pickford et al., 2002). However, crucially for our purposes here, it has been argued that muscle insertions on the humerus bone, and the degree of curvature of phalanges, are best associated with tree climbing and arboreal activity (Richmond and Jungers, 2008). This has led Thorpe and colleagues to suggest that bipedalism in *O. tugenensis*, and perhaps within the hominin clade as a whole, evolved in similar contexts to that seen in orangutans discussed above (Thorpe et al., 2007). The low rounded molars and small canine teeth have been used to argue that this hominin primarily ate leaves, fruit, seeds, roots, nuts, and insects, which would not refute such a theory (Thorpe et al., 2007), while fauna from the Lukeino Formation suggest *O. tugenensis* lived in a region dominated by dry evergreen forest (Pickford, 2012). That said, arguments for the significance of terrestrial bipedalism, and a lack of stable isotope data or directly associated faunal information, for *O. tugenensis* urge caution in such strong claims.

The Miocene hominins that have been most associated with arguments for tropical forest origins are the members of the genus *Ardipithecus* (White et al., 2009; White, 2014) (Figure 3.7). Discovered by Yohannes Haile-Selassie and Tim White in the Middle Awash region of Ethiopia, *Ardipithecus kadabba* was initially considered to be a subspecies of the previously discovered *Ardipithecus ramidus* hominin (Haile-Selassie, 2001). However, based on dental morphology and wear it has since been argued that *A. kadabba* is a species in its own right, dating to between 5.8 and 5.2 Ma (Haile-Selassie et al., 2004). The broad and robust nature of a toe bone also associated with this species has been used to argue that *A. kadabba* walked bipedally. The second member of *Ardipithecus*, *A. ramidus*, or 'Ardi', is actually a Pliocene species dating to *c.* 4.4 Ma (White et al., 1994). The pelvic morphology and divergent large toe have been used to suggest that this species was also bipedal (Lovejoy et al., 2009; White et al., 2009). However, its lower limb proportions, foot, and hand morphology have also been associated with tree climbing (Kimbel et al., 2014) (Figures 3.8 and 3.9). Fauna uncovered from

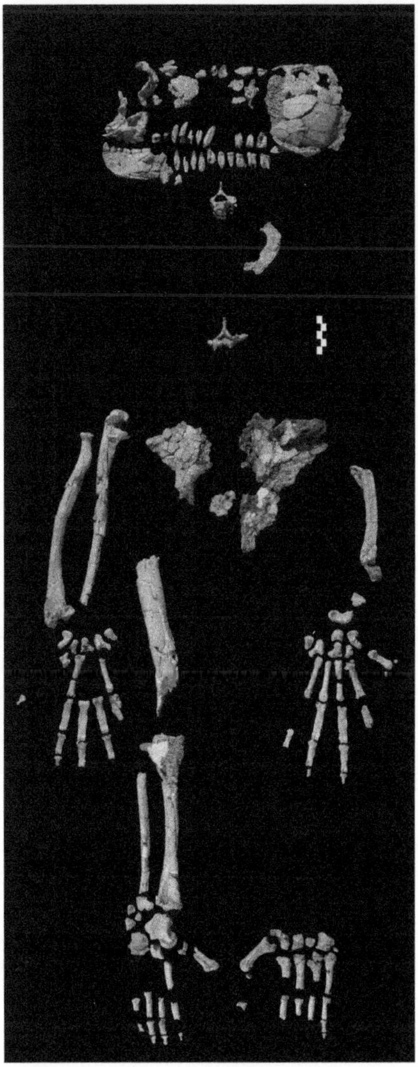

Figure 3.8. *Ardipithecus ramidus* fossil skeleton housed in the National Museum of Ethiopia, Addis Ababa. Photograph taken by Tim D. White (humanoriginsphotos.com). (c) 2009 Tim D. White. With thanks to David L. Brill and Tim D. White.

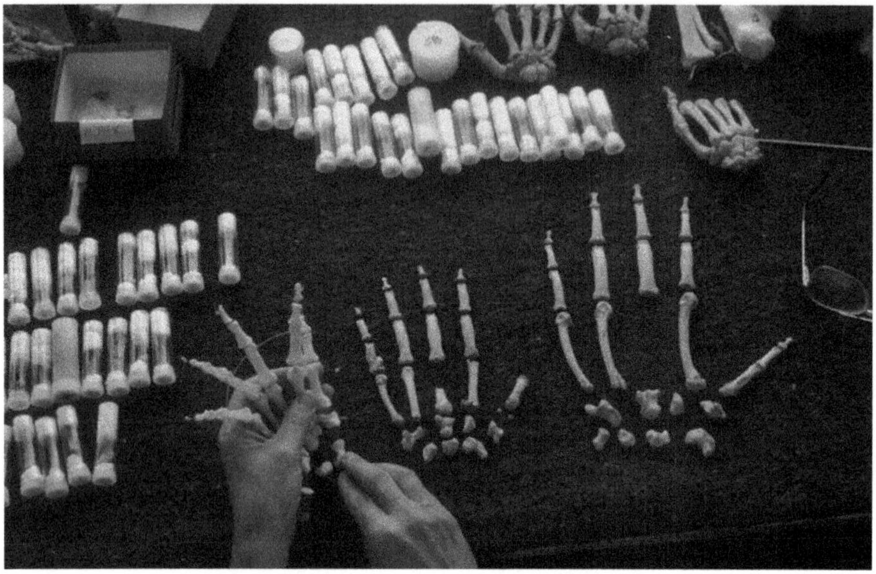

Figure 3.9. Comparative analysis of the 'Ardi' hand bones (centre) with a modern human (left) and chimpanzee (right) at the FSF-sponsored Paleoanthropology Laboratory, National Museum of Ethiopia, Addis Ababa, 1995. The fossils show that the hand of Ardipithecus ramidus was neither chimpanzee nor human in its proportions and articulations. Photograph taken by David L. Brill. (c) 1995 David L. Brill. With thanks to David L. Brill.

the same contexts as *A. ramidus* were used to argue that it lived within closed-canopy tropical forest (WoldeGabriel et al., 1994, 2001), although stable carbon isotope analysis of the tooth enamel of associated fauna was suggested to, in fact, demonstrate that *A. ramidus* and *A. kadabba* both had a preference for open grassland environments (Levin et al., 2008). Stable isotope analysis of *A. ramidus* teeth themselves suggest a mixed forest–woodland–grassland preference that, alongside morphological information, implies at least some adaptation to arboreal settings (White et al., 2009; WoldeGabriel et al., 2009).

The environmental association of the earliest hominins in Africa remains something of an unknown. This is in part a result of the fact that debate exists as to whether *S. tchadensis* and *O. tugenensis* are definitively hominins. This may never be resolved given the difficulty of taxonomically comparing sparse fossil records in such deep time. Another issue is that the dominant environments associated with these taxa are difficult to ascertain with only a few data points and estimations of past faunal ecological preferences. What is clear, however, is that all of the proposed early hominins from the Late Miocene retained morphological adaptations associated with tree climbing and arboreal living, even as bipedal traits developed. Whether Thorpe et al. (2007) are right in their argument that hominin bipedalism evolved through the navigation of tree branches, tropical forests, either as part of variable environments or settings dominated by closed canopy, clearly played an important role in hominin emergence. Finally, it is also worth noting that uncertainties of taxonomic affiliation and genetic divergence data mean that the chimpanzee split from hominins could have occurred *after* the existence of *S. tchadensis* and *O. tugenensis*. Given the forest preference of chimpanzees today this provides a little-discussed, contrasting potential environmental end-point for Late Miocene hominin taxa, away from Pliocene and Pleistocene hominins whose association with 'savanna' settings has too often been used to suggest a linear march towards open environments.

FOREST VERSUS 'SAVANNA' IN PLIOCENE HOMININ EVOLUTION: AN OPEN DRIVER OR A SHADIER PICTURE?

As noted above, from the Pliocene onwards, savanna environments take centre stage in studies of hominin evolution and dietary preferences. However, while expanding grasslands clearly did provide a backdrop to hominin diversification in East and South Africa during these periods, the 'savanna hypothesis' misses a considerable degree of evolutionary variability. For example, although, as discussed above, *A. anamensis* has morphology suggestive of habitual bipedality, the lack of postcranial skeletal information inhibits an understanding of the relative importance of climbing and terrestrial bipedalism. Much discussion in this regard is therefore forced to extrapolate from its successor, *A. afarensis*. While the angle of the leg beneath the pelvis in the latter hominin has been used to suggest constant upright bipedality, and footprints at Laetoli often associated with this hominin suggest the same (Stern and Susman, 1983; White and Suwa, 1987; Raichlan et al., 2010), the limb and scapular morphology of *A. afarensis* has been used to suggest that its locomotor repertoire included a substantial amount of climbing (Green and Alemseged, 2012) (Figure 3.10). Furthermore, the apparent existence of a wrist-locking mechanism suggests that *A. afarensis* may have utilized knuckle walking in terrestrial situations (Richmond et al., 2001). The exact reliance of this hominin on terrestrial bipedalism therefore remains debated and a more complex picture of morphological adaptation and evolution seems likely.

The existing palaeoenvironmental and palaeodietary information for *A. anamensis* and *A. afarensis* also highlight complexity. Based on stable carbon isotope analysis of *A. anamensis* teeth, and that of associated fauna, it has been argued that this hominin existed along riparian forest corridors, or within more open forest biomes (Schoeninger et al., 2003; Cerling et al., 2010; Cerling et al., 2013). As can be seen in Figure 3.11, *A. anamensis* $\delta^{13}C$ suggest similar adaptations to the forest–woodland–grassland mosaics inhabited by *A. ramidus*. Neither document significant inputs of C_4 resources. Dental microwear evidence from *A. anamensis* is indicative of a diet similar to modern gorillas, supporting the isotopic information suggesting reliance on C_3-based resources (Ungar et al., 2010), though the precision provided by this form of analysis is debated (Sponheimer et al., 2013). Stable carbon isotope data from *A. afarensis* tooth enamel indicate that this hominin lived in much more varied habitats than *A. anamensis* and *A. ramidus*. While one individual shows clear preference for pure C_3, and perhaps forest, habitats, other individuals show clear C_4 resource inputs overlapping with *P. boisei*

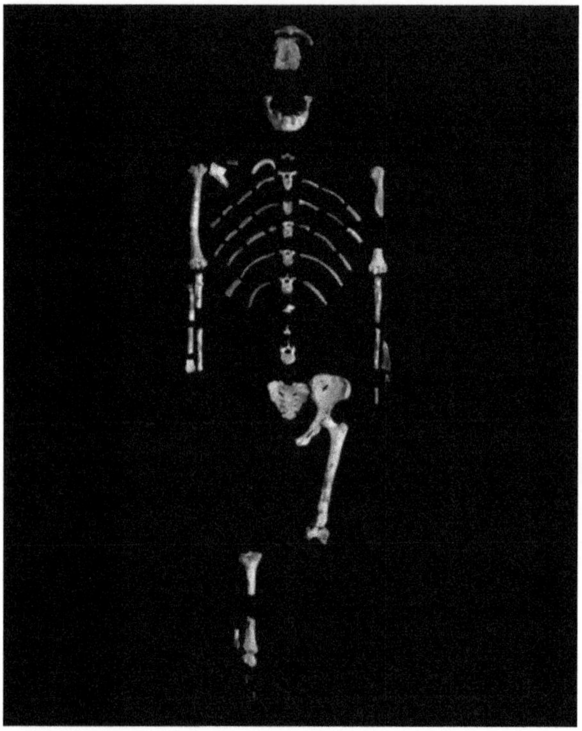

Figure 3.10. *Australopithecus afarensis* 'Lucy' original fossil skeleton housed in the National Museum of Ethiopia, Addis Ababa. Photograph taken by David L. Brill in 1985 (humanoriginsphotos.com). (c) 1985 David L. Brill. With thanks to David L. Brill.

(Reed, 2008; Sponheimer et al., 2013). This variation may be structured by location and Su and Harrison (2008) have suggested a distinction between the dense wooded and mesic conditions inhabited by *A. afarensis* at Hadar versus woodland–shrubland–grassland mosaics at Laeotoli, Tanzania. Based on chimpanzee ecology, the authors suggest that *A. afarensis* populations would have actually been more successful in wooded locales (Su and Harrison, 2008).

A broad dietary shift, from a focus on C_3 resources in *A. ramidus* and *A. anamensis*, to an increasing engagement with C_4 resources in *A. afarensis*, *A. africanus*, and *P. boisei* is clear. Yet, variability still exists within this trend. *Homo* spp., in both eastern and southern Africa, do not appear to specialize in C_4 resource use to the same extent as *P. boisei*, while *Australopithecus sediba* (*c.* 2 Ma) demonstrates a focus on C_3 resources (Sponheimer et al., 2013; Roberts et al., 2016; Robinson et al., 2017). Sequential analysis of *A. africanus* and *P. boisei* tooth enamel has even highlighted dramatic seasonal or interannual variability in the contribution of C_3 and C_4 resources to their diets (Sponheimer et al., 2006b; Lee-Thorp et al., 2010). Laser ablation on four *P. robustus* teeth crowns demonstrates that certain individuals switched between a C_3-dominated diet to a C_4-dominated diet on intra- and interannual scales (Sponheimer et al., 2006b). Similar variability was also noticed for *A. africanus* individuals from Sterkfontein

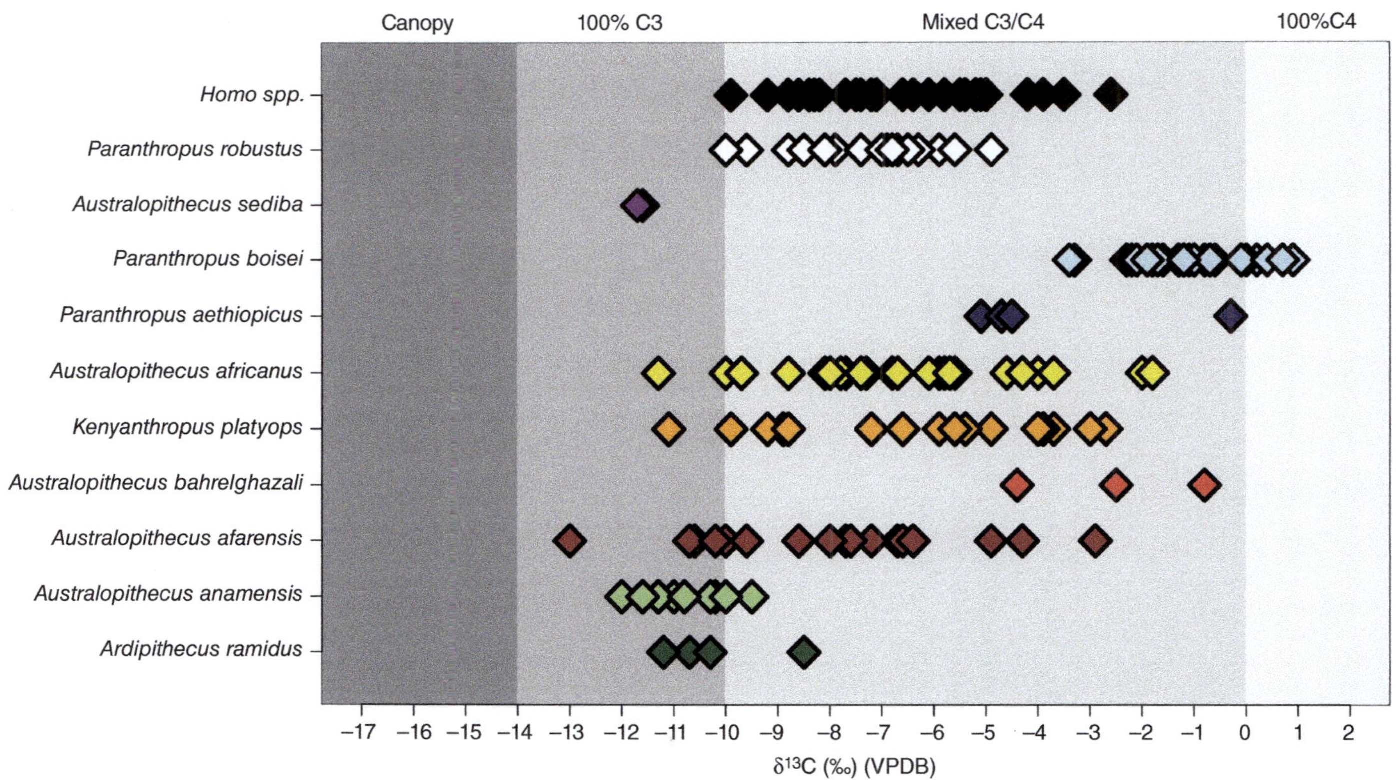

Figure 3.11. Stable carbon isotope ($\delta^{13}C$) of hominin tooth enamel from southern, eastern, and Central Africa plotted by genus/species (as per. Levin et al., 2015). Hominin taxa are arranged from oldest (bottom) to youngest (top) (as per. Levin et al., 2015). Data from Lee-Thorp et al. (1994), Sponheimer and Lee-Thorp (1999), Lee-Thorp et al. (2000), van der Merwe et al. (2003), Sponheimer et al. (2005, 2006b), van der Merwe et al. (2008), White et al. (2009), Cerling et al. (2011), Henry et al. (2012), Lee-Thorp et al. (2012), Cerling et al. (2013), Sponheimer et al. (2013), Wynn et al. (2013), Levin et al. (2015). The data is shown against estimates of 100% canopy resource consumption, 100% C_3 resource consumption, a mixed C_3/C_4 resource diet, and 100% C_4 resource diets highlighted in the text.

Member 4 (Lee-Thorp et al., 2010). One individual indicated a more or less uniform C_3 preference, while another showed *c.* 6‰ variance. This, alongside the fact that the exact C_4 resources used remain unknown, even by *P. boisei/robustus* that appear reliant on these envrionments, make it difficult to uniformly associate hominin adaptations with savanna settings at this point in time. Clearly some degree of selectivity and flexibility was maintained on an intra- and inter-taxonomic basis.

Even in the Late Pliocene and Early Pleistocene, forests, or at least localized arboreal settings, apparently left an important evolutionary legacy. *A. sediba*, (*c.* 2 Ma) demonstrates a mosaic of morphological features that include evidence for terrestrial bipedalism as well as climbing (Kivell et al., 2011). Similarly, biomechanical studies of *H. habilis* limb bones demonstrate high humeral relative to femoral strength, which suggests at least partial arboreality relative to *Homo erectus*, with the more derived pattern of obligate bipedalism not evolving until after 1.8 Ma (Ruff, 2009). The hominin remains dated to c. 2.6-2.0 Ma at Ishango in the Democratic Republic of Congo have also been argued to represent interaction with tropical forests at this time, although directly associated palaeoenvironmental evidence is currently lacking (Crevecoeur et al., 2014). While further cranial expansion, habitual tool-use, and derived bipedal locomotion, akin to *H. sapiens*, in *H. erectus* (1.9 to 0.2 Ma) has been persistently linked to open savanna settings (Dennell and Roebroeks, 2005), our understanding of the environments of the earliest members of the genus *Homo* remains very poor. A similar situation exists for the emergence of stone tool technology. The greater time-depth of this practice, extending back to 3.3 Ma (Harmand et al., 2015), means that more than one hominin species, both prior to and after the evolution of *Homo*, could have been manufacturers of stone tools. Pedogenic carbonate stable isotope evidence has also been used to suggest that Early Pleistocene toolmakers had a preference for wooded rather than savanna environments in the Turkana Basin (Quinn et al., 2013). It is also worth noting that the earliest 'tools' were likely made from wood and other organic materials that will not preserve in the fossil record, particularly if this practice began in tropical forest settings (Panger et al., 2002).

The evidence for persistent variability in hominin morphology and adaptations is also perhaps less surprising when increasingly resolved regional and local palaeoenvironmental studies of the Pliocene and Early Pleistocene are taken into account. Although a general trend towards increasing presence of C_4 grassland and more arid environments in East, and most likely also South, Africa from 10 Ma to 2 Ma certainly exists based on regional records, subtleties often remain undiscussed. Local diversity has now been discerned in the timing of the onset of C_4 vegetation dominance in different lake basins. For example, in contrast to other regions in eastern Africa this only appears to occur after 2.9 Ma in the Lower Awash Valley and Omo-Turkana region (Cerling et al., 2011; Robinson et al., 2017). Furthermore, the trend towards increasing C_4 vegetation was not unilinear. Feakins et al. (2005, 2013) have argued, on the basis of geochemical analysis of East African marine cores, that there is increasing humidity and forest presence *c.* 4 Ma. Moreover, observations at local and regional scales have been used to suggest that in places such as Olduvai Gorge, open C_4 grasslands could abruptly transition to C_3 forests within several hundred to thousands of years as late as 2 Ma (Magill et al., 2013; Maslin et al., 2014). Elements of woodland and forest habitats have also been postulated at Pliocene and Pleistocene contexts at Laetoli (Kingston and Harrison,

2007) and Koobi Fora (Reed, 1997; Elton, 2006), within a context of increasingly open settings in eastern and southern Africa more broadly (Elton, 2007).

One of the key points that is often missed in discussions of hominin evolution is the fact that animals, including our ancestors, have agency. The dominance of a C_4 landscape or savanna does not necessarily lead to pure savanna reliance or a complete neglect of tropical forest habitats. If the above suggests anything it is that 'off-site' palaeoenvironmental records, stable isotope analysis of hominin teeth, dental microwear, and associated faunal habitat preferences can provide conflicting evidence for overall ecological context and that they are best used together in the construction of a more complex picture of adaptation. In addition, microhabitats can drive evolution, and the persistent combination of climbing and bipedal morphologies, as well as mixed C_3 and C_4 preferences, suggest that new narratives and hypotheses should shift towards local, contextual explanations of change and adaptation. There is also the point that even if the consumption of C_4 resources by hominins is demonstrated, it is not currently possible to distinguish whether they are grasses, sedges, underground storage organs, insects, or large animals. The evolutionary implications, and the supposed importance of the savanna, however it is defined, will evidently be different in each case (Sponheimer et al., 2013; Paine et al., 2018). As a result, while the evolution and diversification of the hominin lineage, specialized bipedalism, and stone tool production, occur against a broad backdrop of C_4 tropical grassland expansion, the timing, mechanism, and importance of particular resources in these habitats, relative to those available in tropical forests and tropical forest patches, remain difficult to determine.

TROPICAL FORESTS AND EARLY-MIDDLE PLEISTOCENE HOMININ EXPANSIONS OUT OF AFRICA

Hominin resource selectivity and complex relationships with tropical forest habitats also characterize the Early-Middle Pleistocene period. By 1.8 Ma there is fossil evidence for the expansion of the genus *Homo* beyond Africa and into Eurasia at Dmanisi cave, Georgia (Gabunia et al., 2000; Garcia et al., 2010). By 1.2–0.8 Ma *Homo antecessor* had also reached Atapuerca in Spain (Bermúdez de Castro et al., 1997; Falguères et al., 1999) and by 0.9–0.7 Ma a hominin species had made it to Happisburgh and Pakefield in the United Kingdom (Parfitt et al., 2005; Ashton et al., 2014). This period is also characterized by the appearance of distinctive pear-shaped hand-axes, commonly associated with *H. erectus*, across Africa (from 1.7 Ma), West Asia, South Asia (1.5 Ma), and Europe (0.9 Ma) (Goren-Inbar et al., 2000; Bar-Yosef and Belfer-Cohen, 2001; Scott and Gilbert, 2009). This technocomplex is seen as a direct descendant of the Oldowan flake and core technology, with early Acheulean assemblages including many Oldowan pieces. The production of these hand-axes has been associated with the attainment of increased cognitive or linguistic ability (Isaac, 1976; Wynn, 1991), the development of a social perception of form and craft (Gamble, 1997), or potentially new forms of tool-use and efficiency (O'Brien, 1981). Whatever their exact function or meaning, they are part of the first stone tool complex to exist beyond Africa even if they do not occur at all sites.

The traditional association of savanna environments with the evolution of our genus has also held true for these Early Pleistocene migrations of *Homo*. A prominent hypothesis on the first Out of Africa process has centred on the expansion of open savanna environments (Dennell and Roebroeks, 2005). Dennell and Roebroeks (2005) have argued that the Late Pliocene and Early Pleistocene expansion of a corridor of grasslands ('savannastan'), that occurred in relation to changing atmospheric CO_2 concentrations and orbital characteristics of the Earth, drove the expansion of hominin populations from East Africa into Central Asia. This hypotheses of grassland expansions has also been extended to Southeast Asia (Ciochon, 2009; Marwick, 2009). Grasslands across Eurasia at this time are seen as favouring adaptations to long-distance ranging and the pursuit of large mammals argued to be characteristic of the genus *Homo* (Dennell and Roebroeks, 2005). This also fits with suggested competitive relationships between hunting and competitive scavenging by species of *Homo* and large mammalian carnivores in the Early Pleistocene (Lewis and Werdelin, 2007, 2010). Foraging theory applied to *H. sapiens* in Sahul would suggest that if hominins rank food by net calorific gain, always collect the highest ranked prey encountered, and favour habitats where return rates are higher relative to search and processing costs, then tropical forests would be less likely targets for colonization (Cosgrove et al., 2007; O'Connell and Allen, 2012).

Nevertheless, as part of this hominin expansion Out of Africa, by 1.5 Ma there is also fossil evidence within the 'Sangiran' geological formation that *H. erectus* had reached Java in tropical Southeast Asia (Larick et al., 2001; Zaim et al., 2011). The 'Bapang' and 'Trinil' Formations have yielded further fossil evidence for *H. erectus* in excess of 1 Ma in this region (de Vos et al., 1994; Larick et al., 2001) (Figure 3.12). Some have suggested that rainforest was the dominant vegetation type in Southeast Asia between 2.6 and 1 Ma (Sémah et al., 2002; Sémah and Sémah, 2012; Sémah et al., 2016), linked to ideas of the stability of evergreen rainforest in this region since the Miocene (Whitmore, 1998). This would imply that *H. erectus* may have occupied and exploited complex tropical forest settings rather than savanna grasslands. Changes in sea-level of atmospheric CO_2 have also been argued to have driven significant landmass, and rainforest, fragmentation and isolation in Southeast Asia during the Quaternary (Heaney, 1991; Gathorne-Hardy et al., 2002). This has led others to argue that it was the forest–grassland mosaics produced by tectonic and climate processes that were crucial for the expansion of *H. erectus* into these regions (Bettis et al., 2009; Ciochon, 2009). Ethnographic and ecological observations that tropical forests represent disadvantageous habitats for hominin thermoregulation, navigability, and protein and carbohydrate sourcing (Chapter 1), and a paucity of archaeological sites with organic remains, have made it difficult to confirm *H. erectus*' tropical forest occupation in this region.

A lack of well-dated, high-resolution palaeoenvironmental records, and detailed archaeological data, has made it challenging to distinguish between the use of savanna corridors or tropical forests in Southeast Asia. Where stone tools do occur, organic residue or use-wear information about their use is currently absent (though see Xhauflair et al., 2017). Some have argued that a paucity of stone tools in Southeast Asia may suggest that *H. erectus* made use of bamboo or wooden tools, thus potentially implying utilization of tropical forest mosaic environments (Bar-Yosef et al., 2012).

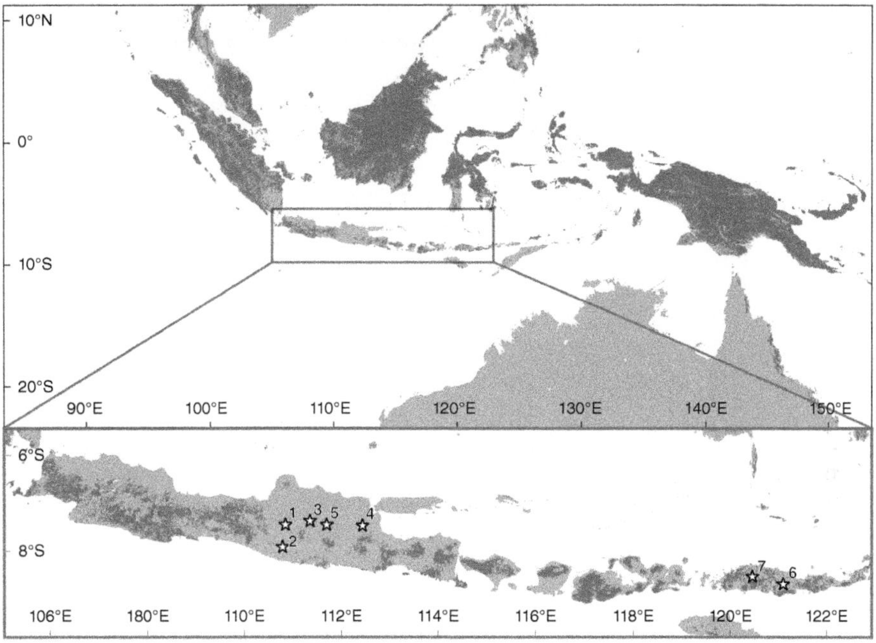

Figure 3.12. Map of major Pleistocene hominin sites in Island Southeast Asia discussed in the text plotted against tropical forest distributions estimated on the basis of the MODIS (Moderate Resolution Imaging Spectroradiometer) Land Cover MCD12Q1 majority land cover type 1, class 2 for 2012 (spatial resolution of 500 m). Downloaded from the US Geological Survey Earth Resources Observation System (EROS) Data Center (EDC). 1. Sangiran, 2. Ngandong, 3. Bapang and Trinil, 4. Mojokerto, 5. Kedung Brubus, 6. Mata Menge, 7. Liang Bua. With thanks to Ulrike Thuering.

Nevertheless, no such implements have been found, nor are they ever likely to be, in Early to Middle Pleistocene sites in this region. It is also worth noting that while increasingly rich subsistence information for *H. sapiens* is being located in cave and rock shelter sites in tropical forest settings (Chapter 4), well-preserved sequences of this nature are largely absent for Early and Middle Pleistocene sites. Whether through preservation or active preference on the part of early *Homo* for open, lake, and riverine settings, this may mean that the archaeological record is biased against observations of interactions between members of our genus with tropical forest habitats at this point in time (Roberts et al., 2016). Furthermore, while it has been argued that the 'Movius' line separates more complex technological production in Africa and western Eurasia (Movius, 1948) from more simple practices in eastern Asia, evidence for symbolic behaviour—in the form of deliberate geometric etchings found on freshwater shell—by *H. erectus* at Trinil *c.* 540–450 ka and frequent bifacial production (Joordens et al., 2015; Dennell, 2016) suggests that it is unlikely that Early-Middle Pleistocene hominins in Southeast Asia lacked the cognitive abilities to effectively exploit tropical forest resources.

Yet, we must rely on the evidence we have to assess the interaction of Early and Middle Pleistocene hominins with Southeast Asian tropical forests. The existing palaeoenvironmental data highlights an apparent preference for lake-edge and marsh environments, with sedges, water-tolerant ferns, and some tropical forest fragments, by *H. erectus* (Tonkunaga et al., 1985; Marwick, 2009). Fauna such as pygmy hippo (*Hexaprotodon*), crocodile (*Crocodylus*), tortoise (*Geochelone*), turtles, and fish (de Vos et al., 1994), as well as open woodland-grassland fauna including pigs (*Sus brachygnatus*), cervids (*Axis lydekkeri*), and bovids (*Bubalus palaeokerabau*) (Rozzi et al., 2013), are commonly associated with this hominin in this region 1.8–1.0 Ma. From 1 Ma, lithic and fossil evidence suggests that *H. erectus* may have moved east of Wallace's line, displaying an ability to cross ocean and perhaps adapt to more isolated forest settings (Van den Bergh et al., 1996b; Brumm et al., 2010). A mixture of Philippine brown deer (*Cervus* cf. *mariannus*), freshwater turtle (Geoemydidae), and monitor lizard (*Varanus salvator*) found in association with a *Rhinoceros philippinensis* skeleton with signs of hominin butchery 0.7 Ma in the Philippines indicates similar habitat variability (Ingicco et al., 2018). Fossil discoveries dated to *c.* 0.7 Ma at Mata Menge, Flores, represent a hominin population even smaller than *Homo floresiensis* (discussed below) (Figure 3.12). The development of small body phenotypes in *H. sapiens* has been argued to be linked to environmental factors, including a potential tropical forest association, such as isolation and lack of resources (Chapter 4) (Brumm et al., 2016; van den Bergh et al., 2016). However, zooarchaeological, pollen, phytolith, and stable isotope proxy data demonstrate that these fossils are associated with dry climates and grassland akin to hominin preferences on the Southeast Asian mainland (Brumm et al., 2016).

As a result, the expansion of Early Pleistocene *Homo* into Flores from 0.7 Ma, and likely from *c.* 0.8 Ma based on recovered lithic assemblages on the island (Brumm et al., 2010), was apparently linked to declines in sea-level and the expansion of mosaics of forest and grassland known to have occurred during this time, rather than the development of specialized subsistence strategies suited to dense tropical rainforest settings (Gathorne-Hardy et al., 2002; Marwick, 2009). At every location in mainland or Island Southeast Asia in which *H. erectus* fossils, or associated flake industries and Acheulean-like technology, are found in the Early and Middle Pleistocene, the faunal record is dominated by a 'Stegodon' fauna (Marwick, 2009; Meijer and Awe Due, 2010; Ingicco et al., 2018). Stegodons are extinct proboscideans first documented in a large-bodied form, *S. trigonocephalus*, in the middle Sangiran Formation alongside *H. erectus c.* 1.6–1.5 Ma (Aimi and Aziz, 1985). These animals, closely related to modern elephants, dispersed throughout Island Southeast Asia and, during the Early Pleistocene, appear to have evolved into small-bodied forms in certain isolated islands such as Flores (*S. sondaari*) and Sulawesi (van den Bergh, 1999). There are several associations of Stegodon fossils with early hominin fossils in Island Southeast Asia, implying a potential subsistence association or, at least, similar habitat preferences (Marwick, 2009). Pleistocene Stegodon appear to have been C_3 browsers making use of a variety of open and more closed habitats (Bocherens et al., 2017), suggesting drier woodland-grassland, riverine, and lacustrine settings rather than dense tropical forest (Rozzi et al., 2013).

One further potential, non-human, contender for Pleistocene hominin tropical forest exploitation is *H. floresiensis* or the 'Hobbit'. Now dated to 190–50 ka (Sutikna et al., 2016), these hominins that measured up to approximately one

Figure 3.13. Photographs of a) a view out of Liang Bua Cave, Flores and b) *Homo floresiensis* fossils recovered from Middle and Late Pleistocene levels at Liang Bua. Photograph credit to the Liang Bua Team and with thanks to Matt Tocheri.

metre in height were found in Liang Bua Cave, Flores, Indonesia in 2003 (Brown et al., 2004; Morwood et al., 2005) (Figure 3.12). The morphology of these fossils appears to derive from *H. erectus* populations that arrived on Flores *c.* 0.7 Ma, and became rapidly dwarfed (van den Bergh et al., 1996a, 2016), although other taxonomic affinities have also been argued (Henneberg et al., 2014; Argue et al., 2017). Beyond small body and brain size, *H. floresiensis* appears to be closely related to *H. erectus* and, as above, it has been argued that food-limited island environments on Flores resulted in insular dwarfism, that is also witnessed in stegodon dwarfing to *Stegodon florensis insularis* (Brown et al., 2004) (Figure 3.13). As at Mata Menge, it is also possible that dense tropical rainforest environments could have provided a 'food-limiting' setting based on some anthropological arguments (Bailey et al., 1989). Indeed, Westaway et al. (2009) argue that *H. floresiensis*' occupation intensity at Liang Bua increases during increasingly wet and humid conditions and the expansion of tropical forest (Westaway et al., 2009). On the other hand, associated faunal and botanical information, including a reliance on the exploitation of *S. f. insularis* which is associated with stone tools dated to 90–50 ka at the site, implies a preference for woodland and grassland habitats (Westaway et al., 2009; Meijer and Awe Due, 2010).

Evaluation of the existing, albeit patchy, evidence for Early and Middle Pleistocene hominin environments in Southeast Asia indicates that both *H. erectus* and *H. floresiensis* were part of a dry tropical forest–woodland–grassland mosaic fauna that expanded across mainland and Island Southeast Asia during this period. *H. erectus* in many places appears to have selected patches of lacustrine and riverine tropical forests (de Vos et al., 1994; Rozzi et al., 2013), but always as part of a wider ecological mosaic in which more open, drier woodlands and grasslands dominated. While dense tropical evergreen rainforest may have persisted in parts of Southeast Asia through the Early and Middle Pleistocene (Sémah et al., 2002; Sémah and Sémah, 2012; Sémah et al., 2016) there is currently no direct evidence that *H. erectus* or *H. floresiensis* either inhabited or utilized these settings. Indeed, perhaps tellingly, these hominin species disappear, alongside the rest of the forest–woodland–grassland fauna that became extinct, *c.* 125 ka with the appearance and expansion of modern rainforest fauna (Westaway et al., 2007; Marwick, 2009). This 'Punung' fauna, characterized by a geological deposit in that region, includes *Pongo* spp. (orangutan), *Hylobates* sp. (gibbon), *Helarctos malayanus* (sun bear), and other species, that form the tropical forest fauna of the region into the Holocene. *H. sapiens* is the only hominin found in association with any of this Late Pleistocene fauna, indicating the novel tropical forest adaptations and adaptive plasticity that was to characterize our species.

TROPICAL FORESTS AS PART OF AN EVOLUTIONARY MOSAIC

'Savanna' and 'tropical forest' have been traditionally treated as opposite ends of the hominin evolutionary spectrum. 'Savanna' has frequently been considered the driver of the hominin trajectory towards specialized upright bipedalism and stone tool manufacture while 'tropical forests' are associated with extant non-human great apes and early hominid and Miocene hominin ancestors. Nevertheless, the above review highlights that palaeoanthropological definitions of savanna extend beyond simply grassland (see also Domínguez-Rodrigo, 2014). There is certainly a general, uncontested shift from closed C_3 forest environments to open C_4 tropical grassland environments in East, and also South, Africa between 10–2 Ma, as well as increasing evidence for hominin dietary reliance on C_4 resources from 5 Ma to 2 Ma. Yet, the C_4 resources in hominin diets could include sedges growing alongside rivers and lakes, grasses in riverside and xeric settings, insects consuming C_4 plants, or medium to large game. Indeed, the exact nature of C_4 resources used also appears to have varied within an individual through time, between locales, and between hominin taxa, with context-specific adaptive implications that were far more complex than the basic broad narrative of unilinear 'savanna' selection. Furthermore, while savanna ecosystems may have provided an important backdrop to the evolution of our lineage, this does not exclude tropical forest patches, or even entire ecosystems, from being actively selected as ongoing contexts of subsistence and occupation.

We saw in Chapter 2 that, like savanna, the term 'tropical forest' captures considerable habitat variability, ranging from tropical evergreen rainforest to dry forest–woodland–grassland mosaics. While tropical rainforests have been considered unattractive for hominin exploitation (Gamble, 1993; Domínguez-Rodrigo, 2014; Finlayson, 2014), given that chimpanzees are known to hunt tropical forest prey such as colobus monkeys (Stanford et al., 1994) there is no *de facto* reason why early *Homo*, or indeed Pliocene and Pleistocene hominins in Africa, could not have occupied, or at least used, dense tropical forest resources. However, as with Early Pleistocene hominins in Africa, *H. erectus* is best associated with environments formed of a mixture of tropical forest, woodland, lacustrine and riverine settings, and grassland (de Vos et al., 1994; Rozzi et al., 2013), usually with the greatest emphasis on the latter two elements (Ciochon, 2009; Marwick, 2009; van den Bergh et al., 2016). At Liang Bua the diminutive *H. floresiensis*, whose body size has been linked to thermoregulatory and nutritional adaptation to dense rainforest habitats, is actually associated with a heavy hunting or scavenging reliance on a Stegodon species that have their own complex relationship with open grassland and woodland settings (Westaway et al., 2009). Rather than an extension of the savanna, therefore, it appears that Pleistocene migration into the tropics beyond Africa was driven by a similar selection of variable forest–woodland–grassland mosaics.

Although focus on savanna contexts in the study of the evolution and expansion of our genus into the tropics clearly misses significant complexity there is also, as of yet, no evidence for the reliance of a hominin on closed canopy forests from the Miocene onwards. While the extinct ape genera *Sivapithecus* and *Gigantopithecus* show stable isotope evidence for dense forest reliance (Nelson, 2003; Loudon et al., 2016; Bocherens et al., 2017), akin to extant non-human great apes today, the broad C_3 focus of *A. anamensis* or *A. ramidus* does not indicate pure forest specialization. Although tropical evergreen rainforests likely existed in Southeast Asia during the presence of *H. erectus* and *H. floresiensis* in this part of the Early Pleistocene world there is no evidence that either of these hominins relied upon such habitats. Stable isotope analysis of *H. erectus* and *H. floresiensis* tooth enamel may conclusively refute this in future, though both of these hominins appear to go extinct in Southeast Asia at around the time when fauna characteristic of modern-day evergreen and semi-evergreen rainforest began to expand across the region (Roberts et al., 2016). It could be argued that this is a result of bias against the preservation of open Early and Middle Pleistocene hominin sites in dense tropical forest. Yet, the fact that non-human species of *Homo* can be associated with well-preserved cave records, such as at Wonderwerk in South Africa (Berna et al., 2012) and Liang Bua in Flores (Sutikna et al., 2016), the clear preservation of Early-Middle Pleistocene *Homo* fossils, stone tool-use, and symbolic behaviour in the tropics, as well as good preservation at forested sites for early primate taxa (Nelson, 2003; Elton, 2008) suggest that taphonomy cannot be the only force at play.

Thus, it appears that, rather than a unilinear selection for savanna, it was habitat variability and selectivity that accompanied Miocene-Pleistocene hominin evolution and dispersal (Potts, 1998, 2013). While there is a broader trend towards more open environments, periods of extreme Pliocene and Pleistocene environmental variability, as well as complex local and regional environmental structuring, likely played an important role in the development of bipedalism, brain size, and stone tool manufacture, all of which could have acted as buffering strategies in the face of changing circumstances (e.g., Trauth et al., 2007; Maslin and Trauth, 2009; Grove, 2011; Magill et al., 2013; Potts, 2013; Maslin et al., 2014). This adaptive flexibility, however, reached new heights with the emergence of our own species, *H. sapiens c.* 300 ka (Hublin et al., 2017; Richter et al., 2017; Brooks et al., 2018; Potts et al., 2018). While our species almost certainly continued to live within mosaic forest, woodland, and grassland settings in Africa and beyond, it also commenced a new, more specialized adaptation to tropical forest settings that had not been seen amongst other hominins. This was part of a flexibility that allowed it to specialize in a number of diverse environments that included desert (Groucutt and Petraglia, 2012) and sub-arctic conditions (Pitulko et al., 2016). It seems to be this plasticity, in the face of environmental and Middle and Late Pleistocene climatic variation (Rabett, 2012), that defines our species, the last surviving hominin.

4

Into the Woods

Early *Homo sapiens* and Tropical Forest Colonization

> He would eventually have to pass through the forest, but he felt no fear. Of course—the forest was inside him, he knew, and it made him who he was.
>
> —Haruki Murakami (2002)

Popular philosophical associations of tropical forests, and forests in general, with an inherent ancestral state, away from the stresses, pollution, and technosphere of modern life, are nicely summarized by Murakami's quote above (2002). Given the probable origins of the hominin clade in tropical forests, this quote is also apt from an evolutionary standpoint. Yet, somewhat surprisingly, tropical forests have frequently been considered impenetrable barriers to the global migration of *Homo sapiens* (Gamble, 1993; Finlayson, 2014). As was the case with the focus on 'savannastan' in facilitating the Early Pleistocene expansion of *Homo erectus* discussed in Chapter 3 (Dennell and Roebroeks, 2005), the movement of *H. sapiens* into tropical regions such as South Asia, Southeast Asia, and Australia has tended to be linked to Late Pleistocene periods when forests contracted and grasslands expanded (Bird et al., 2005; Boivin et al., 2013). Alternative narratives have focused on the importance of coastal adaptations as providing a rich source of protein and driving cultural and technological complexity, as well as mobility, in human populations during the Middle and Late Pleistocene (Mellars, 2006; Marean, 2016). The evidence of early art and symbolism at coastal cave sites such as Blombos in South Africa (Henshilwood et al., 2002, 2011; Vanhaeren et al., 2013) and Taforalt in North Africa (Bouzouggar et al., 2007) is often used to emphasize the role of marine habitats in the earliest cultural emergence of our species. Indeed, for the last decade, the pursuit of rich marine resources (Mellars, 2005, 2006) has been a popular explanation for the supposed rapidity of the 'southern dispersal route', whereby humans left Africa 60 ka, based on genetic information (e.g., Macaulay et al., 2005), to reach the Pleistocene landmass that connected Australia and New Guinea (Sahul) by *c.* 65 ka (Clarkson et al., 2017). In both of these cases, the coast or expanses of grassland have been seen as homogeneous corridors, facilitating rapid expansion without novel adaptation.

Nevertheless, discoveries of early *H. sapiens* fossils in North (Hublin et al., 2017; Richter et al., 2017) and East Africa (White et al., 2003; McDougall et al., 2005), as well as technological correlates in West, Central, and southern Africa (McBrearty and Brooks, 2000; Mercader, 2002a), from 300 ka suggest that our species may have emerged in a number of diverse African environments, not just in open grassland plains or on the coast. In reviewing this, and growing evidence that humans left Africa closer to 200–100 ka (Groucutt and Petraglia, 2012; Blinkhorn et al., 2013; Bae et al., 2017; Hershkovitz et al., 2018), I argue that our species evolved and migrated within a context of climatic and environmental variability. The evaluation of the existing evidence for Late Pleistocene tropical forest use by *H. sapiens* in Africa, South Asia, Southeast Asia, Melanesia, Australia, and South America indicates that tropical forests were a clear part of this adaptive plasticity from at least 45 ka, and most likely 200 ka—although the evidence for this earlier period currently remains patchy and inconclusive (Roberts and Petraglia, 2015). In each of these regions, evidence for the complex exploitation of tropical forest flora and fauna is apparent, often from the first arrival of human populations, albeit with clear differences between geographies and tropical forest formations. In a way then, relatively rapid expansion of *H. sapiens* was perhaps facilitated by a constant adaptation from Africa to the Americas. However, this adaptation was one of a 'generalist specialist' (Roberts and Stewart, 2018), an animal able to adapt to a number of environments but also to specialize in each of those environments. Taking this approach, tropical forests lie at the heart of the potential ecological origins of our species as a global, colonizing force that has made us who we are.

THE GENETIC, MORPHOLOGICAL, AND PHYSICAL ENVIRONMENT OF HUMAN ORIGINS

Defining our own species in the fossil record is more complex than it is often considered to be beyond palaeoanthropology. The term 'human' has been used to define both any hominin within the genus *Homo* as well as just *H. sapiens*. Demonstration of interbreeding between *Homo neanderthalensis*, *H. sapiens*, and the Denisovans (Green et al., 2010; Reich et al., 2011; Vernot et al., 2016) has complicated this picture further given that, in many biological definitions of a 'species', breeding to produce fertile offspring would lead to all three of these hominins being classed as *H. sapiens* (Figure 4.1). Even if we do isolate *H. sapiens* from other members of the genus *Homo*, as a taxon with enlarged cranial capacity, a modern cranial vault, reduction in the size of the supraorbital torus (a bony ridge above the eye sockets), and a modern face (as per Stringer, 2016), determining the exact point at which it emerged is reliant on statistical comparisons with a relatively poor fossil dataset (e.g., White et al., 2003; Hublin et al., 2017). This is even more difficult when there are huge temporal and geographical gaps in the African hominin fossil record. Nevertheless, a growing database of genetic information (including revised methods and modelling), Middle Pleistocene fossils from various regions within Africa, palaeoenvironmental datasets, and more nuanced assessments of technological and cultural complexity are beginning to provide more detailed insights into the evolutionary emergence of *H. sapiens*.

For a long period of palaeoanthropological history there have been two competing views relating to the origin of our species: a recent African origin versus a multi-regional origin. The multi-regional origin model, developed by Milford

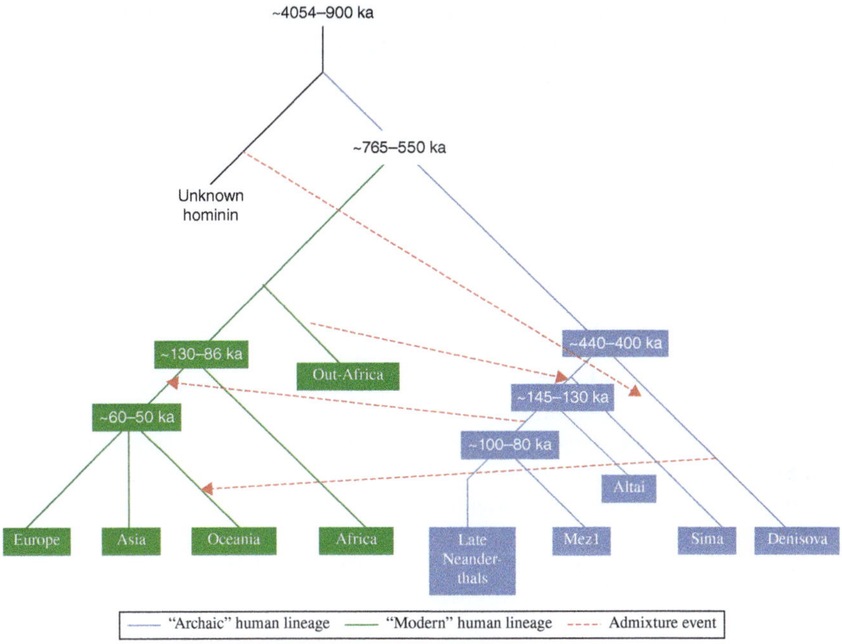

Figure 4.1. State of knowledge of genetic divergence dates of *Homo sapiens*, *Homo neanderthalensis*, and the Denisovan hominins, as well as genetic admixture and interbreeding events between 'archaic' and 'modern' hominin taxa. Date intervals represent population split times in thousands of years before present, assuming a nuclear DNA mutation rate of f 0.5E-9 per bp per year. Inferred phylogeny and major admixture events come from Kuhlwilm et al. (2016), Meyer et al. (2016), Prüfer et al. (2014, 2017), and Posth et al. (2017). With thanks to Cosimo Posth and Hans Sell.

Wolpoff and colleagues (1988), argued that convergent evolution from 2.5 Ma has led to the repeated appearance of *H. sapiens* in different parts of the world at different times. This view was largely considered disproven since mitochondrial DNA of modern human populations suggested the existence of a 'Mitochondrial Eve', a woman from whom all living humans descend. In 1987, Cann et al. suggested that all living human mtDNA originated from a single population living in Africa between 200 and 140 ka. This was based on a 'molecular clock' that utilized estimated mutation rates in extant non-human great apes to equate variation in human genomes to dates of past divergence. Other studies have used slightly different models or modern Y-chromosome data to produce estimated ages of *c.* 234–152 ka (Endicott et al., 2009) and *c.* 148–99 ka (Poznik et al., 2013), respectively. As will be seen in the context of genetic assessments of human migrations below, it is important to note that 'molecular clock' estimates often involve errors of tens of thousands of years. Moreover, the concept of a 'Mitochondrial Eve' does not mean that there was only one woman alive at the time. Rather, the genetic contribution of other female members of our species living *c.* 200 ka is no longer visible as a result of population extinctions and later replacement.

Genome-wide nuclear genetic studies of modern populations have confirmed these trends, and provided yet more detailed insight into human origins and interaction with other hominin species (Campbell and Tishkoff, 2008; Green et al., 2010). In a study of modern African genetic diversity, Tishkoff and colleagues (2009) found that San populations in South Africa had the greatest genetic diversity of 113 populations sampled. This was used to suggest that the origin of modern human migration, and populations alive today, was actually southwestern Africa (Tishkoff et al., 2009). While geographical locations estimated purely from modern DNA should be treated with caution, studies such as this demonstrate the level of demographic detail that can now be gleaned from modern genetic material. Further revelations have occurred with the discovery that 'ancient' DNA is preserved over considerable time-depths. Not only have genome sequences now been obtained for *H. neanderthalensis* (Green et al., 2010) and a mystery 'Denisovan' hominin (Meyer et al., 2012) but these taxa have also been shown to have interbred with *H. sapiens* as it moved beyond Africa. It could be argued that the genetic interactions of these groups indicate that at least some elements of the multi-regional perspective may be accurate, particularly with regards to the biological similarity of Middle and Late Pleistocene members of the genus *Homo*. However, modelling genetic and morphological variation is largely considered to maintain that the variation exhibited in these populations is greater than that seen between modern human populations today.

Despite these advances, not only will extinct populations of *H. sapiens* not be reflected in the modern genetic record, or ancient populations to which they did not contribute, but it is almost impossible from genetic information alone to pinpoint the geographical, temporal, and ecological origins of our species due to later migrations masking past geographical distributions of human populations. For this, the most solid datasets we have are the hominin fossils themselves. Morphologically, *H. sapiens* is distinctive. The oldest accepted fossils with traits associated with our species, including large brain volume (*c.* 1350–1400 ml volume), short, retracted face (Lieberman et al., 2002), and short and globular brain-casing (Schwartz and Tattersall, 2002) have, for a long time, been the Omo Kibish 1 individual re-dated to 195 ka (McDougall et al., 2005) and the Herto fossils dated to 160 ka (White et al., 2003), in Ethiopia (Figures 4.2 and 4.3). Although variation in these fossils is considered to fall beyond that of modern humans, they demonstrate that the distinctive morphology of *H. sapiens* had emerged in East Africa by *c.* 200 ka. The Omo fossils have been indirectly associated with palaeoenvironmental data from speleothems that suggest the presence of a relatively dry environment fed by an active Omo River (Vaks et al., 2007). Fauna discovered from geological members associated with the Herto remains similarly indicate the proximity of aquatic and grassland habitats (Clark et al., 2003). As such, the appearance of *H. sapiens* on the grasslands of East Africa appears to cap the savanna-focused evolutionary trajectory discussed in Chapter 3.

Indeed, the Herto and Omo fossil data, alongside the genetic information, have been used to argue for a single origin of *H. sapiens* in the grassland habitats of East Africa (Clark et al., 2003) (Figure 4.3). Nevertheless, the recent dating of new remains excavated at Jebel Irhoud, Morocco to *c.* 300 ka (Richter et al., 2017) complicates this picture. While these fossils have relatively 'primitive' brain-case morphology, their facial, mandibular, and dental traits all fall well within morphological variation in modern humans (Hublin et al., 2017). A fragmentary fossil

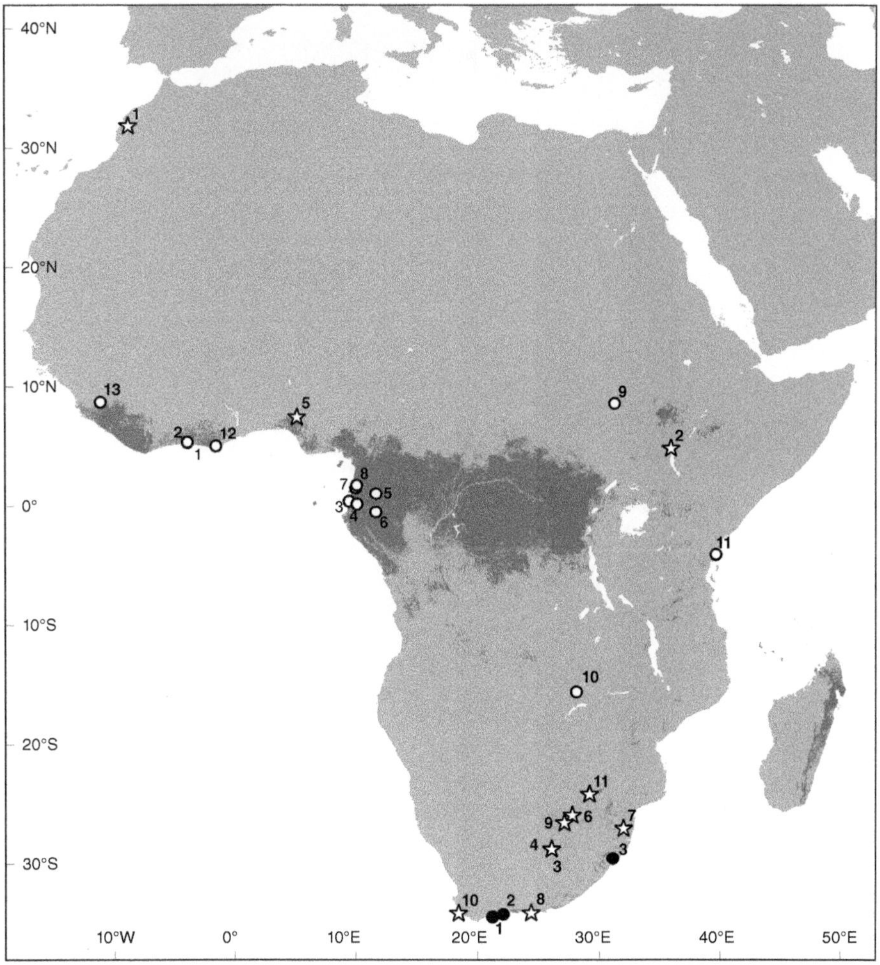

Figure 4.2. Map of major Middle and Late Pleistocene archaeological and fossil sites in Africa discussed in the text plotted against tropical forest distributions estimated on the basis of the MODIS (Moderate Resolution Imaging Spectroradiometer) Land Cover MCD12Q1 majority land cover type 1, class 2 for 2012 (spatial resolution of 500 m). Downloaded from the US Geological Survey Earth Resources Observation System (EROS) Data Center (EDC). White stars denote Middle and Late Pleistocene *Homo* fossils discussed in the text: 1. Jebel Irhoud, 2. Omo Kibish, 3. Herto, 4. Florisbad, 5. Iwo Eleru, 6. Rising Star, 7. Border Cave 5, 8. Klasies River Mouth, 9. Boskop, 10. Fish Hoek, 11. Cave of Hearths. Black dots denote archaeological sites in South Africa associated with the origins of behavioural complexity: 1. Blombos Cave, 2. Pinnacle Point, 3. Sibudu. White dots denote archaeological sites postulated as representing tropical and subtropical forest adaptations: 1. Bete, 2. Guabuo, 3. Médoumane, 4. Kango, 5. Okala, 6. Okanda, 7. Ogooué Basin, 8. Mosumu, 9. Kalambo Falls, 10. Twin Rivers Kopje, 11. Panga ya Saidi, 12. Bosumpra, 13. Yengema Cave. With thanks to Ulrike Thuering.

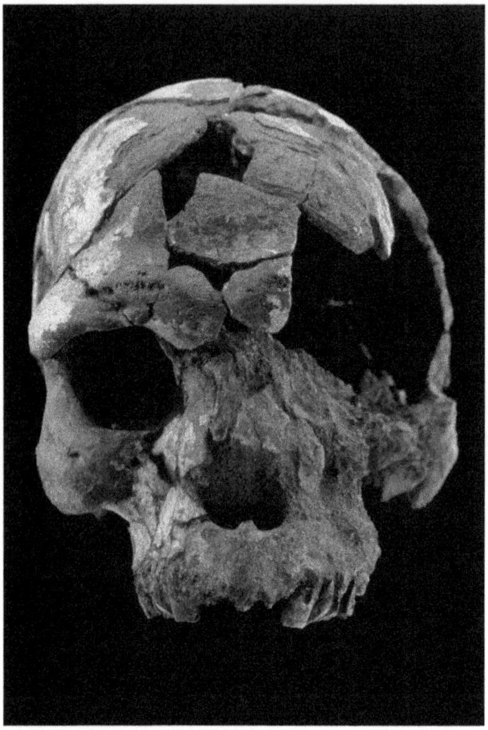

Figure 4.3. *Homo sapiens idaltu* fossil (BOU-VP-16/1), Bouri Herto, adult. Housed in the National Museum of Ethiopia, Addis Ababa. Photograph taken by David L. Brill (humanoriginsphotos.com). (c) David L. Brill 2001. With thanks to David L. Brill.

specimen from Florisbad may also demonstrate a series of derived *H. sapiens* features *c.* 260 ka in South Africa (Grün et al., 1996). Moreover, although the earliest *H. sapiens* fossil in West Africa dates to only 16–12 ka at Iwo Eleru, Nigeria, morphometric analysis of cranial features of these remains indicates deep population substructure in this part of Africa and a potentially complex evolutionary process for our species in different parts of the continent (Harvati et al., 2011). With additional morphological variation seen among specimens from Border Cave 5, Boskop, Fish Hoek, Klasies River Mouth, and perhaps also the Cave of Hearths (Tattersall and Schwartz, 2008), and the fact that the 'archaic' hominin *Homo naledi* of Rising Star Cave, South Africa is now dated to *c.* 335–236 ka (Dirks et al., 2017), it is becoming increasingly apparent that our species emerged in a complex evolutionary setting, in a mosaic manner (*c.* 300–100 ka) across the different regions, and potentially environments, of Africa (Figure 4.2).

The paucity of fossil remains makes it difficult to confidently develop more detailed understandings of the environmental adaptations of early members of our species and its temporal and spatial emergence. As a result, many archaeologists have attempted to use characteristic forms of lithic production and use and evidence for 'complex' cultural artefacts, such as beads, portable 'art', and rock art, to date the presence of *H. sapiens* in Africa (McBrearty and Brooks, 2000;

Henshilwood et al., 2011; Shea, 2011). Although some types of cultural artefacts, particularly non-lithic evidence, will be inherently rarer further back in time (Shea, 2011), there is now evidence for engraved ochre dating to *c.* 100 ka in southern Africa (Henshilwood et al., 2002, 2011), shell beads *c.* 90 ka in southern and northern Africa (Bouzouggar et al., 2007; d'Errico et al., 2009; Vanhaeren et al., 2013), and the appearance of complex lithic and osseous technology, including complex projectiles, in southern Africa from *c.* 80 ka (Backwell et al., 2008; Lombard and Phillipson, 2010). In southern Africa these developments have tended to be found in coastal or open grassland contexts, such as at Blombos Cave, Pinnacle Point, and Sibudu Cave (Henshilwood et al., 2001; Henshilwood, 2008; Marean et al., 2007; Backwell et al., 2008) (Figure 4.2). Site location and faunal datasets have led some to suggest that the complex behaviour of our species was driven by access to protein-rich marine foods (Marean, 2016). Others, by contrast, have argued that the development of projectile technology is better linked to open grassland/woodland settings and the more efficient hunting of medium meat packages (Ambrose, 1993; Shea and Sisk, 2010).

Nevertheless, research in the last decade has now revealed that some of the earliest members of our species likely inhabited more diverse environmental settings than previously assumed. For example, work in Lesotho and the Ethiopian Highlands suggests that *H. sapiens* may have occupied high-altitude settings from as early as 90 ka in the case of the former (Brandt et al., 2012; Stewart et al., 2012, 2016). These habitats were highly sensitive to climate change (Roberts et al., 2013; Stewart et al., 2016) and would have required new adaptations to cold temperatures and highly-seasonal resources. Yet, well-preserved MIS 5 to 3 (130–30 ka) sequences of human occupation of these environments continue to emerge. Similarly, desert settings in Namaqualand, southern Africa (Dewar and Stewart, 2016), North Africa, and across the Sahara (Scerri, 2017), have yielded evidence for early human occupation of highly arid regions during MIS 5 (130–80 ka). In particular, in the Sahara, changes in the Earth's orbit seem to have triggered significant northward expansion of savanna environments into the Sahara during interglacial periods in association with a multitude of lake and river networks (Drake and Breeze, 2016). Palaeoenvironmental information associated with the emergence of Middle Stone Age technologies *c.* 300 ka at Olorgesailie in southern Kenya has also recently been used to argue that the climatic and environmental flux potentially framed the emergence of technological characteristics associated with our species (Brooks et al., 2018; Potts et al., 2018).

It should be apparent that current data make it increasingly difficult to pinpoint a single point of origin for *H. sapiens*. Genetic evidence can only ever be related to surviving lineages and particular fossils that have yielded ancient DNA, meaning that a 'single origin' determined by these means misses a significant, extinct portion of the human story. Morphological data suggest that our species appeared in a gradual, mosaic manner between 300 and 200 ka, with large Middle Pleistocene variation in the genus *Homo* existing across Africa. Similarly, increasing excavation and analysis is revealing that 'symbolic' or 'complex' technological behaviour associated with our species does not simply emerge in one single environment. Furthermore, it was a gradual and cumulative process, found across many human populations at different times and frequencies. Indeed, it appears to be the plastic ability of our species to adapt to the increasingly dramatic and frequent climatic fluctuations associated with the Middle-Late Pleistocene and the diverse multitude of Africa's environments that truly defines *H. sapiens* (Grove, 2015; Grove et al., 2015; Potts et al., 2018). While *H. neanderthalensis* has demonstrated evidence for personal ornamentation and social signalling, alongside morphological and genetic similarities, it is only our species that appears to have been able to adapt to the diversity of the terrestrial environments the world has to offer. Some of these are, of course, the tropical forests.

AN ORIGIN 'UNDER THE CANOPY'?

Desmond Clark (1963, 1971, 1988) was the first to suggest that humans may have occupied tropical forests in West and Central Africa during the Middle and Late Pleistocene. However, it was Larry Barham (2001, 2002), and later Julio Mercader (2002a, 2002b), who really drove home the point that there is no reason why our species could not have occupied tropical forest habitats from the moment it evolved in Africa. In particular, Barham and Mercader have argued that the discovery of Sangoan lithic forms throughout the equatorial regions of Central Africa, West Africa, western East Africa, and South-Central Africa (Barham and Smart, 1996; Barham, 2001, 2002; Mercader, 2002a, 2002b) indicated that humans did not, at any point, avoid tropical forest habitats (Figure 4.2). The heavy duty elements of this industry, including large bifacial sub-triangular 'pick' forms, massive core-scrapers, and coarse Acheulean-like bifaces, have led some authors to suggest that this is a pre-*H. sapiens* complex (Lioubine and Guede, 2000; McBrearty and Brooks, 2000). Indeed, its age and technological affinities, although strongly debated, potentially reach back to 400–300 ka, placing it just beyond limits of the oldest fossil appearance of our species (Barham, 2000). Its stratigraphic positioning below the Lupemban at Kalambo Falls (dated at nearby Twin Rivers, Kopje to 265 ka and older—Barham, 2002) further supports this (Taylor, 2011). Yet, while the archaeological evidence of the Sangoan, supported by palynological and geological inference (Jahns et al., 1998; Dupont et al., 2000), potentially hints at the very early interaction of our species with tropical forest environments, there is huge variability in terms of what people call the Sangoan and archaeologists now question its existence as a meaningful technocomplex (Taylor, 2011, 2016).

The Sangoan has been reported from the evergreen rainforest region of Ivory Coast, at Bete and Guabuo, dated by thermoluminescence to as early as *c.* 254 ka (Lioubine and Guede, 2000; Mercader, 2002a) (Figure 4.2). Palynological records from Atlantic coast marine cores have indicated open forest conditions at Bete and closed forest at Guabuo (Mercader, 2002a). However, Mercader (2002a) reports these dates without their likely-significant error margins. Furthermore, the marine cores used to reconstruct local palaeoenvironments are up to *c.* 200 km away (Jahns et al., 1998; Dupont et al., 2000), inherently 'off-site', and of unknown relevance to site-specific habitats. They are also difficult to correlate chronologically with West and Central African Sangoan archaeological records that have been subject to considerable depositional re-working that prohibits the development of detailed chronological frameworks (Basell, 2008; Taylor, 2011, 2016; Blome et al., 2012). The Sangoan has also been documented from the lowland forests of Gabon at Médoumane, Kango, Okala, and Okanda, as well as in southern Cameroon (Mercader and Martí, 1999) (Figure 4.2). Sparse palaeoenvironmental evidence has been argued to indicate proximate rainforest (Bayle des Hermens et al., 1987; Clist, 1993, 1995) but this evidence is again indirect and difficult to compare both chronologically and spatially with periods of occupation. In all cases the lithics that are being caught under the umbrella 'Sangoan' can be highly variable (Taylor, 2011). Furthermore, there are also no directly-associated fauna or flora remains at any of these sites and so human ecological adaptations remain largely unknown and based on indirect assumptions.

The Lupemban industry, characterized by bifacially flaked lanceolate points, is another lithic technocomplex that has been argued to provide evidence for early tropical forest colonization by *H. sapiens* (Clark, 1963, 1971, 1988; Barham, 2000, 2001, 2002) (Figure 4.4). Core-axes have been associated with heavy-duty digging activities for tropical roots and tubers (Ichikawa, 1996), as well as wood-working (Clark, 1966), while Lupemban lanceolates, points, backed blades, and tranchets have been linked to the hunting of terrestrial and arboreal fauna (Barham, 2001). As Nicholas Taylor (2016) notes, the distribution of Lupemban sites, focused around the Congo Basin and its eastern and southern margins (Clark, 1967; Cole, 1967), has been superimposed onto contemporary distributions of rainforest and woodland ecosystems (Mayaux et al., 2004) to suggest a close relationship between

Figure 4.4. Photograph of two bifacial Lupemban stone tools from the site of Mosumu in Equatorial Guinea. Photograph taken by, and reproduced with permission from, Julio Mercader.

this technocomplex and the southern edge of the Central African rainforest zone. Similarly, pollen records have been used to associate sites of Lupemban finds, such as those within the Ogooué Basin and at Mosumu in Equatorial Guinea, dated by radiocarbon to as old as *c.* 30–20,000 cal. years BP (but are almost certainly older), with modern tropical forest or reconstructed tropical forest for the recent past (Mercader and Martí, 2002) (Figure 4.2). Yet, as with the Sangoan, these palaeoenvironmental records can only be indirectly related to periods of Lupemban occupation (Taylor, 2016).

Overall, there is currently no direct empirical evidence that links Lupemban stone tools to the exploitation of tropical forest resources (Cornelissen, 2002). Post-depositional disturbance of the majority of Lupemban sites has led to considerable difficulties in dating the technocomplex, with estimated dates converging on the range 40–12 ka, resulting in coarse palaeoenvironmental correlations (Clark and Brown, 2001; Cornelissen, 2002; Taylor, 2011). Contractions in the Central African rainforest during the last glacial period would have resulted in an advance and admixture of seasonal, semi-deciduous forests and savanna grasslands at its fringes (Thomas, 2000; Basell, 2008; Blome et al., 2012). Consequently, as recently comprehensively reviewed by Taylor (2016), the main distribution of Lupemban sites, sitting on the current border of the region's rainforests, may not have been located in rainforest during periods of occupation. Where 'on-site' palaeoenvironmental records are available for the Lupemban, deposits tend to be disturbed, leaving palaeoenvironmental reconstruction limited to comparison with modern contexts or the more recent past (Mercader and Martí, 2002; Mercader et al., 2003). At those sites where archaeological sequences are associated with palaeoenvironmental information, including Kalambo Falls, northern Zambia (J. D. Clark, 1969; J. D. Clark and Brown, 2001; Taylor et al., 2001), Twin Rivers Kopje in Zambia (Barham, 2002), and Mufo, Angola (J. D. Clark, 1963) (Figure 4.2), the Lupemban generally corresponds with deciduous woodland and open savanna ecosystems rather than rainforest, though re-working remains a consistent problem.

Modern tropical forest distributions or 'off-site' marine cores cannot be used as a reliable proxy for local tropical forest extent in Pleistocene Africa. Where good terrestrial palaeoenvironmental datasets exist, or historical modelling has been undertaken, it is clear that the rainforests of Africa are some of the most globally vulnerable to reduction in the face of cooler temperatures, decreased rainfall, and lower atmospheric CO_2 concentration in glacial periods (Maley and Brenac, 1998; Malhi and Wright, 2004; Blome et al., 2012) (Figure 4.5). Therefore, while the Sangoan and Lupemban hint at potentially early human encounters with tropical forests, poor palaeoenvironmental resolution, chronological correlation, and a considerable lack of direct subsistence or technological use information (Taylor, 2009) means that there is currently a limited understanding of local Pleistocene environments in these parts of Africa (Roberts and Petraglia, 2015) (Figure 4.5). There is also no Middle Pleistocene fossil evidence in this region meaning that the hominin producers of these industries remain unknown. Nevertheless, based on growing datasets from other regions (below), and the clear adaptive capacity of *H. sapiens*, it seems highly likely that our species could, and probably did, occupy the tropical forests of Africa from *c.* 200 ka. Future work in the evergreen and semi-evergreen rainforests of Central and West Africa,

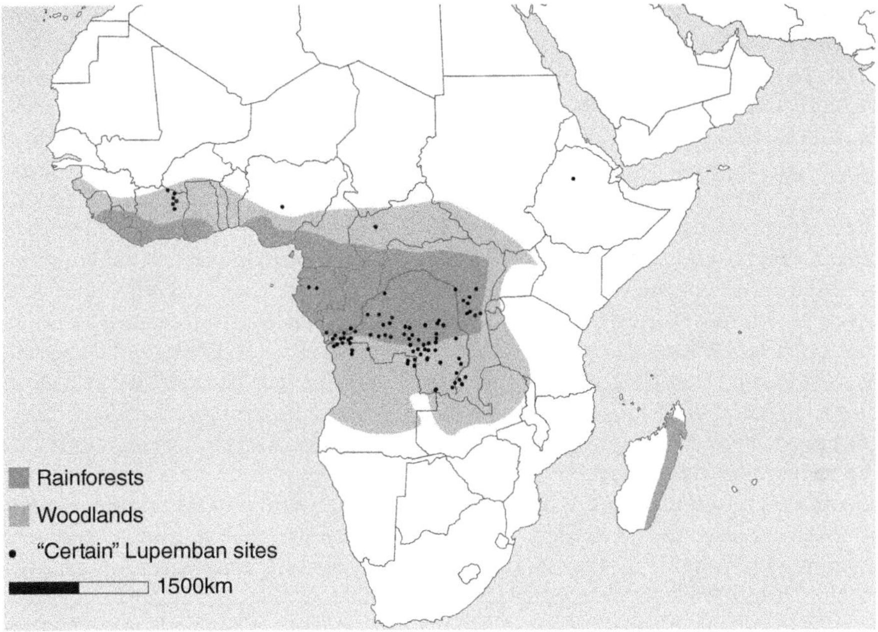

Figure 4.5. The distribution of Lupemban sites, after Clark (1967) and Taylor (2011), plotted against the contemporary extent of rainforest ecosystems as presented by Mayaux et al. (2004) for the year 2000. 'Rainforests' incorporates Mayaux et al.'s categories of: closed evergreen lowland forest, degraded evergreen lowland forest, submontane forest, montane forest, swamp forest, mangroves, mosaic forest/croplands and mosaic forest/savanna. 'Woodlands' includes the categories closed deciduous forest and deciduous woodlands. Modified with permission from Taylor (2011). With thanks to Hans Sell.

particularly focused on the discovery of new cave and rock shelter sites, should definitively demonstrate this, though public health and political issues have made such work impractical in recent times.

In the meantime, increasingly systematic palaeoenvironmental studies of Middle Stone Age sites on the fringes of the African tropics are providing some insight into human tropical forest use. Such records are increasingly suggesting that Middle Stone Age human groups tracked the formation of forest, woodland, and open mosaics during periods of increased climatic aridity (Basell, 2008; Blome et al., 2012), albeit with potentially increased specialization in dry tropical forest and woodland biomes during wetter periods (Mercader et al., 2013). This fits earlier suggestions that both the Lupemban and Sangoan industries demonstrate strong technological continuity across postulated forest habitats and open environments (Van Noten et al., 1972; McBrearty, 1988). Similarly, a newly published site in the tropical coastal forest of Kenya, Panga ya Saidi, demonstrates that from 80 ka to *c.* 1 ka human hunter-gatherers were persistently exploiting tropical forest, as well as grassland, environments (Shipton et al., 2018). Interestingly, in contrast to suggestions that marine adaptations drove human expansion from East Africa into the Middle East and Levant (e.g., Mellars, 2006), there is no evidence at this site for marine resources making up a considerable part of human subsistence. In line with sites in North Africa, it appears that these resources were more significant for symbolic expression (Bouzouggar et al., 2007; Scerri, 2017).

From the Later Stone Age of Africa (*c.* <40 ka) clear evidence emerges for more extensive and specialized human settlement and use of Africa's tropical forests, including rainforests (Mercader, 2002a, 2002b). Many of the Central African forest regions were demonstrably settled prior to the Holocene (Mercader et al., 2003). In the Ituri region, three phytolith sequences indicate high percentages of arboreal rainforest taxa in association with cultural remains, dating across the last 18,000 years, and demonstrating sustained hunter-gatherer settlement of the lowland forests of the northeast Congo Basin (Mercader et al., 2000). Almost all twelve countries that contain the African rainforest demonstrate dense evidence for Later Stone Age occupation. The sites of Bosumpra (Ghana) and Yengema Cave (Sierra Leone), as well as fifty sites characterized as Later Stone Age in Central Africa, yield archaeological evidence in lowland evergreen, seasonal, dry, and mosaic forest contexts prior to the Holocene (Coon et al., 1968; Smith, 1975; Mercader and Martí, 2002). Furthermore, palynological sequences from Atlantic Central Africa and West Africa, as well as phytolith data from the eastern Democratic Republic of Congo, indicate that Later Stone Age hunter-gatherers continued to inhabit rainforests throughout the climatic and vegetation changes of the Terminal Pleistocene/Early Holocene (Mercader, 2002a, 2002b).

PASSING THROUGH FORESTS: A VERY HUMAN OUT OF AFRICA

Two broad categories of recent models, outlined by Groucutt et al. (2015), exist for the dispersal of *H. sapiens* beyond Africa. These are, an early onset dispersal model where our species moved out during Marine Isotope Stage (MIS) 5 (130–80 ka) and a late dispersal model where this only occurred post-80 ka. The earliest known human fossils beyond Africa actually come from Misliya Cave (194–177 ka), Skhul (120–90 ka), and Qafzeh (100–90 ka) in Israel (Valladas et al., 1988; Stringer, et al., 1989; Grün and Stringer, 2000; Hershkovitz et al., 2018). However, these are often considered to be a failed, early dispersal event (e.g., Shea and Bar-Yosef, 2005; Stringer and Galway-Witham, 2018), and the next oldest fossils in the Levant come from *c.* 55 ka at Manot Cave (Hershkovitz et al., 2015) or, more securely, from 43 ka onwards (Douka, 2013). Nevertheless, growing finds of human fossils in Saudi Arabia *c.* 85 ka (Groucutt et al., 2018), and China *c.* 139–111 ka at Tongtianyan Cave (Shen et al., 2002), 106 ka at Zhirendong (Liu et al., 2010), and *c.* 100 ka at Fuyan Cave (Liu et al., 2015), and new early dates of *c.* 65 ka from Australia (Clarkson et al., 2017) support sustained MIS 5 expansion. Although the Tongtianyan Cave and Zhirendong fossils appear to have a combination of *H. erectus* and *H. sapiens* features, and the Fuyan Cave chronology remains hotly debated (Michel et al., 2016), growing fossil datasets appear to offer a tantalizing hint that human populations had begun to expand into Eurasia during MIS 5, perhaps as a result of climatic amelioration in the Levant and Saudi Arabia (Rabett, 2012; Groucutt et al., 2015).

Genetic evidence has started to play an increasingly prominent role in discussions of the timing and process of human dispersal. For a long time it has been argued that the spatial arrangement and estimated mutation chronology of modern human mtDNA and Y chromosome trees indicate that dispersal only occurred from 60–50 ka onwards (Soares et al., 2012; Mellars et al., 2013). In particular, mtDNA haplogroups M and N were argued to have separated from African origins 65 ka to expand into South Asia and Sahul by 45 ka (Macaulay et al., 2005). However, a revised estimate for the nuclear genomic mutation rate of *H. sapiens* (Scally and Durbin, 2012) as well as increasing high-coverage whole-genome demographic studies allow for, or even support, a MIS 5 dispersal >75 ka (e.g., Schiffels and Durbin, 2014; Pagani et al., 2016). Modern DNA reconstructions also have the major limitation that they do not account for now-extinct human populations and dispersals. For example, while ancient DNA has pinpointed one occurrence of *H. sapiens* and Neanderthal interbreeding to 60–50 ka based on a clear signature it left in the ancient genomes of Siberian individuals (Fu et al., 2014), a recent study has argued that an early group of humans came into contact with Neanderthals as early as *c.* 270 ka (Posth et al., 2017). This group then became extinct leaving their DNA in later generations of Neanderthals (Posth et al., 2017).

Paul Mellars and colleagues have tended to argue that coastal resources and routes fuelled the dispersal of humans into southern and southeastern Asia. To support this they have argued that the distribution of microlithic and geometric technologies, as well as elements of symbolic material culture, around the Indian Ocean rim indicate the rapid, single dispersal of our species *c.* 60–50 ka (Mellars et al., 2013). Not only does this perspective link *H. sapiens* to a single technological form (see Lewis et al., 2014 for clear refutation of this claim) but it also neglects the fact that a lot of the sites they plot onto large-scale maps to indicate 'coastal' distributions are, in practice, located many kilometres inland and have little to no evidence for marine-focused subsistence (Groucutt et al., 2015). By contrast, others have argued that an earlier, MIS 5 phase of dispersal beyond Africa was associated with a Middle Palaeolithic lithic technology (Clarkson et al., 2012; Blinkhorn et al., 2013). This includes suggestions of an affinity of 'Nubian' Levallois technology in Northeast Africa to that in Arabia in MIS 5 (Groucutt and Petraglia, 2012; Usik et al., 2013). Similarly, a continuity of Middle Palaeolithic industries in South Asia from late MIS 5 through MIS 3 has been used to argue for hominin population continuity in this region, through tephra markers of the Sumatran Toba super-eruption dated to *c.* 74 ka, implying that MIS 5 technologies were indeed linked to *H. sapiens* (Petraglia et al., 2010).

The weight of evidence thus now suggests that there were significant MIS 5 human dispersals, even if they did not contribute to modern populations due to later replacement or extinction (Bae et al., 2017; Hershkovitz et al., 2018; Stringer and Galway-Witham, 2018). It does also appear to be the case that there is a significant dispersal signature from 60 ka onwards, but it can no longer, if it ever could, be related to a single environmental adaptation. Firstly, *H. sapiens* emerged and dispersed within diverse demographic and geographical environments 300–200 ka in Africa itself, so suggested limitations to the coast or grassland in later dispersal are unlikely to be parsimonious. Secondly, it is clear that different dispersals of our species reacted to intensifying climatic variability that would have had extreme environmental and ecological effects in some regions, such as the Sahara and Arabia (Drake et al., 2011; Groucutt and Petraglia, 2012; Breeze et al., 2015), and more subtle, spatially variable impacts elsewhere (e.g., Roberts et al., 2017b). Hunter-gatherer societies were able to persist through this in a variety of different environmental settings from at least *c.* 60 ka to 12 ka. Finally, and most importantly, growing evidence from the 'northern' routes into Eurasia now indicates that Late Pleistocene *H. sapiens* could also adapt to palaeoarctic tundra (Fu et al., 2014; Pitulko et al., 2016) and high-altitude (BaoYin et al., 2007) conditions. On the 'southern' route, meanwhile, tropical forest settings are yielding evidence for a diversity of specialized adaptations from South Asia to Sahul.

South Asia

Until recently, South Asia has been rarely discussed in the context of early human tropical forest habitation. This is despite the fact that the earliest human fossils from this region (dated to *c.* 38,000–36,000 cal. years BP) were found within the Wet Zone lowland tropical rainforest contexts of Sri Lanka, at the rock shelter sites of Fa Hien-lena and Batadomba-lena (Deraniyagala, 1992; Wijeyapala, 1997; Kennedy, 2000; Perera, 2010; Perera et al., 2011). Human fossils in the evergreen and semi-evergreen rainforest of this region were accompanied by some of the first microlithic and osseous technologies anywhere in South Asia (Deraniyagala, 1992; Perera et al., 2011; Lewis et al., 2014; Perera et al., 2016). This early occupation, as well as remarkable technological stability, is sustained from *c.* 38,000–36,000 cal. years BP, through the Last Glacial Maximum, the Terminal Pleistocene, and into the Holocene, until the adoption of farming in the region *c.* 3,000 cal. years BP (Deraniyagala 2007; Roberts et al., 2015a). Indeed, human occupation of Sri Lanka's Wet Zone rainforest occurs long prior to any firm evidence for human occupation of other ecologies across the island (Figure 4.6). To date, the earliest definitive evidence from the Intermediate and Dry Zone areas of Sri Lanka is found in the form of human remains and microliths at Balangoda Kuragala and Bellan-bandi Palassa, respectively, during the Terminal Pleistocene *c.* 12,000 cal. years BP, prior to the appearance of shell midden sites on the coast *c.* 6,000 cal. years BP (Deraniyagala, 1992; Perera, 2010; Roberts et al., 2015b).

Throughout the Late and Terminal Pleistocene, the available faunal records from the Wet Zone cave and rock shelter sites of Sri Lanka indicate that forest-dwelling, semi-arboreal, and arboreal primates consistently make up between 70 and 80 per cent of the mammalian assemblages (Wijeyapala, 1997; Perera, 2010). Such a proportion is unheard of in both the archaeological and anthropological literature, with one exception (Fortier, 2009), even with the availability of rifle technologies in many parts of the tropics (ProWildlife, 2007; Lu, 2010; Papworth et al., 2013). Other resources exploited at Batadomba-lena include mouse deer, giant squirrel, mongoose, jungle cat, and civet, as well as *Canarium* sp. nuts and starchy rainforest plants (Perera et al., 2011), all of which imply dedicated rainforest subsistence. It has been suggested that Sri Lankan microlithic technologies were used in projectile hunting of arboreal and semi-arboreal fauna, while bone toolkits were used in the shooting and trapping of these fauna, or in accessing freshwater molluscs (Perera et al., 2011, 2016) (Figure 4.7). Trapping may be further suggested by the presence of nocturnal fauna, such as the Indian

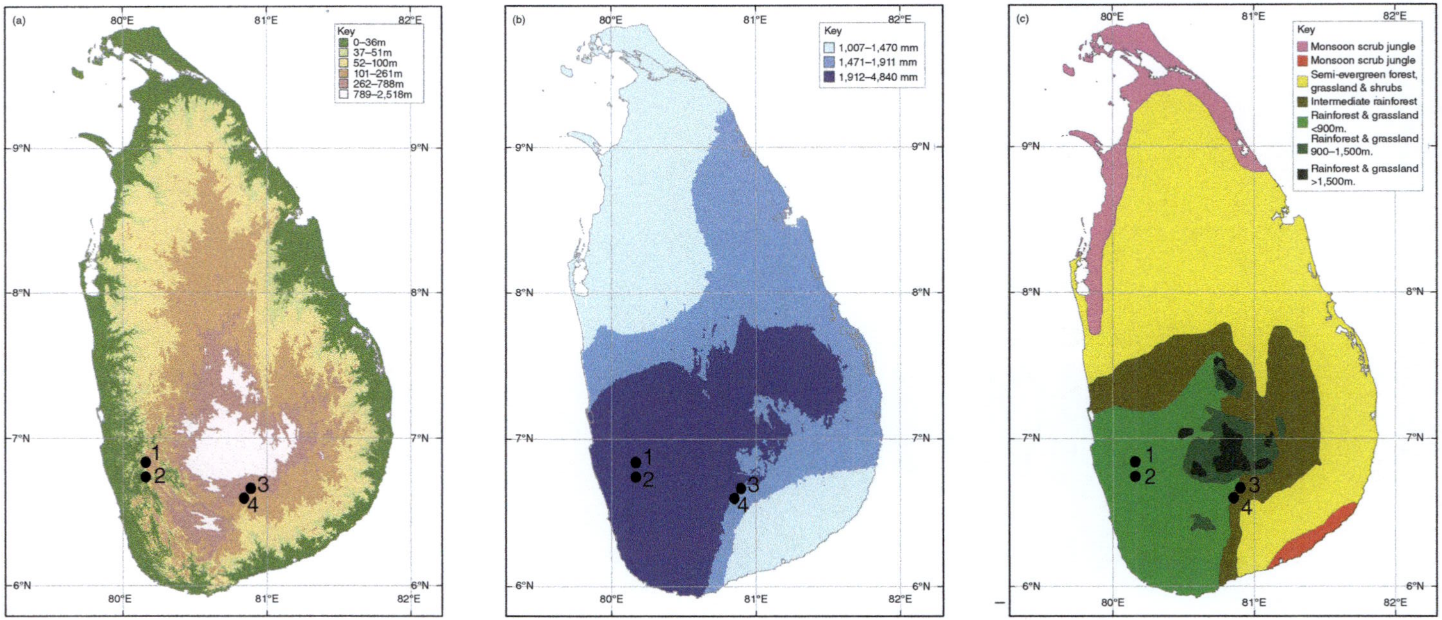

Figure 4.6. The location of key archaeological sites in Sri Lanka discussed in the text (1. Batadomba-lena, 2. Fa Hien-lena, 3. Balangoda Kuragala, 4. Bellan-bandi Palassa) relative to a) elevation, b) precipitation (modelled using Worldclim 1.3 records of international precipitation, and shown in the three arithmetic groups discussed here, 'Wet' Zone, 'Intermediate' Zone, and 'Dry' Zone—per Roberts et al., 2015b) and c) map of vegetation redrawn using data from Dittus (1977) and Erdelen (1988) With thanks to Hans Sell and Jana Zech.

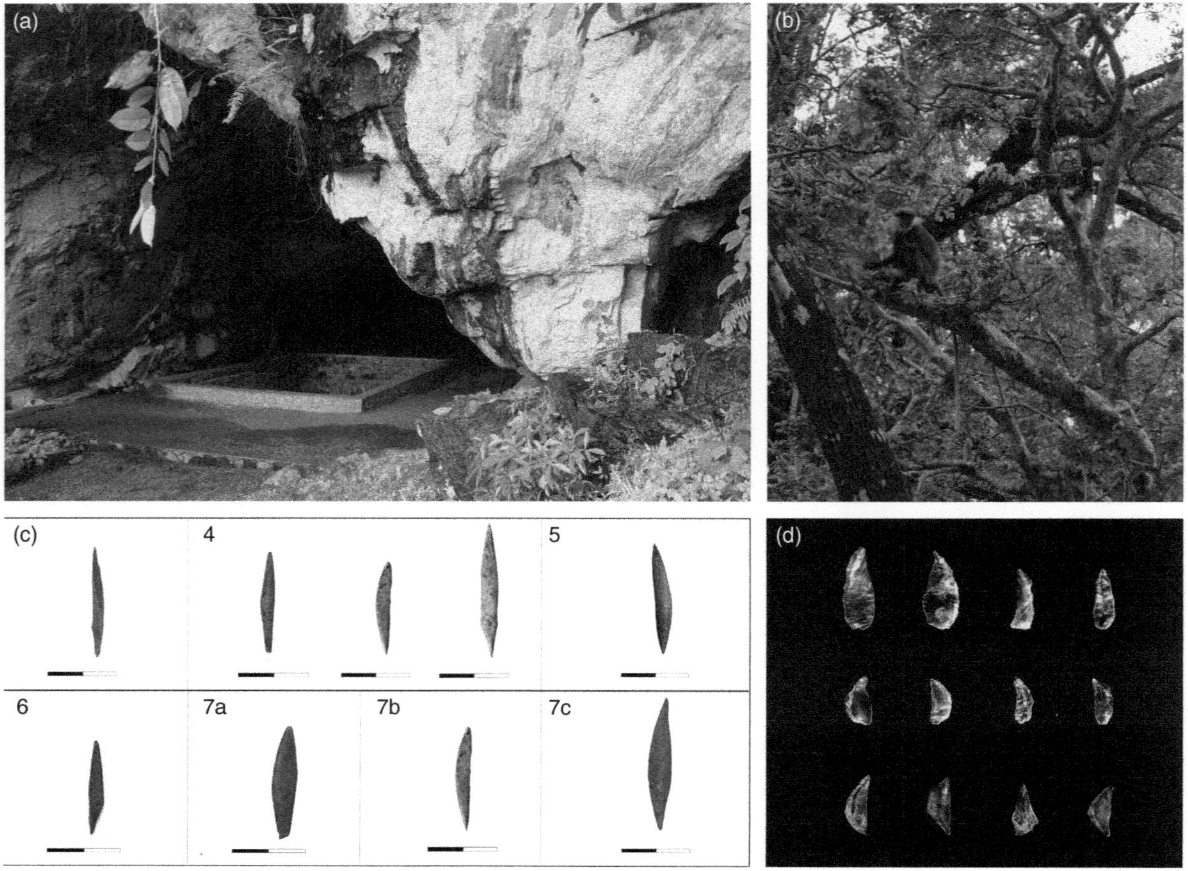

Figure 4.7. Photographs of a) the site of Batadomba-lena in the Wet Zone of Sri Lanka taken by the author, b) a purple-faced leaf monkey (*Trachypithecus vetulus*) taken by the author hunted by Late Pleistocene humans at the site of Batadomba-lena, c) double-pointed bone tools from the site of Batadomba-lena adapted from Perera et al. (2016), and d) microlith technologies from the site of Batadomba-lena taken by, and reproduced with permission from, Laura Lewis.

porcupine, at these sites (Perera et al., 2011; Roberts et al., 2017b). Additional use-wear studies of the microliths and bone tools from the Sri Lankan cave and rock shelter sites may help to clarify this further.

The reconstructed ecologies of faunal and floral resources from Late Pleistocene Sri Lankan sites is supported by 'on-site' palaeoenvironmental information, from molluscan and macrobotanical evidence, that suggests the consistent presence of tropical forest in the immediate vicinity of Batadomba-lena through the Last Glacial Maximum and into the Terminal Pleistocene (Perera et al., 2011). Similar ecological contexts appear to be demonstrated at the cave and rock shelter sequences of Fa Hien-lena (Wijeyapala, 1997; Perera, 2010) and Kitulgala Beli-lena (Kourampas et al., 2009) though systematic multidisciplinary work at these sites remains limited. The persistence of rainforest in the Wet Zone of Sri Lanka through this time period is supported by 'off-site' peat swamp palynological records from the Horton Plains of the Highland Plateau (Premathilake and Risberg, 2003; Premathilake, 2012) and broad-scale regional vegetation models (Boivin et al., 2013), although both records suggest more open tropical forest in the Wet Zone during the Last Glacial Maximum (*c.* 20 ka). Micromorphology at the site of Kitulgala Beli-lena has also provided evidence for periods of decreasing and increasing humidity that correlate with those seen in the Horton Plains Highlands (Kourampas et al., 2009). Nevertheless, further palaeoenvironmental studies across Sri Lanka, and off its coasts, are desirable if the changing extent of the tropical forest is to be studied across the island as a whole (Roberts et al., 2015b).

Sri Lanka has also provided the first direct insights into early human dietary reliance on tropical forest resources, demonstrating that they were not utilized as part of seasonally variable subsistence strategies. Using stable carbon and oxygen isotope analysis of human and faunal tooth enamel from the Terminal Pleistocene and Early Holocene sequences of Fa Hien-lena and Balangoda Kuragala, as well as layers dated from 36,000–20,000 cal. years BP at Batadomba-lena, Roberts and colleagues (2015a, 2017b) demonstrated that Sri Lankan human foragers were almost completely reliant on tropical rainforest foodstuffs, from their first appearance in this part of the world until the introduction of agriculture. The human and faunal isotope data suggest that the Wet Zone rainforests were more open from 36,000 cal. years BP and during the Last Glacial Maximum, though humans maintained a general reliance on forest resources (Roberts et al., 2015a, 2017b) (Figure 4.8). The presence of marine shell beads and shark teeth at the rainforest sites suggests there may have been links with groups living on the coast of Sri Lanka (Perera, 2010; Perera et al., 2011). However, it remains clear that in Sri Lanka, early humans were able to subsist within and occupy tropical forests without recourse to other environments, despite the existence of Sri Lanka's modern ecological transect of Wet Zone lowland rainforest, Intermediate Zone rainforest, and Dry Zone grassland existing from at least 12 ka, and perhaps 36 ka (Roberts et al. 2015b, 2017b).

Southeast Asia

Southeast Asia is potentially home to some of the earliest fossils of our species in tropical forest contexts. Four teeth at Panxian Dadong in South China, dated to 300–130 ka, based on biostratigraphy, U-series, and Electron Spin Resonance

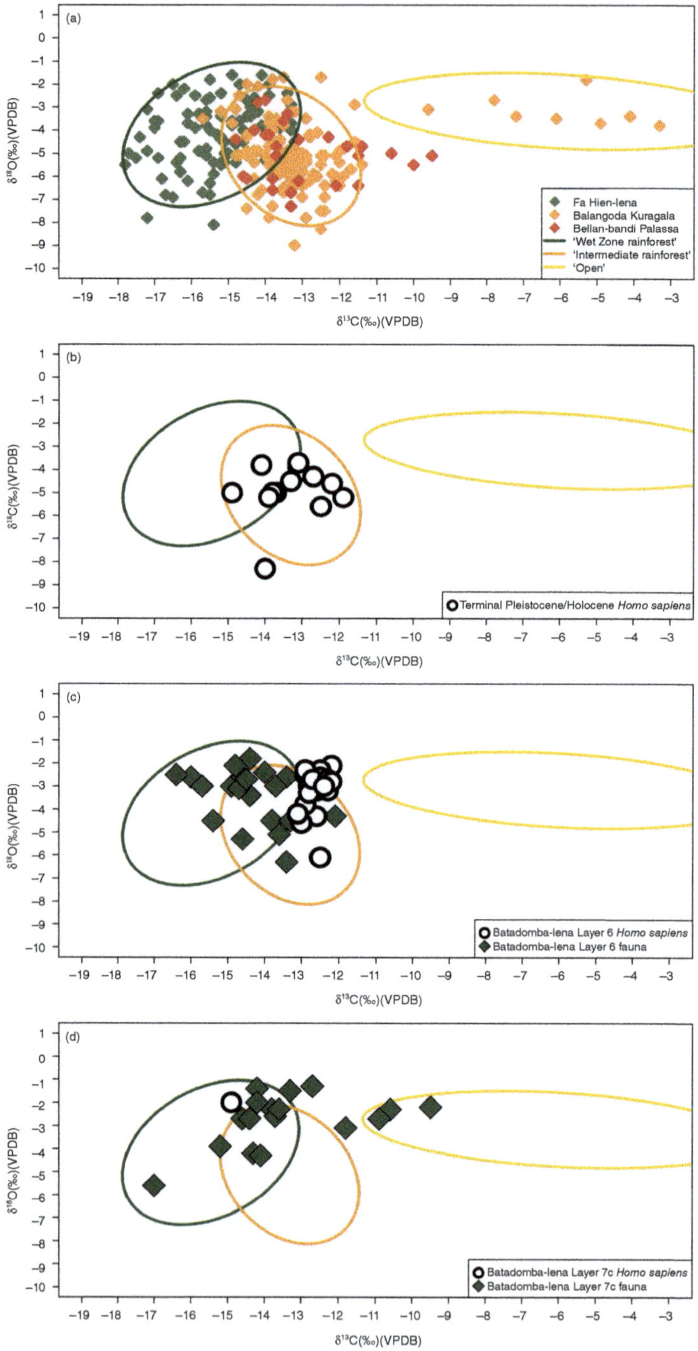

Figure 4.8. a) Stable carbon ($\delta^{13}C$) and stable oxygen isotope ($\delta^{18}O$) measurements of fauna from the Terminal Pleistocene/Holocene deposits of Fa Hien-lena, Balangoda Kuragala, and Bellan-bandi Palassa with 90% confidence ellipses representing 'Wet Zone' tropical forest, 'Intermediate' tropical forest, and 'Open' environments in Sri Lanka at this time, b) Terminal Pleistocene/Early Holocene human $\delta^{13}C$ and $\delta^{18}O$ measurements from Balangoda Kuragala and Fa Hien-lena, c) $\delta^{13}C$ and $\delta^{18}O$ values of human and faunal specimens from Layer 6 dated to *c.* 20,000 cal. years BP at Batadomba-lena, d) $\delta^{13}C$ and $\delta^{18}O$ values of human and faunal specimens from Layer 7c dated to *c.* 36,000 cal. years BP at Batadomba-lena. Data from Roberts et al. (2015a, 2017b).

dating (Shen et al., 1997; Rink et al., 2003; Schepartz and Miller-Antonio, 2004), have been identified as *H. sapiens* (Schepartz et al., 2000), and associated with fauna indicative of mixed woodland, bamboo forest, and grassland (Schepartz et al., 2000). However, the species attribution of the teeth, and their chronological association, remains debatable (Liu et al., 2013). Confirmed *H. sapiens* has been found at Fuyan Cave in the same region in association with mixed tropical rainforest, woodland, and grassland fauna (Liu et al., 2015) though, as noted above, their date remains disputed (Michel et al., 2016) (Figure 4.9). A premolar

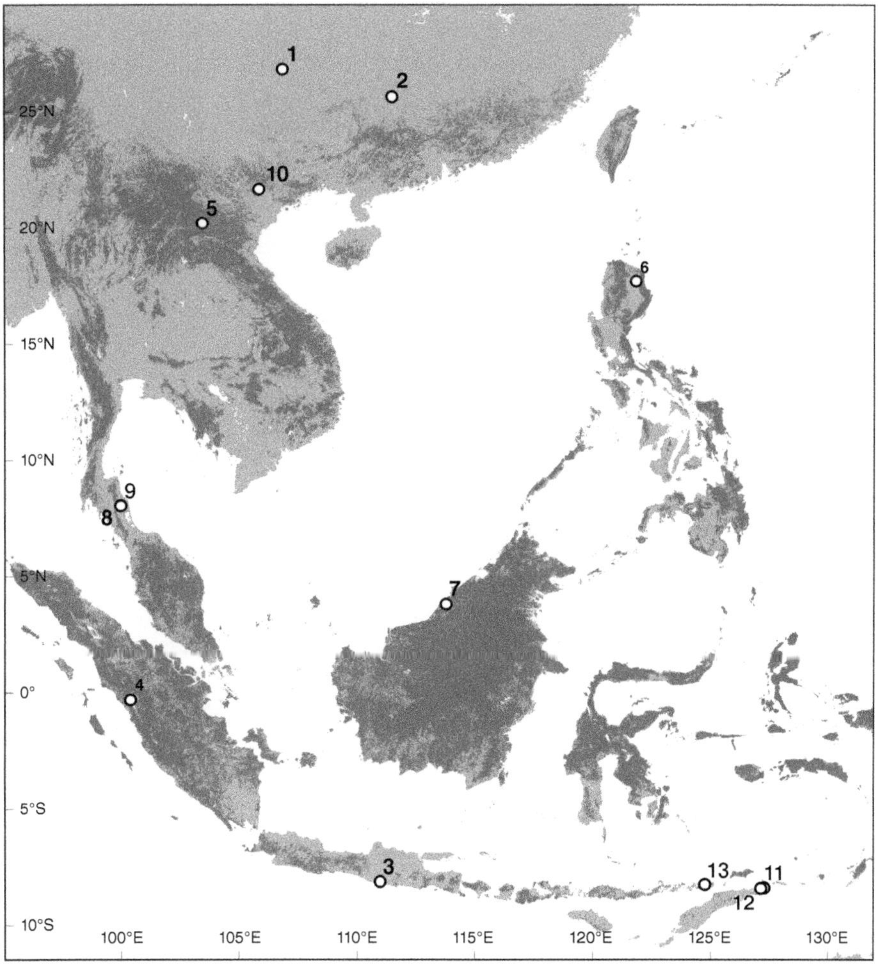

Figure 4.9. Map of major Late Pleistocene archaeological and fossil sites in Southeast Asia discussed in the text plotted against tropical forest distributions estimated on the basis of the MODIS (Moderate Resolution Imaging Spectroradiometer) Land Cover MCD12Q1 majority land cover type 1, class 2 for 2012 (spatial resolution of 500 m). Downloaded from the US Geological Survey Earth Resources Observation System (EROS) Data Center (EDC). 1. Panxian Dadong, 2. Fuyan Cave, 3. Punung Cave, 4. Lida Ajer, 5. Tam Pa Ling, 6. Callao Cave, 7. the Niah Caves, 8. Lang Rongrien, 9. Moh Khiew, 10. Nguom rock shelter, 11. Jerimalai, 12. Matja Kuru, 13. Makpan. With thanks to Ulrike Thuering.

possibly dating to earlier than *c.* 100 ka on the island of Java, associated with Punung tropical rainforest fauna, has been argued to represent early human rainforest presence in Indonesia (Storm et al., 2005; Storm and De Vos, 2006; Westaway et al., 2007). However, its identification and dating remain provisional (Storm et al., 2005; Barker et al., 2007; Bacon et al., 2008). Finds of *H. sapiens* associated with rainforest fauna at Lida Ajer in Sumatra *c.* 73–63 ka (Westaway et al., 2017), at Tam Pa Ling, Laos *c.* 60–46 ka associated with some elements of tropical forest microfauna (Demeter et al., 2015), and a human metatarsal found within a rainforest river valley environment at Callao Cave in the Philippines (Mijares et al., 2010), potentially also represent early human presence in the tropical forests of Southeast Asia (Figure 4.9).

Although the dating and environmental context of many of these early fossil finds remain disputed, the Niah Caves located in Sarawak, Borneo, have yielded some of the most detailed evidence for Late Pleistocene human rainforest subsistence the world over (Barker et al., 2007; Barker, 2013) (Figure 4.9). These caves are home to another early human fossil, the 'Deep Skull', in this region (*c.* 37,000 cal. years BP) (Barker et al., 2007; Higham et al., 2009; Hunt and Barker, 2014; Pike, 2016), associated with human activity now dated to between 53,000 and 34,000 cal. years BP (Barker et al., 2007; Barker and Farr, 2016). Furthermore, multidisciplinary analysis of sedimentology, pollen, phytoliths, macrobotanical remains, starch grains, and vertebrate remains has indicated that human foraging activities were tailored to a tropical forest environment. These include the collection and processing of toxic plants and a specialized hunting focus on wild boar, as well as some arboreal primates, using some of the earliest bone technology found in Southeast Asia *c.* 45,000 cal. years BP (Rabett et al., 2006; Barker et al., 2007; Rabett and Barker, 2007; Barton et al., 2009, 2016; Hunt et al., 2012, 2016; Piper and Rabett, 2014; Barton, 2016; Rabett, 2016) (Figure 4.10). Hunt and colleagues (2012, 2016) have also suggested that during periods of maximum forest extent, human groups at the Niah Caves deliberately practised vegetation burning and disturbance in order to promote the presence of open habitats. The maintenance of mosaic forest habitats would have, in turn, attracted wild pigs to new plant growth.

Indeed, unlike Sri Lanka, from the evidence available at present, most early human occupation in both mainland and Island Southeast Asia appears to have focused on mosaic tropical habitats rather than uniform evergreen rainforest (Rabett, 2012; Pawlik et al., 2014a). For example, the earliest layers of occupation at the Lang Rongrien rock shelter on the Thai side of the Thai–Malay Peninsula, dated to between *c.* 44,000 and 32,000 cal. years BP, are associated with open savanna-like conditions, with the presence of some open-canopy forest on the basis of molluscan and mammalian assemblages (Anderson, 1997; Mudar and Anderson, 2007) (Figure 4.9). Similarly, although the nearby site of Moh Khiew shows increased exploitation of arboreal taxa *c.* 30,000 cal. years BP, the palaeoenvironmental and faunal data are indicative of a mosaic of open and forest habitats immediately surrounding the rock shelter (Pookajorn, 1996). Mosaic and shifting environments have also been inferred for the Punung Cave sites in central and eastern Java from 45,000 years ago (Sémah et al., 2002), as well as pre-27,000 cal. years BP and 27,000 cal. years BP layers at the Nguòm rock shelter in Vietnam (Reynolds, 1990) (Figure 4.9). While caves in the Philippines (Détroit et al., 2004; Mijares et al., 2010) have been argued to show Late Pleistocene human

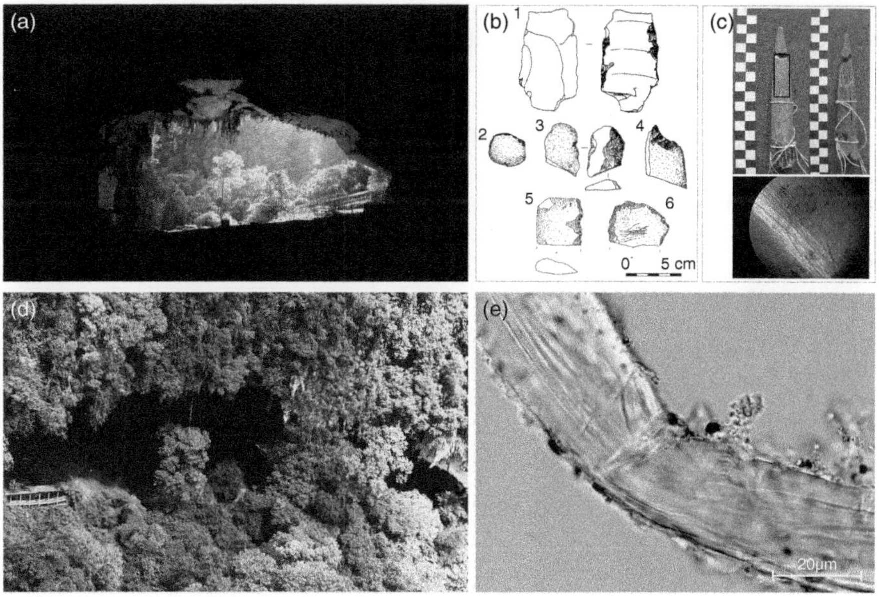

Figure 4.10. Images of a) view out of the Great Mouth Cave of the Niah Caves, b) lithic material from the Great Mouth Cave of the Niah Caves, c) a bone tool with use-wear from Lobang Hangus Cave of the Niah Caves, d) a view of the environment surrounding the Niah Caves, and e) a close-up of starch granules found on lithics at the Niah Caves. Used with permission of Graeme Barker, Huw Barton, Ryan Rabett, Tim Reynolds, and the National Museum of Sarawak, Borneo. With thanks to the Niah Caves Team.

presence in tropical forest environments, palaeoenvironmental evidence suggests changeable mosaic ecologies on the islands of Palawan and northern Luzon (Pawlik et al., 2014a).

In Wallacea, particularly the islands of Timor and Alor that were never connected to Sahul during the Late Pleistocene, tropical forest use has been considered an unviable source of subsistence for the earliest colonists given depauperate terrestrial forest environments (O'Connor et al., 2011; Samper Carro et al., 2016) (Figure 4.9). Instead, as was once the case for the whole of the southern dispersal route, sea-faring abilities and evidence for some of the earliest pelagic fishing has been emphasized at the sites of Jerimalai and Matja Kuru 1 and 2 (O'Connor et al., 2011) (Figure 4.9). However, giant rat taxa, with postulated preferences for forest environments, are also present at a number of Late Pleistocene sites (Aplin and Helgen, 2010), while other sites, such as Makpan on Alor, are located inland from the coast (Julien Louys pers. comm.). On the whole there is clear evidence for tropical forest use across mainland and Island Southeast Asia from at least 45 ka onwards, though often combined as part of mosaic subsistence strategies. Rabett (2012) has argued that more specialized tropical forest use only occurs from the Terminal Pleistocene as precipitation increased and evergreen rainforest expands (though other researchers have suggested rainforest fragmentation—Cannon et al., 2009). It is at this point when complex bone technologies, a diversification of lithic assemblages, and increased selective exploitation of arboreal and semi-arboreal fauna seem to simultaneously occur across this region (Piper and Rabett, 2009; Rabett et al., 2009; Rabett and Piper, 2012).

Sahul and the Bismarck Archipelago

The earliest evidence for human tropical forest occupation in Melanesia comes from the Ivane Valley region of Papua New Guinea (Figure 4.11). Here, at an

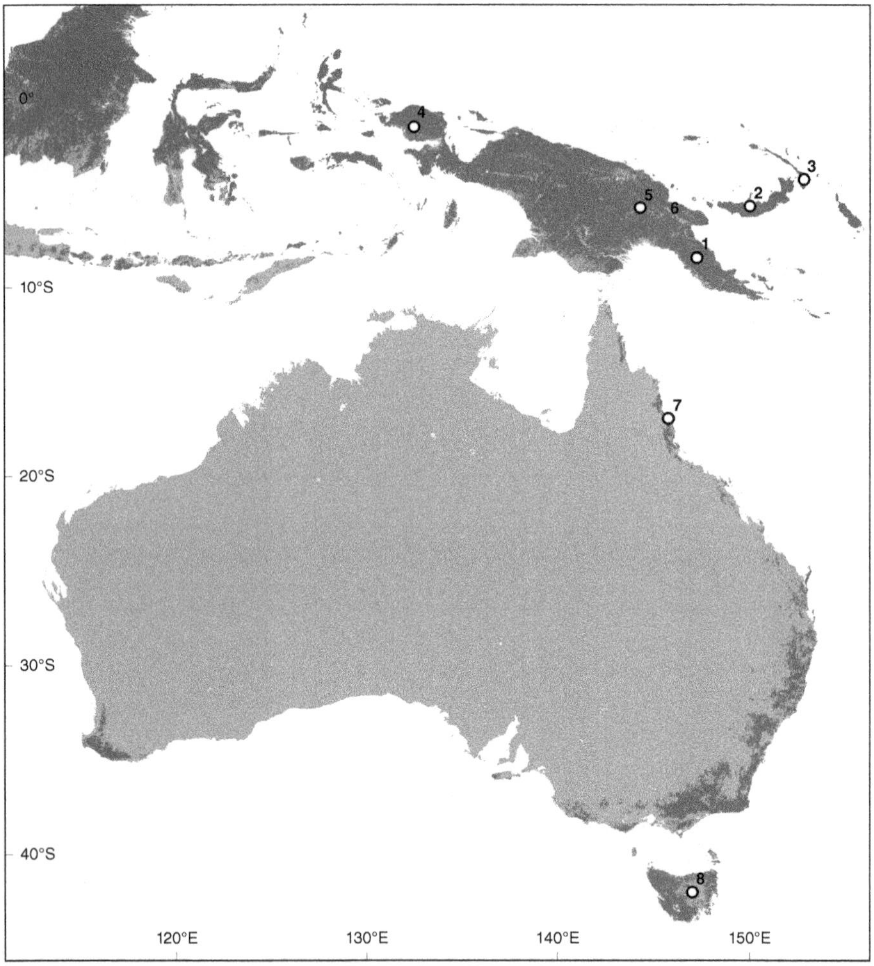

Figure 4.11. Map of major Late Pleistocene archaeological and fossil sites in Melanesia and Australia discussed in the text plotted against tropical forest distributions estimated on the basis of the MODIS (Moderate Resolution Imaging Spectroradiometer) Land Cover MCD12Q1 majority land cover type 1, class 2 for 2012 (spatial resolution of 500 m). Downloaded from the US Geological Survey Earth Resources Observation System (EROS) Data Center (EDC). 1. The Ivane Valley sites, 2. Yombon, 3. Matekupkum, Buang Merabek, Panakiwuk, Balof, 4. Bird's Head, 5. Nombe, 6. NFX and Wañelek, 7. Sites in the northeast Queensland rainforest, 8. Tasmania. With thanks to Ulrike Thuering.

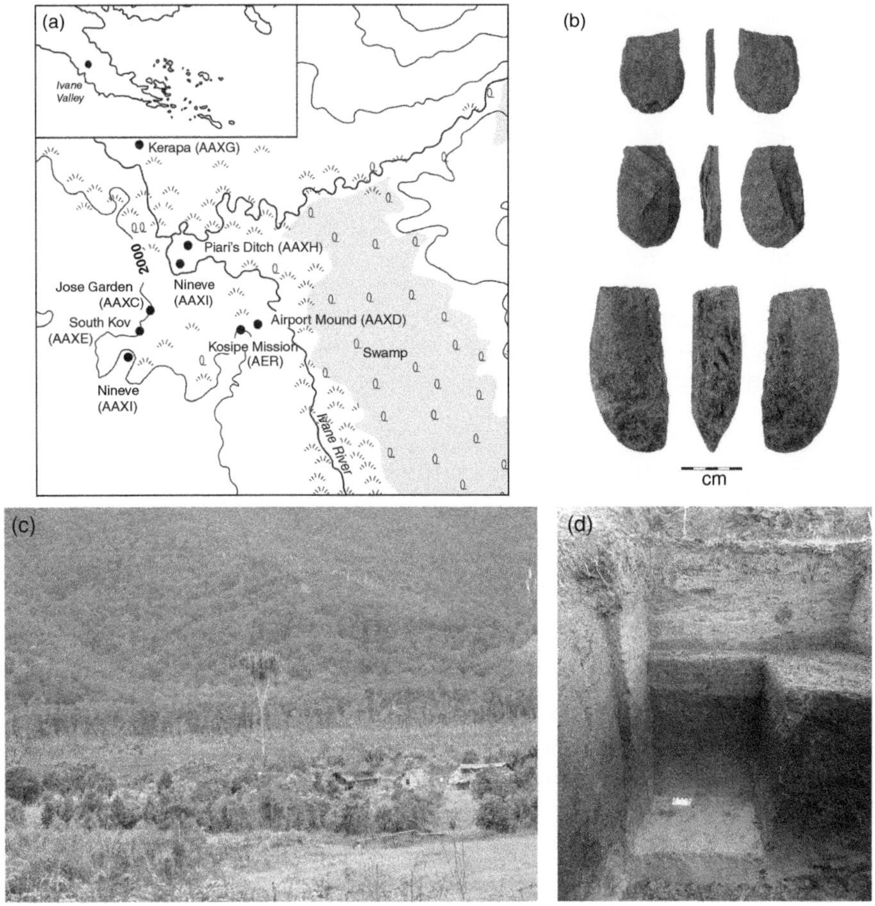

Figure 4.12. Compilation of a) Map of the Ivane Valley sites in the highlands of Papua New Guinea, b) Photograph of lithics found at the site of Kosipe in the Ivane Valley, c) Photograph of montane rainforest in the Ivane Valley today, d) Photograph of stratigraphic section of Kosipe in the Ivane Valley. Taken by, and used with permission from, Glenn Summerhayes.

elevation of 2,000 metres, archaeological remains have been radiocarbon dated to between 49,000 and 43,000 cal. years BP at the site of Vilukuav (Summerhayes et al., 2010). Waisted stone axes from the early layers of Airstrip Mound, South Kov, and Kosipe (dated to *c.* 40,000 cal. years BP) have been interpreted as implements used to modify forest environments (Groube, 1989). Starch grains of *Dioscorea* yam species and charred *Pandanus* nuts from radiocarbon-dated Ivane Valley sites also indicate the exploitation of forest products from as early as 49,000–43,000 cal. years BP (Summerhayes et al., 2010) (Figure 4.12). Comparably early evidence for the human occupation of modern-day tropical forests in the Melanesian portion of the Late Pleistocene continent of Sahul has been found elsewhere in Papua New Guinea (Summerhayes et al., 2017) and the Yombon region of West New Britain in the Bismarck Archipelago where the organized utilization of tropical forest resources is seen *c.* 35,500 cal. years BP (Pavlides and Gosden, 1994; Pavlides, 2004; Summerhayes et al., 2017). Late Pleistocene evidence for human occupation in the tropics of this region has also been documented on the offshore island of New Ireland back to *c.* 40,000 cal. years BP (Leavesley, 2002, 2004, 2005).

In the Ivane Valley of New Guinea, as well as at Late Pleistocene sites in New Ireland such as Buang Merabek, tropical forest foragers have been argued to have been small, highly mobile groups, flexibly exploiting local raw materials as part of a seasonal subsistence strategy that made use of a wide variety of environments and ecologies (Pavlides, 2004; Leavesley, 2007; Gosden, 2010; Summerhayes and Ford, 2014; Summerhayes et al., 2017) (Figure 4.11). This is not only based on patterns of lithic exploitation and manufacture (Pavlides, 2004; Leavesley, 2007; Summerhayes and Ford, 2014) but, in the Ivane Valley at least, also the fact that cold winter temperatures would have made it unlikely that humans lived there permanently (Gosden, 2010). Although Late Pleistocene cases of adaptations to high-altitude environments are increasing, as noted above, human activity appears to have completely ceased in this region almost entirely during the Last Glacial Maximum, when mean annual temperatures may have fallen by as much as 5°C (Summerhayes et al., 2010). While there is evidence for the exploitation of montane tropical forest resources at the Ivane Valley, limited forest-specific subsistence evidence is available from the mosaic tropical ecologies of West New Britain and New Ireland. Furthermore, detailed reconstruction of local environmental contexts during the Late Pleistocene is almost completely lacking across West New Britain, New Ireland, and the Ivane Valley.

After *c.* 26 ka, and into the Terminal Pleistocene (*c.* 18 ka), more intensive human occupation and manipulation of tropical rainforests is visible in Melanesia. Two long-term cave sequences in the Bird's Head region of western New Guinea have been argued to provide evidence for rainforest hunting and gathering strategies during the last 26,000 years (albeit with these behaviours only really intensifying during the Middle Holocene) (Pasveer, 2004). Further sites across the rest of New Ireland (including Buang Merabek, Panakiwuk, and Balof) and mainland Papua New Guinea (Nombe, NFX, and Wañelek) also show increasing evidence for human tropical forest presence at this time (Pavlides, 2004; Leavesley, 2007). Evidence from the site of Matenkupkum, New Ireland, indicates that post-Last Glacial Maximum (*c.* 20–18 ka) people began to introduce wild plants and animals into forest ecosystems, naturally poor in animal protein, that they were now persistently inhabiting over the long-term (Gosden and Robertson, 1991; Gosden, 1995; Barton and Paz, 2007) (Chapter 5) (Figure 4.11). Future work in the Solomon Islands may reveal further tropical forest adaptations at this time. The increasingly intensive ecological manipulation of tropical forest resources during this period culminates in the emergence of horticulture in Papua New Guinea during the Early Holocene *c.* 9 ka, with intensive clearance and the progressively specialized exploitation of endemic plants and animals (Golson, 1977; Gosden, 2010) (Chapter 5).

Despite Australia being joined to the rest of Sahul until the end of the last glacial cycle (Allen et al., 1977), firm evidence for Late Pleistocene human occupation of its rainforests has yet to be established. A thin strip of fragmented rainforest in the far north of Queensland is buffered from fire-prone sclerophyll vegetation by wet sclerophyll forests, characterized by a sclerophyll overstory and fire tolerant rainforest species underneath (Head, 1996; Hilbert et al., 2001) (Figure 4.11). It has been argued that frequencies of charcoal and pollen in sediment sequences, including those from Lake Euramoo, can be associated with the onset of anthropogenic exploitation of the fire-prone sclerophyll vegetation from *c.* 30,000 years ago (Kershaw, 1985, 1986; Haberle, 2005; Haberle et al., 2010). However, inferences of early human presence in these forests are based on indirect association of burning with human action (Cosgrove et al., 2007). Evidence for human presence in the temperate rainforests and glaciated highlands of Tasmania *c.* 34 ka (Pike-Tay and Cosgrove, 2006) implies that such adaptations were not beyond Australia's earliest inhabitants. Nevertheless, at present, clear evidence for human occupation and utilization, in the form of the detoxification of toxic forest nuts and the hunting of forest marsupials, of this region's rainforests occurs only with the Holocene expansion of rainforest from *c.* 8,000 years ago (Cosgrove et al., 2007; Haberle et al., 2010; Ferrier, 2015).

Central and South America

With the highly dubious claim of a 130 ka arrival of *H. sapiens* in America aside (Holen et al., 2017), archaeological and genetic evidence demonstrates that our species first arrived in the New World *c.* 24–13 ka (Goebel et al., 2008; Bourgeon et al., 2017; Dillehay et al., 2017; Moreno-Mayar et al., 2018). This evidence demonstrates that humans were present in the Americas over nearly two thousand years prior to the traditional 11.5 ka date of population explosion linked to the Clovis technocomplex (Waters and Stafford, 2007). Humans appear to have rapidly made it to South America through North and Central America, with early evidence for arrival demonstrated at Los Toldos Cave in Patagonia *c.* 12.6 ka (Cardich et al., 1973; Dillehay, 2000), Tagua-Tagua in Chile *c.* 11.4 ka (Montané, 1968), Monte Verde in Chile *c.* 18.5–14.5 ka (Dillehay, 2000; Dillehay et al., 2008), Huaca Prieta in Peru 15 ka (Dillehay et al., 2017), Taima-Taima in Venezuela 13 ka (Ochsenius and Gruhn, 1979; Oliver and Alexander, 2003), and, more controversially, at Pedra Furada, and elsewhere (Boëda et al., 2016), in Brazil *c.* 50–20 ka (Guidon and Arnaud, 1991; Meltzer et al., 1994) (Figure 4.13). While the MIS 3 dates are largely dismissed given the lack of any other evidence for dispersal at this time, and the nature of the 'tools' or origin of the charcoal at Pedra Furada are argued to be the result of natural processes, the early chronology of Monte Verde (Meltzer et al., 1997; Dillehay et al., 2008, 2015), is now generally accepted, supporting a rapid movement of people from North to South America by at least 14.5 ka and perhaps as early as 18.5 ka.

As with many of the regions discussed, it has been argued that such a rapid human dispersal across the Americas could only have been facilitated by a marine adaptation and navigation (Dillehay, 2000; Erlandson et al., 2007). There is, however, growing evidence for the exploitation of different interior environments including, perhaps, tropical forests. At Monte Verde, regional environmental records suggest that from 18 ka to 14 ka an open, dry Patagonian forest transitioned to a dense, temperate

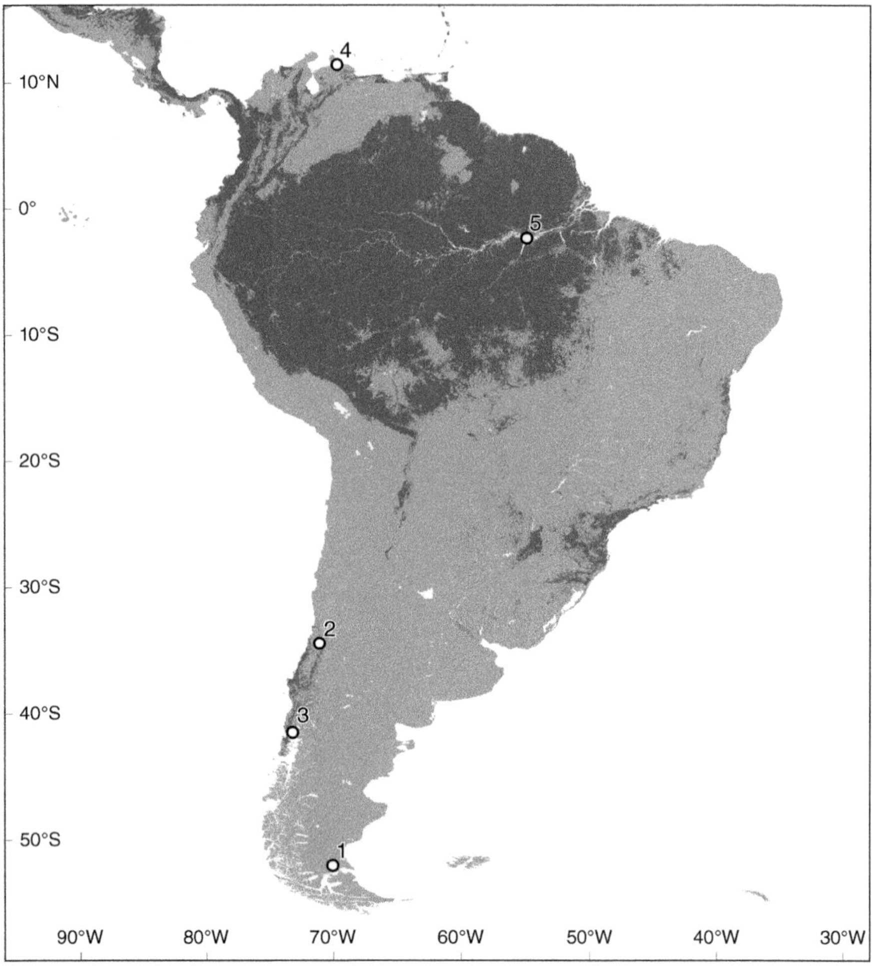

Figure 4.13. Map of major Late Pleistocene archaeological sites in South America discussed in the text plotted against tropical forest distributions estimated on the basis of the MODIS (Moderate Resolution Imaging Spectroradiometer) Land Cover MCD12Q1 majority land cover type 1, class 2 for 2012 (spatial resolution of 500 m). Downloaded from the US Geological Survey Earth Resources Observation System (EROS) Data Center (EDC). 1. Los Toldos Cave, 2. Tagua-Tagua, 3. Monte Verde, 4. Taima-Taima, 5. Amazon Basin sites. Created by Ulrike Thuering.

rainforest (Dillehay, 1989; Heusser et al., 1999) (Figures 4.13 and 4.14). Although these are beyond the tropical constraints imposed in Chapter 2, their dense canopies would have presented similar challenges to protein and carbohydrate sourcing and thermoregulation. Faunal and plant remains from the site indicate that humans obtained resources from a mosaic of environmental zones ranging from these forests to the Pacific coastline, including forest nuts, grass seeds, specific medicinal plants, seaweeds, deer or small horses, and elephant-like gomphotheres (Dillehay et al., 2015). In southern North and Central America, while most

Figure 4.14. Photograph of the temperate evergreen rainforest in the vicinity of Monte Verde, Chile. Photograph taken by, and reproduced with permission from, Tom Dillehay.

Pleistocene research has been focused in the dry and cool highlands (Ochoa, 2012), human occupation of lowland, seasonal tropical forests, as well as early squash and maize cultivation, dates to at least 9 ka (Piperno et al., 2009; Ranere et al., 2009) (Chapter 5). Early human presence in the Amazon Basin 11–9 ka (Roosevelt et al., 2002; Neves and Hubbe, 2005; Roosevelt, 2013) almost certainly occurred at a time of increased precipitation and tropical evergreen rainforest expansion (Rothhammer and Dillehay, 2009).

Additional excavation of rock shelter and cave sites in the Amazon Basin and other tropical forest regions of South America, as well as the tropical forests of Guatemala and Mexico, is likely to reveal more detailed insights into early human tropical forest hunting and gathering in Central and South America, including the different resources utilized. Indirect genetic evidence that certain human populations became isolated in South America following an initial rapid dispersal (Zegura et al., 2004; Skoglund et al., 2015; Skoglund and Reich, 2016) perhaps indicates that different environmental territories and specializations were developed. Further indirect evidence of local human subsistence strategies may also be found in the Late and Terminal Pleistocene extinctions of South American megafauna that significantly altered biodiversity, nutrient cycling, and forest structure in the region (Doughty et al., 2016). However, the potential role of humans in this process is unclear and yet to be fully explored (Rossetti et al., 2004). Regardless, what is clear is that some of the earliest humans in the Americas appear to have very quickly occupied and exploited the tropical forests of Central and South America *c.* 13 ka (Piperno, 2011a) (Figure 4.13). In North and Central America, in particular, intensive manipulation of tropical forest plants, perhaps from earliest arrival, led to some of the earliest global examples of tropical cultivation (Chapter 5).

TROPICAL FOREST PREHISTORIES AND THE DEFINING OF A PLASTIC PIONEER

While the above demonstrates that our species was able to adapt to tropical forest environments in Africa, South Asia, Southeast Asia, Sahul, and South America it would be a mistake to think these adaptations were uniform. Different tropical forest formations in different regions certainly shaped these adaptations. For example, the montane and evergreen-deciduous forests of the New Guinea Highlands are prone to burning, and large openings frequently form as a result of cyclones (Whitmore, 1998). This may explain the early evidence for human burning in this region that would have stimulated the availability of a diversity of tropical floral resources and their consumers (Groube, 1989; Denham and Barton, 2006), but made the tracking of fauna through resurgent undergrowth difficult. This, and the susceptibility of these montane and subalpine rainforests to changes in temperature across the Late Pleistocene–Holocene transition, likely encouraged the highly mobile, multi-ecotonal lifeways documented for these populations, which may have involved the exploitation of grassland megafauna (Fairbairn et al., 2006; Summerhayes et al., 2010). Although not strictly tropical, human adaptations to the varied forest types in montane regions can also be seen at the Monte Verde site in Chile, where hunter-gatherers appear to have used a complete altitudinal transect of resources ranging from montane forest, temperate lowland rainforest, and coastal settings (Dillehay et al., 2015).

Spatial variation in tropical forest formations need not be altitudinal, however, and Late Pleistocene human occupations in Southeast Asia clearly made continued use of a multitude of different forest types. For example, the presence of peat swamp forests, and accompanying soil acidity, in the vicinity of the Niah Caves may be the reason behind the focus there on wild boar and semi-arboreal mammal hunting. Poor soils would have resulted in limited floor-level plant resources, though some techniques were evidently developed to process hardy, toxic plant foods, and a combination of other forest types were also exploited and even promoted through burning (Cranbrook, 2000; Barker et al., 2007; Hunt et al., 2012; Rabett, 2012). In Wallacea, if tropical forests were used in the Late Pleistocene, this was combined with a contribution of marine resources (O'Connor et al., 2011). At Panga ya Saidi in Kenya humans consistently utilized tropical coastal forest and grassland resources over a period of 76,000 years without evidence for coastal subsistence (Shipton et al., 2018). Sri Lanka provides a clear exception to the use of multiple forest types and semi-evergreen and evergreen rainforest was apparently utilized from 36–3 ka with a persistent focus on arboreal and semi-arboreal primate resources as a source of protein (Roberts et al., 2017b). That said, even here, riverine freshwater snails provide a hint of microhabitat variability (Perera et al., 2011).

Climate change has also played a major role in evolving tropical forest adaptations during the Late Pleistocene. Based on modern data, of all the regions where early human tropical forest contact has been documented, Africa appears to be the most susceptible to rainforest reduction as a result of climatic impacts, likely leading to the relatively low floral diversity in African rainforests today (Maley, 1996; Malhi and Wright, 2004; Malhi et al., 2013). During the last glacial period, the Central African rainforest contracted, resulting in an advance of seasonal, semi-deciduous forests and savanna grasslands at its fringes. This may explain why the main distribution of Lupemban sites skirts the border of the Central African rainforest as it sits today and both Taylor and Cornelissen have emphasized the importance of forest ecotones as the key environmental context for this industry (Cornelissen, 2002; Taylor, 2011, 2016; Cornelissen, 2013). Indeed, palaeoenvironmental and subsistence evidence from around the Central African tropical forests is beginning to suggest that Late Pleistocene *H. sapiens* exploited semi-evergreen and deciduous forest edges and moved with them as climate changed their position (Basell, 2008; Blome et al., 2012). Similarly, in Sri Lanka, as the Wet Zone semi-evergreen and evergreen rainforest became more open leading up to the Last Glacial Maximum, *H. sapiens* utilized more open Intermediate semi-evergreen and dry 'Intermediae' tropical rain forests (Roberts et al., 2015a, 2017b). Nevertheless, in contrast to Africa, there is clear evidence for the continued specialization in dense tropical forest resource use from peoples' first arrival in this part of the world.

Changes in climate, across the Last Glacial Maximum and Terminal Pleistocene/Holocene transition in particular, also did not have homogeneous global effects. Indeed, while conditions warmed and precipitation generally increased in the tropics after the Last Glacial Maximum, significant sea-level rise (from a minimum of *c.* -116m) had hydrological impacts that led to a continued aridity and persistence of mosaic and open tropical forest formations in Southeast Asia (Kienast et al., 2001; Lambeck et al., 2002; Bird et al., 2005; Wurster et al., 2010). The effects of climate change will also have been felt very differently between groups living at the centre of an evergreen rainforest versus those living on tropical forest-savanna fringes or in montane rainforests. As noted in Chapter 2, many temporal fluctuations in forest types and species are recognized in the Amazon Basin, the Andes, Africa, and Southeast Asia for the Terminal Pleistocene and Holocene (e.g., Early to Middle Holocene arid phases and claims of the 'Younger Dryas' in Amazonia—Mayle et al., 2004—and Africa—Maley, 1996; Garcin et al., 2007; Bayon et al., 2012) suggesting we are missing finer-scale changes in forest structure and extent during the Late Pleistocene. What is clear, however, is that Late Pleistocene human rainforest adaptations should be viewed on a regionally and chronologically specific basis. It is only through detailed documentation of rainforest 'prehistories' that local exploitation strategies can be properly documented.

With this, it is also important to note that humans themselves have altered tropical forest structure and extent, sometimes with lasting ramifications, in different settings. Mounting evidence indicates deliberate burning in the tropical forests of Southeast Asia from *c.* 45 ka (Hunt et al., 2012). In Australia, fluctuations in plant species proportions have been correlated with human burning from 40 ka, including the decline of *Araucaria* and the rise of Eucalypts and *Casuarina* (Kershaw et al., 2007; Bird et al., 2013). Human landscape impacts have also been documented in the montane tropical forests of Highland New Guinea from 45–35 ka, even slowing vegetation re-colonization in the region following the Last Glacial Maximum (Fairbairn et al., 2006). Late Pleistocene humans also potentially had impacts on tropical forest animal populations. While discussions of megafaunal extinctions have been relatively limited in these settings, they possessed diverse megafauna (Malhi et al., 2014; Doughty et al., 2015). In the New Guinea Highlands there is evidence for extinct marsupials (such as *Maokopia ronaldii* and *Thylogale hopeii*), at West Balim River *c.* 30 ka and at Nombe *c.* 25 ka, whose gradual demise occurred following human presence and burning (Fairbairn et al., 2006). While in this, and in other cases, the role of humans remains debated, such extinctions could have massive environmental implications.

H. sapiens is certainly not the only Middle-Late Pleistocene hominin species to have adapted to extreme shifts in climate. *H. neanderthalensis* survived through glacial cycles and extreme environments in Europe *c.* 300–40 ka (e.g., Barton et al., 2011) while *H. erectus* appears to have persisted in Southeast Asia from *c.* 1.5 Ma to *c.* 100 ka (Marwick, 2009) (Chapter 3). In a tropical forest context it is also interesting that there is limited use of organic technology in Southeast Asia with the arrival of *H. sapiens* until an explosion in this form of material culture in the Terminal Pleistocene (Rabett, 2012) and *H. erectus* has also been shown to have potential symbolic capacity. However, what is clear is that *H. sapiens* is the only hominin able to adapt to the huge variety of environments, ranging from palaeoarctic to tropical forest, through increasing Late Pleistocene climate variability that posed threats to the continuity of the genus *Homo* (Rabett, 2012). *H. sapiens* was able to do so for three primary reasons: (i) it could specialize in the exploitation of varied local environments of a given type (in this case different tropical forest formations) and develop diverse strategies for long-term continuity, (ii) where necessary it could change the environments it inhabited to promote favoured resources that was to eventually lead to agriculture (Chapter 5), and (iii) as Rabett puts it, 'only *H. sapiens* evolved palaeoculture that became capable of adapting fast enough and successfully enough to keep pace with the conditions of the Upper Pleistocene' (Rabett, 2012: 280).

Finds of symbolic culture, technological complexity, and broad resource usage among other hominins have made these designations insufficient in the definition of our species. Instead, it is becoming increasingly argued that it is the plastic ability of our species to biologically and culturally adapt to different environments and changing climates that makes us human (Rabett, 2012; Grove, 2015). Rabett (2012) has additionally argued that our species shows a unique capacity for the 'colonization' of new environments, which now reach beyond our own planet, through a process of contextual cultural alteration (or transference), diversification, innovation, and divergence. To this I would add the proposal that our species is the first ecological 'generalist specialist' (Roberts and Stewart, 2018). As a generalist it can thrive in a variety of environmental conditions and make use of different resources. As a specialist it can also excel in a narrow range of environmental conditions with a limited diet. This is exemplified in studying Late Pleistocene tropical forest adaptations. The diversity of tropical forest environments inhabited by our species, alongside other extreme environments, demonstrates its plastic or generalist nature. However, when situated within very particular tropical forest environments it is able to specialize in locally specific cultural and ecological strategies. It is not necessarily so much that *H. sapiens* adapted 'fast enough' but rather that its adaptation, from its origins in Africa to the final stages of the Late Pleistocene colonization process in the New World, was to change and variability.

5

Tropical Bounties

The Emergence of Tropical Forest Agricultures

One touch of nature makes the whole world kin.
—William Shakespeare (2008)

The transition from the Terminal Pleistocene to the Holocene (*c.* 12–8 ka) witnessed increasingly intensive human manipulation of plant and animal resources that resulted in genetic and phenotypic changes in various species as part of what has been termed the 'origins of agriculture'. This process has been cited as one of the most significant ecological occurrences in human evolutionary history (Bocquet-Appel, 2011; Larson et al., 2014), representing a shift in human interactions with the natural world with global environmental ramifications (Fuller et al., 2011a; Boivin et al., 2016). Martin Jones (2007) has also discussed the cultural and social changes resulting from the new spatial and practical proximity of domesticated plants and animals that made them effectively 'family' or 'kin'. The tropics have, for a long time, been left out of discussions of this process, with poor preservation conditions considered unlikely to produce incipient crop or animal domesticate remains and some even arguing that the wet and acidic soils of tropical forests were too poor to support agriculture (Meggers, 1971, 1977, 1987; Grollemund et al., 2015). Nevertheless, emerging datasets from Melanesia, North and Central America, South America, and Africa are demonstrating that cultivation and, to a lesser extent, herding practices also emerged indigenously in these regions and, in some cases, perhaps as early as the traditional focus point of the 'Fertile Crescent' in the Near East. Moreover, these examples are having significant impacts on the way we conceptualize the emergence of 'agriculture' and the adaptive and social changes required (Denham et al., 2004, 2009; Barton and Denham, 2011).

Here, I explore the distinctive nature of early agricultures in tropical forest environments. I also evaluate their predecessors in the form of human management including forest burning to stimulate faunal and floral growth and diversity, the deliberate movement of faunal species into tropical forest environments, and the emergence of arboriculture cultivation. In doing so, I document how the species and strategies involved in these processes differ globally with varying tropical forest formations, ranging from a focus on long-term forest interaction, drainage system construction, and tree-cropping in Melanesia (Denham et al., 2003; Denham, 2011) to diverse hunting, fishing, and cultivation strategies in the Amazon (Roosevelt, 2000;

Meggers and Miller, 2002). The development and impacts of these 'indigenous' agricultures also differ starkly to the introduction of exogenous agricultural lifeways into tropical forests, including the spread of the Austronesian language speakers and their associated domesticates and commensal species into Southeast Asia and the Pacific and the relationship between Bantu speakers and tropical forest communities in Africa. I argue that this multi-regional, multi-temporal perspective on human–rainforest interactions provides a platform for questioning definitions of 'agriculture' more broadly. Moreover, the themes raised link into broader debates regarding the desirability of tropical forests for human subsistence (Chapter 1).

A BREAK FROM THE PAST?

When discussing the 'origins of agriculture' it is best to start with definitions. Price and Bar-Yosef (2011) argued that the term 'agriculture' should only be applied where the management of domesticated plants and animals is the major activity and subsistence source for a particular society. Domesticated taxa are those that have faced human selection that has had genotypic and phenotypic ramifications so that, in many cases, they can no longer survive without human input. For example, domesticated animals tend to be more docile or smaller in stature than their wild counterparts, while domesticated crops tend to have larger grains or fruits, and reproductive strategies that are often tuned into human behaviour such as sickle harvesting (Smith, 1999). It has been argued that these taxa are often predisposed to human selection and domestication (Diamond, 2005). Domesticated animals tend to either live in herd structures that can be easily led by humans or live as commensal species with human groups. Domesticated plants often grow naturally in dense stands, reproduce quickly, and have obvious calorific (and often carbohydrate) benefits (Smith, 1999; Diamond, 2005). Agricultural definitions that focus on the dominance of domesticated resources in the diet of a given society work well in terms of the traditional archaeological focus on the Near East in the search for agricultural origins. Here, domesticated cattle, sheep, goats, and pigs combined with barley, emmer and einkorn wheat, and a series of legumes to become major subsistence staples familiar to most Europeans today (Smith, 1999; Price and Bar-Yosef, 2011).

Such an approach has traditionally framed tropical forests as unlikely candidates for becoming 'agricultural centres'. Low light on the forest floor has meant that natural stands of dense grasses are unlikely and productive tree resources are often highly spaced throughout the forest. Acid soils and high water flow have also led to organic remains being rarely preserved and so whether taxa are wild or domesticated, let alone the relative reliance on these plants, is almost impossible to ascertain. In the case of animal taxa there are few, if any, natural 'herd' species in tropical forests, where animals are often small, live in small groups, are nocturnal, difficult to pen and, overall, not lent to domestication (see Smith, 1999). Definitions such as Price and Bar-Yosef's also draw a clear line between hunter-gatherers and farmers. Moreover, even where productive tropical crops, such as maize or yams, have been identified, poor soils that are prone to erosion have meant that the development of sedentary, 'farming' societies familiar to a Eurocentric mind-set have been considered unsustainable (Meggers, 1971, 1977, 1987; Grollemund et al., 2015). For example, the hunter-gatherer usage of domesticated dogs, bottle gourds, and squash in northern America as 'utility' species from at least *c.* 8,000 years ago would be considered beyond agriculture (Erickson et al., 2005; Larson et al., 2014; Mitchell, 2017). Similarly, active plant and animal management strategies by mobile groups or hunter-gatherers exploiting a diversity of subsistence strategies in tropical forests have not been considered 'agricultural' in the traditional sense.

More recently, however, there has been something of a shift in global archaeological definitions of 'agriculture' that has, at least in part, been stimulated by the documentation of new types of intensive plant and animal selection in tropical settings (Figure 5.1). For example, in the traditional case study of the Near East, the development of farming societies was seen as a sudden 'revolution' (Childe, 1936). Yet, it is now being increasingly recognized that Near Eastern foraging societies gradually experimented with the domestication of different animals and plants, in a diversity of regions and environments, that were only later combined into a 'package' staple for farming communities (Brown et al., 2009; Fuller et al., 2010, 2011b, 2014). In Price and Bar-Yosef's (2011) definition only those groups relying on this package would be considered as part of the agricultural story, which clearly misses a significant proportion of the human agency behind this process. As a result of this conceptual reconfiguration, it is now more popular to globally explore and document the increasingly intensive ways in which humans brought plants and animals into their own cultural and adaptive systems, and the corresponding impacts this had on various taxa. For example, cultivation is 'a human action that enhances the survival,

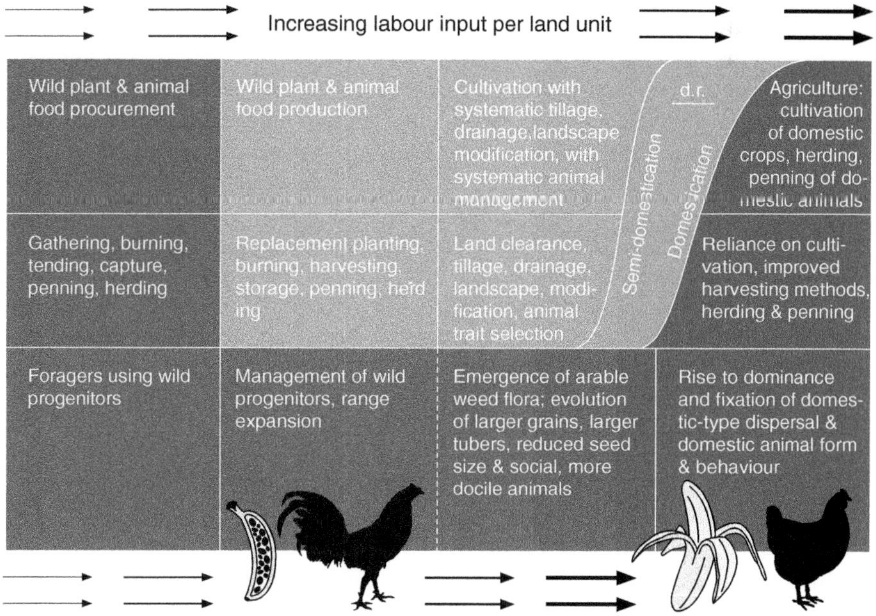

Figure 5.1. Schematic of the process of domestication and agricultural practice. Adapted from Harris (1989) and Fuller (2007a) with thanks to Hans Sell.

reproduction, or growth of certain plants' (Fuller and Hildebrand, 2013: 508) while herding is a human practice that defines the mobility and population structure of certain animals. Domestication is the genetic and morphological result of the selective pressures 'imposed by cultivation [and herding]' (Fuller and Hildebrand, 2013: 509, my addition) (Figure 5.1).

As such, in this chapter, I will evaluate the origins and continuity of agricultural *practice* in tropical forests by documenting evidence for cultivation and herding, and subsequent domestication. Yet, before going further, it is worth noting that a focus on these processes and outcomes can still miss significant cases of human plant and animal management in tropical forest environments. Firstly, as was touched upon in Chapter 4, Late Pleistocene humans from 45 ka did not just simply use tropical forest resources but also modified them in fundamental ways, with significant outcomes that remain present today. Forest burning to maintain ecotonal environments and promote understory plant growth has been documented in the montane tropical forests of Highland New Guinea, the evergreen rainforest of Borneo, and potentially the dry and wet tropical forests of Australia from 45–40 ka (Kershaw, 1985, 1986; Summerhayes et al., 2010; Hunt et al., 2012). While in some of these cases it can be difficult to differentiate deliberate human burning from natural fires, practices that enhance the survival and growth of particular plant types have also been documented in Holocene groups. For example, the first foraging colonists of the eastern Caribbean in the Middle Holocene engaged in modification and management of tropical ecosystems that is reflected in significant shifts in pollen and phytolith datasets (Siegel et al., 2015). Similarly, as seen in Chapter 2, historical and ethnographically documented hunter-gatherer groups in Australia used burning to promote different tropical forest habitats and their associated plant taxa (Hill and Baird, 2003).

There is also clear evidence for the translocation of plant and animal taxa by hunter-gatherer groups in tropical forests. In Melanesia, people translocated small mammals for reliable protein from 20 ka (Allen et al., 1989; Flannery and White, 1991; Gosden and Robertson, 1991). The result is that species like bandicoot (*Echymipera* spp.) and cuscus (*Phalanger* sp.) are now widely distributed across Melanesian islands, including the Bismarck Archipelago, where they are not endemic. Such practices certainly impacted the mobility and demography of these species, albeit not in a direct manner, following initial translocation. Greater yam (*Dioscorea alata*) has been argued to present on both sides of the Wallace Line by 40 ka, implying human agency (Barton and Paz, 2007), though its identification on the basis of starch grains and implications remain questioned (Denham, 2011). By the Terminal Pleistocene and Early Holocene, economically important plants, including the sago palm (*Metroxylon sagu*), yams (*D. alata* and *D. hispida*), taro (*Colocasia esculenta*), and swamp taro (*Cyrtosperma merkusii*), appear to have been widely dispersed, most likely with some human intervention, from Mainland Southeast Asia to the coastlands and islands of Southeast Asia, the Philippines, and Wallacea, and possibly also into North Australia (Denham et al., 2009; Barker et al., 2011; Denham, 2011; Hunt and Premathilake, 2012) (Figure 5.2). The modification of the distribution and density of edible and economic tree species has also been observed among Amazonian hunter-gatherers (Balée, 1994). These examples demonstrate that a focus solely on taxa that were eventually 'domesticated', or on cultivation and herding in a traditional sense, may miss relevant human behaviours.

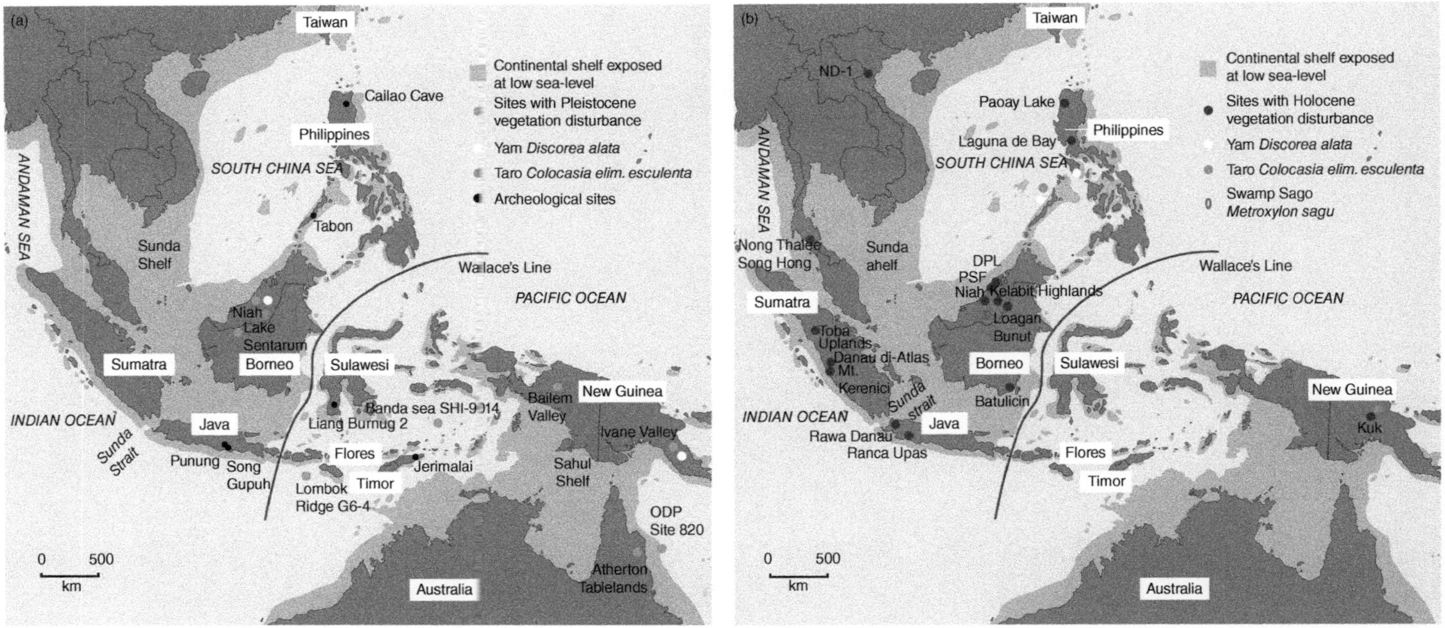

Figure 5.2. Tropical Australasia showing sites with reasonably certain human presence, sites where there is vegetation disturbance by fire atypical of the longer Pleistocene record, or where humans are directly implicated, and locations with evidence for economically useful plants found both sides of the biogeographical discontinuity of the Wallace Line. a) Late Pleistocene (45–11 ka), b) Early-Middle Holocene (11–5ka). The figure is compiled based on data from Paz (2005), Storm et al. (2005), Moss and Kershaw (2007), Westaway et al. (2007), Mijares et al. (2010), Summerhayes et al. (2010), Barker et al. (2011), Denham (2011), Hunt et al. (2012), Hunt and Premathilake (2012), Hunt and Rabett (2014), Pawlik et al. (2014b), Marwick et al. (2016). Adapted from Roberts et al. (2017a). With thanks to Chris Hunt and Hans Sell.

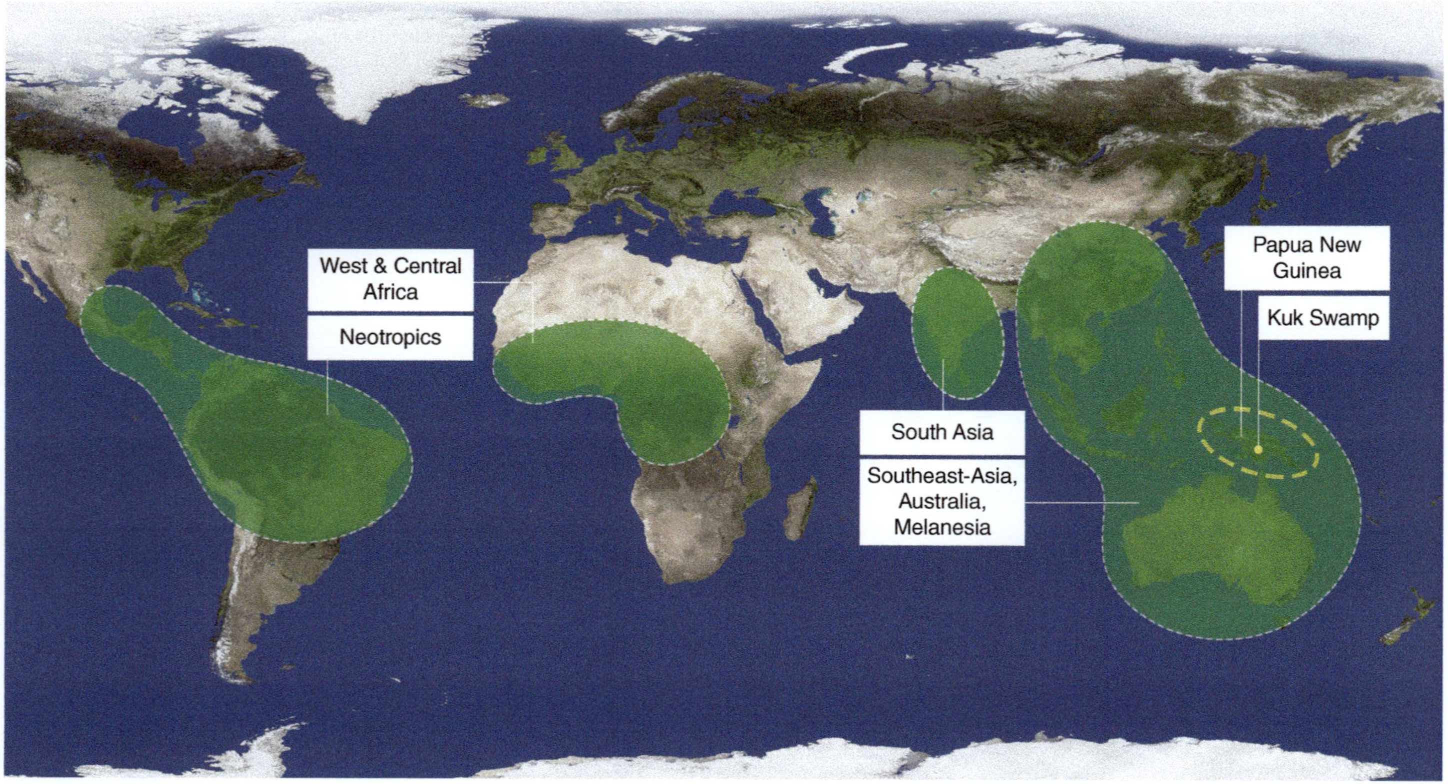

Figure 5.3. Map of main regions discussed in Chapter 5. With thanks to Hans Sell.

Finally, as noted in Chapter 4, Late Pleistocene hunter-gatherer groups were able to hunt particular tropical forest fauna for long periods of time, even though these resources are often considered to be limited (Bailey et al., 1989) and vulnerable to over-exploitation (Harrison, 2011). Increasingly detailed zooarchaeological assemblages from Late Pleistocene Sri Lanka *c.* 36–3 ka (Perera et al., 2011), Southeast Asia *c.* 45–12 ka (Barker et al., 2007; Piper and Rabett, 2014), and East Africa *c.* 70–1 ka (Shipton et al., 2018) are highlighting that human foragers could sustainably exploit particular tropical microhabitats and even tropical taxa, including primates often considered particularly susceptible to hunting pressures (Mittermeier, 1987), for millennia. Ecologically it has been noted that hunters often target species that live sympatrically and that when the numbers of one species are reduced others increase with the removal of competition (Bodmer et al., 1997; Peres and Dolman, 2000; Refisch and Koné, 2005). Moreover, ethnographic hunter-gatherers have been shown to shift the forest localities they exploit as primate and other small- and medium-sized mammal resources become depleted in one, allowing faunal recovery (Van Vliet et al., 2009). Late Pleistocene *Homo sapiens* thus apparently had a clear understanding of the demographic thresholds of tropical forest fauna (Roberts, 2018), even if they did not directly control them.

Practices like these have given tropical forests an important place in emphasizing the continuum between the behaviours and perceptions of foragers and farmers (e.g., Harris, 1989; Roscoe, 2002). Clearly humans do not have to rely on domesticated resources to manipulate plants and animals in ways similar to cultivation and herding. It is also incredibly clear that it is not only farmers that have treated plants and animals as a close part of their kin or cultural classification systems (e.g., Lévi-Strauss, 1963, 1966; Balée, 1994). While in many cases burning, translocation, or population management did not lead to formal domestication, this does not mean that such examples should be discarded as irrelevant to the origins and practice of 'agriculture', in tropical forests as well as other environments. Rather, these practices demonstrate the significant temporal and geographical extent over which humans have experimented with new cultural and adaptive relationships with the natural world (Allaby et al., 2016, 2017). For the sake of space and focus I limit discussion in this chapter to those instances in which deliberate (albeit perhaps not initially) intervention in tropical forest plant and animal lifecycles, through cultivation or herding as defined above, led to new genotypes and phenotypes in a variety of domesticated species familiar to us today (Figure 5.3). Many of these instances, however, occur alongside continued management of hunted taxa or the translocation and nurturing of plant taxa that were never to become fully domesticated.

EMERGING FROM THE SWAMP: INCIPIENT CULTIVATION, A TROPICAL PERSPECTIVE

The story of deliberate, independent cultivation and domestication in tropical forests begins in the Wahgi Valley of Highland New Guinea. It is here, at the site of Kuk Swamp, that ideas surrounding indigenous tropical forest domestication began to be taken seriously and have a major influence on global discussions of the 'origins of agriculture' (Denham et al., 2003, 2004; Denham and Barton, 2006) (Figure 5.4). Jack Golson and colleagues, in particular Tim Denham, have archaeologically investigated this site from the 1970s through to the 1990s. From the more than two hundred trenches that have been excavated Golson identified artificial drainage ditches and house structures that he argued represented agricultural practice at the site in a series of six, increasingly intensive, phases (Golson, 1977, 1989, 1991). Phase 1, now dated to 10,220 to 9,910 cal. years BP (Denham et al., 2003; Golson et al., 2017), consists of a palaeochannel and structural elements within a palaeosurface that have been seen as part of the first anthropogenic attempts to drain the local wetland and promote the growth of carbohydrate-rich plants (Golson, 1977, 1989; Golson and Hughes, 1980). This culminated with the likely cultivation, and domestication, of taro (*Colocasia esculenta*) and banana (*Musa* spp.) in Phase 2, from 6,950 to 6,440 cal. years BP (Denham et al., 2003; Haberle et al., 2017; Lentfer and Denham, 2017) (Figure 5.4). If the longevity of this cultivation processes is correct, this would not only make the montane tropical forests, and their neighbouring grassland, an independent centre of agricultural practice, but also one of the oldest such centres in the world.

Nevertheless, as even Golson himself has acknowledged (Golson, 1991; Hope and Golson, 1995), the direct evidence for plant cultivation or domestication *c.* 10,220 to 9,910 cal. years BP is largely lacking, with such assertions being based on analogy with more recent prehistoric, and ethnographic, practice and indirect assumptions based on ditch structures (Smith, 1999). Detailed evaluation of the course and morphology of the palaeochannel from this phase suggests that it may not, in fact, be artificial (Denham, 2004). Similarly, while the presence of stake holes and artefacts from the associated palaeosurface demonstrate human activity, and there is a presence of *Musa* spp. phytoliths in an environment in which they would not usually appear, these cannot be unequivocally linked with agricultural practice and it is not known if the bananas represented were edible or how they dispersed to the site (Wilson, 1985; Denham, 2004; Lentfer and Denham, 2017) (Figure 5.4). Furthermore, there is limited evidence from elsewhere in the New Guinea Highlands that would imply the presence of agriculture at this time (though see Lentfer and Denham, 2017 for Middle and Late Holocene examples). No other wetland sites have yielded evidence for houses or the construction of drainage ditches from this early period, and lithic and faunal assemblages from nearby rock shelters and caves, such as Kiowa, demonstrate a continuity of Pleistocene foraging strategies rather than a clear transition to agriculture (Bulmer, 1966; White, 1972; Mountain, 1991; Gaffney et al., 2015; Golson et al., 2017; Roberts et al., 2017d).

On the other hand, the Terminal Pleistocene and Early Holocene were clearly periods of increasing anthropogenic alteration of New Guinea's montane rainforest (Figure 5.5). Major disturbance and burning events have been shown across a number of large valleys in Highland New Guinea in the form of long-term primary forest decline and increasing secondary forest and grassland in association with growing charcoal frequencies (Haberle et al., 1991; Haberle, 1994; Hope and Golson, 1995; Denham, 2004; Haberle et al., 2017) (Figure 5.5). At Kuk Swamp itself, Haberle notes a clear increase in burning and the development of a grassland, swamp forest, and secondary tropical forest mosaic driven by

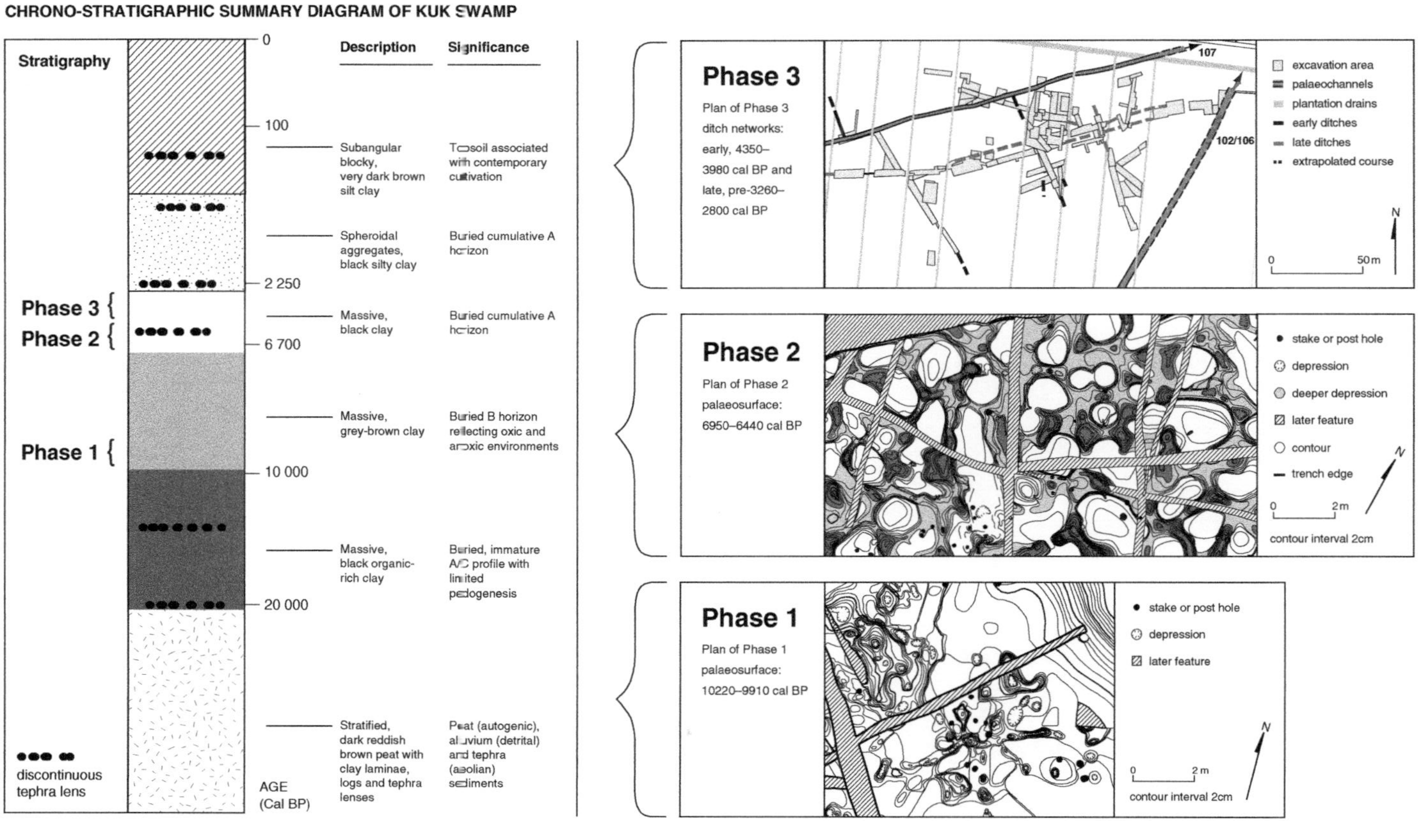

Figure 5.4. Diagram of palaeochannels and landscape uses during Phases 1, 2, and 3 of Kuk Swamp. Only the most relevant prominent features are shown. Reproduced and modified with permission from Denham et al. (2003). With thanks to Hans Sell.

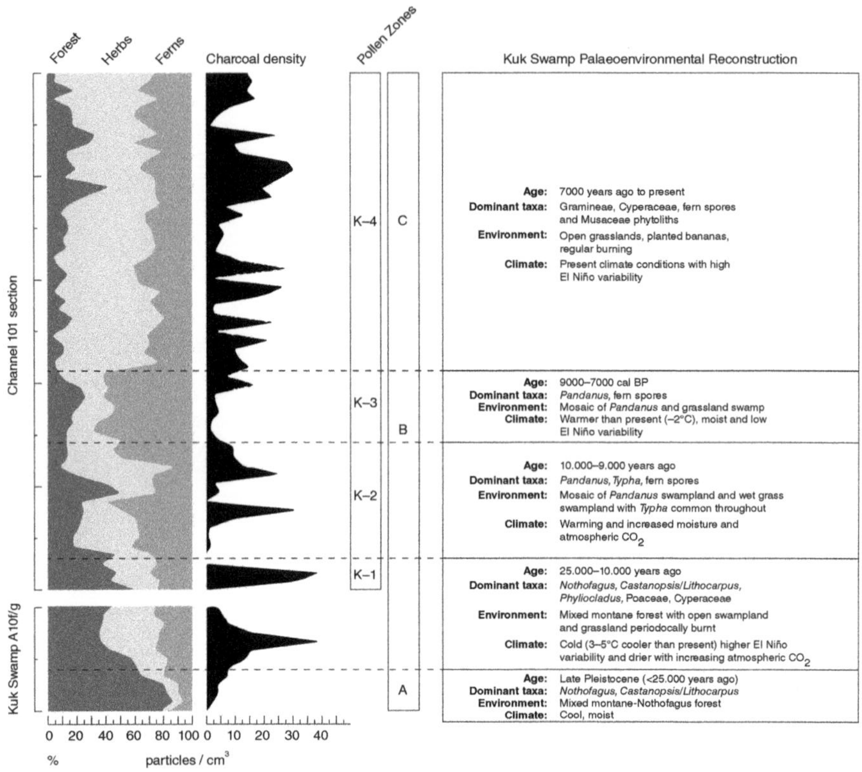

Figure 5.5. Terminal Pleistocene to Late Holocene composite pollen diagram of archaeological and stratigraphic contexts at Kuk Swamp, primarily of Early and Middle Holocene age. The diagram shows selected pollen, spore, and phytolith taxa plotted according to sample age and association with agricultural phases/features. Charcoal density refers to microscopic charcoal particle concentration from pollen samples. Four significant pollen zones are recognized, K-1 to K-4. Drawn by Simon Haberle. Reproduced with permission from Haberle et al. (2017). With thanks to Hans Sell.

anthropogenic firing (Haberle et al., 2010, 2017). Hints of permanent Terminal Pleistocene settlements at NFX (Watson and Cole, 1977) and Wañelek (Bulmer, 1977, 1991), and the increasing focus of populations on plant use from the Early Holocene, have been used to argue that such increasing anthropogenic influence is linked to crop production (Denham, 2004; Lentfer and Denham, 2017). Genetic evidence has also shown that taro, the greater yam, and bananas were likely independently domesticated in Melanesia (Lebot et al., 1998, 2004; Perrier et al., 2011; Chaïr et al., 2016; Arnau et al., 2017), albeit of unknown timing. It has been suggested that such plants, as well as sugarcane, would have been at their ecological limit in the highlands and their persistence there through the Terminal Pleistocene and Holocene transition would have necessitated human intervention (Haberle, 1995; Lentfer and Denham, 2017).

In Phase 2 (6,950 to 6,400 cal. years BP) and Phase 3 (4,350 to 3,980 cal. years BP) clear evidence for anthropogenic drainage and ditched cultivation of taro and bananas is present at Kuk Swamp (Denham et al., 2003) (Figures 5.4 and 5.5). In these later periods sugarcane and yams were also managed (Denham et al., 2003; Denham, 2011). Given that bananas originate in forest settings, taro and yams grow in more open forest floors, and sugarcane prefers open conditions, it seems probable that these practices originated in the maintenance of forest-grassland ecotones from 10,220 to 9,910 cal. years BP. On balance, it appears that the occupants of Kuk were practising varied strategies to enhance plant food production from the earliest Holocene. Yet, while we can note the practice of cultivation from this point, and certainly from 6,950 to 6,400 cal. years BP, identifying the onset of domestication of any of these plant foods is more problematic. As will be seen, preservation conditions in tropical forest settings mean that the identification of organic, domesticated remains, particularly from earlier periods, is challenging. While starch granules and phytoliths may be preserved (as they are at Kuk; Fullagar et al., 2006) it is often impossible to distinguish wild and domesticated taxa using either of these methods. Modern genetic information can be used to construct a 'clock' of domestication but with error margins that are of little use on Holocene timescales. At Kuk Swamp, and in many tropical forest regions to be discussed, we must base discussions of the onset of 'agricultural practice' on human actions (e.g., cultivation and herding) rather than the domesticated results (Lentfer and Denham, 2017).

It is increasingly becoming consensus that the Early Holocene plant management strategies at Kuk Swamp represent a significant change and intensification of human relationships with certain plant taxa (Denham et al., 2017a, 2017b, 2017c; Golson et al., 2017; Lentfer and Denham, 2017). The next question then becomes: what was the reason for this change? Here it is worth focusing on the environmental context of these early forays into tropical agriculture. Located *c.* 1,500 metres above sea-level at the transition between montane rainforest and grassland, the habitats of Kuk Swamp could be drastically altered by climate change. As seen in Chapter 4, Late Pleistocene human occupation of the Ivane Valley in the New Guinea Highlands has been argued to be heavily influenced by temperature, with significant drops during the Last Glacial Maximum rendering many areas inhospitable and necessitating mobile foraging strategies (Gosden, 2010; Summerhayes et al., 2010, 2017). Even at the slightly lower altitudes of Kuk, permanent occupation and demographic increase would necessitate a strategy to buffer potential climate change, particularly during the climatic instability of the Terminal Pleistocene and Holocene boundary. One such strategy would be landscape modification and interference in plant lifecycles that preserved the productive growth of important carbohydrate-rich food sources (Haberle and Lusty 2000; Denham et al., 2003).

Climatic drivers associated with the Terminal Pleistocene–Early Holocene transition have also been a popular explanation for the origins of agricultural practice in the Near East (Richerson et al., 2001; Bar-Yosef, 2011). However, there have been few detailed, well-dated, 'on-site' palaeoenvironmental studies, directly associated with forager behaviour, in the New Guinea Highlands. Recent isotopic palaeoenvironmental reconstruction at the site of Kiowa, which is located at a similar ecological and altitudinal position to Kuk Swamp, demonstrated that tropical highland forest and open mosaics persisted, alongside a focus on the hunting of wild animals such as cuscus (*Phalanger carmelitae*), ringtail possums,

(*Pseudocheirus corinnae*/*P. cupreus*/*P. forbesi*), and macropods (*Dendrolagus*/*Thylogale*) (Gaffney et al., 2015), from 12–0.5 ka at this site (Roberts et al., 2017d). This implies that climate change may not have been a major driver of subsistence in the region. However, palaeoenvironmental records from elsewhere in the Central Highlands suggest sub-montane tropical forest encroachment under the warmer and wetter conditions of the Holocene (Haberle 1998, 2003; Haberle et al., 2001). As a result, reliance on taro, yam, and sugarcane at Kuk Swamp may have necessitated increasing drainage and forest clearance, as well as local cultivation of clusters, through periods of changing climate. Indeed, it is also possible that humans actively maintained the mosaics inhabited by animals hunted at Kiowa.

The emergence of cultivation at Kuk Swamp was the final stage of a diversity of stable, successful tropical forest foraging adaptations in the region that extend back to 45 ka (Summerhayes et al., 2010, 2017), through the Last Glacial Maximum (Pavlides, 2004; Leavesley, 2006) and into the Holocene (Denham et al., 2003, 2004). During the Terminal Pleistocene and Holocene there is evidence for more frequent burning of forests to stimulate plant growth (Fairbairn et al., 2006) and the deliberate translocation of animal protein between tropical forest islands (Allen et al., 1989; Gosden and Robertson, 1991). Instead of a dramatic, forced event, plant management at Kuk Swamp was part of long-term relationships between hunter-gatherer communities and tropical forest habitats that characterized this part of the world from the Late Pleistocene. Nevertheless, while this blurs traditional dichotomies of 'hunter-gatherers' and 'farmers' (Harris, 1989; Denham et al., 2003; Denham and Haberle, 2008; Denham, 2011), it is clear that deliberate landscape modifications and plant cultivation at Kuk represent the formation of new human relationships with tropical forests. More sedentary and/or growing populations from the Terminal Pleistocene and Early Holocene here, and in other parts of the world, required increasing local control of resource availability (Lentfer and Denham, 2017). As part of this process, plants and animals were to become a permanent fixture of human domestic life with social energy, politics, and economic networks all being invested in their reproduction and sustenance.

INDIGENOUS TROPICAL FOREST AGRICULTURES

From the Early Holocene onwards, similar examples of agricultural practice, mainly in the form of cultivation leading to domestication, can be seen across the tropics. The diversity of agricultural practices, developed in the different tropical forest formations highlighted in Chapter 2, demonstrates that humans in these regions had an intimate knowledge of and connection with those species that were becoming increasing parts of their cultural lives and economic reliance.

Southeast Asia, Melanesia, and Australasia

At present there are no other Early or Middle Holocene records of plant cultivation in New Guinea that compare to the dataset available from Kuk Swamp, although rectilinear ditch networks appear in the Wahgi Valley and at Tambul 4,000 years ago (Golson, 1996; Haberle et al., 2017). Evidence from the Late Holocene demonstrates that Highland New Guinea groups continued to practise the deliberate planting, vegetative propagation, and drainage witnessed at Kuk Swamp (Denham, 2011). Based on areas of greatest genetic diversity and ancestral wild forms (Yen, 1985, 1991), it has been argued that a range of highland and lowland plants were domesticated in New Guinea between the Early and Late Holocene, including pandanus (*Pandanus julianettii*/iwen/brosimos complex), *Canarium* spp., and sago palm (*Metroxylon sagu*; Barrau 1955: 46). As mentioned above, bananas, greater yam, sugarcane, and taro have all been argued to have a genetic origin in New Guinea (Lebot et al., 1998, 2004; Malapa et al., 2005; Perrier et al., 2009). The lack of organic remains and ancient DNA has hindered the definitive identification and dating of the earliest examples of these 'domesticated' forms, though continued cultivation and human selection from the Early to Late Holocene is clear (Denham, 2011). In all these cases the altitudinal ecotone from lowland evergreen forest to montane rainforest and grassland provided an important resource to human foragers in New Guinea. This is clearly identified in the form of the combination of active drainage to open up areas for tuber propagation and active tree crop planting to maintain a horticultural 'mosaic' (Denham, 2011).

In contrast to New Guinea, neighbouring tropical Australia has often been framed as the 'continent of hunter-gatherers' (Lourandos, 1997) and there is very little evidence for indigenously domesticated plants or animals before colonial arrival. Nevertheless, many ethnographic records report plant selection strategies akin to cultivation documented in New Guinea, albeit perhaps less obvious. For example, phenotypic changes in tubers, including repeated harvests growing larger, have been shown to revert to wild type with no lasting genetic or phenotypic change once harvesting pressure relaxes (Denham, 2008). Similarly, there are many accounts of Aboriginal groups noting that viable portions of tubers will regrow if buried (Berndt and Berndt, 1993; Denham, 2011). Ethnographically documented burning has been shown to have the effect of maintaining mosaics of plant availability across different seasons (Hill and Baird, 2003). It has also been tentatively argued, on the basis of the distribution of botanically and genetically feral forms of taro, yam, and banana, that deliberate cultivation, connected to that seen in the New Guinea Highlands when Sahul existed as a single landmass, was practised in tropical northern Australia during the Early Holocene (Denham et al., 2009). These strategies were then abandoned at some point prior to the arrival of British explorers and colonizers (Jones and Meehan, 1989; Denham et al., 2009). Definitive proof of domesticated plants in this part of Australia in the Early and Middle Holocene remains to be found, however.

Island Southeast Asia has also traditionally been argued to be bereft of agriculture until the arrival of Austronesian speaking populations in the Late Holocene (Bellwood, 2005). Yet there are a number of documented cases of the dispersal of cultivated plants from New Guinea into the region between the Early and Late Holocene (Denham, 2011). Swamp sago palms were spread by humans from New Guinea and the Moluccas across the Indonesian and Malaysian archipelagos to the most eastern parts of India (Gangwar and Ramakrishnan, 1990; Flach, 1997; Kjær et al., 2004; Abbas et al., 2009; Blench, 2013) (Figure 5.6). The varieties in these regions have all been shown to be clones of one species (Kjær et al., 2004;

Figure 5.6. Photograph of a sago grove at Pa'Dalih, Borneo. Photograph taken by, and reproduced with permission from, Chris Hunt.

Abbas et al., 2009), despite phenotypic variation in starch production and colour (Kjær et al., 2004), implying deliberate selection (Denham, 2011). Similarly, paper mulberry can now be found across Island Southeast Asia and Oceania (Whistler and Elevitch, 2006; Seelenfreund et al., 2010). The Oceanian cultivars at least exhibit almost no genetic variability, likewise pointing towards deliberate vegetative dispersal by humans (Seelenfreund et al., 2011). Again, these practices seem to have focused on managing plant resources along open-closed gradients of tropical forest, and maintaining these gradients, in a similar way that is documented ethnographically in the region today (Brosius, 1986; Barton and Denham, 2011) (Figure 5.6).

Few domesticated animals have tropical forest origins but the domesticated water buffalo (*Bubalus bubalis*) may be one contender. Amazingly, more humans depend on the *c.* 130 million members of this species today than on any other domesticated animal (Scherf, 2000). This is likely due to the fact that its modern distribution centres on the densely-populated regions of South Asia, Southeast Asia, and China. Two extant types of domesticated water buffalo are thought to exist based on morphological and behavioural criteria (MacGregor, 1941; Barker, 2014). The river buffalo (*Bubalus arnee spp. bubalis*), found in South Asia, the Balkans, and the Mediterranean, prefers wallowing in deep water. The swamp buffalo (*Bubalus arnee spp. carabensis*) is found from east India, through Southeast Asia, to southern China, and prefers to wallow in swamp settings (Cockrill, 1974, 1977) (Figure 5.7). Both types feed primarily on aquatic plants and rely on waterlogged settings for thermoregulation. The environmental origins of these species are currently unknown but, on the basis of recent phylogenetic research, the swamp type appears to have originated in southern China or Southeast Asia

Figure 5.7. Photograph of swamp water buffalo (*Bubalus arnee* spp. *carabensi*) in eastern India. Taken by Neelsky and reproduced with permission from shutterstock.com.

c. 4,000 years ago, while the river type likely originated in South Asia *c.* 5,000 years ago (Yang et al., 2008). While more research, particularly with regards to the morphological identification of domestic forms is required, it is possible that the increasingly intensive management of these tropical animals formed part of tropical forest–swamp–river mosaics utilized by human populations in Southeast Asia during the Holocene.

South Asia

In the last two decades a wealth of archaeological data from archaeobotany, archaeozoology, and archaeological artefacts has enriched understandings of the development of agriculture in South Asia (Fuller, 2006a, 2011). A series of independent trajectories towards agricultural economies have now been documented across this region, including within the ecological zones of tropical India (Fuller, 2003, 2006b, 2007b; Southworth, 2005). Potential centres of agricultural origins in Gangetic India and Orissa may have seen the indigenous domestication of cucumbers (*Cucumis sativus*), snake gourd (*Trichosanthes cucumerina*), and possibly Indian rice (*Oryza sativa* ssp. *indica*), Kodo millet (*Paspalum scrobiculatum* L.), browntop millet (*Brachiaria ramosa* (L.) Stapf.), yellow foxtail millet (*Setaria pumila*), and pigeonpea (*Cajanus cajan*) along a gradient from rivers and lakes to tropical dry deciduous forest settings (Fuller, 2003, 2006a; Saraswat, 2004). Tuber crops, such as taro (*Colocasia esculenta*) and various yams (*Dioscorea* spp.) are significant elements of subsistence strategies in Orissa today and it is possible that wild forms of these taxa are native to Orissan tropical forests

(Fuller, 2006a). Likewise, wild types of the balsam apple (*Momordica dioica*) and bitter gourd (*Momordica charantia* L) are found in the forests of eastern India, where they may have been indigenously cultivated and cropped. Wild okra (*Abelmoschus ficulneus*) populations that contributed to modern domesticated forms may also have been subject to cultivation in this region (Fuller, 2006a).

The indigenous trajectory towards agriculture in the tropical forest mosaics of India is also supported by distinctive terminology for tree species, found in tropical dry deciduous forests and of economic importance in prehistory and today, in the South Indian Dravidian language family (Fuller, 2006a, 2007b, 2011; Asouti and Fuller, 2008). Flotation from the sites of Senuwar (Saraswat, 1992, 2004), Mhagara (Harvey et al., 2005), and Lahuradewara (Tewari et al., 2003, 2005) indicates the indigenous cultivation of rice and small millets, by populations primarily reliant on wild animals for protein (Joglekar, 2004), prior to the introduction of crops such as wheat and barley from Southwest Asia (Fuller, 2006a, 2006b). By the mid-third millennium BC, and potentially the mid-fourth millennium BC, there is indisputable evidence for sedentary, agricultural villages alongside rice-tempered pottery (Possehl and Rissman, 1992; Singh, 1997; Fuller, 2002). Nevertheless, beyond rice, and potentially associated millets, there is relatively little direct evidence for early cultivation in the Gangetic Plain and Orissa. Even in the case of rice, debate relating to the reliability of the morphological identification of Indian rice and a general paucity of direct, AMS radiocarbon dates, leaves trajectories of cultivation and domestication somewhat uncertain in this region. Nevertheless, it seems very likely that these two highly agriculturally fertile regions today represent instances of humans manipulating tropical forest–grassland–riverine mosaics to maximize resource potential from the Middle Holocene onwards.

South Asia is also home to perhaps the most reliable instance of animal domestication in a tropical forest setting. The domestic chicken (*Gallus gallus*) is descended from the red jungle fowl (*Gallus gallus*) with which it can still interbreed (International Chicken Polymorphism Map Consortium, 2004). Some genetic input into modern domesticates, at least in the form of the gene for yellow skin, has also occurred through hybridization with the grey jungle fowl (*Gallus sonneratii*) (Eriksson et al., 2008). It was previously argued on the basis of modern genetic data that a single domestication event occurred in Thailand (Fumihito et al., 1994). However, re-evaluation, with a much larger dataset, points towards multiple maternal origins for the domestic chicken. Indeed, the clade found in the Americas, Europe, and Africa appears to have originated from the Indian subcontinent, where a large number of unique haplotypes occur, and where the two wild species of jungle fowl are found today (Liu et al., 2006) (Figure 5.8). Ethnographic tropical forest hunter-gatherers commonly practise the hunting of land-bird species and it seems likely that, in the past, they were identified as a productive food source that could be effectively penned. Interestingly, while Sri Lanka is often neglected in discussions of chicken domestication, the long-term records of tropical rainforest use from 36,000 to 3,000 years ago mentioned in Chapter 4 all contain intensive use of jungle fowl, alongside primate populations, for subsistence as well as the manufacture of bone tools (Perera, 2010; Perera et al., 2011, 2016).

Sri Lanka has also hosted more controversial claims for agricultural origins in tropical forest settings. Based on a pollen record in the Horton Plains Highlands,

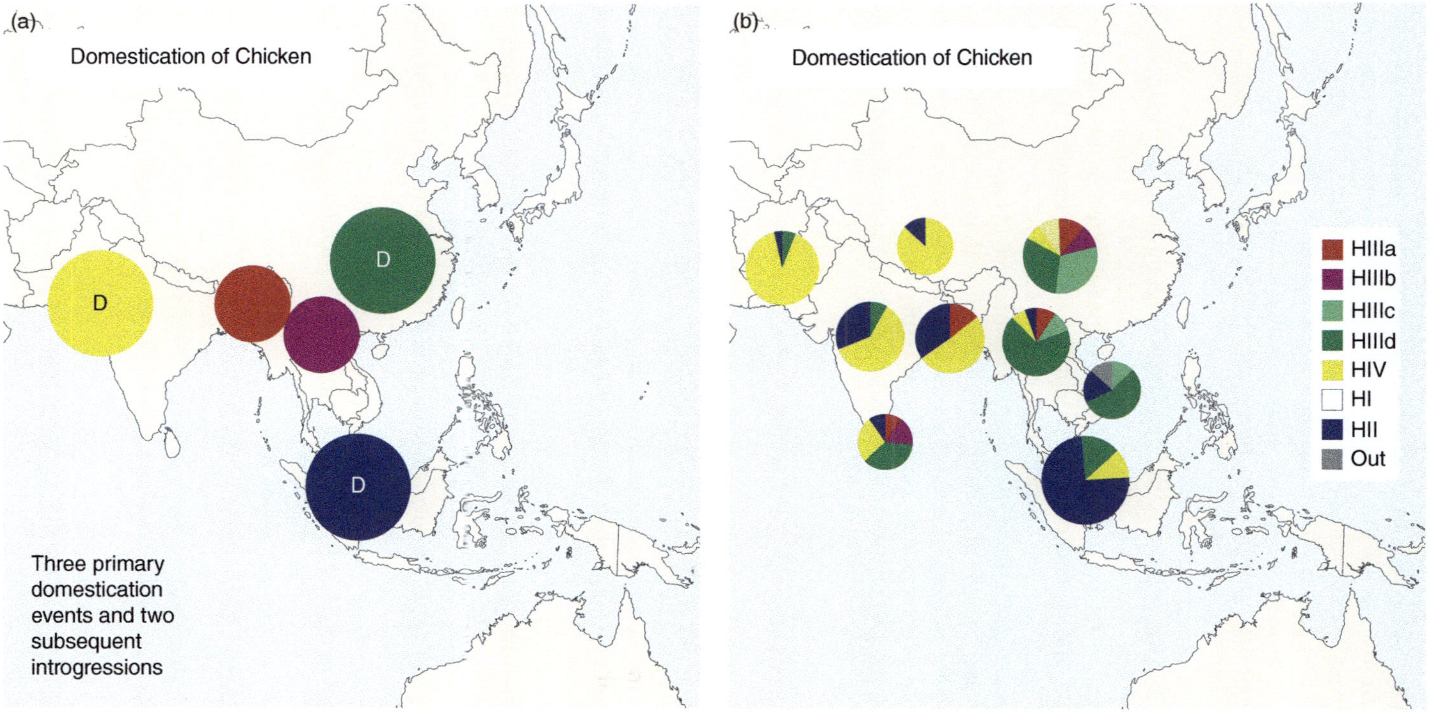

Figure 5.8. a) Three centres for domestication of chickens indicated by three dominant colours, green (Yellow River, China), yellow (South Asia), blue (Indonesia). b) Map of prominent mtDNA haplogroups across main centres of indigenous chicken domestication. Based on genetic data from Liu et al. (2006), Gongora et al. (2008), and Xiang et al. (2014). With thanks to Hans Sell.

Premathilake (2012) has argued that evidence for increased burning (*c.* 18,000–16,000 years ago) coincides with a period when the Wet Zone rainforest was expanding, implying anthropogenic influence. This has been linked to pollen characteristic of disturbance colonizers (e.g., *Macaranga* cf. *peltata* and *Ficus* sp.) and, alongside grass pollen with similarities to oats, rice, and millets, has been used to argue for early incipient cultivation (Premathilake, 2012). Premathilake has then claimed the presence of pollen grains with 'some' morphological similarities to Sri Lankan domesticated rice (cf. *O. sativa*) and millet (*P. miliaceum*) at *c.* 13,800 years ago (Premathilake, 2012). However, while the Horton Plains likely do record human influence, including burning to maintain habitat mosaics, evidence for cultivation and early domestication remains purely speculative based on uncertainties, and even impossibilities in the case of rice, in domesticate pollen identification. A similar scenario applies to the 'domesticated' banana phytoliths identified at Middle-Late Holocene levels at Fa-Hien lena (Premathilake, 2015). Although wild banana (*Musa* spp.) and jackfruit (*Artocarpus* sp.) phytoliths have been identified at the cave site of Batadomabalena *c.* 12,000 years ago (Perera et al., 2011), and wild banana seeds at Beli Lena *c.* 11,500–13,500 years ago (Kajale, 1989), suggesting they may have formed a part of hunter-gatherer subsistence, discussions of cultivation and domestication are not currently warranted.

Africa

As characterized by Marshall and Hildebrand (2002), Africa has often been seen as a continent of 'cattle before crops'. Indeed, plant-based agriculture, particularly that involving cereals, appears along the western fringes of the Sahara–Sahel border 3,000 years later than the first appearance of domesticated animals in the eastern Sahara (Manning, 2010). This open, arid location for the first African crops, and the lack of domesticable animals in tropical forest contexts, has meant that these ecotones have been neglected in discussions of the origins of agriculture in Africa. Linguistic evidence suggests that many of the cultivated plants found in the Benue-Congo languages of Nigeria, including pearl millet (*Pennisetum glaucum*), cowpea (*Vigna unguiculata*), and fonio (*Digitaria* sp.), have Bantu reflexes indicating a West African origin (Blench, 1996, 2007). Pearl millet is a drought-tolerant domesticate first identified on the southern fringes of the western Sahara at Dhar Tichitt *c.* 1,800 BC and the Tilemsi Valley, Mali by 2,000 BC (Amblard and Pernès, 1989; Manning, 2008; Manning et al., 2011). As will be seen below, its movement south of the Niger River to Birimi in northern Ghana in the early second millennium BC (D'Andrea et al., 2001), south of the River Niger and northern Ghana by the second millennium BC (D'Andrea et al., 2001), and into the Cameroonian rainforest by the mid-first millennium BC (Neumann et al., 2011) is generally linked to the expansion of Bantu farmers and climatically- or anthropogenically-driven tropical forest retreat (Willis et al., 2004; Bayon et al., 2012; Grollemund et al., 2015).

Fonio is today cultivated throughout much of West Africa though solid archaeobotanical evidence for its utilization appears relatively late. It has tentatively been identified dating to *c.* 850 BC at Kolima Sud Est (Takezawa and Cissé, 2004), though the most reliable early find comes from Cubalel in Senegal at AD 500 (Murray et al., 2007). Cowpea is one of the most important food legumes in the semi-arid tropics of Africa, being drought tolerant and shade resistant. Phylogenetic studies and linguistic evidence indicate tropical West Africa to be the point of origin of its domestication (Blench, 1996; Coulibaly et al., 2002) and the earliest domesticated cowpea currently dates to *c.* 1,700 BC in central Ghana only a few hundred years after the earliest confirmed domesticated pearl millet (D'Andrea et al., 2007). African rice (*Oryza galberrima*) is today grown only in West Africa with the earliest evidence of domesticated forms dating to AD 200 at the site of Jenné-Jeno on the Niger River (McIntosh and McIntosh, 1980). As with millet, the cultivation of these crops is linked to tropical savanna and river settings, although stable tropical forest biomes to the south, alongside expanding desert to the north, *c.* 4,000–2,000 years ago may have encouraged increased plant manipulation by large populations trapped between environmental extremes. Manning (2010) has argued that these indigenous cereal domestications should be affiliated with pastoral communities moving from the north during this period of climate change.

Although none of these taxa have an explicitly tropical forest origin, early finds of pearl millet at Kintampo in northern Ghana occur alongside oil palm, yams, and cowpea within a multi-cropping tropical forest package (D'Andrea et al., 2001, 2006). This evidence suggests potentially more complex ecological contexts for the emergence or, at least, the assimilation of these cereals in different parts of West Africa. Indigenous cultivation of various African plants has also been argued as occurring *within* the West and Central African rainforests, alongside pollen and micro-charcoal evidence for deliberate clearance of a diversity of tropical forest formations from the second millennium BC onwards. Forest yams constituted a stable food alongside groundnut (*Vigna subterranea*), castor bean (*Ricinus communis* L), and gourds (*Lagenaria siceraria*) (Fuller and Hildebrand, 2013). Tree-crops were also planted or at least tended, including the oil palm (*Elaeis guineensis*) and the raffia palm (*Raphia monbuttorum*) (Hamilton et al., 2016) (Figure 5.9). The banana (*Musa* spp.) may have been introduced into such indigenous tropical forest cultivation strategies from Southeast Asia as early as 1,000 BC, though the phytolith findings at Nkang, Cameroon remain debated (Neumann and Hildebrand, 2009). This all suggests that tropical forests may have played a significant role in early food production in Africa, though poor archaeological visibility of domesticated root and tree crops, and difficulty of excavation mean this process currently remains far less well understood than the later Bantu expansion.

Neotropics

It is archaeological investigations of Neotropical forests in central and southern America that have really shaped our understanding of forest manipulation and agricultural origins in tropical forest settings. It has long been argued that tropical forest agriculture began in the Neotropics long prior to colonial arrival (Harris, 1972), and that current patterns of tree species distribution may have resulted from past human management (Balée, 1988). However, it is only recently that solid archaeobotanical and chronological evidence, alongside DNA-based studies, has begun to provide a more detailed geographical and temporal picture of this

Figure 5.9. a) Oil palm (*Elaeis guineensis*) tree on the Tshuapa riverbanks, Democratic Republic of Congo and b) Oil palm macroremains from the upper layers of an excavation in the Democratic Republic of Congo. Photographs taken by, and reproduced with permission from, Barbara Eichhorn.

process. The major staple of the Americas, maize (*Zea mays*), with a total modern production surpassing that of wheat and rice (Franklin, 2013), is often considered an ideal arid-adapted food, appearing in *Interstellar* (Nolan, 2016) as humanity's last hope on a drying planet. However, genetic and archaeobotanical data have now demonstrated maize to have its origin in the lower, warmer, and moister habitats of Guerrero and Michoacán states of Mexico, where seasonal tropical forest is the native vegetation (Matsuoka et al., 2002; Doebley, 2004; van Heerwaarden et al., 2011; Piperno, 2011b). The earliest identified record of maize comes in the form of starch grains and phytoliths from Xihuatoxtla rock shelter in the Central Balsas River Valley (Piperno et al., 2009). This is broadly supported by molecular clock estimations determined from modern DNA for maize domestication between 12,000 and 6,400 years ago (Matsuoka et al., 2002) (Figure 5.10). Although the Xihuatoxtla data are indirectly dated, and the earliest directly dated cob comes from Guilá Naquitz, Oaxaca, *c.* 6,500 years ago (Piperno and Flannery, 2001), a dry tropical forest-grassland-riverine ecotone origin for this crucial global staple seems probable.

Early instances of maize have also been identified, on the basis of phytolith datasets, in humid tropical forest settings in Panama (*c.* 8,000 years ago),

Figure 5.10. Photographs of a) Xihuatoxtla rock shelter in the Balsas River Basin of Mexico and b) ground stone tools from the Xihuatoxtla rock shelter taken by, and reproduced with permission from, Dolores Piperno.

Colombia (*c.* 8,000–6,000 cal. years BP), Ecuador (*c.* 8,000–7,000 years ago), and now even Bolivia and French Guiana in the Amazon Basin (*c.* 5,000 cal. years BP) (Piperno, 2011b), though the indirect dating of these silica entities is problematic (Langlie et al., 2014) (Figure 5.11). That said, while direct AMS dates from cobs tend to give younger ages (e.g., 5,000 years ago for maize in Ecuador—Piperno et al., 2009; Piperno, 2011b), microbotanical data from Panama indicate the presence of maize *c.* 8,000 cal. years BP (Piperno, 2011b), while starch evidence has been used to suggest maize cultivation in the southern Caribbean at *c.* 7,790–7,670 cal. years BP (Pagán-Jiménez et al., 2015). Such an early appearance has been further supported by the discovery of macrobotanical and microbotanical evidence for maize at Huaca Prieta and Paredones in Peru *c.* 6,700 cal. years BP (Grobman et al., 2012). Early phytoliths of domesticated squash (possibly *Cucurbita argyrosperma*) have been identified at Xihuatoxtla *c.* 9,000 cal. years BP (Piperno et al., 2009), the Oaxaca Valley *c.* 8,000 cal. years BP, as well as in Panama (*Cucurbita moschata—c.* 8,500 years ago), Ecuador (*Cucurbuta ecuadorensis—c.* 10,000 years ago), and Colombia (*c.* 9,000–8,000 years ago) (Smith, 1999; Piperno and Stothert, 2003; Piperno, 2011b). Direct dates on squash (*Cucurbita moschata*) (10,402–10,253 cal. years BP), peanut (*Arachis* sp., 8,630–8,580 cal. years BP), coca (*Erythroxylum novagranatense*, 7,950 cal. years BP), and cotton (*Gossypium* sp., 6,301–6,133 cal. years BP), alongside finely-ground stone hoes, also reveal early cultivation in the humid Zaña Valley of northern Peru (Dillehay et al., 2007; Piperno and Dillehay, 2008).

In addition to the well-known staples of maize and squash, some of the earliest crops in Central and South America include tree crops and tubers such as arrowroot (*Maranta arundinacea*), leren (*Calathea allouia*), peanuts (*Arachis* sp.), avocado (*Persea americana*), and pacay (*Inga fuellei*), indicating selective management of a mixture of particularly productive seed, tree, and root crop tropical forest plants (Piperno, 2011b). As in New Guinea and Southeast Asia the systematic human utilization of these plant foods has broadened archaeological perspectives beyond Eurocentric cereal- and herding-based agricultural narratives.

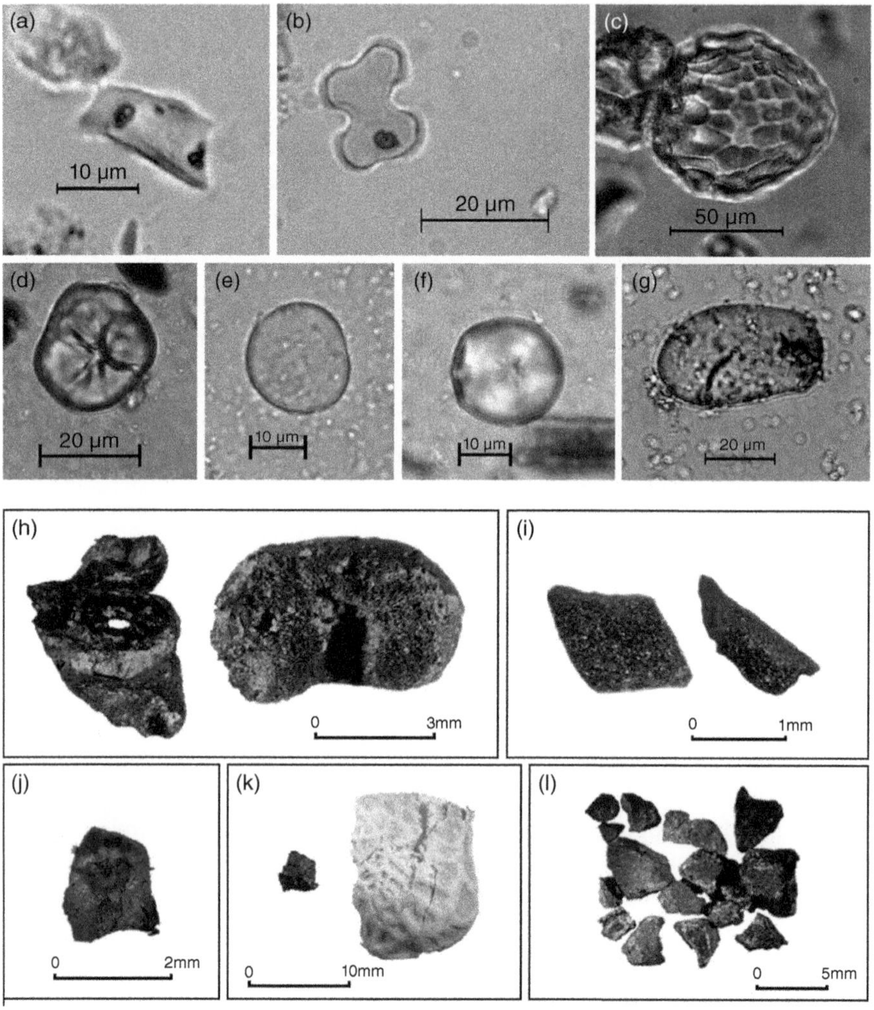

Figure 5.11. Selection of botanical remains recovered from raised agricultural fields and associated domestic sites in South America. a) Wavy-top roundel phytolith from maize cob (*Zea mays*) from site K-VIII, French Guiana, b) Cross-shaped phytolith Variant 1 from the maize leaf (*Zea mays*) from the Bois Diable site, French Guiana, c) Scalloped sphere derived from squash rind (*Cucurbita* spp.), Loma Salvatierra, Bolivia, d) Maize starch grain (*Zea mays*), Loma Salvatierra, e) Chilli pepper starch grain (*Capsicum* sp.), Loma Salvatierra, f) Manioc starch grain (Manihot esculenta), Loma Salvatierra, g) Yam starch grain (Dioscorea sp.), Loma Salvatierra, h) Fragment of maize cob and grain (*Zea mays*), Loma Salvatierra, i) Fragment of squash rind (*Cucurbita* spp.), Loma Salvatierra, j) Cotton seed base (*Gossypium* sp.), Loma Salvatierra, k) Fossil and modern fragments of peanut (*Arachis* sp.), Loma Salvatierra, l) Fragments of palm endocarp (Arecaceae) Loma Salvatierra. Photographs all taken by, and reproduced with permission from, José Iriarte.

Figure 5.12. Photographs of a) a manioc (*Manihot esculenta*) plot in the Versalles Itonama Indigenous community, Itenez River, Bolivia and b) silicified secretory cells from manioc (*Manihot esculenta*) Triunfo Amazonian Dark Earth, Itenenz, Bolivia. Photographs taken by, and reproduced with permission from, José Iriarte.

Indeed, in contrast to previous suggestions that South American rainforest soils and dense forests could not support dense agricultural communities (Meggers, 1971, 1977, 1987), the Amazon Basin has also emerged as a centre of agricultural origins and large sedentary human populations. Here, manioc (*Manihot esculenta*), a woody shrub that is today extensively cultivated for its starchy tuberous root, was domesticated at some point during the Early and Middle Holocene, no more than 10,000 years ago (Olsen and Schaal, 1999, 2006; Olsen, 2002) (Figure 5.12). Chilli peppers (*Capsicum baccatum*) were probably also domesticated just south of the Amazon (Perry et al., 2007), and have been identified in southwestern Trinidad *c.* 7,790–7,670 cal. years BP (Pagán-Jiménez et al., 2015), while the aforementioned leren and peanuts also seem to have been domesticated in this region (Kochert et al., 1996). More recently, the domestication of wild rice has been documented *c.* 4,000 cal. years BP at the Monte Castelo shell mound in the Brazilian Amazon (Hilbert et al., 2017).

The globally recognized pineapple (*Ananas comosus*) is also an Amazonian domesticate, alongside a number of species of palm that are valued for their fruits (Pickersgill, 1976; Clement, 2006; Clement et al., 2010). Of these the peach palm (*Bactris gasipaes*) was likely one of the earliest domesticates in Amazonia (Clement et al., 2009; Galluzzi et al., 2015), and its starchy fruits are cooked and seeds made into flour in many regions of the Amazon rainforest today (Clement, 1988, 2006; Clement et al., 2010). Cacao, a firm favourite with chocolate lovers, may also have been domesticated, or semi-domesticated, in the Amazon Basin (Clement et al., 2010), and certainly was domesticated and used in a variety of products in Mesoamerica by 3,800 cal. years BP (Powis et al., 2011). In this, and other cases, the exact date and geographical location of Neotropic domesticates, and particularly those exploited for tubers and fruits in the Amazon Basin, have often been difficult to pin down due to issues of identifying domestic forms and the lack of preservation of organic remains. Furthermore, while we do have a better picture of prehistoric plant management practices in this part of the world than in many tropical forest regions, key questions also remain in terms of management practices, the scale of land-use, and the diversity of pre-Columbian resource use, settlement, and geomorphological modification across the largest river basin in the world.

The growth of manioc in the interior of the Amazon Basin, and maize when it was introduced, has tended to be associated with mobile, shifting, or 'slash and burn' cultivation—with patches of forest being burned to make clearings for crop growth prior to being allowed to regenerate (Heckenberger et al., 2003, 2007). This perception has been challenged, however, and pre-Columbian agricultural systems have been more recently characterized as polyculture agroforestry through which 'gardens' were intercollated with managed forests in different stages of succession (McMichael et al., 2012; Neves, 2013). In areas of modern-day grassland on the edge of the Amazon Basin, the growth of sweet potato and manioc has been linked to labour-intensive engineered ditch systems, such as those seen in the humid lowlands of Bolivia at the Llanos de Moxos (Lombardo and Prümers, 2010) and the Guyanas (Rostain, 2014). Iriarte et al. (2012) have argued that these raised field systems allowed for fire-free land-use and forest clearance, with significant biomass burning only occurring later. In the western Amazon, by contrast, earthwork construction has been linked to ritual gatherings and village site placement (Schaan et al., 2013). Regardless, the ubiquity of human investment in modifying tropical forest landscapes to promote food production in the Amazon Basin is increasingly apparent. In particular, the formation of fertile anthropic soils known as *terras pretas* and *terras mulatas* has left a visible legacy across the region, from the east coast of South America to well into the western Amazon, from the first millennium BC (Neves et al., 2001; Woods et al., 2009; Arroyo-Kalin, 2010, 2012).

Recent ecological and genetic datasets are also demonstrating that the distribution of trees in the Amazon rainforest today is strongly linked to human selection (Steege et al., 2013). For example, stands of Amazon (or Brazil) nut (*Bertholletia excelsa*) have been shown to closely map onto ancient human settlements (Shepard and Ramirez, 2011; Thomas et al., 2015). While this did not necessarily lead to 'domestication' in the genetic sense it certainly represents clear, deliberate human management akin to 'cultivation', occurring alongside the more traditional ideas of domesticated crops, agricultural clearings, and raised earthworks discussed above. Finally, as will be seen in Chapter 6, settled Amazonian communities even 'farmed' turtles in corrals placed within the rivers on the banks of which they resided and relied upon for soil fertility (Hemming, 2009). In sum, while much work remains to be done in terms of dating and locating the origins of different domesticated crops in the tropical forests of the Americas, particularly within the Amazon, it is clear that these habitats have provided some of the major foodstuffs in the world today. They also document millennia of human 'cultivation-like' practices that have left a lasting impression on the natural heritage of the region. The extent of these behaviours across the Amazon rainforest in particular is now supporting calls for this once supposedly 'pristine' environment to be part of the push to place the 'Anthropocene' era long before the industrial revolution and nuclear bomb-testing (Roosevelt, 2013; Lewis and Maslin, 2015; McMichael et al., 2017) (Chapter 8).

INVASIVE AGRICULTURES

The above discussion has focused on domesticates and agricultural practices that originated within tropical forest ecosystems. However, significant cultural and environmental changes have also taken place where crops and animals domesticated elsewhere were moved into these habitats, either through processes of migration or exchange. These introductions often coincided with environmental or climatic change or significant human alteration of the existing forest landscape through clearance and burning. In Central Africa, mature evergreen trees were rapidly being replaced by dry forests and grasslands during the first millennium BC (Maley and Brenac, 1998; Maley, 2002; Ngomanda et al., 2009; Garcin et al., 2018), at the same time as Bantu-speaking peoples migrated southwards from the Nigeria–Cameroon border (Grollemund et al., 2015). Evidence for early Iron Age farming in Cameroon indicates the incorporation of pearl millet into indigenous cultivation strategies including oil palm, and likely also yams and plantains (Casey, 2013; de Maret, 2013; Fuller and Hildebrand, 2013). Evidence for early Iron Age farming south of Cameroon currently remains limited, however (though see Neumann et al., 2011; Livingstone Smith et al., 2017). Genetic analysis of forest hunter-gatherers, practising indigenous tree-cropping, shows introgression with these new, populous communities (Verdu et al., 2009). Local and regional records, including a near-shore marine core, demonstrate significant, unprecedented human land-use intensification at this time in the form of soil erosion, charcoal production, and deforestation linked to increasingly dense, sedentary populations using pearl millet in more open, drier conditions (Willis et al., 2004; Ngomanda et al., 2009; Bayon et al., 2012; Garcin et al., 2018).

One of the better-documented tropical agricultural expansions is that associated with the expansion of Austronesian language families from Taiwan *c.* 4,000 BC throughout much of Southeast Asia and the Pacific (Bellwood, 1993, 2005). The spread of rice from China and Taiwan into the tropical forests of this region has been notoriously difficult to study as a result of major preservation issues and excavation practices (Castillo, 2011). At present, the earliest rice in Mainland Southeast Asia comes from coastal Thailand *c.* 2,000–1,500 BC (Thompson, 1996), but it is reliably dated in the interior to only 1,000 BC, and after the earliest evidence for the cultivation of millet (Weber et al., 2010). In Island Southeast Asia rice husks have been dated in pottery to *c.* 2,200 BC in Gua Sireh, Sarawak, Borneo (Beavitt et al., 1996). Rice also seemingly made it to Luzon in the Philippines by 1,400 BC (Snow et al., 1986) where it eventually led to forest clearance and the construction of the well-known Banaue terraces (Acabado, 2010). The arrival of rice and millet agriculture in the tropical forests of Southeast Asia is generally associated with large-scale forest clearance, particularly within the more deciduous forests to the north of the equatorial belt in Mainland Southeast Asia which would have been easier to burn (Bellwood, 1993) (Figure 5.13). However, in some cases early rice appears to be combined with local cultivation of yams (*Dioscorea alata*) and taro (*Colocasia esculenta*) (Paz, 2002) and some forest hunter-gatherers also initially resisted its adoption (Krigbaum, 2003; Barton and Denham, 2011).

Rice, alongside millet, is absent in the Pacific from Indonesia onwards, countering traditional ideas relating to the inexorable expansion of cereal agriculture. More prolific domesticated expansions come in the form of the pig across Southeast Asia and the Pacific. The wide distribution of native wild pigs in Southeast Asia makes pinpointing domestication difficult (Larson et al., 2007, 2010). A domesticated pig appears to be present on Luzon *c.* 2,000 BC (Piper et al., 2009), prior to the main influx of the so-called Pacific pig throughout Oceania, perhaps representing a local domestication (Bellwood, 2011; Larson, 2011). The

Figure 5.13. Photograph of a rice padi field in Bario, Borneo taken by, and reproduced with permission from, Chris Hunt.

latter 'Pacific' clade went on to expand from Southeast Asia out across Near and Outer Oceania (Larson et al., 2007). Similarly, the domesticated chicken, either from South Asia or from a separate, local domestication, also became part of the wider expansion of populations, ancestral to Polynesians, carrying Lapita style pottery, from the Bismarck Archipelago to Hawaii, Easter Island, and New Zealand (1,600 BC to 500 BC) (Kirch, 2000, 2010). Interestingly, these populations added these domesticated animals to a reliance on root crops and tubers, including taro, yams, bananas, and breadfruit, domesticated in the tropical forests of Melanesia and Southeast Asia. In some instances, such as the Chatham Islands, the introduction of these crops was combined with a management of local forest tree crops (Maxwell et al., 2016). This suggests an intimate knowledge of different island environments that perhaps belies a rapid, destructive agricultural expansion.

Nevertheless, in contrast to the apparently highly-sustainable cultivation of tuber and tree crops in Melanesia and Southeast Asia for millennia, the agricultural practices of Lapita communities appear to have led to significant landscape changes in many cases. For example, on Tonga, tree species declined in abundance following Polynesian colonization (Fall, 2010). This could be related to the disturbance by large-scale populations and the use of vegetation for structures, boats, and animal pens (Kirch and Ellison, 1994; Spriggs, 1997). Many alluvial plains in the Pacific islands also show evidence for human-accelerated soil erosion (Spriggs, 1997). Transported pigs may also have played a role (Kirch, 2001) and feral pigs in Hawaii have been noted to reduce the abundance of native plant species, enhance conditions for the establishment of invasive plant populations, negatively impact native forest bird species, and also increase soil erosion (Nogueira-Filho et al., 2009). However, perhaps the main agent of these changes was the commensal species, the Pacific rat (*Rattus exulans*) that inadvertently accompanied, or was deliberately transported by, these agricultural groups. The consumption of eggs and tree seeds appears to have had major impacts on local tropical environments, and avian extinctions after Lapita colonization (but prior to European arrival) numbered in the hundreds if not thousands of species (Steadman, 2006; Swift et al., 2017a, 2017b).

A similar process is also seen in the Caribbean. Two initial waves of colonization at 4,000–2,000 BC in Cuba, Haiti, the Dominican Republic, and the Lesser Antilles and 2,000–500 BC in Puerto Rico and once more in the Lesser Antilles, appear to have been primarily driven by hunter-gatherers who made use of shellfish, wild plants, and hunted animals (Wilson, 1999, 2007) (Figure 5.14). It was, however, during two later waves of occupation, by Saladoid horticulturalists (500 BC–AD 600) and Taíno chiefdoms (AD 600–1200), that communities began to introduce a series of new plant and animal resources to these islands (Fitzpatrick and Keegan, 2007; Wilson, 2007). Agriculturalists brought with them wild avocado (*Persea americana*), manioc (*Manihot esculenta*), as well as dog (*Canis lupus familiaris*) and guinea pig (*Cavia porcellus*) (Fitzpatrick and Keegan, 2007). On the island of Carriacou, for example, five species were introduced between 1,000 and 1,400 years ago including peccaries, armadillos, guinea pigs, opossum, and agoutis (Giovas et al., 2012). While some have suggested that the earliest occupants forced a number of hunted animal taxa to extinction (Wilson, 2007), this largely remains unverified. Fitzpatrick and Keegan (2007) argue that it is only during the arrival of the later horticultural and agricultural groups that observable changes to island environments occurred as a result of forest clearance, over-exploitation of terrestrial and marine resources, increasing populations, and the introduction of commensal taxa.

Similar alterations concordant with agricultural 'invasion' can be seen in the Wet Zone rainforests of Sri Lanka where a greater than 30,000 year old record of tropical forest foraging only changes with the introduction of millet and the arrival of iron technology 3,000 years ago (Deraniyagala, 1992; Roberts et al., 2015a). Across the tropical regions of India, incoming cattle, sheep, goats, wheat, barley, and pulses also led to large-scale clearance along the tropical forest regions of the western Ghats and Orissa that continued during the colonial period (Tian et al., 2014) and is only intensifying today (Joshi and Singh, 2003). The impact of large, grazing animals on tropical forest landscapes can also be seen following the relatively recent introduction of cattle into the tropical forest zones of North, Central, and South America. In both areas, the requirements of pastureland have led to unsustainable burning, deforestation, and soil erosion (Nicholson et al., 1995; Geist and Lambin, 2002). Meanwhile, the grazing of cattle hinders the regrowth of tropical forests, breaking managed succession cycles that had persisted for millennia. Indeed, this remains one of the primary conservation issues in the Amazon today, as will be seen in Chapter 8 (Hecht et al., 1988; Hecht, 1992; Veiga et al., 2004; de Sy et al., 2015). Similarly the introduction of plantations of sugarcane, soybean, and the monoculture cropping of rubber trees (Hemming, 2009), has led to deforestation and a huge reduction in plant diversity within the Amazon rainforest in a pattern increasingly witnessed across the global tropics in the 21st century (Ghazoul and Sheil, 2010).

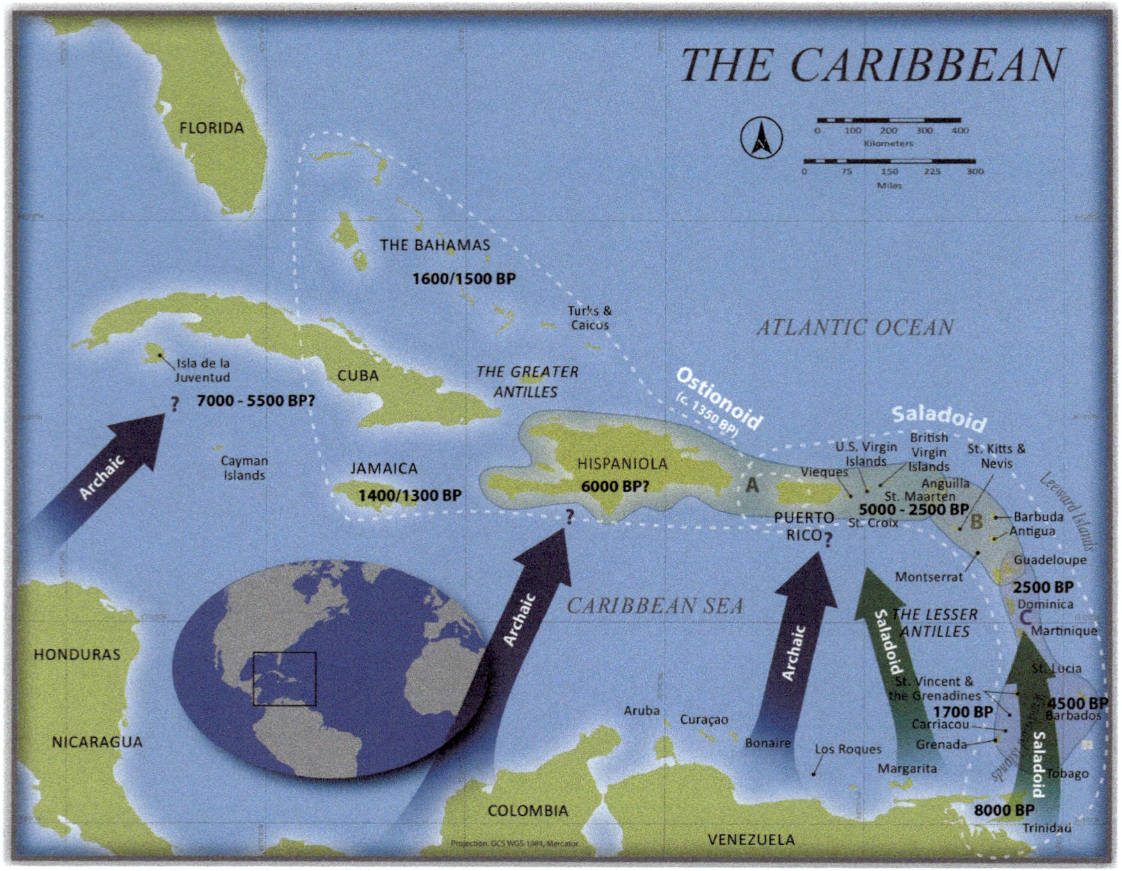

Figure 5.14. Map of the Caribbean showing major prehistoric population dispersals and some of the ceramic style zones referred to in Fitzpatrick (2015). Dates in bold (in calendar years before present) indicate earliest known dates or date ranges for the settlement of particular areas or islands in the region. Created by, and used with permission from, Scott M. Fitzpatrick.

TROPICAL AGRICULTURES: A COUNTERPOINT TO TRADITION

The above review demonstrates the growing wealth of knowledge that is now available relating to the emergence of indigenous tropical forest agricultural 'practices' during the Holocene period. From being considered 'pristine' wildernesses untouched by farming communities, or even human impacts, prior to the colonial period it is now becoming clear that humans cultivated tropical forest plants, and potentially herded or kept animals in these habitats, for many millennia, leading to a series of domestications, across a series of different tropical forest regions and formations. While poor organic preservation, a reliance on phytolith and pollen evidence that is often not specific, and issues of chronology mean that the exact timing of many of these domestication events remains lacking, it is clear that the Holocene witnessed the intensification of tropical forest plant and animal management that began during the Late Pleistocene. It is also evident that the indigenous agricultural developments were often very different, and had very different impacts, to agricultural practices introduced from different biomes and cultures. Together, these data have had three major ramifications for studies of agriculture in archaeology: these relate to the definition, causation, and sustainability of prehistoric agricultural practices, not just in tropical forests but worldwide.

Definition

Firstly, agricultural definitions focused on the level of contribution of domesticated plants and animals to human subsistence (Price and Bar-Yosef, 2011) miss the long-term human management (including cultivation and herding) that led to domestication. Furthermore, they dichotomize hunter-gatherers and farmers, culturally and economically, when, in many cases, agricultural practices and the use of domesticates continue alongside existing hunting and gathering strategies (e.g., Harris, 1972; Roberts et al., 2017d). As will be seen in Chapter 6, it is certainly the case that societies reliant on domesticated resources undergo demographic and cultural shifts that distinguish them from more mobile horticulturalists and communities complementing the foraging of wild foods with domesticated resources. However, in many cases, sedentary tropical forest farmers also continued to make use of the 'gardens' and natural dynamics of the forests and their wild resources to live within these environments. Tropical forests force us to reassess what our definitions of 'agriculture' mean and to view long-term and diverse 'agricultural practices', including cultivation and herding but also tree-cropping and penning, as crucial pathways towards the final domesticated outcome. This work is now influencing studies of agriculture in traditional zones such as the Near East and Europe (Langlie et al., 2014; Fuller and Lucas, 2017), which are showing more diverse geographical and temporal origins, including often gradual and discontinuous development, of agricultural subsistence and farming communities.

Causation

The diverse environmental origins of agricultural practices within tropical forests also provide insights into questions relating to the causes of increasingly intensive human management of plant and animal resources. Firstly, monocausal explanations for the origins of agriculture are clearly inappropriate. For example, while rapid climate forcing may have extreme impacts on arid grassland systems, as hypothesized in the Middle East (Bar-Yosef and Belfer-Cohen, 1989; Richerson et al., 2001), or in high-altitude tropical forest areas such as in the case of Kuk Swamp, in humid lowland tropical forests, such as in Amazonia or the Balsas River Valley of Mexico, any effects are likely to be dampened or more drawn out. Similarly, while increases in population density may have been behind the adoption of pearl millet, fonio, and cowpea cultivation in the space between desert and tropical rainforest in West Africa, such growth is unlikely, and undocumented in the region of Kuk Swamp during early phases of ditch creation or prior to the adoption of rice agriculture in Southeast Asia. The fact that tropical forests can even be identified as centres of origins for agriculture practices, alongside environments such as the 'Fertile Crescent' of the Near East and the temperate river valleys of northern America, suggests that monocausal explanations are always likely to be insufficient.

This is even more the case when we consider the level of variation within 'tropical forest' formations and the subsequent impacts on indigenous domestication. It may even be that it was this variety that was crucial to the domestication process in tropical forests. One apparent commonality between all of the cases of indigenous tropical forest domestications, at least with regards to plants, is ecological mosaics. Whether it is an altitudinal gradient from montane rainforest to high-altitude grassland, or lowland gradients between river settings, savanna, dry deciduous forests, and wet rainforests, indigenous cultivation and tree-cropping appear to occur as part of diverse management of various environments, with diverse corresponding resources. Active maintenance and manipulation of such mosaics is common practice among ethnographic communities inhabiting tropical forests today (Chapter 2), and it is clear in some of the earliest examples of tropical forest management in Southeast Asia and Melanesia. In these mosaics, while rapid, short-term impacts of climate change or population growth would likely have been buffered by ecological diversity, gradual, long-term changes in the balance of different habitats would have increasingly threatened the delicate balance of subsistence and required more intensive manipulation, including translocation, burning, and irrigation, that led, eventually, to domestication.

Sustainability

It is also telling in this regard that the maintenance and occupation of these mosaic habitats facilitated sustainable agricultural practices for thousands of years, without any obviously adverse effects. It is only with the development of large, sedentary communities, often with monumental construction or established political elites, and increasing reliance on a few stable crops that some tropical forest farmers began to push the boundaries of sustainability and abandon mosaic habitats. However, even then, as will be seen in Chapter 6, many urban groups continued to function through the diverse use of forest products and resources, often in the form of the maintenance of forest 'gardens' through which natural plant diversity was maintained. By contrast, the 'invasion' of agricultural forms from outside often led to rapid landscape clearance and homogeneity, with massive impacts on soil erosion and the regional diversity of plant and animal taxa. Indeed, these 'invasions' often seem to be stimulated by large-scale climate-change impacts, such as in the case of the opening of savanna corridors in Central Africa (Grollemund et al., 2015; Garcin et al., 2018), or population migration and transmission stimulated by demographic growth, as is perhaps the case in the context of the Lapita expansion into the Pacific (Kirch, 2000, 2001, 2010). This significant conversion continued apace during the colonial period and industrial monoculture and extensive cattle ranching have introduced rampant clearance, biodiversity reduction, soil erosion, and increased susceptibility to wild fires, some of the greatest dangers facing tropical forests today.

6

'Ruins' of the Forest

Social Complexity and Tropical Cities

> If we may find a city, where one was considered impossible to exist, it may well write a whole new chapter in human history.
>
> —Charlie Hunnam as Colonel Percival Fawcett in *Lost City of Z*, James Gray (2016)

The above quote from a recent Hollywood film presentation of Colonel Percival Fawcett's obsessive early twentieth-century search for the remains of the Lost City of Z (Gray, 2016) highlights the effort that it has taken to convince the academic world and the public alike that large urban forms can be developed in tropical forest settings. While the film, and the book by David Grann (2009) upon which it was based, grossly overplay the exploration credentials, respect for Indigenous peoples, and scientific abilities of Colonel Fawcett (Hemming, 2017), this quote encapsulates the difficult working conditions and environmental determinism in western thought that have led to perceptions of 'impossibility' of extensive settlements and social complexity in tropical forests. Beyond searches for debated 'lost' cities, even where the clear ruins of ancient urban sites have been found in tropical forests, as with the Classic Maya in North and Central America and the Khmer Empire in Southeast Asia, their collapse has been seen as almost inevitable given necessary forest clearance, soil erosion, and population pressure on these delicate environments (Webster, 2002; Diamond, 2005; Chen et al., 2014; Lentz et al., 2014). In particular, the intensive agriculture seen as necessary to fuel the 'urban revolution' (Childe, 1950) and the development of cities and elite structures familiar to most archaeological definitions of cities (Adams, 1981; Postgate, 1992), has been considered impossible on the fragile, low nutrient soils of tropical forest habitats (Meggers, 1954, 1971, 1977, 1987). Other, less-discussed threats include natural disasters, such as mudslides and mass-flooding, that continue to trouble tropical regions prone to high annual or seasonal rainfall (Larsen, 2017). Nevertheless, new methodologies and theoretical shifts are highlighting the clear emergence of social complexity and extensive human populations prior to the arrival of European settlers in many of the world's tropical forest settings.

Here, I review the growing dataset of past 'urban' forms in tropical forests. As with 'the origins of agriculture' in Chapter 5, tropical forests have been crucial in demonstrating that traditional ideas of 'urbanism' in archaeology–namely 'compact', bounded, and dense populations documented in early Mesopotamia and the

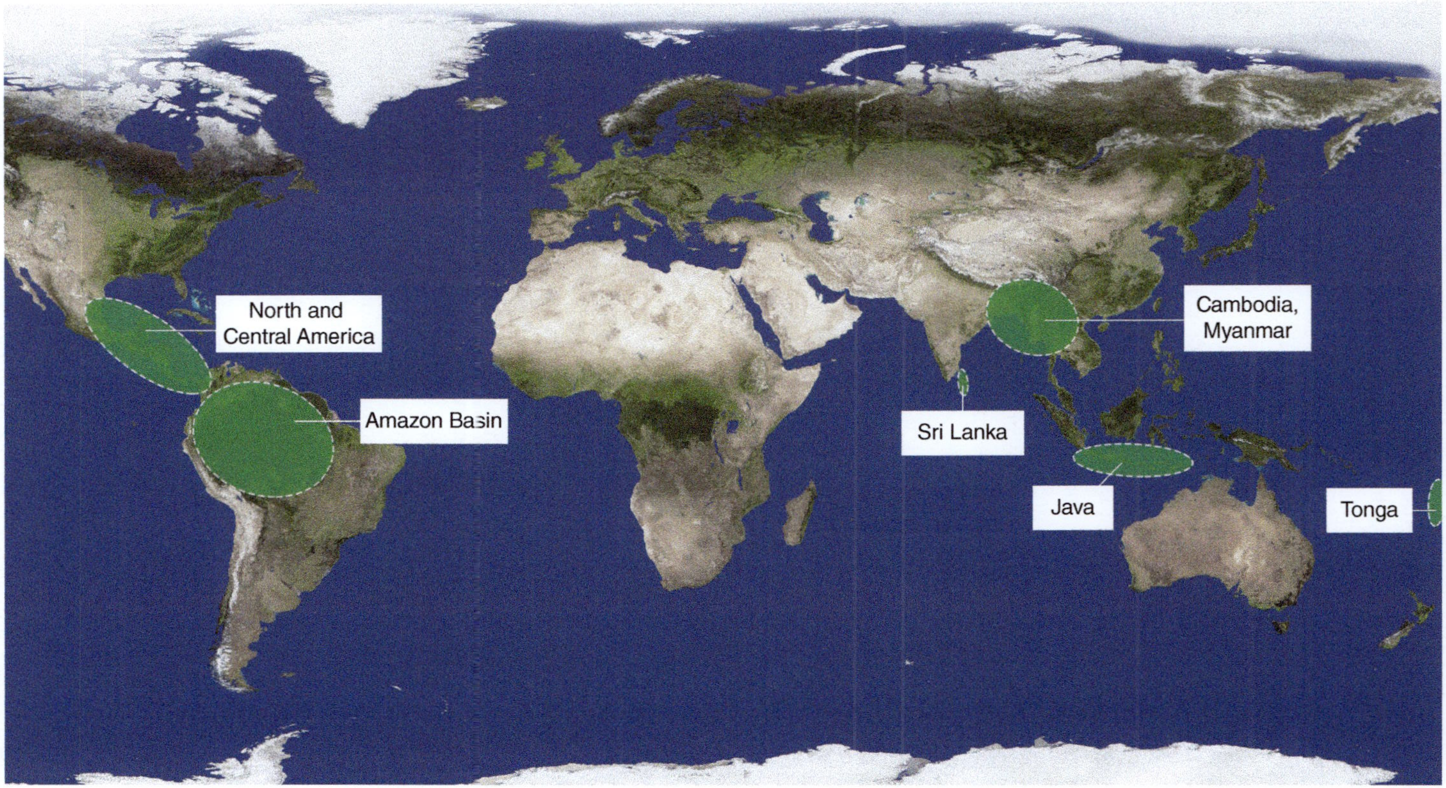

Figure 6.1. Map of main areas discussed in Chapter 6. With thanks to Hans Sell.

Mediterranean, and that dominate European thought—do not capture the whole wealth of 'urban' diversity and settlement networks that began to develop from the Middle Holocene. In particular, the fact that these models were developed for temperate or arid environments has made it challenging to apply them in tropical settings. Indeed, recent archaeological research focused on tropical forest urbanism has argued for the definition of a completely new urban trajectory, that of 'low-density agrarian settlements' (Fletcher, 1995, 2009, 2010, 2012). Here, I explore to what extent this trajectory towards dispersed settlement, incorporating large amounts of open space and water management networks in an extensive sprawl, is consistently characteristic of prehistoric social complexity in tropical forests by comparing examples from North and Central America, Southeast Asia, South Asia, South America, and Oceania (Figure 6.1). Furthermore, I evaluate to what extent the tension between increasingly expansive populations, and central political and bureaucratic foci, predisposed these urban leviathans to eventual dispersion, disintegration of central power, and the movement towards smaller, more-contained settlements across the formally occupied region, what Lucero et al. (2015) term an 'urban diaspora'.

DEFINING URBANISM AND THE CHALLENGE OF 'LOW-DENSITY AGRARIAN-BASED URBANISM'

A proliferating problem for the global comparative study of 'urbanism' in archaeology is that there are no 'precise well-understood additional characteristics for societies so described' (Adams, 1981: 81). As a result, archaeological discussions of 'urbanism' have tended to follow the biases or perceptions of a particular scholarly tradition working in a particular region at a particular point in time (see also Fletcher, 1995, 2004). In the 1950s, 1960s, and 1970s archaeologists, drawing upon neo-evolutionary theory, placed the city at the final end of a trajectory in social complexity (White, 1949; Service, 1975). Gordon Childe (1950), one of the most influential archaeologists of the twentieth century, characterized urbanism, based on finds from the fourth millennium BC in the Near East, as representing a 'revolutionary' increase in population density, craft specialization and surplus production, monumental public buildings, an elite, writing, the origins of science, regular stylistic uniformity, long-distance trade, and 'state organization' in the form of hereditary control over political, religious, and economic bureaucracy. This ancient Near Eastern 'urbanism' was framed as the antecessor for the cities of the Classical, Medieval, and then Industrial worlds of Europe through a principle of *ex oriente lux* (Childe, 1950; Trigger, 1989).

Although no exact limits were placed on the population density required for something to be 'urban', examples from the Mesopotamian region (that includes most of Iraq and Kuwait, eastern Syria, southeastern Turkey, and parts of Saudi Arabia) initiated a focus on 'compact' definitions of a city. Such cities have rarely exceeded 100 km^2 and are characterized by dense planning with a general lack of open space (Fletcher, 2004, 2009, 2010). In the case of Mesopotamia, this form of organization fit with state methods of control that co-opted surplus that had accumulated in temple buildings at the centre of settlements.

During the Uruk period (4,000–3,100 BC), that takes its name from the famous biblical city of Ur, hierarchies of settlements, increased labour input and centralized control of monumental temples and other buildings (Nissen, 1988; Postgate, 1992), craft specialization in districts of cities separate from villages, the erection of walls around dense population areas (Postgate, 1992), and the invention of writing all point towards growing administrative complexity and focused power in the hands of a small section of society on a small section of the landscape. By the end of the Jemdet Nasr period (3,100–2,900 BC) and Early Dynastic period (2,900–2,350 BC) most settled populations had moved into city-states, with established hinterlands, that were often located within 50 kilometres of each other and characterized by competitive relationships (Adams, 1981; Crawford, 2004) (Figure 6.2). The distribution of power between these focal nodes was identified using unique 'city-seals' that highlighted the ideological and administrative identity of a given city. 'Rural' settings and villages were located in specified areas beyond the 'urban' zone, existing as a formalized hinterland of control (Adams, 1981) (Figure 6.2).

The clear demarcation of a dense 'urban' inhabited area opposed to a 'rural' hinterland fit nicely with western histories of Industrial cities such as London in the eighteenth and nineteenth centuries (Fletcher, 1995). What Fletcher terms the

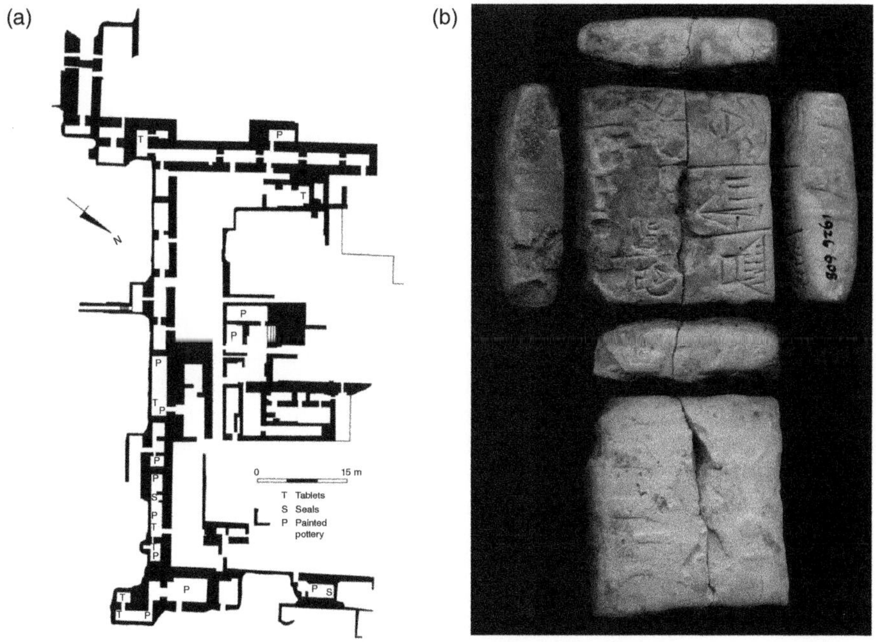

Figure 6.2. A) Schematic of a large building excavated in Mound B of Jemdet Nasr, 100 kilometres south of Baghdad, between 1926 and 1928. Reproduced with permission from Roger Matthews and Cambridge University Press from Matthews (1992) and B) Impression of a city seal representing Jemdet Nasr. AN1926.608 Tablet, Uruk III (*c.* 3,200–3,000 BC), find spot: Jamdat Nasr. Image © Ashmolean Museum, University of Oxford. Reproduced with permission of the Ashmolean Museum, Oxford.

'great pre-industrial capitals of the world', including nineteenth-century Edo in Japan, ninth-century Tang Dynasty Chang-an in China, and Abbasid Baghdad in the seventh century all stayed below 100 km^2 with populations of approximately 1 million people (Fletcher, 1995, 2009, 2010). Second- to fifth-century Rome and most European cities from the sixth to the eighteenth century did not exceed 20 km^2 (Fletcher, 1995, 2009). The idea that this was the 'normal' density and extent for urban forms pervaded comparative archaeology. For example, in ancient Egypt, a key example in the origins of statehood, the small, dispersed pastoral communities that gradually began to congregate along the Nile between 4,000 and 3,000 BC were not considered to represent trajectory towards traditional urbanism. Ancient Egypt was even characterized as a 'civilization without cities' (Wilson, 1958). More recently, nuanced analyses have argued that the focus of craft specialization and mortuary ritual in large central settlements such as Naqada, Hierakonpolis, and Abydos constituted urban activity (Kemp, 1972) even if as symbolic 'cities of the dead' that attracted dense populations at certain points of time (Wengrow, 2006) (Figure 6.3). However, despite often being a focal point for the formation of western identity, ancient Egypt has remained an uncomfortable case study for comparative urbanism due to the fact that state control apparently extended over a territory (Trigger, 2003) rather than dense inhabited cities.

One major blow to such traditional ideas of 'urbanism' in archaeology was the argument that past cities do not necessarily equate to 'states' in the sense of centralized, hierarchical elites. For example, the site of Jenné-jeno, and the smaller adjacent mound of Hambarketolo, in Mali had reached 40 hectares in area and housed between 10,000 and 30,000 people by *c.* 800 AD (McIntosh and McIntosh, 1980; McIntosh, 1995; Connah, 2001). This makes it 'urban' by most prehistoric standards. Yet, there is no evidence for central elite control in the form of rich grave goods. Instead, Jenné-jeno was apparently guided by a heterarchical structure whereby power was spread horizontally between different groups (perhaps representing different crafts) (McIntosh, 1995, 1999). In a tropical context, in his study of traditional Balinese irrigation systems Lansing (2007) demonstrated that the concentration of power in a centralized pyramidal hierarchy was not essential for elaborate water management systems or monumental constructions that could instead be managed cooperatively by a multitude of stakeholders. Coming from the opposite angle, the elites that marshalled control over the construction of Great Zimbabwe, and its access to long-distance Indian Ocean trade, built monumental enclosures and residences in the apparent absence of long-term sedentary, dense populations (Pikirayi, 2001; Huffman, 2009). Together, these examples have suggested that the model of hereditary elites in densely populated, monumental city strongholds is not universally applicable.

Of greater interest here, however, is the instrumental contribution tropical forests have made in delivering a second significant blow to Eurocentric ideas of 'urbanism' in archaeology. As early as the 1960s, settlement and landscape survey approaches demonstrated that the famous ritual centres of the lowland Classic Maya in the tropical settings of North and Central America did not conform to the Mesopotamian urban ideal (e.g., Coe, 1965, 1967). These seventh- to tenth-century AD centres were located in extensive settlements of dispersed living mounds, plazas, and raised field and road systems rather than at the heart of a defined urban area (Sabloff, 1990; Fletcher, 2009). Drawing on this, as well as

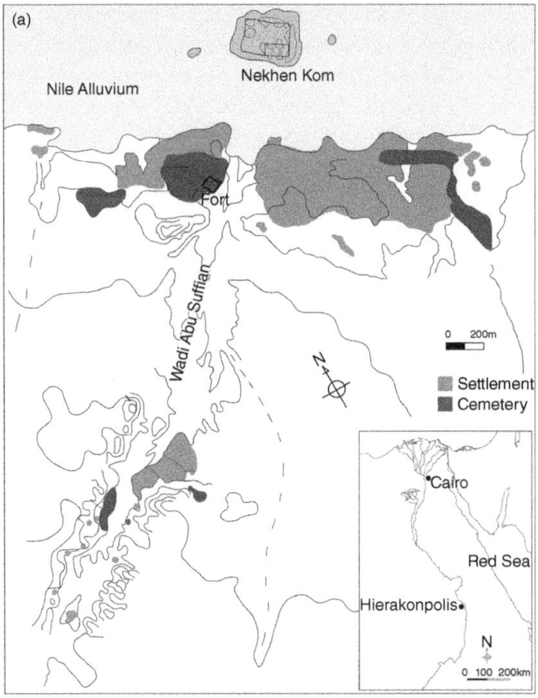

Figure 6.3. a) Map of the location of the Predynastic settlement of Hierakonpolis, Egypt and b) Evidence for a specialized brewery location (HK11C brewery Op B) within Hierakonpolis. Used with permission from Renne Friedman and the Hierakonpolis-online.org team.

examples from tropical forest ecosystems in South and Southeast Asia, Fletcher (2004, 2009, 2010) defined three trajectories of low-density settlement: 'industrial low-density urbanism', 'agrarian-based, low-density urbanism', and smaller agrarian 'No Name' settlements. Of these, the second, demarcating cities between 100 and 1,000 km^2, where monuments, buildings, and large, extensive populations were interspersed with fields, anthropogenic woodlands, and water management systems, have been increasingly highlighted as a common solution to large population organization in the tropics (Lucero et al., 2015). Legitimate classification of such settlements as 'urban' was supported by low-density industrial cases, such as the 'East Coast Megalopolis' of the United States of America, whereby extended conurbations had effectively joined the previously isolated cities of the eastern seaboard (Gottman, 1961). Indeed most industrial cities in the modern world are heading towards extensive suburbs and agricultural inclusion within the urban framework (Angel et al., 2005; Angel, 2012; Fletcher, 2009).

Although what exactly should be contained within the definition of 'urban' in archaeology is likely to remain disputed as a result of regional variation and personal experience (Heckenberger et al., 2008), here I will use Fletcher's 'agrarian-based, low-density urbanism' as a conceptual way into a global comparison across tropical forests. Tropical forests, particularly seasonal tropical forests with large variations in rainfall, have been referenced as having certain characteristics that encourage this particular urban trajectory, although these characteristics, and their variability, have rarely been reviewed in detail on a case by case basis (though see Fletcher, 2009, 2012; Lucero et al., 2015). Here, I review to what degree 'urban' developments in tropical forests in North and Central America, Southeast Asia, South Asia, South America, and Oceania fit with 'agrarian-based, low-density urbanism'. While this review cannot cover every tropical forest example of a prehistoric or historic settlement that has been termed 'urban', I seek to evaluate to what extent local cultural, social, and environmental variations in these more comprehensively-studied tropical regions alter this class of urban trajectory, or even encourage completely different trajectories. I also compare these cases to 'agrarian-based, low-density urbanism' beyond tropical forest environments. Given the modern urban tendency towards low-density, urban sprawls I highlight the challenges faced by 'agrarian-based, low-density' urban forms in tropical forest settings in the past, and how this might be used to inform conservation and urban planning policy in the present.

INDIGENOUS TROPICAL URBANISM: UNIQUE CHALLENGES AND EXAMPLES

As noted by Bacus and Lucero (1999), obstacles of disease, seasonally-constrained field opportunities, poor surface visibility, and issues of preservation, also noted for tropical forest archaeology in general in Chapter 1, have made the long-term, large-scale excavations usually associated with urban sites in archaeology somewhat unattractive. It is, however, the perceived unattractiveness of tropical forest settings to 'compact' forms of urbanism involving dedicated dense population centres and a rural hinterland, and the state control of large agricultural surpluses, that has really tended to discourage research into social complexity in these settings. Delicate balances in soil acidity and nutrient content, forest regeneration thresholds, faunal lifecycles, large or seasonally variable amounts of rainfall, risks of flooding, cyclones, and storms, diseases in warm stagnant water and insects, all pose significant challenges to modern dense urban forms in tropical forest settings (Chapter 8; Ghazoul and Sheil, 2010; Larsen, 2017) (Figure 6.4). In turn, this has

Figure 6.4. Aerial view of urban community on Caribbean coast, Vargas state, north of Caracas, Venezuela showing apartment buildings that were damaged by flood waters and debris flows associated with a rare and massive storm, December 1999 (Larsen and Wieczorek, 2006; Larsen, 2017). Note extensive deposition of sediment throughout the area of apartment buildings as well as offshore, and numerous landslide scars on hill-slopes in the background. Photo taken by, and used with permission from, Matthew Larsen of the Smithsonian Institution.

led to dense populations and monumental construction being considered destined for failure in these environments in the past (Meggers, 1954, 1971; Beach, 1998; Beach et al., 2002, 2006; Diamond, 2005). Nevertheless, humans demonstrate considerable adaptive flexibility and, when we move away from traditional European *intensive* ideas of urbanism, by which I mean a focus on a settlement of human population (50–100 persons per hectare) within less than 100 km^2 (following Fletcher, 2009, 2010), large populations and social complexity seem more at home in tropical forests.

'Agrarian-based, low-density urbanism' represents the *extensive* (1–10 persons per hectare) spread of urban infrastructure, residential populations, and agricultural fields, plots, and open space, over an area of approximately 100 to 1,000 km^2 (Fletcher, 2012; Lucero et al., 2015). Central areas of ritual, political, or economic importance maintain a 'centripetal' draw for the settlement, while farmsteads and villages draw the population gradually outwards in a 'centrifugal' manner (Lucero et al., 2015). This urban adaptation seems particularly suited to the large and dispersed biodiversity of tropical forests, and the reliance of large human populations on the mosaic of agricultural strategies highlighted in Chapter 5 (Fletcher, 2009, 2012). In addition, in seasonally dry tropical forest (as in monsoon forest) (Chapter 2), where many of the examples discussed below are placed, 'agrarian-based, low-density urbanism' enables small-scale hydraulic landscape modification to protect against flooding and drought to operate alongside central planning of large reservoirs, overall system flow, and the mitigation of stagnant water (Fletcher et al., 2008a, 2008b; Scarborough and Lucero, 2010; Lucero et al., 2015). Although this urban trajectory has been noted in tropical forests in the Old World and the Neotropics, I seek to determine potential differences in local practice and outcome, as well as tropical forest case studies where this trajectory is not seen, that are linked to local topography, climate, forest formation variability, and human cultural and social organization and resilience.

North and Central America: The Case of the Maya

The emergence of the Pre-Classic and Classic Maya Periods in southeastern Mexico, Guatemala, Belize, western Honduras, and El Salvador, has been held up as a case of evident, but unsustainable, urban expansion in tropical forest settings (e.g., Diamond, 2005). Based on the cultivation of the North and Central American staple crops, including maize, beans, squashes, and chilli peppers (Smith, 1999, 2006), Maya urban forms first began to emerge in the middle of the Pre-Classic period *c.* 800 BC (Webster, 2002). At Nakbe in the region of Petén, Guatemala, monumental stone structures are dated to 750 BC (Sharer and Traxler, 2006; Coe and Houston, 2015), while settlements also emerged across the lowland tropical forest areas of Yucatán at this time (Sharer and Traxler, 2006). The first stela, a sculpted monument that often had textual reference to kings or important events, was raised at Tikal in Guatemala *c.* 400 BC (Coe, 1965; Rathje, 1971; Miller, 1999) (Figure 6.5). Alongside the formation of a developed script (Saturno et al., 2006), these city-states show a similar overlap of political power, in the form of divine

Figure 6.5. Photograph of Temple 2 at the Great Plaza of Tikal, Guatemala, with Temples 3 and 4 in the background. Photograph taken by, and used with permission from, Macduff Everton.

kingship, and a dense, single city, to that seen in Mesopotamia (Figure 6.5). Tikal emerged as a large city *c.* 350 BC but was still dwarfed by the vast El Mirador in the same region (Matheny, 1986). Urban sites also began to emerge in the Guatemalan Highlands at Kaminalijuyu (Love, 2007) and on the Pacific coastal plain at Chocolá (Sharer and Traxler, 2006). Many of these early city-states were, however, abandoned in the first century AD (Matheny, 1986; Martin and Grube, 2008; Estrada-Belli, 2011).

The Classic Maya period (AD 250–900) witnessed the dramatic growth of certain centres in the power vacuum left by Pre-Classic abandonment. This period is also associated with the famous Long Count calendar and increasing numbers of stelae and dated monuments, as well as large-scale construction, urbanism, long-distance exchange, and artistic expression (Coe, 1999). At this point, divine kingship became solidified in symbolism and writing, with alliances and wars being recorded as elites developed and lost spheres of influence (Martin and Grube, 2008; Masson, 2012). The cities of Tikal, Uaxactun, and Calakmul, with populations between 50,000 and 120,000 people, developed particularly extensive systems of vassals (Marcus, 1993). These networks were also influenced from the edge of the tropics and the vast city of Teotihuacan in the distant Valley of Mexico deposed Maya rulers in the fourth century AD to install a loyal dynasty (Demarest, 2004; Martin and Grube, 2008). Kings used rituals and religion to boost their status, involving the decapitation of high status captives to appease the deities during major building projects or the instalment of a new ruler, or to boost agricultural harvest (Sharer and Traxler, 2006). The elite also controlled long-distance exchange networks, with a particular focus on jade and *Spondylus* shells (Miller, 1999; Masson and Freidel, 2012), as well as craft specialization, in the form of masks, murals, sculptures, and elaborate ceramics. The degree of centralized control is evident in the scale of monumental elite residences and temples (Foster, 2002).

The clear display of elite power and kingship in the form of monumental architecture, ritual, and symbolism provides some important parallels between the Mesopotamian and Maya city-states. Indeed, the most powerful Classic Maya centres in the southern Maya lowlands, such as Tikal and Naranjo in Guatemala, and Calakmul in Mexico, are located on pockets of particularly rich soils for maize agriculture (Fedick, 1996; Lentz et al., 2014, 2015). Nevertheless, it was surveys of the Classic Maya by Gordon Willey and William Coe, as early as seventy years ago, that first began to demonstrate extensive forms of urban settlement that did not fit with existing ideas of urban populations and organization (Sabloff, 1990; Fletcher, 2012; Chase and Chase, 2017) (Figure 6.6). Now two of the best-studied Classic Maya centres, Copan and Tikal are both seen has having a clear tendency towards 'agrarian-based, low-density urbanism' (Fletcher, 2009, 2012). Although Copan is only 25 km^2, with a ceremonial centre close to the river, residential buildings of timber and thatch, surrounded by agricultural land, are dispersed evenly along the alluvial plain and up into the hill-slopes (Fletcher, 2012). This patterning can be found in a grander form at Tikal where survey has indicated a network of moats, residences, reservoirs, and pyramid clusters extended out from a central hilltop for 200 km^2 with population densities of 6 people for every hectare (Webster et al., 2007). This included the joining of various 'sub-centres' into a larger dispersed urban conglomeration that survived until the ninth century AD.

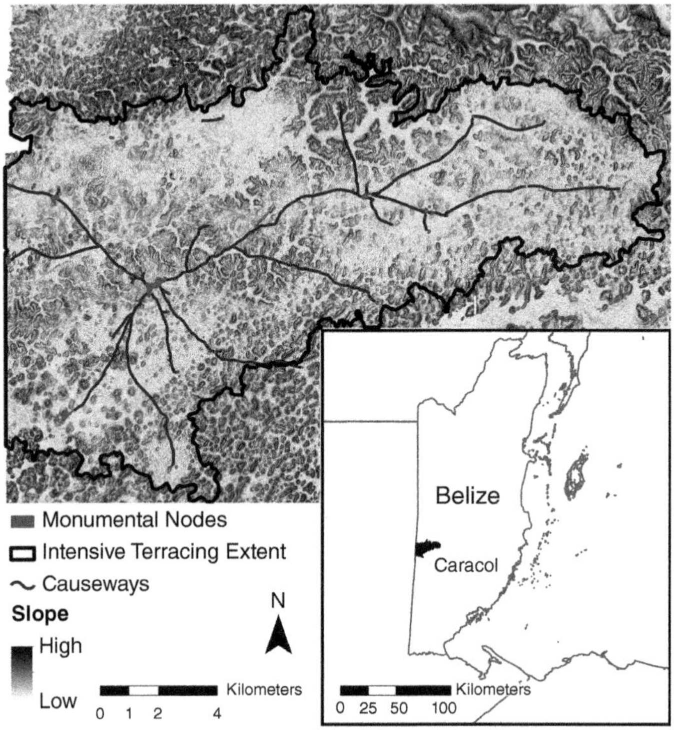

Figure 6.6. Map of the extensive terracing and causewayed network of monumental nodes at the Classic Maya site of Caracol, Belize. Reproduced and adapted with permission from Chase and Chase (2017).

This dispersed settlement pattern was encouraged by Pre-Classic and Classic Maya subsistence strategies that go well beyond an often-assumed intensive reliance on maize, squash, and beans. A huge network of locally-managed raised fields can still be observed near Maya centres today (Adams and Jones, 1981), alongside patches of 'garden' and managed fallows that facilitated the maintenance of an environmental mosaic and ensured security in unpredictable rainfall conditions (Sharer and Traxler, 2006; Ross, 2011). Plants such as sunflowers, tomatoes, and manioc contributed to this more dispersed agricultural strategy (Foster, 2002), while historical ecology studies have been used to argue that the abundance and distribution of tropical forest plant species indicate a close management of wild products by the Classic Maya (Ross, 2011). The Maya certainly made use of non-domesticated resources, including the penning and fattening of wild turkeys and deer (Foster, 2002). Although in some instances the Maya did focus on particularly fertile mollisols, these were often tracked in an extensive and winding manner along rivers and into slopes (Fedick and Ford, 1990; Lucero, 1994). It has even been argued that many of these urban farmers would have been mobile, relocating houses seasonally (Lucero, 2006). That these dispersed agricultural networks were locally managed is highlighted by the fact that small-scale farming

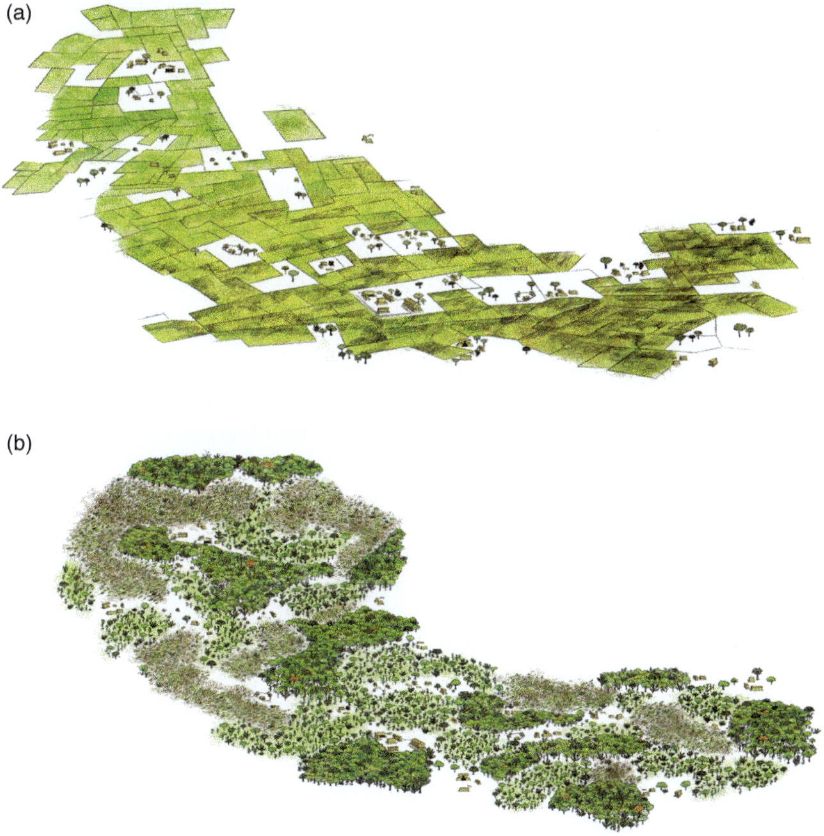

Figure 6.7. Illustration showing a) traditional Eurocentric ideas of Maya landscape modifications and b) a 'Forest Garden' landscape whereby 'milpa' agriculture takes into account natural forest succession cycles (Ford and Nigh, 2015). Reproduced and adapted from Ford and Nigh (2015). Created and owned by Anabel Ford at the Meso American Research Center.

continued at Classic Maya centres well beyond the demise of the political institutions synonymous with this period (Webster, 2002).

Ford and Nigh (2015) have put forward a particularly interesting subsistence model of 'Forest Gardens' in discussing sustainable cultivation in the tropical forests of Mesoamerica for eight millennia (Figure 6.7). Here, maize can still form a crucial part of subsistence as part of perennial, multi-cropping swidden cultivation otherwise known as the *milpa* (Yukatek *kol*). This system rotates annual crops, such as maize and manioc, with stages of shrub and tree growth that culminate in the re-establishment of mature canopy trees (Nations and Nigh, 1980; Hernández et al., 1995; Nigh, 2008). 'Forest Gardens' also integrate agricultural practices within tropical forest succession where nearly all of the dominant tree species can be utilized (Campbell et al., 2006). On a wider scale, this system can also be adapted to a transect of diverse tropical forest environments, ranging from well-drained, drier upland forests with palm resources, to lowland and wetland forests that provide raw materials for dyes, fruit, fire wood, construction, and a habitat for utilized animals (Schulze and Whitacre, 1999). As Ford and Nigh (2015) argue for the region of El Pilar near Tikal, this mosaic was resiliently managed by extensive populations of up to 90,000 in the wider Tikal region (Valdés and Fahsen, 2005), through periods of rise and decline at the ceremonial centres. This not only demonstrates a clear understanding of the water cycle, biodiversity, and soil fertility but also that wide-scale forest clearance was unnecessary (Figure 6.7).

Local independence and management within larger urban forms was also a resilient adaptation to the large regional variation (between 1,350 and 3,700 mm between July and December) in precipitation as well as the un-retentive karstic landscape in southern North and Central America (Scarborough, 1993). This resulted in a reliance on natural rain-fed sinkholes (*aguadas*) and swamps (*bajos*) (Ford, 1986; Fedick and Ford, 1990), as well as anthropogenic manipulation of the landscape. From 100 BC micro-reservoirs diverted and stored water through the drier periods of the year (Scarborough and Gallopin, 1991). By the time of the Classic Maya, large-scale networks comprised of a variety of locally-managed reservoirs, dams, and constructed channels, recently highlighted by the application of Light Detection and Ranging (LiDAR) survey at Caracol, dispersed across major lowland Classic Maya urban centres (Chase et al., 2011; Scarborough et al., 2012). The Classic Maya also utilized particular hydrophytic and macrophytic plants within these water systems in order to guard against the build-up of stagnant water and water-borne parasites and diseases (Lucero et al., 2011). The intersection of small-scale communities and elite central power in water management can be found in the proliferation of the water lily on reservoir surfaces as well as within royal symbolic contexts (Lucero, 1999).

The difficulty of maintaining large, extensive populations in a landscape where water can be hard to come by and suitable soil for agriculture exists in limited areas appears to have told in the Post-Classic period (950–1539 AD). While it is unlikely that a rapid drying event caused political collapse across the entire Maya heartland (Gill, 2000; Diamond, 2005), evident increases in seasonality, drought frequency, as well as long-term subtle changes in precipitation could be enough to tip the delicate water balance in different regions (Webster, 2002; Lucero et al., 2011; Dunning et al., 2012; Medina-Elizalde and Rohling, 2012). Indeed, this may explain a redistribution of urban centres to the northern lowlands, the Gulf of Mexico, and the Maya Highlands where permanent rivers and lakes were available (Masson, 2012). The abandonment of a number of Classic Maya sites in the southern lowlands from the tenth century AD, including Copán and Tikal, has also been linked to population over-reach and soil erosion as a result of intensive staple agriculture on increasingly marginal soils (Beach et al., 2002, 2006; Diamond, 2005). Lake sediment records reveal that maize in particular was cultivated in association with deforestation in the region (Colunga-García Marín and Zizumbo-Villarreal, 2004). With falling agricultural returns, exacerbated by local droughts, the power of the elite to provide for the remainder of society came into question, leading to political challenges, the cessation of monumental construction, and the abandonment of ceremonial centres (Lucero, 2002; Fash, 2005; Martin and Grube 2008).

Southeast Asia: Cases from Burma, Cambodia, and Java

North and Central America was the first region in which extensive ancient urbanism was documented. However, it is Southeast Asia that has provided some of the most emphatic cases of 'agrarian-based, low-density urbanism'. Urbanism developed in this part of the world from the first millennium BC in the Ayurawaddy basin (O'Reilly, 2007) and in the Mekong delta in the early first millennium AD (Higham, 2002). In the case of the former, 'Pyu' towns, such as Sri Ksetra, reached up to *c.* 20 km^2 in extent, incorporating substantial hydraulic management features, such as reservoirs, as well as moats and walls (Fletcher, 2012). In the eastern Mekong region, smaller towns, also encircled by walls and moats, including Angkor Borei, reached up to *c.* 5 km^2 and included canals and water management systems (Stark, 2006). From the sixth century AD the trajectory towards dispersed, low-density urbanism that was to become a characteristic feature of social complexity in Mainland Southeast Asia had begun at sites such as Sambor Prei Kuk (Ichita et al., 1999). Networks of shrines, huge water management features and networks, and extensive, dispersed residential areas and agrarian spaces were to combine in various ways, leading to the formation of the two best-studied, but also different, urban cases of Bagan (90 km^2) and Angkor (1,000 km^2) in the early second millennium AD and late first millennium AD, respectively (Fletcher, 2012).

Bagan lies in the Bagan Plains on the eastern bank of the Ayurawaddy River in an arid region of Myanmar. Based on historical texts the city was founded in the mid to late ninth century AD (Lieberman, 2003), though a long history of settlement stretches back to the earliest 'Pyu' towns in this region (Hudson, 2001, 2004). From 1000–1300 AD Bagan went on to become the political and economic capital of the Pagan Kingdom (Hudson, 2001) (Figure 6.8). In contrast to the

Figure 6.8. Photograph of the central area of Bagan, Myanmar. Photograph taken by, and used with permission from, Ben Dharmendra.

previous 'Pyu' towns, a central area, apparently composed of brick palaces, was surrounded by Buddhist shrines and monasteries that extended from the banks of the river across a total area of 80 km^2, with no observable boundary or limit (Hudson, 2001) (Figure 6.8). Interspersed between these religious monuments were residential areas for the service staff and slaves required to support them. The Pagan state relied upon these religious and academic centres, as well as maritime access to exotic goods, to justify their central place within this system, and Bagan was to become a multi-cultural hub for medicine, law, religion, and language, whose pull reached as far as South Asia (Ishizawa and Kono, 1989; Lieberman, 2003; Rao, 2013). However, Bagan is actually located in an area of low agricultural productivity and there is limited evidence for agrarian activity in the vicinity of Bagan (Aung-Thwin, 1985; Ishizawa and Kono, 1989). The maintenance of this urban system was therefore reliant on connectivity to, and control over, agricultural producers from as far away as 200 km along the Ayurawaddy River (Aung Thwin, 1990).

The famous centre of Angkor in Cambodia shows urban dispersion of a different nature and scale that made it the most extensive urban complex of the pre-industrial world (Evans et al., 2007) (Figure 6.9). Eighth- and ninth-century Angkor was actually represented by two unbounded, low-density major centres with complex water management systems, Ak Yum to the west and Hariharalaya to the southeast (Pottier, 2006). In the late ninth century a new capital, Yasodharapura, was built, incorporating the previous two centres within 'Greater Angkor' (Figure 6.9). This new capital was also unbounded and was contemporary with the construction of a huge reservoir now known as the East Baray (Pottier, 2000). Administrative centres and palaces varied in location within what was to become an urban network of over 1,000 km^2, until the development of a walled, moated central enclosure, Angkor Thom. As at Bagan, this centre of power was associated with religious control administered through state Buddhist and Brahmanical temples. Unlike Bagan, however, the urban area of Angkor incorporated sprawling domestic mounds, smaller shrines (*prasat*), and water tanks (*trapeang*) connected through a massive network of channels and ditches. While some residential areas are found within large temple enclosures, and likely reflect a religious support staff and labour, fields of the major staple, rice, and domestic refuse can be found through Angkor along the sides of channels, mounds, and embankments in a more unregulated manner, including adjacent to the world-famous tourist attraction of the Angkor Wat temple (Evans et al., 2007; Fletcher, 2012).

As in North and Central America, remote sensing has proved incredibly advantageous in mapping the sheer extent of the Angkor urban complex. The densely built central area of Angkor Thom has now been mapped using LiDAR, which has stripped away the seasonal forests of Cambodia to reveal this gridded area (Evans, 2016). From this centre, RADAR has revealed an engineered landscape spanning 3,000 km^2 and residential landscape of over 1,000 km^2 (Evans et al., 2007, 2013; Evans, 2016) (Figure 6.10). As seen at Tikal, while giant reservoirs and markets served to draw people into the temple and palace centres, rice farming and water control systems, including smaller scale canals and reservoirs, extend across the lowlands and the surrounding hills (Evans et al., 2007). In addition to rice, the dispersed agrarian urban mass that lived alongside and managed this water control network also exploited palm trees, seasonally available forest fruits in orchards

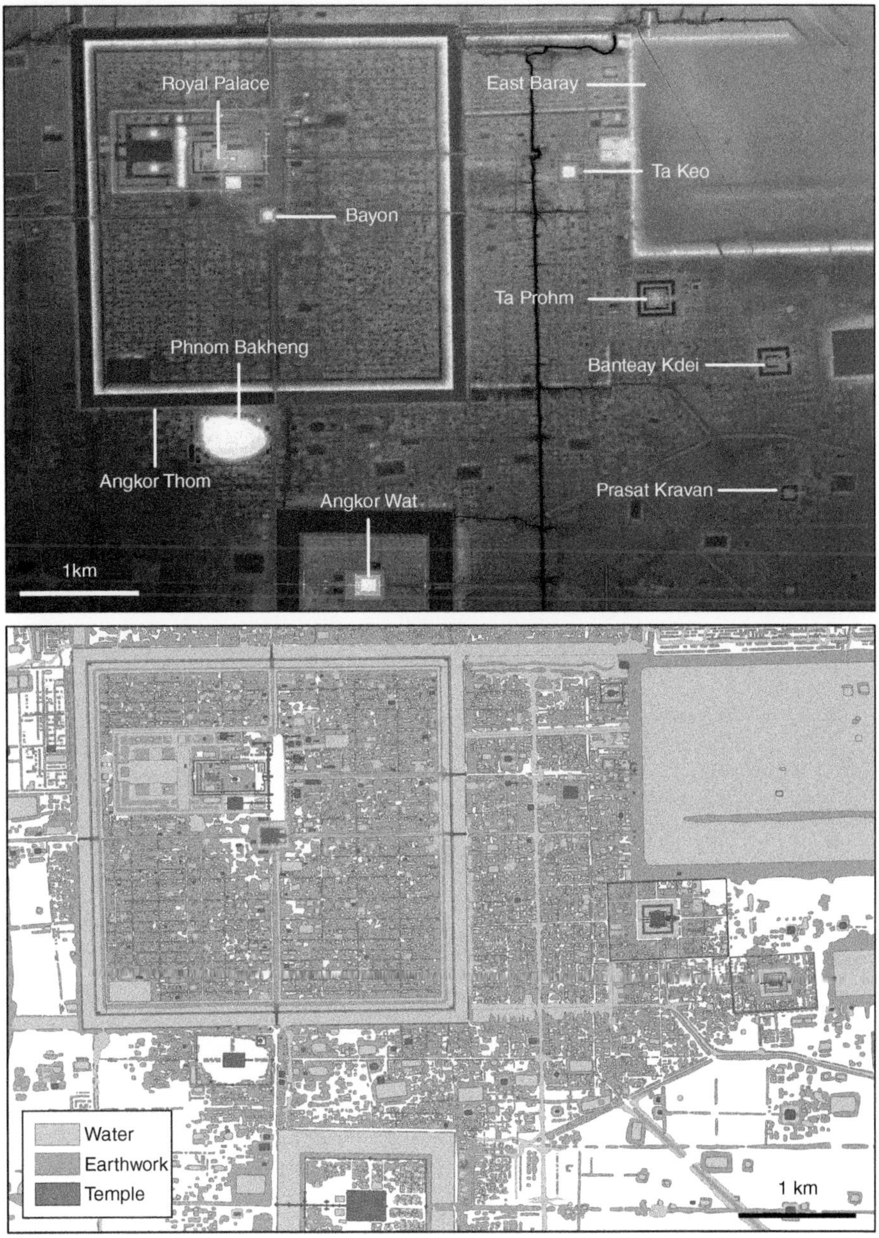

Figure 6.9. Map of the urban core of Greater Angkor in northwestern Cambodia. The top panel shows the LiDAR terrain model for the central zone, while the map below is the result of 150 years of mapping in that same area using various techniques, including LiDAR imagery, since 2012. With thanks to Damian Evans, Christophe Pottier, Pelle Wijker, and Sarah Klassen.

Figure 6.10. Oblique perspective showing the ability of airborne laser scanning, or LiDAR, to virtually strip away land cover including villages, swidden agriculture, and dipterocarp forest to reveal archaeological topography, including elements of the medieval city of Mahendraparvata, in the Greater Angkor region. Created by, and reproduced with permission from, Damian Evans.

that thrived during the wet season, and vegetables. In addition, numerous rivers and ponds were manipulated into freshwater fisheries that provided a major source of protein in the form of roasted or steamed fish products. Pigs, cattle, and chickens were also kept in the vicinity of domestic settings and provided an additional source of nutrition (Higham, 2001; Daguan, 2007). Although anthropogenically induced changes to the landscape in the form of deforestation and soil erosion are clear (Evans et al., 2007), the dispersed spread of Angkor allowed small-scale farmers to extensively manage an entire river catchment, and a mosaic of resources, rather than intensively utilize a single location.

In response to the seasonal dry periods, the population of Angkor had also effectively created their own anthropogenic catchment. This avoided reliance on a single reservoir, and encouraged the constant, controlled flow of water that helped to avoid tropical, water-borne diseases that remain highly problematic in the region today (Miksic, 1999; Fletcher et al., 2008a, 2008b). The overall arrangement of this hydraulic system implies some degree of state control. East to west running embankments trapped water flowing from the north, north to south channels delivered water to large reservoirs (the *barays*) and large temple moats, embankments and channels running northwest to southeast distributed water back towards the east, and channels running to the southwest could dispose of water rapidly from the system (Fletcher et al., 2008a, 2008b). The policy of using sand mixed with clay as a means of controlling water flow in different parts of the system may also be reflective of some central instruction (Fletcher et al., 2008a). The immense scale and extreme stability of this system in providing water to the Angkor urban sprawl throughout repeated dry seasons is indicative of the longevity of this centre despite broader political fluctuations in the Khmer Empire as a whole prior to the thirteenth century AD. However, the direct role of the state in the everyday management of this system is up for debate and evidence for small-scale, *ad hoc* repairs, highlighted by LiDAR imaging, suggests agrarian stakeholders may have had a greater role.

A tension between the centripetal desires of the state and centrifugal forces of small-scale farmers attempting to adapt to change seems evident during the demise of Angkor. The seasonal dry tropical forest region of Angkor was highly vulnerable to a series of both protracted droughts and deluge rain events during the mid to late fourteenth century AD (Buckley et al., 2010, 2014). The rupturing of parts of the network demonstrates the destruction during natural disasters (Evans, 2016), and led to masses of cross-bedded sand washing into the lower canals creating blockages and dysfunction (Fletcher et al., 2008a, 2008b). After this time smaller, more compact towns were established along the southern side of the Tonle Sap lake and the banks of the Mekong River (Lucero et al., 2015). The elite meanwhile had shifted to new urban centres in the Phnom Penh region from the sixteenth century AD (Thompson, 2004). This process seems to highlight the agency of the agrarian populace within a dispersed urban form that, when faced by external challenge, will quickly neglect centralizing forces. Interestingly, at Bagan, the fact that the agricultural produce required to sustain this centre came from 50–200 km away in monsoon-fed settings apparently enabled greater central longevity (Fletcher, 2010, 2012). There was no need for the state to fight the disaggregation of small-scale farming communities, as they existed at a considerable distance in settings less vulnerable to climatic changes. The decline of Bagan as a population centre in the thirteenth century AD appears to be rather linked to a political shift towards new parts of Myanmar.

The centre of Borobudur in Island Southeast Asia demonstrates a different trajectory again. Built between AD 780 and 825, Borobudur includes the world's largest Buddhist temple, located in the tropical uplands of Java between two volcanoes and two rivers (Soekmono, 1976). The construction of this temple is thought to have taken 50 million hours of labour to complete (Miksic, 1990) and involved huge artistic and sculpting expertise. As at Angkor and Bagan then, the centre of Borobudur, built under the Sailendra Indonesian Dynasty, utilized control over religion to legitimize its authority. However, its strategic location also led to dynastic influence, not only over Java, but also into the neighbouring islands of Sumatra and Bali. This was done through a thalassocracy, a term that denotes state control over the sea, with primarily maritime realms, and built upon long-distance exchange networks that extend back to at least 2,000 BC in Southeast Asia (Hall, 1985; Hung et al., 2007; Zakharov, 2012; Stark, 2015). Unlike Angkor, however, there is no evidence of any areas of settlements in Java engaged in non-ritual or non-agricultural craft specialization until mid-way through the second millennium AD (Miksic, 1999). Moreover, Borobudur appears to be a more 'compact' phenomenon, with no evidence for buildings beyond formal shrine or palace enclosures. As such, it appears that Borobudur relied upon the elite dominance of maritime connections and religious spheres of influence. The lack of a self-sufficient agrarian base at any degree of proximity may explain the short-lived nature of this centre.

South Asia: Sri Lanka

The rise of urbanism in Sri Lanka in the first millennium BC is chronicled by the *Mahavamsa* at the site of Anuradhapura in the northern Dry Zone of the island *c.* 500 BC (Bandaranayake, 2003) (Figure 6.11). The *Mahavamsa* documents that the Sinhalese King Pandukabhaya founded his capital at this site with a focus on its strategic position between the western and eastern coasts of the island for trade and the development of monastic complexes. Monumental structural evidence and imported material culture from the third century AD onwards supports this (Bandaranayake, 2003). Like the Southeast Asian examples, this Sri Lankan urban site is focused on a walled centre surrounded by massive monasteries (including the largest brick structure in the world—Bandaranayake, 2003). From this centre there is an extension of ritual sites for 15 km to other religious sub-centres such as Mihintale (Coningham et al., 2007). Together, the Anurdahapura network of state-controlled temples dominated the political and religious landscape of north, and at times even the entirety, of Sri Lanka between the ninth and tenth centuries AD (Coningham, 1999; Coningham et al., 2007; Fletcher, 2010, 2012). The capacity of Anuradhapura to do this is evident from estimations that the three great monasteries at its centre could have accommodated 30,000 Buddhist pilgrims, with the stone rice containers in each feeding over 1,000 monks at a single sitting (Coningham and Gunawardhana, 2013).

Although the knowledge of residential areas at Anuradhapura remains limited compared to the examples discussed above, pioneering survey work by Coningham has documented a variety of occupation locales of similar form and extent, composed primarily of timber and thatch structures, across the Anuradhapura urban landscape (Coningham, 1999; Coningham et al., 2007) (Figure 6.11). It is estimated that there were at least 250,000 (Bandaranayake, 2003) people within a potential area of as much as *c.* 500 km^2 at the height of Anuradhapura in the ninth and tenth centuries AD (Coningham et al., 2007). This population was reliant on a large spread of irrigated rice agriculture throughout the complex (Coningham, 2006). As Anuradhapura is located in the 'Dry Zone' of Sri Lanka that receives an annual rainfall of 1,300–1,450 mm between December and March, with a marked dry season in the summer months (Coningham, 1999), natural water sources were seasonal. As a result, in addition to religious control, the elite also justified its status through the construction of massive water tanks from the fourth century BC onwards to provide water for the irrigation of bunded rice fields (Coningham, 1999). These reservoirs were developed and expanded with demand. Tissawewa (6 km^2) was constructed in the third century BC and Nuwarawewa (9 km^2) was built in the first century AD. By the fifth century AD, timber and stone dams and canals up to *c.* 90 km long were built to link the centre to water sources in the hills via storage tanks such as the Kalawewa (25.8 km^2) (Lucero et al., 2015).

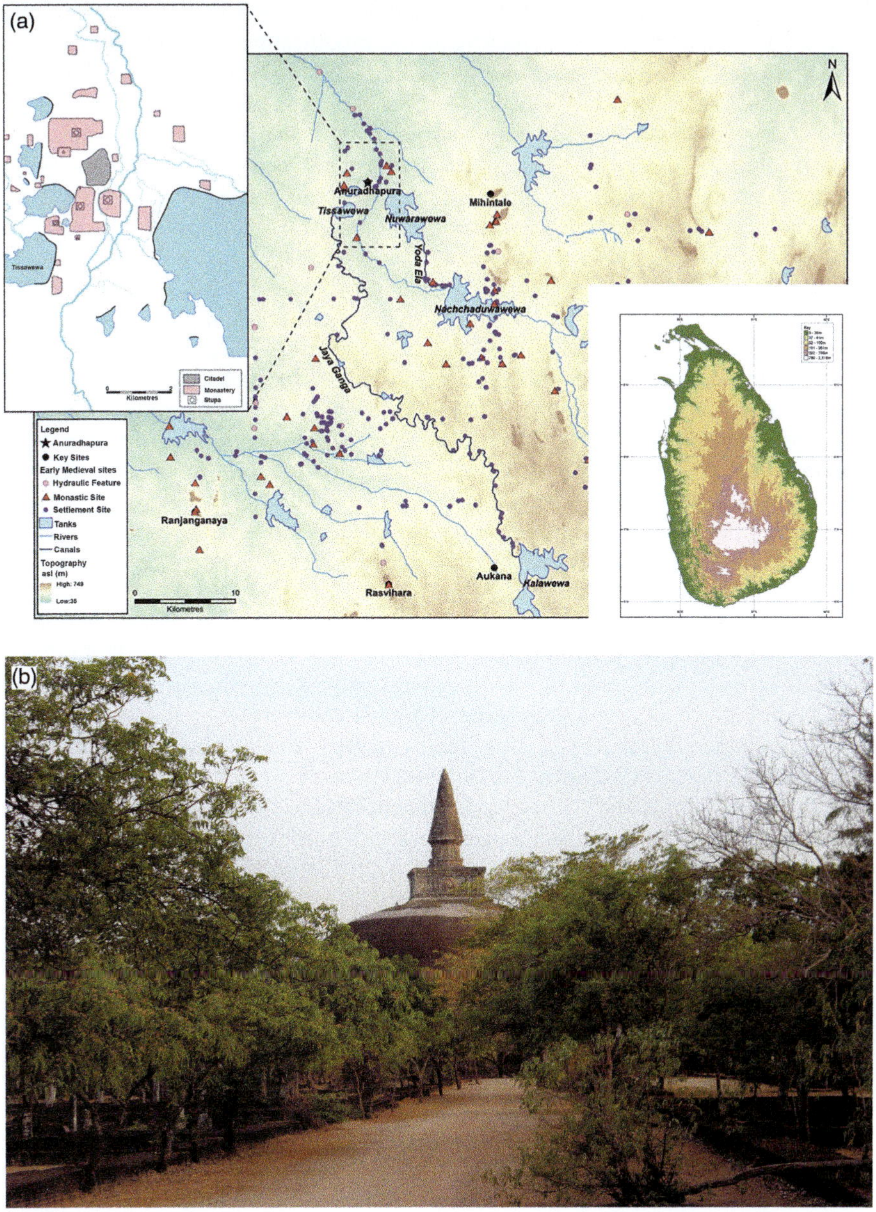

Figure 6.11. a) Map showing the extensive distribution of monastic sites and settlements in the Anuradhapura region in the Dry Zone of Sri Lanka. Modified from Coningham et al. (2007) with thanks to Jana Zech. Acknowledgement also to Robin Coningham and Durham University's 2014 UNESCO Chair in Archaeological Ethics and Practice in Cultural Heritage, and b) Photograph of a monumental Buddhist stupa at the centre of a monastic area in Anuradhapura, Sri Lanka, taken by the author.

Nevertheless, in addition to these state-coordinated efforts, there was a significant amount of independent land management and modification by smaller stakeholder groups and families. Individual domestic gardens and the swidden cultivation of dry tropical forest prevalent in the region, known locally as *chena*, supplemented rice agriculture (Coningham, 2006). Furthermore, extensive landscape modification occurred independently to larger state developments across the urban spread of Anuradhapura (Coningham, 1999). Rivers were redirected and cascade systems were developed throughout the hinterland, managed by multiple domestic foci as well as smaller monastic areas (Coningham, 1999). In addition, a number of smaller storage reservoirs were built adjacent to monastic and domestic locales, linked by smaller-scale canal networks (Coningham, 1999; Coningham et al., 2007; Lucero et al., 2015). This local agency is evident in the partial reoccupation and monument refurbishment that took place in Anuradhapura during the later twelfth century, after the Sinhalese capital had moved elsewhere, and the continued small-scale farming and water management in Anuradhapura documented by British colonial administrators in the nineteenth century (Seneviratna, 1994; Gilliland et al., 2013).

The number of central monumental sites in the inner 15 km^2 of Anuradhapura declined from 292 to 11 by the eleventh century AD (Coningham and Gunawardhana, 2013). This has been associated with the establishment and growth of a new capital in Polonnaruwa in the eleventh century AD, associated with the construction of a tank 87 km^2 in extent (Gilliland et al., 2013). This was short-lived and abandoned by the thirteenth century AD when the court moved into compact urban forms within the evergreen rainforests of the Wet Zone of Sri Lanka at Dambadeniya, Kurunagala, and then finally Kandy (Coningham, 1999). It has been argued that the abandonment of Anuradhapura was stimulated by high-amplitude fluctuations in the operation of the Southwest monsoon, bringing severe drought and cyclonic storms in the eleventh and twelfth centuries AD (Sinha et al., 2007; Lucero et al., 2015). Similarly, shifts into the Wet Zone have been linked to reduced monsoon rainfall in South Asia in the thirteenth century AD (Sinha et al., 2007; Lucero et al., 2015). It should be noted that there currently exist no suitable, high-resolution palaeoenvironmental records in the Dry Zone of Sri Lanka to test this proposition. However, as at Angkor, such external pressures are likely to have increased the tension between the religious and political pull of the elite centre and the local interests of smaller agrarian and monastic stakeholders. The maintenance of dispersed farming settlements in Anuradhapura and Polonnaruwa until the nineteenth century AD, and the elite shift to compact cities with demarcated hinterlands and tribute centres in the Wet Zone, seems to add further support to this idea.

The Amazon Basin

The Amazon Basin represents a completely different tropical forest context to the dry, seasonal formations discussed so far. Indeed, the dominance of evergreen and semi-evergreen rainforest, and correspondingly high temperatures, humidity, and poor soils, in this region led Betty Meggers to prominently argue that sedentary agriculture and the development of urbanism was impossible (Meggers and Evans, 1957).

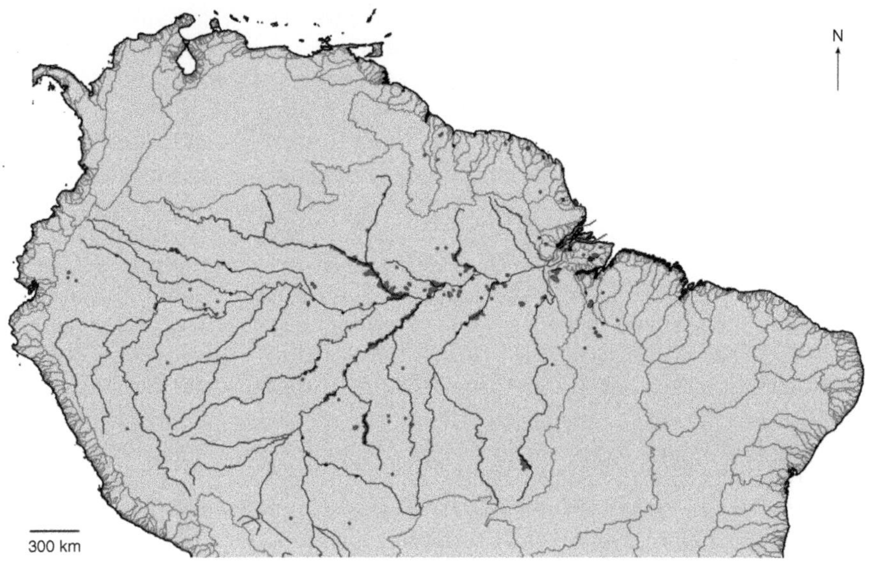

Figure 6.12. Map of the distribution of identified *terras prettas* sites in the Amazon Basin. Adapted, with permission from Manuel Arroyo-Kalin, from Arroyo-Kalin (2017).

This was thought to be supported by the fact that most ethnographic accounts for the lower Amazon Basin in the 1950s depicted small Indigenous communities living in independent villages with little to no political integration (Meggers, 1954, 1971). These communities practised shifting horticulture, not monoculture, and had little evidence for art, permanent monumental architecture, or social hierarchy (Meggers, 1971; Roe, 1982). Given that the proposed environmental limitation was theoretically constant, it seemed likely that this was also the case in the past. Yet, this perspective tended to discount the fact that Amazonian cultures and populations faced massive dislocation and mortality in the face of European colonial arrival (Roosevelt, 1993, 1994; Heckenberger, 1996; Hemming, 2009). Moreover, it also ignored the fact that large populations can be sustained on agricultural practices more attuned to tropical forest environments (Chapter 5), as well as the long-term management of non-domesticated resources (Balée, 1994, 2013; Clement, 1999; Hilbert et al., 2017; Levis et al., 2017). Such activities and their widespread presence across the Amazon Basin in the Middle and Late Holocene—for example in the form of anthropogenic soils (Figure 6.12)—are hinting at an ever increasing pre-Columbian human presence and impact in this region.

Ethnohistorical accounts made during European exploration and colonization document the presence of dense Indigenous populations undertaking monumental landscape alteration and living within settlement networks in the Amazon (Medina, 1934; Bettendorff, 1990; Hemming, 2009). In contrast to the invading Europeans, these Indigenous groups were clearly able to navigate tropical forests and maximize the resource potential of these environments (Hemming, 2009; McMichael et al., 2012, 2017). Increasing archaeological survey and evidence in the last two decades has even bolstered claims of pre-Columbian 'urbanism' in the Amazon Basin. Survey and geophysical studies, alongside large-scale excavation and chronometric dating in the Upper Xingu river basin (Heckenberger, 1996; Heckenberger et al., 2003, 2008, 2013), at Santarem and Marajó Island (Bevan, 1989; Roosevelt, 1991, 1993, 1994), and Acré State (Schaan et al., 2013) in Brazil, the Llanos de Moxos, Bolivia (Denevan, 1966; Lombardo and Prümers, 2010; Rostain, 2017), and in the Upano area of eastern Ecuador (Salazar, 1998; Rostain, 2008, 2014), have revealed an abundance of densely clustered structures and landscape features that have been argued to be part of what Heckenberger et al. (2008) call 'Garden Cities'. These 'Garden Cities' peaked between AD 1250 and 1650, prior to colonial arrival. In this model, based on excavations in the Upper Xingu, larger towns formed a central point in an urban network, of potential ritual importance, and were surrounded by monumental wooden walls and ditches. These, in turn, were surrounded by satellite villages, connected by roads and pathways cleared through the forest. Between these settlements lay a patchwork of fields of manioc, regions utilized for tree-cropping, and manipulated river dams and ponds used to farm turtles and fish (Heckenberger et al., 2008; Heckenberger and Neves, 2009). Although Fletcher (2012) would classify these entities as a different 'No Name' category of low-density settlement, smaller than the 'low-density agrarian urban' forms discussed above, such clusters may have been home to tens of thousands of people and demonstrate evidence for planning and the integration of an extensive, dependent settlement network.

Marajó Island, at the mouth of the Amazon, is another region with evidence for increasingly extensive human population and social organization. Mobile settlement utilizing estuarine middens from the Middle Holocene (Roosevelt, 1995, 2006) is replaced by evidence for sustained occupation of mound sites, such as Teso dos Bichos and Guajara, by populations with large domestic structures, cemeteries, and middens (Roosevelt, 1991) from AD 0 (Figure 6.13). Lasting until *c.* AD 1300, the mounds of this Marajoara phase, characterized by elaborately decorated pottery, were composed of successive earth platforms containing domestic floors, clay hearths, and cemetery urn fields. The population, estimated at around 100,000 (Roosevelt, 1999), practised a mixed system of cultivation and foraging in an area of periodically open savannas surrounded by gallery forests (Roosevelt, 1991; Smith et al., 2011). Humans appear to have used water-loving palms, root crops, and tree fruits, and demonstrate little reliance on maize agriculture (Roosevelt, 1991; Meggers 2003; Hermenegildo et al., 2017). The bones of small fish and small turtles dominate faunal assemblages (Roosevelt, 1991), demonstrating the importance of rivers as a source of protein in the Amazon Basin. Interestingly, there is no material evidence that hierarchical political groups administered these settlement complexes. Graves do not differ in terms of their content or architecture, and only slightly in size, and all contain elaborately decorated pottery as well as plain ceramics. Similarly, the domestic floors contain mainly domestic ceramic wares. Images of humans in funerary and feasting art seem mainly shamanic, and demonstrate closer links to the forest environment than to anything overtly political (Roosevelt, 1991, 1999) (Figure 6.13).

Another area where significant settlement investigation has taken place is the Santarem/Monte Alegre region at the mouth of the Tapajos River in Brazil. Associated with elaborately incised and modelled pottery (de la Penha, 1986), occupation of this site appears to begin around AD 500 before peaking *c.* AD 1500–1600. Settlement

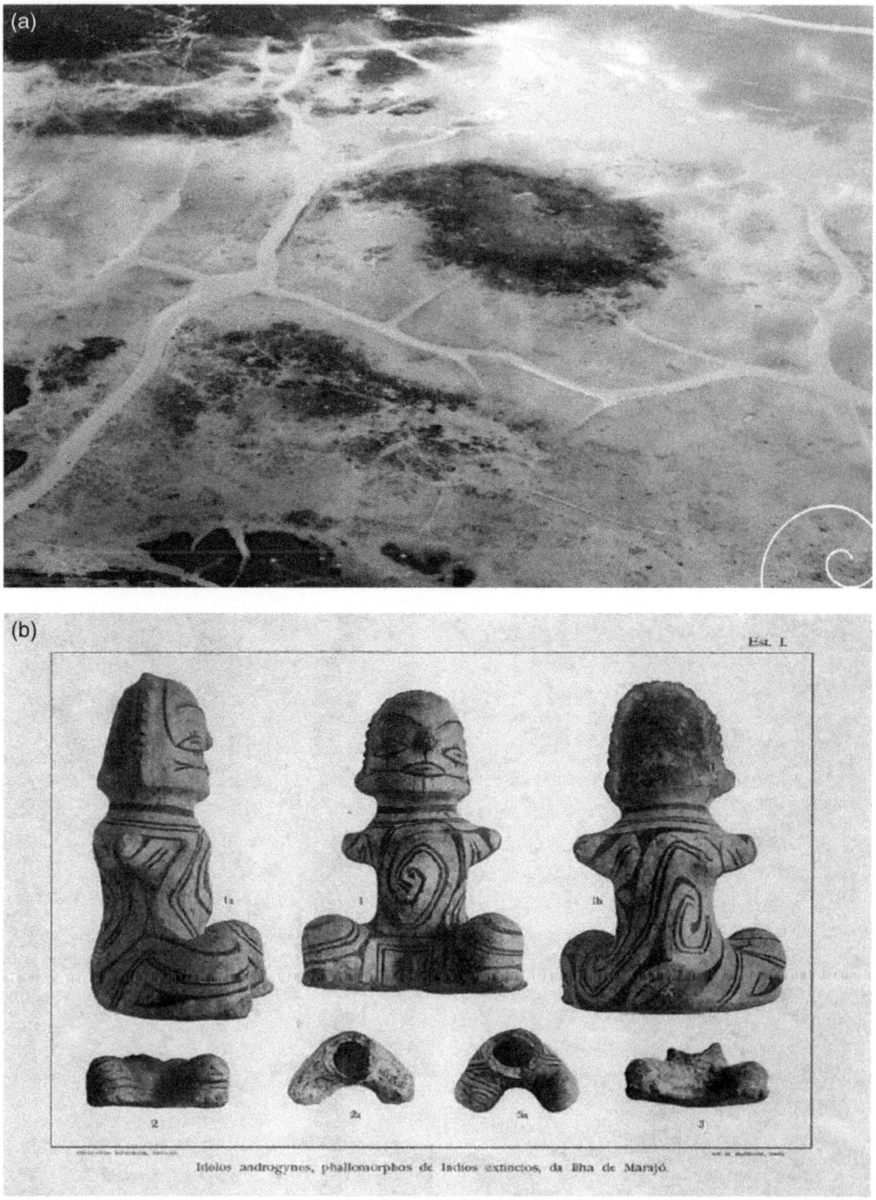

Figure 6.13. a) Aerial photograph of Marajó Island, Brazil taken by Wagner Souza e Silva. The mound at the centre is 'Teso dos Bichos', b) Plate showing a classic rattle figure combining male and female elements discovered during excavations at Marajó Island, Brazil. Drawn by Charles Hartt. Reproduced with permission from Eduardo Neves and the Museu de Arqueologia e Etnologia Department, Universidade de São Paulo.

was apparently more dispersed than on Marajó Island, with numerous, large areas of dense *terras pretas* soils (including charred bones and ceramics left over from occupation) and structural remains linked together in a similar way to that described for the Upper Xingu (Nimuendaju, 1949). As at Marajó Island, the subsistence of populations in this region appears to have been focused on tree fruits and fishing, alongside the cultivation of gourds and manioc (Roosevelt, 1999). Maize was rare and may have formed part of ritual ceremonies based on ethnohistoric data (Nimuendaju, 1949; Roosevelt, 1999). Interestingly, satellite settlements, such as Caverna da Pedra Pintada, show evidence for a wider range of forest fauna (Roosevelt, 1999). Social stratification may have been more pronounced here than on Marajó Island. Ethnohistoric records refer to 'community leaders' or 'nobles' separate from those working for subsistence (Bettendorff, 1990), and large ceramic figurines of women and records of war bands at the time of colonization (Nimuendaju, 1949) imply some form of centralized control. Nevertheless, elaborate dishes, female figurines, musical instruments, stone and jade ornaments, and other ritual objects again appear to have been widely available across the sites (de la Penha, 1986; Roosevelt, 1999; Gomes, 2001), while all burials seem equally highly ritualized (Roosevelt, 1999).

In Bolivia, extensive field earthworks and mounds, such as the Llanos de Moxos, have been revealed (Denevan, 1966; Lombardo and Prümers, 2010; Rostain, 2017). These landscape features appear to have been in place since AD 1000, and document clear knowledge of flooding risks to agricultural settlements (Lombardo and Prümers, 2010; Rostain, 2014). Large mound systems focused on a ceremonial complex have also been discovered at Faladas, in the Ecuadorean Amazon uplands dating to 100 BC–AD 1000 (Salazar, 1998; Rostain, 2008). Increasing Late Holocene populations across the Amazon Basin appear to have resulted in an increase in the *terras pretas* soils mentioned above, that represent the accumulation of habitation debris, and *terras mulatas* soils, documenting increasingly intensive cultivation and near surface burning to maintain soil fertility (Neves et al., 2003; Arroyo-Kalin, 2008, 2014, 2017) (Figure 6.12). While these are largely found in the eastern Amazon, recent work is suggesting that this may simply be a result of research bias (Arroyo-Kalin, 2012; Schmidt et al., 2014; Arroyo-Kalin, 2017). The extent of these soils highlights the dispersed, urban networks characteristic of social complexity in the pre-colonial Amazon. Yet, rather than being purely agrarian, satellites engaged in these networks were potentially mobile, returning to different plots following regrowth, and often utilized the hunting and gathering of key wild resources. While the size of these networks, and evidence for hierarchy, clearly varies between examples (Heckenberger and Neves, 2009; Neves, 2013), these dispersed urban settlements were remarkably resilient, persisting until the arrival of colonial powers in many cases.

More debated is the scale and nature of environmental impacts of pre-Columbian human populations in the Amazon Basin. Beyond the key sites mentioned above, the continuous extension of earthworks along a vast region of southern (de Souza et al., 2018) and eastern Amazonia (Heckenberger et al., 2003, 2007) implies large-scale human modifications. Furthermore, the number of pre-Columbian plant domestications, huge extension of anthropogenic soils caused by human burning or settlement (Arroyo-Kalin, 2012, 2017), and lasting impacts on forest structure and species composition (Levis et al., 2012; Steege et al., 2013;

Thomas et al., 2015) (Chapter 5) all imply large populations with the potential to have environmental impacts on a continental scale (Lewis and Maslin, 2015) (Chapter 8). Nevertheless, palaeoenvironmental studies in the Amazon Basin have remained more equivocal. Carson et al. (2014) argue, on the basis of pollen and charcoal evidence from lake sediments, that 'earthwork builders' in north-eastern Bolivia made use of open savanna forming under natural, drier conditions prior to 2,000 years ago. Similarly, Watling et al. (2017a) argued that there was no evidence for sizeable clearings lasting any significant period of time around the geoglyphs of Acré State, Brazil in southwestern Amazonia. Yet these palaeoenvironmental interpretations remain debated (Piperno et al., 2017; Watling et al., 2017b) and the degree to which pre-Columbian human disturbance in the Amazon Basin had widespread environmental and climatic repercussions remains a central question of archaeological research in the region.

The Pacific: Island-Based Tropical Statehood

The final case I wish to mention in detail is urbanism in the insular tropical forests of the Pacific Islands. Between 3,000 and 1,000 BC speakers of Austronesian languages appear to have migrated from Taiwan, through Island Southeast Asia, and into island Melanesia (Kirch, 2000). The extension of this process into Polynesia has been associated with the Lapita culture, so-characterized by its distinctive geometric dentate-stamped pottery (Kirch, 2000, 2010, 2012), which reached Samoa by 1,000 BC (Green and Leach, 1989) and Tonga by 900 BC (Burley et al., 2010). This massive expansion was supported by domestic pigs, dogs, and chickens, as well as horticulture based on the root crops of taro and yam, and tree crops of coconuts, bananas, and breadfruit (Kirch, 2000). Although Lapita pottery has not been found further east of Samoa, these domesticated crops soon extended into the more remote Polynesian islands as well (Kirch, 1997, 2012; Summerhayes, 2000). These settling groups also supplemented their subsistence with fishing, mollusc gathering, and the utilization of local and translocated tree crops within unique systems of arboriculture and agroforestry (Maxwell et al., 2016). Settlements tended to be small, often focused on beach sites in the west of the Pacific and more inland areas towards the east of this region. However, larger settlement complexes and states did begin to emerge during the early second millennium AD, particularly within the Tongan and Hawai'ian archipelagos, representing insular tropical polities.

The Tu'i Tonga Empire projected control over long-distance exchange in Oceania from the Tongan archipelago between AD 950 and 1500, and is considered unique in a Pacific context in the way that it aggregated an entire archipelago under a single political system (Clark et al., 2014). Based on historical and oral history approaches it seems that the ancient Tongan state was headed by the paramount Tui Tonga or Lord of Tonga who administered other islands through related noble families and a series of political and economic networks (Kirch, 1984; Neich, 2006). The first monumental centre of this state, with evidence for hierarchical organization of dwellings, emerged at Heketa on the island of Tongatapu *c.* AD 1,300 (Clark and Reepmeyer, 2014; Clark, 2016). Power then appears to have relocated to Lapaha between AD 1350 and 1400 at which point

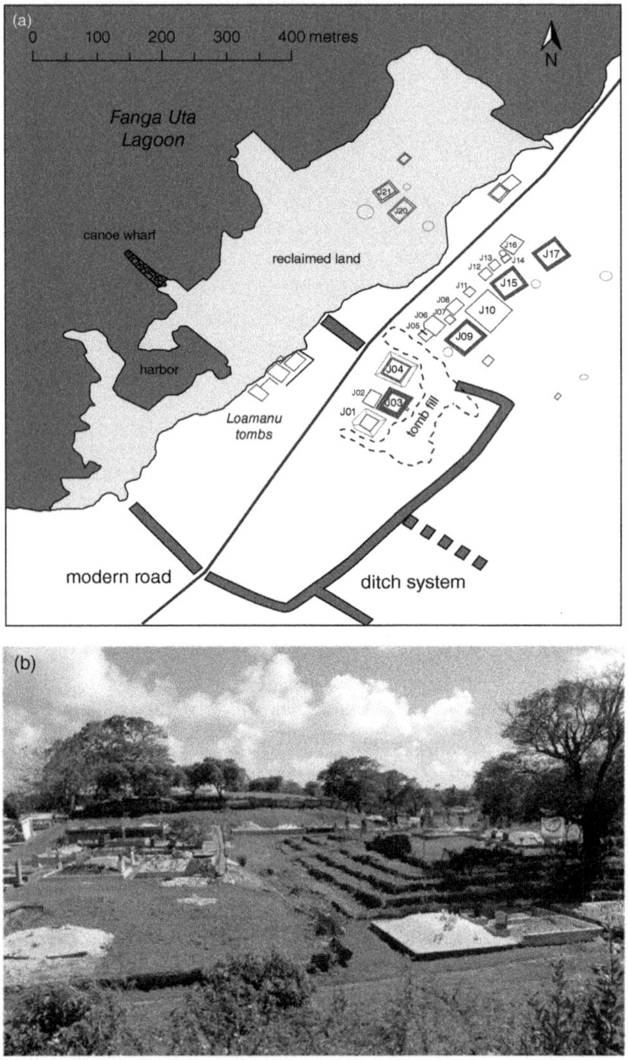

Figure 6.14. a) Map of Lapaha showing the location of the royal tombs and prehistoric features of the ancient Tongan state. Tombs are identified by a 'J' number as some tombs have multiple and conflicting names. b) View of the royal tombs (langi). In the foreground J02, foreground right J03, and in the distance is J04. The tombs are surrounded by graves in the Catholic cemetery of Lapaha. Images produced by, and reproduced with permission from, Geoff Clark.

the state reached its zenith (Figure 6.14). A monumental central place of over 50 hectares, including royal tombs of reef limestone, ditch systems, roads, sitting platforms, bathing complexes, standing stones, and a canoe harbour and wharf, emerged here and lasted until the early eighteenth century (Clark et al., 2014; Clark and Reepmeyer, 2014; Clark, 2016). Recent LiDAR research, as well as ethnohistorical data, indicates that Tongatapu provided a redistributional centre for subsistence products, such as yams, wood, sea birds, fish, and pearl shells, and craft specialization products, such as mats and fishing technology, for the entire Tongan archipelago (Freeland et al., 2016; Parton et al., 2018).

The primary source of power for this state appears to have been monopolization of regional and long-distance exchange. Ethnographic and ethnohistoric sources suggest that the majority of imports at the centre of this political trade would have been perishable materials that are unlikely to preserve in tropical contexts (Kaeppler, 1978; Clark and Martinsson-Wallin, 2007). However, petrological and geochemical analyses of stone tools found in elite burial contexts at Lapaha have provided more detailed insights into the scale and nature of Tonga's economic sphere of influence at this time. This research has documented trade from not just other volcanic islands in the Tongan archipelago, but also Fiji, Samoa, and Tahiti, 2,500 km east of the central Tongan island of Tongatapu (Clark et al., 2014). The numbers of lithic tools, particularly in the form of adzes, from Samoa within elite tombs at Lapha appear to have increased through time, and seem to demonstrate influence over developing nucleated workshops in Samoa that produced high-quality adzes which were widely distributed across the Pacific from AD 1200 onwards (Winterhoff, 2007). While preservation issues remain, it seems that the compact monumental urban agglomerations themselves may have been of limited importance to the Tongan state. Instead, local redistribution, as well as the domination of religious rituals, was superseded by the ability to develop a vast network of craft specialization and long-distance exchange as the real source of power.

In contrast to Tonga, the emergence of kingdoms in Hawai'i between AD 1200 and 1819 was focused on internal redistribution and symbolic control of *local* economic resources, rather than long-distance trade networks. Following occupation at some point in the first millennium AD, gradual political centralization occurred in Hawai'ian society from the thirteenth to nineteenth centuries AD during a time of population increase and expansion into all of the environmental zones of the archipelago (Kirch, 1990). Different island polities began to compete with each other for influence and power. For example, in Maui, during the Consolidation period (*c.* AD 1400–1500) the compact centres of Hāna and Kaupō were part of an East Maui polity while Wailuku and Kula aligned with West Maui (Kolb, 1999). By the time of the Classic period (AD 1650–1820) a more integrated Maui competed with other islands including Hawai'i and O'ahu (Hommon, 1986; Kirch, 1990). Oral histories state that chiefs traced their lineage back to the gods (Wichman, 2003; Kirch, 2012), dictating the harvest and land rights for all below them, and power seems to have been focused on monumental temple centres, such as those found at Loaloa and Popoiwi at the chiefly centre of Kaupo, Maui (Kolb, 1999). These temples appear to have operated as elite residences but also as ceremonial centres for warfare, ritual, and agricultural redistribution (Kolb, 1993, 1999; Kolb et al., 1997).

Kolb (1999) documents the key role of these temples in centralizing local economic resources, particularly in the form of the pig that has ritual importance across a number of Pacific island societies today, under elite authority. During the Formation and Consolidation period (AD 1200–1500) pig remains were minimal in temple contexts, with wild resources such as sea birds and fish being more frequent. During the Unification and Classic periods (AD 1500–1820) there was a dramatic increase in burnt domesticated pig bones. By contrast, fish, wild birds, and domesticated chicken remained the primary protein sources in domestic settings. Kolb (1999) argues that in a tropical environment with limited resources the gradual adoption and monopolization of pig feasting in elite and religious, settings enhanced the ideological and political value of pigs and made it a symbol of their power (Kolb, 1994). Large-scale specialization in pig farming and proscribed rules for pig treatment and distribution (Luomala, 1960) demonstrate how the Hawai'ian elites effectively developed their own system of craft specialization, exchange, ritual, and redistribution around a common resource in an insular tropical setting. Although there is a difference in scale in exchange networks, as in the case of Tonga, a compact monumental and ritual centre formed more of a symbolic focus for dispersed agrarian populations in island and archipelago settings rather than a 'lived-in', extensive urban environment.

AGRARIAN-BASED, LOW-DENSITY SETTLEMENT: A UNIFORM TROPICAL FOREST ADAPTATION?

Building on the nature of indigenously-developed agricultural strategies in tropical forests noted in Chapter 5 it is perhaps unsurprising that low-density settlement patterns are an attractive solution for the maintenance of large populations within an administrative system in tropical forest settings. Sedentary, 'compact' populations reliant on the intensive cultivation of a few staple plants would have had major impacts on often-fragile tropical soils (Beach et al., 2002). Moreover, in seasonally dry tropical regions, the congregation of a dense population on one or two major water sources would represent a risky strategy in the face of droughts or floods (Larsen, 2017). By contrast, the extension of low-density agrarian populations in a sprawl away from a centre, be it ceremonial or otherwise, often engaging in local hydraulic management and subsistence control, would place less pressure on resources, and allow diverse combinations of different agricultural crops, tree crop management, and domestic or wild animal corralling to continue to provide for ever-growing communities. This strategy can be clearly seen in the 'Forest Gardens' of Maya urban populations as well as the sprawling agrarian network of Angkor and enabled the persistence of extensive populations over centuries, and even millennia, when the elite centres themselves are excluded. Indeed, the extensive urban fringes around many ancient urban centres in tropical forests are drawing interest within present-day urban planning in these environments (Simon and Adam-Bradford, 2016).

Nevertheless, the above review clearly highlights differences in the 'agrarian' basis for low-density settlement, the role of a central state or elite, or even the existence of this trajectory altogether in different tropical forest formations and locales. Firstly, although occupants of many lowland Maya centres, Greater Angkor, Bagan, and Anuradhapura were reliant on particular agricultural staples (Coningham, 2006; Fletcher, 2012; Ford and Nigh, 2016), proximity of these resources to the religious and political centre varied. In the Classic Maya case, for example, a lack of major river transportation, no beasts of burden, and no metal tools likely resulted in independent, local groups. Indeed, at Tikal, a large population of independent small-scale farming groups existed well beyond the fall of the Classic Maya elite at the site (Webster, 2002; Webster et al., 2007). By contrast, a closer meshing of elite and agrarian interests around the large reservoir of Anuradhapura, Sri Lanka led to a significant decline in both these social groups in this region with the relocation of the court to Polonnaruwa and, later, southern Sri Lanka. In Southeast Asia the integration of residential farmers within the urban sprawl itself ranges from complete immersion throughout the 1,000 km^2 landscape of Greater Angkor, to a separation of dispersed shrine and palatial enclosures at Bagan from the agricultural heartlands by 50 to 200 km along river basins. This placed different strains on each associated landscape and also presented different challenges to the elite maintenance of large populations in a given political system.

The degree to which the basis of low-density urban forms in tropical forests would be considered uniformly 'agrarian' is also debatable. The fact that domesticated maize, squash, and beans supported Maya urbanism, and domesticated rice sustained Greater Angkor, Bagan, and ancient Sri Lankan urbanism is well-attested (Fletcher, 2009, 2012; Lucero et al., 2015). Certainly at Anuradhapura and Greater Angkor, complex irrigation networks demonstrate the importance of rice cultivation for the occupants of these massive urban networks. Nevertheless, small-scale farmers integrated tropical forest products into their subsistence regimes in both cases. In the case of the Maya 'Forest Gardens', mobile farming groups combined domesticated resources within cycles of forest regeneration and wild resource availability (Ford and Nigh, 2015). This mobility goes beyond what we might associate with usual ideas of agrarian-based urbanism and may also be documented in Sri Lanka. The smaller low-density settlements, or 'Garden Cities', in the evergreen rainforest of the Amazon Basin seem to rely equally on the corralling of freshwater turtles, fishing, the management of certain productive trees, the hunting of various mammals and birds, and the growth of certain domesticates such as manioc (Heckenberger et al., 2008), and carbohydrate-rich crops such as maize seem to have played relatively select roles even when introduced (Hermenegildo et al., 2017). While still fitting with ideas of dispersed resource use, the subsistence strategies of dispersed urban groups are not uniform, globally or locally.

There is also clear variation in the relative importance of a central hierarchical elite between the dispersed urban forms discussed above. The importance of religious and political control at monumental temple centres in the case of the Classic Maya, Greater Angkor, Bagan, Borobudur, and Anuradhapura is clear. In the case of Bagan and Borobudur these elites could even demand agricultural tribute from 200 km away and across oceans, respectively. Yet the control of the elite over different aspects of society clearly varied between cases. State maintenance and construction of reservoirs at Anuradhapura and Polonnaruwa led to a close integration of elite and agrarian society. In the cases of Greater Angkor and many of the lowland Maya centres, however, it remains dubious to what extent the elite held direct control of day-to-day maintenance of water management systems and small-scale farming organization. At the other end of the spectrum, in the Amazon case studies, there is almost a complete lack of evidence for the existence of an elite hierarchy at the centre of an urban network at all. In most cases, the development of the urban nodes, subsistence independence, and lack of grave good distinctions seem to have more of an affinity with the heterarchical form of governance and administration proposed for Jenné-jeno in Mali. Indeed, this may be why the 'Garden Cities' resemble the nodal landscape of this latter example, rather than the broader, continuous spread of Tikal and Angkor. Local dominance of hierarchical or heterarchical systems may also have varied *within* urban systems and Scarborough and Lucero (2010) note that heterarchies can rise as hierarchies disintegrate.

The development of low-density urbanism in other environments further highlights the broader potential advantages of this urban system in seasonal environments or those with a relatively low carrying capacity in terms of soil quality and nutrient prevalence. In the past, sustained low-density, and often ceremonially focused, settlements can also be found at the heart of the Mongol Empire (Fletcher, 1991) on the steppes of Mongolia that are vulnerable to overgrazing and become perilously cold during winter. Technically within the tropics, though in drier settings than those mentioned above, urban instances are also evident among the populations of the Inland Niger Delta, Mali and the Middle Senegal Valley. These populations spread out along rivers, relying upon dry banks and floodplains for cattle, pearl millet, sorghum, and African rice agriculture, as a resilient package against seasonal aridity or prolonged drought, from the third century BC onwards (McIntosh, 1995; Connah, 2001). By 800 AD Jenné-jeno, Mali, together with the smaller adjacent mound of Hambarketolo, had reached 40 hectares in area and housed between 10,000 and 30,000 people (McIntosh and McIntosh, 1980; McIntosh, 1995; Connah, 2001). In the Middle Senegal Valley, dispersed, but connected, settlements dominate the landscape prior to the Takrur Empire from 1000 AD. Nevertheless, the linkage between low-density urbanism and highly seasonal environments does not hold in all cases, particularly as the 'Edgeless City' becomes a characteristic of cities in the eastern coast of North America, western Europe, tropical Africa, and China alike (Lang and LeFurgy, 2003).

It is also clear that even within tropical forest settings pre-industrial urbanism does not always conform to low-density, dispersed models. As noted above, in the case of the states that formed in Tonga and Hawai'i, compact urban centres appear to have formed focal points for craft production as well as ceremonial political power. While populations did live within these settings, the main goal of the elite heads of these spheres was to project power across entire archipelagos and out across the Pacific Ocean in the case of the Tu'i Tonga Empire between 950 and 1,500 AD. It is likely that the delicate, insular ecosystems, that were prone to over-hunting and soil erosion (Kirch, 2000, 2001, 2010, 2012; Steadman, 2006; Swift et al., 2017a, 2017b), made such a basis of power more feasible than the incorporation of a large agrarian population. A similar situation existed at the ceremonial centre of Borobudur, where tribute was demanded across oceans in order to sustain what was primarily a ceremonial and monumental centre. In Sri Lanka, from the second millennium AD, compact urban centres developed in the evergreen rainforests of the Wet Zone, where the control of tribute, exchange, and agricultural hinterlands clearing ever-wet forests made intensive population more sustainable than the northern Dry Zone (Coningham, 1999, 2006; Fletcher, 2012). Finally, the smaller 'Garden Cities' of Amazonia, perhaps highlight the local reliance on natural, rather than anthropogenic, watercourses in the Amazon Basin. These different forms of 'urbanism' each had their own vulnerabilities that become apparent when we look at the declines of tropical forest urbanism in different regions.

AGRARIAN-BASED, LOW-DENSITY URBANISM: A UNIFORM RECORD OF COLLAPSE AND LESSONS FOR THE FUTURE?

Although appearing to be a prevalent trajectory towards urbanism in tropical forest settings, it remains the case that upon colonial arrival in North and Central America, Southeast Asia, South Asia, Oceania, and Africa, monumental urban centres and continuous extensive population had either disappeared from these regions completely or had been replaced by 'compact' urban forms (Fletcher, 2004; Lucero et al., 2015). For example, as Fletcher (2004) notes, in 1525 Hernán Cortés passed within a few kilometres of Tikal but did not deem the remnants significant enough to mention in his accounts (Webster, 2002). Similarly, in the nineteenth century, John Davy remarked 'Annoradapoora, so long the capital of Ceylon, is now a small mean village, in the midst of a desert' (Davy, 1821). Although, as will be seen in Chapter 7, the European powers, and the pathogens they brought with them, may have had a hand in some of the declines of urban centres in tropical forest settings, there do appear to have been vulnerabilities inherent to agrarian-based, low-density urbanism. In a recent paper, Lucero et al. (2015) highlight that the two biggest factors in the decline of these settlement forms in tropical forests were severe climate change and instability and internal tensions within a complex political system. In turn, these challenges may have important implications for the tendency towards these forms of settlement in the tropics, and beyond, today (Fletcher, 2009, 2012; Lucero et al., 2015).

The relationship between climate and social change is not simple, and social agency and resilience will play a considerable role in buffering external change while the impacts of climate change on local environments can be difficult to discern (Turner and Sabloff, 2012). Nevertheless, the decline of dispersed urbanism in Greater Angkor, northern Sri Lanka, and North and Central America appears to be associated with increased climatic, and environmental, instability (Lucero et al., 2015). Eight prolonged droughts between *c.* 800 and 930 AD discerned in a speleothem in Yucatán, Mexico apparently affected the availability of water for drinking and agriculture throughout the Maya lowlands, including within anthropogenic reservoirs (Medina-Elizalde et al., 2010; Lucero et al., 2015). At Greater Angkor and Sri Lanka a combination of severe droughts and deluges, highlighted by tree-ring data (Buckley et al., 2010, 2014) and a stalagmite from central India (Sinha et al., 2007), respectively, led to the drying-up and blocking of hydraulic channels, as well as the breaking of banks and catastrophic flooding of different regions (Lucero et al., 2015; Evans, 2016; Roberts et al., 2017a). This made the extensive agriculture required to support such large populations in these seasonally dry areas untenable, and the failure of the state to fulfil obligations of water provision led to the abandonment of ceremonial centres (Lucero et al., 2015).

Yet, as Lucero et al. (2015) highlight, in none of these cases was there a total social 'collapse'. In the Maya case some population congregations around areas of intensive agriculture did lead to soil erosion and deforestation (Beach et al., 2006), and a lack of access to permanent water sources left the southern Maya lowlands far more vulnerable to subtle climate change than the north, the coast, or higher altitudes with access to running rivers and standing lakes. In these latter regions, particularly in the highlands, 'compact' urban forms, influenced by

Classic Maya traditions, developed at strategic points on watercourses in the Post-Classic phase (Sabloff, 2007; Graham, 2011; Masson and Freidel, 2012; Lucero et al., 2015). Moreover, even in the southern lowlands, previously under-appreciated mosaic 'garden' approaches to agriculture and small-scale water management continued. After Greater Angkor and Anuradhapura, smaller, dispersed agrarian towns and villages developed in Cambodia and Sri Lanka, pursuing existing management and subsistence strategies, as the elite centres headed for more compact forms with reliable rainfall sources (Fletcher, 2012; Lucero et al., 2015). This variable dislocation and relocation of urban populations and traditions, rather than complete collapse, is what Lucero et al. (2015) call changes in the 'urban diaspora'.

In fact it is in this reorganization that we can find one of the major internal vulnerabilities of agrarian-based, low-density urbanism in tropical forests, the tension between an elite seeking to centripetally attract tribute and labour to a political centre and small-scale agrarian stakeholders centrifugally adapting to the impacts of external forces on their local subsistence concerns. Such tension is clear in the nature of reorganization in the Maya lowlands, Greater Angkor, and Anuradhapura. In all cases, often extensive agrarian settlement continues beyond the abandonment of an elite centre indicating that occupation is still viable in these locales. Bagan is an exception that really proves the rule. The agricultural landscapes this centre relied upon were 50–200 km away, in areas with more reliable water access, meaning that local soil erosion and environmental change were not pressing. In this case, the centre moved as a result of political reorganization and the waning ability of Bagan to demand tribute from so far away. It is also noteworthy that the new forms of urbanism that appeared after agrarian-based, low-density urbanism in Cambodia, Sri Lanka, and Central America come in a more compact form. This suggests that the concern of the elites had shifted to a more defensible, delimited area, where a reliable agricultural hinterland can be commanded effectively, and where external forces, such as changes in rainfall, are less damaging.

The unique challenges facing agrarian-based, low-density urbanism can also be seen when we look at the other, more 'compact' forms of urbanism in tropical forest settings. For example, the Tu'i Tonga Empire thrived as a result of the control of external trade connections and craft specialization, facilitating the rise of a single noble family to dominate the entire Tongan archipelago (Kirch, 1984; Neich, 2006). Nevertheless, such a means of control, as well as an urban form dedicated to its ceremonial legitimation, is vulnerable to internal disputes between competing elite families as well as any external shifts in trade relationships (Kirch, 1984; Neich, 2006). The political competition was even more fierce in Hawai'i between 1200 and 1800 AD as numerous island polities competed for the same island resources and ceremonial influence, with none able to bring enough political centrality to dominate exchange and ideology over the huge distances that separate the Hawai'ian archipelago from the rest of the Pacific (Kirch, 1990). Similarly, although Borobudur was an extensive ceremonial centre, the lack of a resident agrarian populace, and the reliance on maritime relationships for subsistence as well as political power may have contributed to its demise and the shift in power to elsewhere in Southeast Asia (Miksic, 1999; Stark, 2015). In these cases, the fall of the urban centre was based on the fragility of the dominant political and economic elite. The small-scale farmers remained in their independent villages throughout these changes, often at some great distance away.

The tendency towards eventual population dislocation and political shifts to more compact forms has been taken as a warning that agrarian-based, low-density urbanism was not viable over the long-term. For example, it is notable that rugged, compact urban networks have dominated the Near East and Europe from 5,000 years ago (Fletcher, 2009, 2012). While particular settlements may rise and decline, the network never fails and is not completely abandoned. Indeed, the majority continue to play a prominent role in the urban and political landscapes of their respective nations today (Fletcher, 2012). For example, Rome, while facing multiple invasions and abandonment as a political centre from mid-way through the first millennium AD, is now the capital city of Italy, seemingly refuting its part as the centre of a Gibbonian 'fall' (Wickham, 2009). Similarly, while urbanism in Roman Britain is often perceived as a fleeting phenomenon (Frere, 1987; Reece, 2002) it is notable that Roman towns and even encampments are at the heart of major cities in the United Kingdom today. By contrast, dispersed urban settlements in environments beyond the tropics, located in difficult seasonal environments and facing issues of political integration, such as the Mongolian capital Karakorum, were often short-lived, leaving few remains on the landscape today beyond 'ruins' and tourist sites (Kato, 1997; Roth et al., 2002; Rogers et al., 2005). As such, recent trends towards extensive low-density urbanism in the modern world are not necessarily seen as beneficial (Fletcher, 2012; Lucero et al., 2015).

Nonetheless, it is worth noting that the dispersed urban centres reviewed above existed in tropical forests for many centuries, and even over a millennium (Figure 6.15). This is currently far longer than any industrial city in a tropical region, making any qualitative judgement perhaps premature. Anuradhapura, for example, dominated the political and urban landscape of Sri Lanka for over 1,000 years, while Greater Angkor and Tikal lasted for 700 and 900 years, respectively (Fletcher, 2010). In the case of Tikal, extensive agricultural populations persisted well beyond this, and small-scale farmers persist in the region today. Similarly, in Greater Angkor, settlement redistributed into dispersed towns along the edge of the previous urban sprawl. In the case of the Amazonian Basin, extensive urban populations lived in 'Garden Cities' right up to the arrival of colonial powers and European diseases that, as will be seen in Chapter 7, decimated millions worldwide (Heckenberger et al., 2008; Heckenberger and Neves, 2009; McMichael et al., 2017). Furthermore, it is clear that the extension of population over a wider landscape and the integration of forest lifecycles and intimate local ecological and hydraulic knowledge into the subsistence patterns of extensive urban networks can provide a remarkably stable form of occupation in often delicate, and changeable, tropical forest, and even rainforest, environments (Watling et al., 2017a).

As a result, it is now clear that extensive urban populations can be, and were, frequently sustained in tropical forest settings. Indeed, there is much more work to be done in investigating the extent and landscape modifications of large human populations in the past in all tropical forest regions. The benefits of remote sensing, and particularly LiDAR, in this regard are becoming evident as forest is virtually removed to reveal the land features below (Chase et al., 2011; Evans, 2016). As more and more of the world's human population moves into the tropics over the next century (State of the Tropics, 2016), the resulting, growing urban centres in these regions will face challenges—including seasonal precipitation,

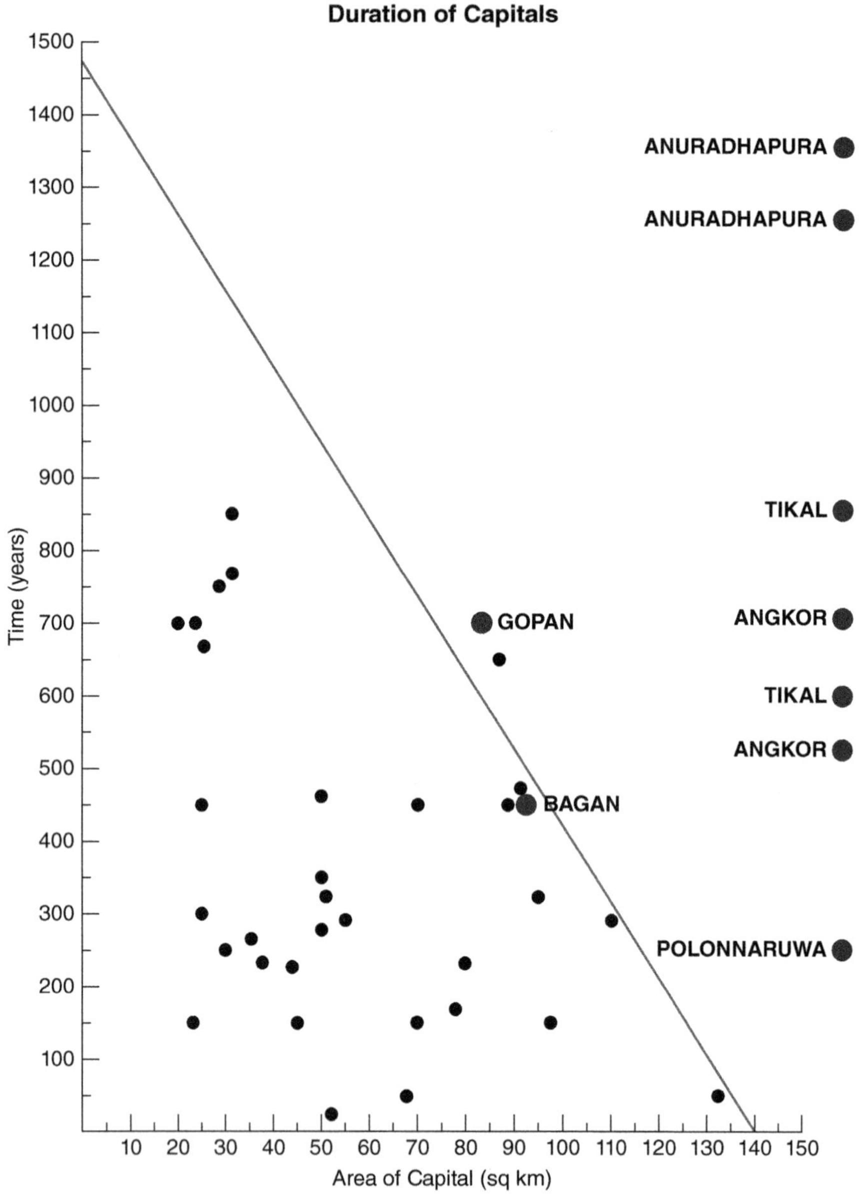

Figure 6.15. Graph showing the duration of urban settlements relative to their area. Key 'low-density agrarian urban' settlements discussed in the text are shown with larger circles. Modified, with permission, from Roland Fletcher (2010).

global changes in temperature and precipitation, climatic variability, disease, and the challenges of sustainable resource use—familiar to the case studies discussed here. The past demonstrates that urban populations can develop highly innovative and complex solutions to these issues, resilient in the face of persistent and extreme climatic variability. Local knowledge and independence on the part of smaller stakeholders is essential if this is to be effective on a region-by-region basis. It will also ensure the longevity of populations through periods of considerable stress. The greatest challenge for our own ideas of governance and bureaucracy, however, is perhaps how to successfully maintain large agglomerations of small-scale domestic, religious, and industrial interest groups united under a common ideology and administrative system in the face of such external challenges.

7

The Last in a Long Line

Historical and Ethnographic Tropical Forest Encounters

> They moved us around and we never stayed in one place for very long. After that I was sent out . . . away from the rainforest.
>
> —Lizzie Wood, *Jirrbal* elder quoted by Åsa Ferrier (2015)

On the basis of the plethora of evidence for tropical forest agricultural practices and urban networks in Chapters 5 and 6 it seems somewhat surprising that these environments, and their human occupants, could ever come to be seen as static, primordial, or 'elusive' (Biesbrouck et al., 1999). Yet, today, Indigenous peoples living in tropical forests are often depicted as being isolated from the outside world, and as equally, passively threatened by external agricultural, economic, and political forces as the habitats in which they reside. As I showed in Chapter 2, the hunting and foraging practices of these groups can provide useful insights into how our prehistoric ancestors may have made a diverse living in environments that have frequently been considered too poor in crucial resources for long-term human occupation. Somehow, however, these parallels and comparisons have also seen these groups framed as relics of some of the earliest members of our species. This has been encouraged by claims that some of these groups genetically represent 'archaic' lineages of *Homo sapiens* that survived in dense forest habitats in different regions (Endicott et al., 2003; Ranaweera et al., 2014; Ranasinghe et al., 2015). Whether this is the case or not, it is now clear from historical and ethnographic information that tropical forest foragers, the world over, have been involved in complex economic and political networks to varying extents at different points in time. Moreover, the present cultural and subsistence systems of many of these groups have been significantly affected by the infiltration of colonial and imperial regimes from the seventeenth century onwards, as well as the more recent, disruptive effects of global capitalism.

This chapter is an attempt to document how Eurocentric concerns with 'exploration', developments in literature, modern conservation movements, and the 'pristine' hunter-gatherer debate have contributed to the removal of tropical forest societies from history and their placement into isolated, primeval conceptions of tropical forest environments. In response to this, I review evidence for historical and ethnographic connections of tropical forest hunter-gatherers, and agriculturalists, with societies in neighbouring territories. In many cases it is clear that these groups have been essential to the functioning of trading networks and economies on a global scale (Morrison and Junker, 2002). It is also evident that colonial forces have shaped the subsistence and geographic situation of many of these groups meaning that their current practices should not be seen as primordial (Headland and Reid, 1989). Indeed, I will also discuss the diverse agency shown by Indigenous tropical forest societies as they engage in, and try to take advantage of, the modern capitalist system that continues to encroach on their territory and values. Nevertheless, while extensive connections and dynamism have clearly been a part of recent human history in tropical forests, this should not detract from the Indigenous knowledge and adaptive solutions such groups have developed for sustainable occupation in tropical forest settings, as well as their ongoing resistance to ways of life that demand intensive forest clearance and economic exploitation. The loss of this knowledge is one of the greatest threats to the human capacity to coexist with these environments today.

DISAPPEARANCE FROM HISTORY

The disappearance of tropical forest peoples from history is inextricably linked to the fact that history, as we know it, is inherently Eurocentric. At the beginning of European ideas of history, the ancient Romans and Greeks classified anything beyond their realms as unknown and populated by sub-human monsters (Brotton, 2014). Even where novel environments in Asia were explored and surveyed, as in the case of Alexander the Great's military movement into India, monsters and demons still played prominent roles as the occupants of landscapes beyond the shores of Europe and the Middle East (Brotton, 2014) (Figure 7.1). This inward focus continued into the Middle Ages with the rise of Christianity that placed Jerusalem at the centre of the world (Brotton, 2014). A Eurocentric focus is also evident in ideologies of social and human development. The Roman writer Lucretius (first century BC) wrote about a progression of technological and social complexity through time from the stone, to the bronze, and then the iron ages. The appropriation of Classical references and values during the Renaissance and Enlightenment periods meant these concepts continued to play a strong role in European ideology and they formed the basis for the nineteenth-century Three Age System of C. J. Thomsen (Stone, Bronze, Iron) that was one of the first European modes for classifying archaeological materials and the prehistoric development of human groups (Daniel, 1975; Trigger, 1989).

One of the biggest challenges to the European psyche was the fifteenth- and sixteenth-century discovery of populations in the New World, and in other areas occupied by colonial powers, that still utilized stone tool technology and predominantly practised hunting and gathering (Trigger, 1989). These people were clearly human but were not Christian, nor, in many cases, did they have metal technology, meaning that they fell outside traditional ideas of human advancement. Some argued that they represented human groups in decline at the far corners of the world in the absence of Christianity while others, such as Herbert Spencer, Lewis Henry Morgan, and Edward Tylor, suggested that they had simply

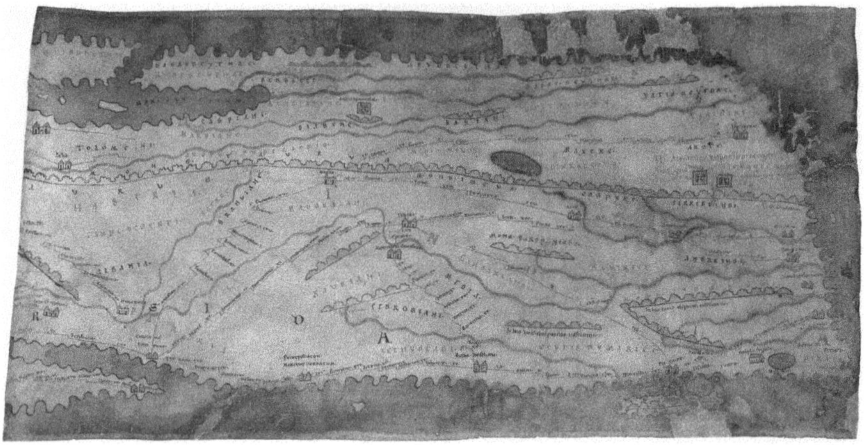

Figure 7.1. Section of the 'Tabula Peutingeriana', an illustrated roadmap showing the layout of the road network of the Roman Empire. The map is a thirteenth-century parchment copy of a possible Roman original. In this panel, the southeastern edge of the known world is shown in the form of India. The edge is inhabited by 'pirates' and beyond there is nothing. Housed and scanned at ÖNB/Wiene and ANL/Vienna, Cod. 324. Reproduced with permission.

progressed more slowly along universal technological and social trajectories than Europeans (Trigger, 1989) (Figure 7.2). In the cases of Morgan and Tyler these groups were placed towards the base of a continuum, between more 'primitive' and 'civilized' societies (Trigger, 1989), defined on the basis of religion, political form, and technology. Morgan and Tyler also suggested that human societies passed through the organizational stages of band, tribe, chiefdom, and state that formed part of the basis of neo-evolutionism and traditional approaches to social complexity in archaeology. As pointed out fairly swiftly by Franz Boas (1911, 1940), however, these perspectives denied these societies their own history, forcing them into categories based on European political and technological values that are often irrelevant in different environmental or cultural contexts.

There was also the additional factor of the emergence of Europe as a global economic power. Eric Wolf, in his book *Europe and the People without History* (1982), famously argued that it was the development of unequal 'tributary' relationships between European states and Indigenous groups in various parts of the Americas, Asia, Africa, and Oceania that really removed their historical agency in ethnographic and historical documentation. He discussed how colonial states created structures, and even 'tribes', in order to control and administer access to precious metals, furs, and slaves in different parts of the world (Wolf, 1982). Thus societies became essentialized by a given name and were considered passive in the face of colonial machinations and economic and political strategy. In a World Systems context (Champion, 1989), Indigenous societies effectively became a 'periphery' that was manipulated and acted on by the European core (Roseberry, 1989). This core extracted raw materials necessary for the growing global capitalist system, while the periphery became enmeshed in increasingly

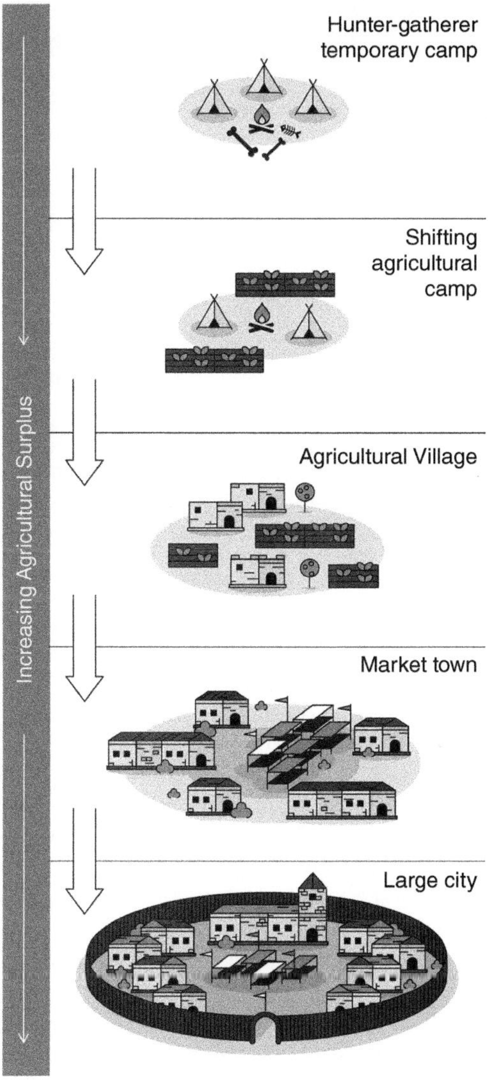

Figure 7.2. Schematic of social evolutionary theory in anthropology and archaeology based on the writings of White (1959) and Service (1975). With thanks to Hans Sell.

unequal economic relationships and lost their voice in western histories. Things happened to them rather than them being considered active participants in new commercial or cultural forms (Wolf, 1982; Morrison, 2018a). A similar line of thinking can also be seen in the gradual development of 'Orientalism' which Edmund Said (1978) describes as the characterization of Middle Eastern, Asian, and North African culture as static and undeveloped relative to the European west.

A clear example of this can be seen in the fifteenth-century carving up of the tropics between Spain and Portugal. On the largest scale, the Treaty of Tordesillas, signed on 7 June 1494, divided newly discovered lands beyond Europe between the Portuguese Empire and the Kingdom of Castile. The Portuguese obtained Africa, as well as much of Asia, and a thin strip of South America because their navigators had already rounded the Cape of Good Hope. Spain on the other hand received almost all of the (still largely undiscovered) Americas (Hemming, 2009; Brotton, 2014). The Indigenous populace of these regions remained unaware of the transition in their political status. Similarly, in western Africa in the eighteenth and nineteenth centuries, the British created a set of territories, including Sierra Leone, the Gambia, Nigeria, and the Gold Coast, with which to control and block trade, and administer the attainment of exotic resources from the interiors of these countries. In many cases Indigenous rulers were supported to govern the interiors, with tribal loyalties coming second to the benefits of patronage in administering an imperial protectorate (Tattersfield, 1998; Mitchell, 2005). Perhaps the greatest dehumanizing example of de-historization is the trans-Atlantic slave trade whereby people from across Africa, albeit building upon indigenous economies of slavery (Mitchell, 2005), were taken away from their traditional lands and families throughout the eighteenth century in order to boost colonial infrastructure projects and productivity in sugar and cotton plantations (Tattersfield, 1998; Christopher, 2006; Lindsay, 2008). History and culture were ignored, as to acknowledge humanity would be to admit to mass-murder.

The removal of Indigenous groups from European history has been particularly stark in tropical forest environments for a few reasons. Firstly, pre-colonial tropical forest groups have produced very few surviving historical records. While religious inscriptions and texts from Sri Lanka and Southeast Asia (Higham, 2001; Morrison, 2018b), monumental dates such as the Maya Long Count system (Webster, 2002) (Figure 7.3), and references made in merchant texts to cities in Africa and Asia provide some insight into past ideologies, events, and social structures (Mitchell, 2005), by and large Europeans have provided the dominant literary records for these regions and environments. In many cases this is because complex oral histories and rituals form the basis of social memory in groups that inhabit tropical forests today (Lévi-Strauss, 1966; Turner, 1967; Basso, 1995), though poor preservation of organics also limits the available textual corpus were it to be present. Where the large monumental structures and works of art discussed in Chapter 6, even in ruinous form, unavoidably demonstrated the long, rich history of tropical communities, greater interest was placed on the discovery of material wealth rather than the achievements of ancient peoples. The classic example of this is Gonzalo Pizarro and Francisco de Orellana's quest for 'El Dorado' that followed from the conquest of the Inka by the former's older brother (Medina, 1934; Hemming, 2009). A subsequent AD 1541 expedition saw Orellana become the first European to navigate the entire Amazon River—in spite of mass mortality and desertion and a follow-up voyage that ultimately led to his death.

Yet, perhaps the main reason these groups have been absent from history, and the reason there has been traditional doubt relating to dense agricultural or urban settlements in the New World, is the fact that the warfare, slavery, translocation, and diseases brought by colonial powers to tropical forest settings in North, Central, and South America decimated millions of human tropical forest

Figure 7.3. Ancient Mayan hieroglyphs on the wall of a building at the Classic Maya site of Copan, Honduras. Photograph taken by Vladimir Korostyshevskiy and reproduced with permission from shutterstock.com.

inhabitants (Watts, 1997; Roosevelt, 1999; Hemming, 2009; Lewis and Maslin, 2015). Similarly, in the Old World, the gradual growth of nation-states and global markets (Deraniyagala, 1992), the dislocation of people from their traditional environments and migratory paths (Ferrier, 2015), the international repercussions of the Atlantic slave trade (Mitchell, 2005; Lindsay, 2008), and the writing of history by state centres and European powers (Morrison, 2014), led to an apparent isolation, foraging-focus, and small-scale of groups living in the tropical forests of South Asia, Africa, and Southeast Asia by the time ethnographers came to document their existence. It is becoming increasingly clear, however, that the settlement patterns, subsistence practices, and population sizes of these Indigenous peoples have had a dynamic history. Furthermore, as archaeologists, historians, and anthropologists delve deeper into the interactions between Indigenous peoples and expanding industrial markets, nation-states, education systems, and state- and European-imposed belief systems, it is evident that there are multiple Indigenous histories that tendencies to focus on universal characterizations of 'noble savages' or 'pristine tribes' have failed to adequately express.

THE ISOLATED 'NOBLE SAVAGE' AND 'PRISTINE' HUNTER-GATHERERS

As early as 1754, the Genevan writer Jean-Jacques Rousseau criticized Hobbesian trajectories of social evolution that framed western society as the pinnacle of civil, religious, and moral development. Instead, Rousseau (1992) argued that humanity is best in its 'primitive state, when placed by nature at an equal distance from the stupidity of brutes and the fatal enlightenment of civil man'. For Rousseau, it was in fact the 'savage' that represented the peak of moral and political advancement in contrast to decadent and immoral 'civilization'. Although Rousseau never used the term, this perspective has been associated with the concept of the 'noble savage' (first used in a play *The Conquest of Granada* written by John Dryden in 1672) (Miner, 1972; Ellingson, 2001) (Figure 7.4). In this line of thinking, Indigenous peoples from around the globe became the 'innate' embodiment of natural morality and the best examples of humanity living harmoniously with its environment. Rousseau's writings had a large influence on anthropologists, including Claude Lévi-Strauss, as Indigenous groups became an important focus of study in

Figure 7.4. Benjamin West's portrayal of the Native American in 'The death of Wolfe' painted in 1770 has been cited as an example of western perceptions of the 'noble savage' (Fryd, 1995). ART Collection/Alamy Stock Photo. Reproduced with permission from Alamy.com.

their own right, with their subsistence, political, classificatory, and ideological systems having much to teach western society (e.g., Lévi-Strauss, 1963, 1966).

In a tropical forest context this perspective was not unique to Rousseau, and many of the earliest Spanish colonial administrators, soldiers, and explorers wrote favourably of Indigenous peoples' morality, ideologies, and adaptive relationship with the environment (Figure 7.4). As noted by Hemming (1995: 5), Pero Vaz de Caminha, a Portuguese explorer writing about Indigenous tropical forest dwellers in Brazil stated 'they are stronger and better fed than we are with all the wheat and vegetables we eat . . . their bodies are so clean and plump and beautiful that they could not more be so'. Amerigo Vespucci, the Italian explorer and cartographer who first demonstrated that South America was part of a truly 'New World', described Indigenous peoples and their bountiful environment as a terrestrial 'paradise' (Morison, 1993; Hemming, 2009). Later, in the African continent the writings of the anthropologist Colin Turnbull (1961, 1965) romantically described the Mbuti 'pygmy' populations as harmoniously living within the Equatorial rainforest. Indeed, many of the Bantu neighbours of so-called 'pygmy' populations have associated them with the 'bush' and the 'forest' or even an 'invisibility' (Schadeberg, 1999). Similarly, although acknowledging semantic issues with the term 'pygmy' and the lack of a unified culture across the African rainforest belt, Hewlett formed a new category of 'tropical forest forager' that was characterized by a minimum period spent hunting and gathering and identification with a preference for forest life (1996).

In archaeology and anthropology, tropical forest hunter-gatherers were a leading part of the 'Man the Hunter' symposium that sought to universally characterize 'hunter-gatherers' as sharing small group size, mobility, and egalitarian social organization (Lee and DeVore, 1968; Leacock and Lee, 1982). A hunter-gatherer 'mode of production' was also promoted in Marshall Sahlins' influential 'Original Affluent Society' idea (1968, 1972). Drawing heavily on Turnbull's writing and datasets relating to Mbuti 'pygmy' subsistence (Frankland, 1999), this highlighted how the emergence of farming or urbanism should not necessarily be considered progress. Hunter-gatherers from around the world have a hugely varied diet, particularly in tropical forest settings as we have seen in Chapter 2. Through intimate knowledge of their resources, hunter-gatherers can transform environments, such as tropical forests, that seem unproductive from a Eurocentric perspective, into bounties. Sahlins argued that the subsistence strategies of these groups are so efficient that they are left with a large amount of 'leisure' time and, by foraging for only their immediate needs and not for a surplus, hunter-gatherer societies avoid the waste and materialism of the western world. 'Pristine' examples were sought, uninfluenced by agricultural communities or capitalist societies, in order to develop a direct behavioural analogue of prehistoric human societies focusing on hunting and gathering (Lee and DeVore, 1968; Sahlins, 1972; Johnson and Earle, 1987).

This perspective, while painting Indigenous foraging communities in a more positive light than previous neo-evolutionary perspectives, still imparted an ahistorical primordiality on these groups. In tropical forests this was compounded by ideas that these environments were impenetrable and even today only home to small-scale societies (Meggers, 1954, 1971). The growing academic and public popularity of the idea of the 'noble savage' stimulated a shift in conservation strategies in tropical forests in the 1960s and 1970s. Perceptions of incompetent land-use by Indigenous groups were replaced by ideas of 'Ecologically Noble Savage[s]' (Redford, 1990). Comparative ecological studies showed high biodiversity being associated with Indigenous groups and low biodiversity associated with non-Indigenous groups (Redford and Robinson, 1987; Borgerhoff and Coppolillo, 2005). During the Earth Summit in Rio de Janeiro in 1992 many international non-governmental organizations (NGOs) used Indigenous groups in the Amazon to advance their own agendas as case studies of cultures with a strong conservation ethic (Hames, 2007) and as 'guardians of the forest' (Conklin, 1997). This image is designed to stimulate donations from wealthy members of the western world that in turn has been influenced over the last half century by the emergence of the 'modern' Environmental and Wilderness Preservation movements (Bosso and Guber, 2006; Carter, 2007). However, in doing so, Indigenous hunter-gatherers are essentialized as part of the tropical forest itself and equally susceptible to threat (Conklin, 1997; Hames, 2007).

In a world of increasing industrialization and urban expansion there has also been a growing public fascination with nature, often epitomized by forests. This is encapsulated in the book *Trees: Reflections and Poems* by the German poet Herman Hesse that was quoted at the start of Chapter 2. He argued that forests provided ideal environments for notions of truth, beauty, and happiness. As Lindsay Baker (2017) notes in a recent BBC article '"tree time" is enjoying a resurgence in popularity—on social media the tree hashtags #treesofinstagram and #lovetrees are rapidly proliferating'. She also points out that an increasingly popular type of health care in Japanese medicine is *Shinrin-yoku* or 'forest bathing' (Figure 7.5). There is also growing interest in more exotic, tropical forests from the western public. Television documentaries such as Bruce Parry's 'Tribe' and 'Amazon' (BBC, 2007), as well as competition shows such as Bear Grylls' 'The Island' (Channel 4, 2014) and 'I'm a Celebrity Get Me Out of Here' (ITV, 2002), seek to paint a similar isolationist idea of tropical forests and the unique, and intentionally shocking, human adaptations required to sustainably survive there. There has also been a similar association of hunter-gatherer 'palaeodiets' with a prehistoric well-being that we have lost sight of (e.g., Wolf, 2010). All of this has fed into the exhibition of Indigenous tropical forest groups as living an affluent life, in tune with the natural environment. Their isolation and protection should be maintained as a showcase for those wishing to return to this earlier state of being.

While this perspective rightly emphasizes the wealth of environmental knowledge Indigenous tropical forest groups often have (Balée, 1994; Hames, 2007), it denies Indigenous groups their history and dynamism in the same way as cultural evolutionism. Firstly, as has been seen in Chapters 4, 5, and 6 tropical forest hunter-gatherers, agriculturalists, and urban dwellers have all modified their environments significantly, including deforestation, topographic re-modelling, and regular forest burning. In many places significant tropical forest communities existed prior to colonial arrival, and the small groups seen today are the product of historical factors rather than instances of a primeval state (Brosius, 1999). Moreover, Indigenous groups do not always intentionally 'conserve' the forest but rather local knowledge, low population density, different technologies, and limited economic markers inevitably lead to a more sustainable existence (Hunn, 1982;

Figure 7.5. '*Shinrin Yoku*' (Forest bathing) was introduced into Japan in the early 1980s as a form of nature therapy. Photograph of a forest pathway in the Shirakami Sanchi World Heritage Reserve, Japan taken by eye-blink and reproduced with permission from shutterstock.com.

Redford, 1990). Perhaps the greatest problem, however, is that these views, no matter how well-intentioned, neglect the agency of Indigenous groups, seeing them as being passively threatened by external forces in their isolated wilderness. Yet, what is clear from a number of ethnographic, ethnohistoric, and historical works is that, far from being isolated, Indigenous tropical forest communities the world over have sought or been part of connections in both pre- and post-colonial settings (Morris, 1982; Junker, 1999; Morrison and Junker, 2002). In addition, populations have agentively adapted or maintained their subsistence strategies as part of wider, changing political and economic contexts.

DOCUMENTING CONNECTIONS

To counter the ideas of tropical forest hunter-gatherer communities as 'pristine' and isolated, Headland and Reid (1989) and Bailey et al. (1989) famously surveyed a series of ethnographic and historical communities inhabiting tropical forest environments. Taking case studies from tropical forests around the world, they demonstrated that supposedly 'isolated' groups had, in fact, been in consistent historical contact with adjacent agricultural societies. For example, the term 'Philippine Negritos' has been used to describe varied linguistically diverse groups in the Southeast Asian island chain. Today, most of these groups practise some form of horticulture and are part of an intense trade of forest products, with rice providing the main staple and wild starch foods sustaining only a small portion of meals (Headland and Reid, 1989). However, some, such as the Agta of eastern Luzon, still reside in small nomadic tropical forest camps and hunt wild boar, deer, and primates with a bow and arrow (Griffin and Headland, 1985; Reid, 1987; Headland, 1988) (Figure 7.6). This led to suggestions that these populations managed to maintain an isolated, prehistoric life that mimicked the first Late Pleistocene humans in this part of the world (Stewart, 1954; Solheim, 1981; Bellwood, 1985). Nevertheless, Headland and Reid used historical records to document symbiotic Agta trade relationships with nearby farming groups dating as far back as the eighteenth century (Figure 7.6). Seemingly, these groups brought forest products to local Christianized farmers in exchange for rice, tobacco, metal tools, beads, and pots (Santa Rosa, 1746, cited in Pérez, 1928; Headland and Reid, 1989) and were far from isolated.

Similarly, Turnbull's (1961, 1965) writings about the Mbuti 'pygmy' groups of Central Africa had framed these forest foragers as affluent and they have often been cited as the original human occupants of the African tropical rainforest (e.g., Mann et al., 1962; Cavalli-Sforza, 1986). Campbell (1983) suggested that these populations were 'independent forest groups'. I used ethnographies of Mbuti populations in Chapter 2 to highlight the diversity of potential hunting techniques utilized by human groups, including trapping and net capture, in bringing down tropical forest prey. Yet, Mbuti pygmy groups also show close socio-economic

Figure 7.6. Photograph of Agta men in the Philippines floating hardwood down river from the Sierra Madre Mountains. The key position of forest-dwelling populations in national and international trading systems can be traced back at least into the Iron Age and historic periods in Southeast Asia. Photograph taken by Jacob Maentz, Alamy Stock Photo. Reproduced with permission from Alamy.com.

relationships with neighbouring populations, including intermarriage (Bahuchet and Guillaume, 1982; Headland and Reid, 1989; Hewlett, 1996). Bailey et al. (1989) also suggested that well over 50 per cent of Mbuti calories come from foods grown by neighbouring agricultural villages (see also Ichikawa, 1983). This African case, and the perceived challenges of dense vegetation, spaced protein and carbohydrate resources, and seasonal flux, even drove arguments by these authors that independent tropical forest life was impossible in the absence of food production based on crops of an ultimately exotic origin (Hart and Hart, 1986; Headland, 1987; Bailey et al., 1989). In particular, Hart and Hart (1986) suggested that for almost half a year no high calorie forest fruits or seeds were available to foragers, vegetative foods were often scarce, and prey animals generally had a low fat content leaving independent forest life essentially impracticable (Bailey et al., 1989).

The same line of argument can also be found in Bailey et al. (1989) and Headland and Reid's (1989) treatment of South Asian 'forest' societies. The Sri Lankan Wanniya-laeto (Vedda), Andaman Islanders, and Nicobarese Islanders were all seen as tropical forest foragers, and ancient remnants of early human occupants of this region. For example, ethnographic analysis suggested that *Dioscorea* yams (Spittel, 1924, 1961), wild date palms (*Phoenix pusilla*; Sarasin and Sarasin, 1893), wild breadfruit (Sarasin and Sarasin, 1893), and other tropical forest plants (Spittel, 1924, 1961) met the majority of Wanniya-laeto starch requirements in the early twentieth century in the semi-evergreen and deciduous tropical forests of the 'Intermediate' rainforest of Sri Lanka (Roberts et al., 2015b). Similarly, bow and arrow hunting and trapping of large reptiles and small- to medium-sized mammals enabled adequate protein to be obtained (Bailey, 1863; Sarasin and Sarasin, 1908; Seligmann and Seligmann, 1911). However, as with the Agta, it was argued that the Wanniya-laeto had economic contacts with agriculturalists as early as the seventeenth century (Knox, 1981; Seligmann and Seligmann, 1911), and traded forest produce such as honey, wax, dried venison, and elephant tusks with local Sinhalese communities for cultivars such as rice and millet, alongside cloth, iron arrowheads, and axes, throughout the nineteenth and twentieth centuries (Sarasin and Sarasin, 1893; Seligmann and Seligmann, 1911; Spittel, 1924).

Although the Andaman Archipelago is composed of an arc of 200 islands in the Bay of Bengal only a few of these, including the Great and Little Andaman islands, are still occupied by Indigenous inhabitants who are known for a distinctive phenotype and supposed affiliation with early human occupants of South Asia (Man and Temple, 1880; Radcliffe-Brown, 1932). Ethnographies from the nineteenth and twentieth centuries (Dutta, 1978; Radcliffe-Brown, 1932; Cipriani, 1966) suggest that they obtained the entirety of their protein and carbohydrate needs from hunting and gathering, including small animals such as snakes and rats (Dutta, 1978), fishing, and the consumption of the seeds of *Cycas* and *Entada phaseolides*, and the pith of *Caryota sobolifera* (Radcliffe-Brown, 1932). *Dioscorea* yams, honey, and *Artocarpus sp.* (jackfruit) (Cipriani, 1966) also formed important resources. Nevertheless, Bailey et al. (1989) argued that it was population displacement from the coastal regions of the islands during the colonial influence and migrations from the Indian subcontinent of the nineteenth century that encouraged the Andamanese to utilize 'forest' resources. This drew upon the work of Cipriani (1966) that suggested that the Jarawa group was forced inland by recent immigrants to subsist on wild pigs obtained with the help of dogs which arrived on the island during the 1800s, as well as some marine resources gathered on occasional beach visits. In addition, recent archaeological investigation shows no evidence for occupation of this island chain prior to 2,000 years ago (Cooper, 2002). An analogous scenario has also been suggested for the Indigenous occupants of the Nicobar Islands (Prasad et al., 2001).

Modern genetic studies of these groups have also provided increasing evidence for contact with other populations. Y-chromosome and mitochondrial DNA (mtDNA) of Filipino populations, for example, does not support a simple dichotomy of 'Negrito' versus non-Negrito, suggesting that 'Negrito' populations such as the Agta were influenced by Holocene migrations in the region and were not isolated (Delfin et al., 2011, 2014). Similarly, genetic studies of mitochondrial, Y-chromosome, and nuclear genomes in 'pygmies' versus other African populations demonstrate asymmetrical gene flow from non-pygmy populations into 'pygmy' populations, with subsequent isolation occurring more recently (Verdu et al., 2009; Verdu and Destro-Bisol, 2012; Verdu et al., 2013) (Figure 7.7). Recent mtDNA work on the Wanniya-laeto communities of Sri Lanka potentially demonstrates similar local admixture with Sinhalese and Tamil populations (Ranaweera et al., 2014), while Indigenous populations of the Great Andaman island show a degree

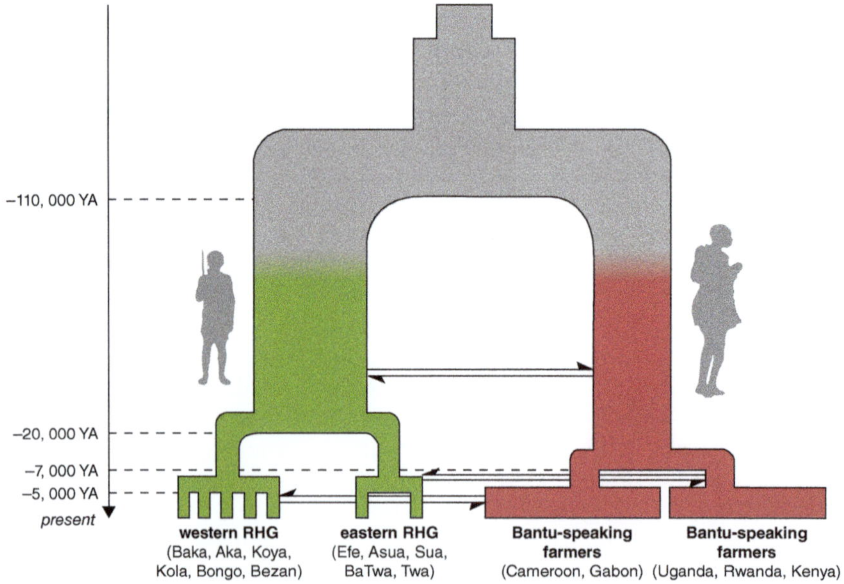

Figure 7.7. The illustration shows a demographic model of 'pygmy' and non-pygmy populations from Central Africa. The model is currently that most supported by genetic studies of autosomal markers discussed in the text. The sizes of branches are proportional to the effective size of populations. Arrows represent admixture between populations. Created by, and used with permission from, Etienne Patin.

of relatively recent genetic admixture from adjacent South Asian populations while also sharing some significant genetic ancestry with Malaysian 'Negrito' groups (Thangaraj et al., 2003, 2005, 2006; Reich et al., 2009; Chaubey and Endicott, 2013). mtDNA variation in Nicobarese Islanders also points towards recent migrations to these islands from Southeast Asia (Prasad et al., 2001).

Together, these data suggest that perceptions of 'pristine', isolated tropical forest forager are wrong in many cases. However, it has also denied these populations adaptive agency in these environments. Bailey et al. (1989) and Bailey and Headland (1991) took this evidence to indicate that no hunter-gatherer population could occupy tropical forests without access to agricultural trade or production and that these habitats were unsuitable for long-term human occupation. In doing so, this 'revisionist' work dichotomized 'hunter-gatherers' and 'farmers' in much the same way as the previous focus on 'pristine' foragers had. The evidence reviewed in the previous chapters demonstrates this to be an invalid over-stretch of the data. Firstly, it is clear that independent tropical forest foraging is possible, not only in the past but also in the present. Here, it is telling that Bailey et al. (1989) and Headland and Reid (1989) do not provide a thorough review of South American tropical forest groups, who have provided an abundance of evidence for specialized tropical forest subsistence (Posey, 1985; Balée, 1989, 1992, 1994, 2013; Posey and Balée, 1989). As reviewed in Chapters 1 and 2, even the Southeast Asian, South Asian, and African populations they do cite have been shown to be able to rely upon tropical forest resources for all of their subsistence requirements. Moreover, as shown in Chapter 5, the boundary between species management practices associated with hunting and gathering and farming is much more murky than Bailey and Headland (1991) would care to suggest.

Thus, rather than indicating passive influence from the outside, evidence for genetic admixture and trade may instead hint at Indigenous agency in the formulation of new subsistence and social frameworks (Bird-David, 1992, 1997). Indeed, although the genetic evidence from these groups tends to document population admixture with adjacent groups, it also hints at long-term tropical forest adaptations. For example, Perry et al. (2014) used genome-wide admixture mapping analysis to identify genomic regions that are significantly associated with the 'pygmy' phenotype in the Batwa, a rainforest hunter-gatherer population from southwestern Uganda. The identified genomic regions have multiple attributes that provide supporting evidence of genuine association with the 'pygmy' phenotype, including enrichments for single-nucleotide polymorphisms (SNPs) previously associated with stature variation in Europeans and for genes with growth hormone receptor and regulation functions. The genomic regions are also enriched in molecular signatures of positive selection in the Batwa, but not in the Baka of Cameroon and Gabon. Although all pygmy populations have separated relatively recently (in the last 20 ka; Patin et al., 2009), these results suggest that the 'pygmy' phenotype has developed independently in western and eastern Central Africa (Figure 7.7). Perhaps the most significant problem with Bailey and Headland's backlash against perspectives of isolated tropical forest foragers is that, while it rightly emphasized the cultural and economic history of these groups, it still painted them as passively forced into new lifeways, new environments, or novel patron–client trading relationships.

DRIVING THE TRADE: TROPICAL FOREST GROUPS AS ECONOMIC AGENTS

Focus on whether tropical forest hunter-gatherers are 'pure' or 'isolated' misses the point that groups are able to employ hunting, gathering, trading, agriculture, and now integration into capitalist wage labour, to different degrees. As was seen in Chapter 5, this has been the case since at least the Early Holocene, and Indigenous tropical forest groups have continued to manipulate different parts of the landscape and different social network opportunities to their own ends. Kathleen Morrison (1997; Morrison and Junker, 2002) and Laura Junker (1990, 1999, 2002a, 2002b) have focused on the long-term histories of tropical forest forager-traders in South and Southeast Asia from 2,000 years ago. These regions were part of the development of a network of commerce and culture that saw the exchange of native crops and plants across the Indian Ocean and Bay of Bengal, from the Bronze Age through to the historic period (Fuller et al., 2011c). Tropical crops such as the banana, and perhaps also mango and citrus, moved from Melanesia and Island Southeast Asia onto mainland Asia, while sandalwood travelled from Island Southeast Asia to the Indian subcontinent (Fuller et al., 2011c). Black pepper moved from the tropical forests of the western Ghats to Sri Lanka and apparently even as far as the Middle East and Classical Mediterranean world (Asouti and Fuller, 2008) (Figure 7.8). The expansion of Buddhism, Hinduism, and Islam tracked these interaction networks (Morrison and Junker, 2002). Perhaps the most interesting part of the growth of this incipient globalization and commercialism was that these maritime transfers were not the work of states or great empires, but rather a multitude of small-scale entrepreneurs that included the inhabitants of tropical forests themselves (Fuller et al., 2011c).

For South Asia, from the Early Historic period (*c.* 300 BC–AD 300), inscriptions relate hunter-gatherers to the 'specialized procurement of forest products' (Morrison, 2002b: 33) that included honey and medicines. Other texts from southwestern India suggest that this activity was part of tribute obligations to lowland kings, with the 'goods' consisting of elephants or the practical guidance of royal hunting parties (Morrison, 2002a). Later historical accounts from this point onwards emphasize the growing interdependence of hill 'tribes' and lowland farmers or urban settlements, in political and economic relationships (Morris, 1977; Murty, 1978; Morrison, 2002a). These relationships intensified in the sixteenth century, as the pan-Indian Ocean trade in spices, including black pepper, cardamom, ginger, and cinnamon in southwestern India and Sri Lanka, became an important resource for the maintenance of local elites through control and access to international entrepôts on the coast (Diffie and Winius, 1977; Morrison, 2002b) (Figure 7.8). A similar set of economic relationships has been described for the Irula and Kurumbas hunter-gatherers who occupied the highland grassland/forest mosaics of the Nilgiri Hills and lowland pastoralists and agriculturalists (Zagarell, 2002). Growing demand for wild products, as well as products that could be cultivated, led to agricultural expansion into many of these tropical forests, and corresponding marginalization of Indigenous foragers as can also be seen in the case of the Wanniya-laeto of Sri Lanka (Deraniyagala, 1992). However, in many cases, tropical forest dwellers maintained a symbolic and culturally powerful role in relationships of interdependence (Morrison, 2002a,b).

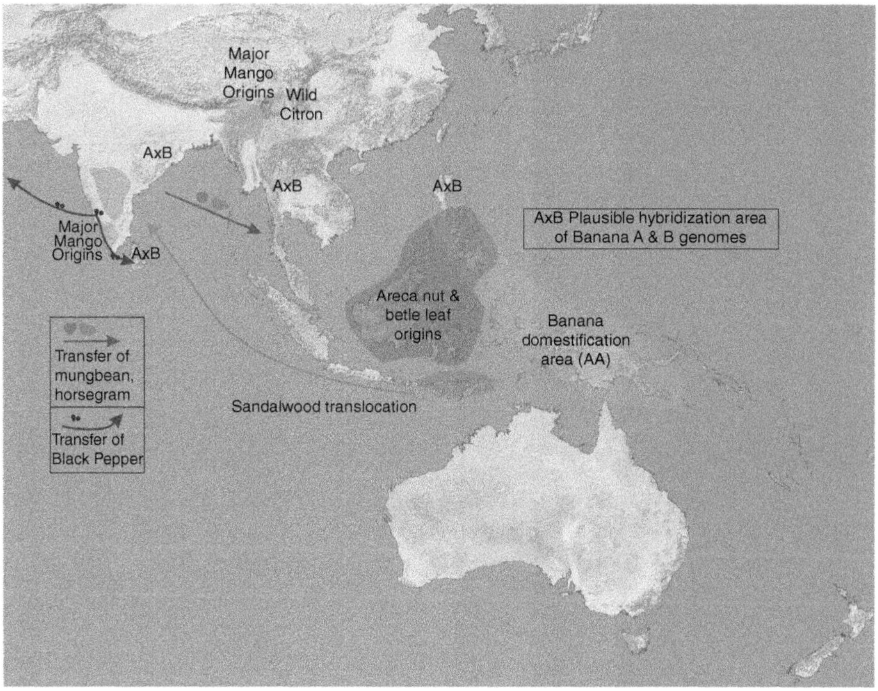

Figure 7.8. Map summarizing major prehistoric translocations of economic plants between Southeast Asia and South Asia, as discussed in the text. Archaeological sites with banana phytoliths are indicated (based on reviews of Lentfer, 2009; Fuller and Madella, 2009). Probable wild distributions of some taxa are indicated based on Fuller et al. (2011c). With thanks to Alison Crowther, Jana Zech, and Hans Sell.

Junker (2002b) argues that the ecological diversity and island fragmentation of Southeast Asia may have led to even greater levels of economic specialization and inter-ethnic exchange between tropical forest foragers, small-scale agriculturalists, and complex chiefdoms and kingdoms focused on maritime trade and intensive farming. Junker (2002b) notes how from the Late Pleistocene and Early Holocene mobile foragers were utilizing a variety of tropical forest and coastal settings. As farming entered the region, there is clear evidence across Mainland and Island Southeast Asia for mixed economies of wild plants and animals with domesticated rice. Linguistic and genetic evidence further suggests intermarriage between 'Negrito' foragers and agricultural populations (discussed above; Reid, 1987; Fix, 2002), while rice agriculturalists were living in close proximity to hunter-gatherers from 3,500 years ago (Hutterer, 1974, 1976). Junker (2002a, 2002b) suggests that these existing frameworks of forager mobility and subsistence experimentation provided the basis for the elite extraction of tropical forest products, such as spices, aromatic plants such as camphor (Donkin, 1999), and tropical bird feathers, during the rise of maritime polities in Island Southeast Asia in the historical period (Boomgaard, 1998; Junker, 2002b) (Chapter 6) (Figure 7.8). The necessity of monopolizing luxury porcelain, silk, bronze and other precious metals from China and Mainland Southeast Asia, led to intense negotiation of trade terms with foraging groups, often with few social or political ramifications for the foragers themselves unless they chose them (Junker, 1999, 2002b).

A similar local agency is even perceptible in the development of brutally extractive economic exchange networks following colonial contact, including the Atlantic slave trade. Although some islands off the Atlantic coast of Africa were settled long before European arrival (Mitchell, 2005), it seems to have been the arrival of the Portuguese in the fifteenth century AD that stimulated extensive trade along the coast of West Africa (Figure 7.9). In search of the gold, slaves, and ivory that were important elements of sub-Saharan regional and international trade (Chapter 6), the elite of Portugal sponsored voyages to Ghana and the kingdoms of Benin and Kongo that opened centuries of European economic

Figure 7.9. Map of Atlantic Coast trading posts in West and Central Africa, some of which are discussed in the text. Adapted and reproduced with permission from Peter Mitchell from Mitchell (2005).

interest in this coastline (Figure 7.9). By the eighteenth and nineteenth centuries this had been formalized into a 'triangular' trade of people to the Americas in exchange for commodity crops such as cotton and sugar (Northrup, 2010). From the beginning, however, the Portuguese colonization of Cape Verde Islands and other regions involved close intermarriage and local political integration, with trade driven by European and African associates (Fage, 1980; Brooks, 2003). No large, indigenous polities existed along the Gold Coast in the late fifteenth century but, established in the fourteenth century, the Kingdom of Kongo already controlled the mouth of the Congo River and its elites were eager to be involved in the benefits of European trade, including the import of firearms, clergy to promote their conversion to Catholicism, and exotic textiles (Hilton, 1985; Mitchell, 2005).

Interestingly, in the case of the Kingdom of Benin, while there was initial interest in the trade of firearms and religious paraphernalia, this state was apparently more indifferent to European contact, demanding higher prices and prohibiting the export of male slaves (Northrup, 2010). European focus on slaves from Africa intensified from the mid-seventeenth century AD as plantations and settlements in the New World drove up the demand for labour. While there should be no question regarding the horrific role of Europeans in driving this inhuman exercise, the broader trade in pan-African goods, Indian cottons, and American tobacco cut across cultural boundaries (Alpern, 1995; Mitchell, 2005; Northrup, 2010). Slave trading was also a long-standing custom of elite economic enrichment in an African context (Goody, 1971; Mitchell, 2005). Moreover, the practical involvement of Europeans in the slave trade itself was often limited beyond the coast and operated along existing African socio-political networks and relying on African traders (DeCorse, 2001). Devastating tropical diseases and a lack of local knowledge meant that Europeans could often only effectively trade 'with African permission and consent' (Mitchell, 2005: 202). At Savi, Benin, European trading lodges were placed adjacent to the existing palace and built from traditional architecture, associating this practice, as well as the access to exotic products, with the local elite (Kelly, 1997; Mitchell, 2005) (Figure 7.10). The settlements of Elmina in modern Ghana and Savi also suggest that European trade did not always result in dramatic cultural changes among Indigenous societies, with new goods and cultural elements adopted, transformed, and rejected as was seen fit (Kelly, 1997; DeCorse, 2001).

African communities also exported manufactured resources, and not just slaves, through the Atlantic network. Mitchell (2005) notes how sixteenth-century 'Afro-Portuguese' ivories from Benin and Sierra Leone were deliberately made for export to Europe (see also Bassani and Fagg, 1988). Benin also exported raffia to France, South America, and Angola while one million mats were exported from Sierra Leone during the eighteenth century (Thornton, 1992: 52–3; Mitchell, 2005). Domesticated African plants such as oil palm (*Elaeis guineensis*), African rice (*Oryza glaberrima*), sorghum (*Sorghum bicolor*), watermelon (*Citrullus lanatus*), and cowpea (*Vigna unguiculata*), among others, also made their way to the Americas as part of the so-called 'Columbian exchange' (Crosby, 1972; Carney, 2001; Nunn and Qian, 2010). Africans acquired European languages and religions, not because they deemed them to be 'superior' but rather to boost their own commercial benefit. Nevertheless, one reason that all of this pre-European socio-political complexity and agency of African societies is often neglected in discussions of the

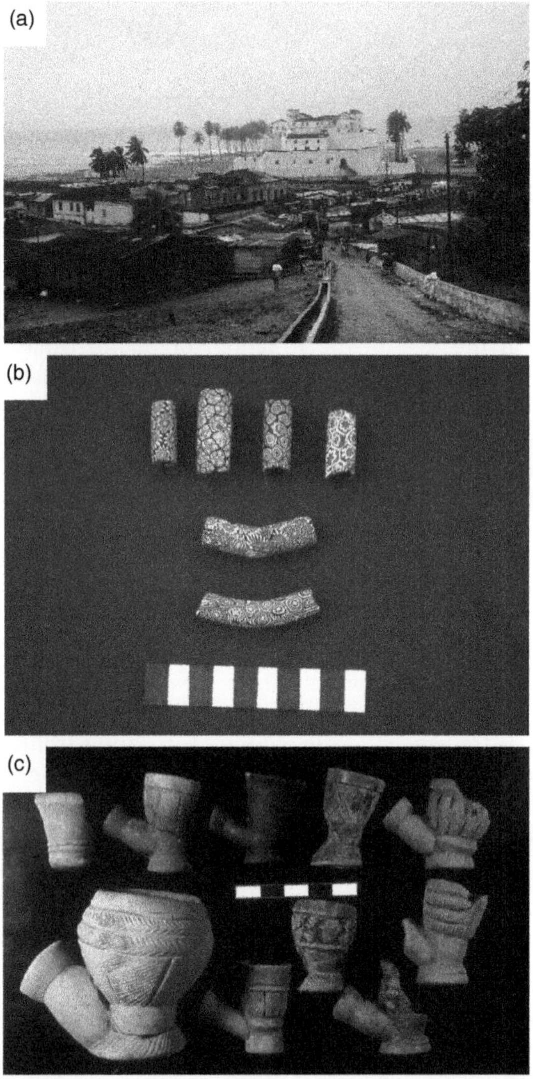

Figure 7.10. Photographs of a) Elmina Castle, Ghana, b) beads excavated at the site of Savi, Bénin, and c) Pipe stems excavated at the site of Savi, Bénin. Adapted with permission from Peter Mitchell from Mitchell (2005).

Atlantic trade network, is the undoubted impact of the environmental, social, demographic, and political landscape transformations wrought by European involvement that persist to this day. Fortified trading posts, including more than sixty built along the Ghanaian coast alone, document the sites where more than 11 million Africans were forcibly removed from their continent of origin from the fifteenth to nineteenth centuries, often dying during capture or transit. This affected household reproduction and political development across tropical Africa, and stimulated relocation into refuge areas, including mountains (McEachern, 2001) and perhaps also dense tropical forest.

A similar large-scale case of social, economic, and political reorganization in tropical forests can be seen in the growth of rubber plantations and trade throughout the eighteenth and nineteenth centuries, particularly in South America. The significant economic production of rubber products began in the early nineteenth century following a series of European and American inventions, including Charles Macintosh's waterproof coats, and Charles Goodyear's 'vulcanization' (Hemming, 2009). It became a key component of the industrial revolution in these regions, being used in pumps, belts, and, perhaps most notoriously, as bike and automobile wheels. By the end of the nineteenth century, Brazil was exporting 21,500 tons of rubber worth £14 million (Hemming, 2009). In Amazonia it was *Hevea brasiliensis* on which this trade was based. Plantations were hindered by insect predators and fungal diseases and so the trade relied on wild-rubber tappers, known as *seringueiros* in Brazil. Indigenous groups increasingly undertook this tapping, with groups such as the Mundurukú voluntarily tapping rubber in exchange for spirits and other western produce and families migrating along tributaries to take part in wage labour (Spruce, 1908; Barham and Coomes, 1994; Hemming, 2009; Vadjunec et al., 2011). However, other tribes, such as the Pakawara and Kariupura in Bolivia, were forced into poor patronage relationships that effectively amounted to slavery. The *seringueiros* were renowned for their brutality, murdering and raping Indigenous peoples at gunpoint with few governmental repercussions until the twentieth century.

The Amazonian rubber trade is undoubtedly a case of unequal exchange and the development of an exploited periphery (Bunker, 1984) that had massive consequences for social and political organization, and population size, in a tropical forest region. In the case of South America, diseases such as smallpox, brought by Europeans as they mined tropical forests more and more extensively, obliterated dense Indigenous occupations that have been described in Chapter 6. Slavery, warfare, and forced religious conversion led to huge relocations and population isolation that fostered perspectives of Amazonian forest groups as inevitably small in scale (Meggers, 1954, 1971; Roosevelt, 1994; Hemming, 2009). Nevertheless, as in Africa, agency on a group and individual level was demonstrated in the negotiation of new capitalist and political structures. First of all, rubber was used by Indigenous groups as a key part of hallucinogenic rituals at the time of European arrival and knowledge of this tree enabled Indigenous populations to gain an advantage in tapping (Hemming, 2009). Indeed, much of the extraction of rubber across the Amazon, as well as the necessary steps of latex hardening and transport, were reliant on Indigenous populations. Tropical diseases such as malaria, and malnutrition due to a lack of knowledge about local resource use, also affected local groups much less than those attempting to subjugate them. To cite an early example of this, starving Spaniards accompanying Gonzalo Pizarro in an ill-fated search for 'El Dorado' in the Peruvian Amazon in 1542 plundered Indigenous manioc fields in the hope of satiation without understanding that consuming bitter manioc unprocessed would lead to poisoning and death (Hemming, 2009).

GIVING BACK A VOICE

The rise of proximate nation-states, and the later influence of European powers, has done much to shape the economic, social, and political histories of peoples inhabiting tropical forests the world over from the sixteenth century, if not earlier (Guha, 1999; Morrison, 2014). From this point, state societies, including European colonists, wrote their histories. Meanwhile, ethnographic and historical discussions of local hunter-gatherers and farmers were based upon societies severely diminished by mass mortality in the Neotropics, or societies that had been politically and economically marginalized in the Old World. As I have tried to demonstrate above, it is becoming clear that Indigenous peoples were central to growing trade networks in the Late Holocene and were not simply passively affected by external forces. This further supports the 'revisionist' contention that hunter-gatherers today are not good 'pristine' models of the past (Bailey and Headland, 1991; Wilmsen and Denbow, 1990). Moreover, it highlights that Indigenous tropical forest groups actively traded and developed subsistence practices through careful negotiation of new cultural and economic influences. However, the remaining small, and often isolated, groups, or tropical forest foragers enmeshed within capitalist wage labour, should not simply be neglected as artificial, damaged remnants of a previous stage of social complexity. Existing Indigenous tropical forest populations have not only continued to agentively negotiate their unique place within a new capitalist world system, but have also continued to act as dynamic reservoirs of knowledge of tropical forest environments and their tipping points. While they may no longer be 'essentialized' as pure hunter-gatherers, they are still often characterized, like their forest homes, as passive victims of capitalism and colonialism.

It is therefore not enough to simply add Indigenous groups into local nation-state or European history but rather we need to actually use their own forms of history and their own value systems to guide historical and archaeological research into their past. A brilliant example of this is Åsa Ferrier's (2015) archaeology of the Jirrbal groups from the Evelyn Tableland of North Queensland. The tropical forests of Australia, and their human occupants, were some of the last to be encountered by Europeans, with intensive interaction with loggers and explorers occurring only during the early twentieth century. The colonizers saw tropical forests as something to be tamed (largely through deforestation and conversion to urban or farmland), and the quote used at the start of this chapter refers to the upheaval of Indigenous forest occupants, such as the Jirrbal, as part of this process. However, Ferrier combines existing historical records of European–Indigenous interactions in this part of Australia with oral histories passed down through Jirrbal elders. As mentioned above, oral histories have been the primary mode of the passing down of knowledge and information in many tropical forest societies and, if we are to really give these people a voice in studies of tropical forest history and prehistory, the inclusion of this information is essential. The correlation of environmental and social changes documented in histories transferred by word of mouth with palaeoenvironmental and archaeological datasets has already demonstrated that they should not be dismissed simply because they do not fit our own mode of historical recording (e.g., Duin et al., 2013; Stump, 2013).

Ferrier uses Jirrbal oral histories, and ethno-historical records of Jirrbal camps, as well as ceremonial and hunting and gathering grounds, in order to identify regions of archaeological importance, such as Urumbal Pocket. Evidence from the excavation of this site fits with broader evidence that the Australian rainforest was only occupied in the Early Holocene *c.* 8,000 cal. years BP, prior to phases of increasingly intensive site use from 5,000–2,000 years ago, and permanent occupation from 1,500 years ago (Cosgrove et al., 2007) (Chapter 4). Interestingly, the processing of toxic rainforest walnuts, remembered in Jirrbal traditions today, is documented at this site from at least 1,500 years ago and this practice most likely extends back to 8,000 years ago (Ferrier, 2015). Excavation of the site of Boignjul shows occupation limited primarily to the contact period, with occupants continuing nut detoxification and traditional forest clearance strategies alongside the use of coins, bone buttons, footwear, and glass styles that provide a material indicator of Aboriginal enmeshment within growing colonial networks. Perhaps the most striking example of this is the European-derived boot being repaired and curated by its Aboriginal owner (Ferrier, 2015: 136). Yet, while the manufacture of glass tools and wearing of European clothes, highlights a culture in transformation, continued hunting of rainforest animals, and visitation of sites such as Boignjul, are argued by Ferrier to demonstrate a form of resistance to this change that is also documented by Jirrbal elders and historical records.

This presents an interesting contrast to Bailey and Headland's arguments that tropical forest use was often forced onto Indigenous groups during periods of colonial interaction. In this case, a combination of oral histories and archaeological data demonstrates that continued tropical forest foraging was an economic and symbolic form of resistance to new power structures. The Sri Lankan Wanniya-laeto (or Vedda) have had to draw on tropical forest subsistence in a slightly more indirect way in recent times. These communities have faced massive upheaval from their traditional hunting grounds during the British colonial phase (Parker, 1909; Spittel, 1924; Deraniyagala, 1992) and the current Sinhalese government continues to encourage their movement away from the forest and into capitalist wage labour and national education systems (Dharmadasa and Samarasinghe, 1990). In response to this, the Wanniya-laeto, with the help of national and international funding, have created a cultural heritage museum at Dambana that tourists can visit to learn about forest subsistence and their cultural practices. Furthermore, the Wanniya-laeto have also been actively involved in scientific research that seeks to understand their relationship to some of the earliest occupants of Sri Lanka as well as the degree of forest reliance of their ancestors prior to colonial interference (Roberts et al., 2018b). The results of this latter study suggest a similar subsistence resistance to wage labour and colonial agriculture to that seen amongst the Jirrbal with continued tropical forest subsistence into the late nineteenth century.

Particularly prominent examples of active Indigenous involvement in political and social developments come from the Brazilian Amazon. As noted above, Indigenous tropical forest groups have often been presented as synonymous with their habitats and tribal groups have used this to their advantage. For example, there are a number of alliances between Indigenous peoples and conservation organizations, of which the A'ukre Kayapo village and the Institution

SocioAmbiental organization are a good example (Schwartzman and Zimmerman, 2005). These groups work together to focus on environmental, education, and political rights in tandem. Similarly, Amazon Watch works with various Indigenous parties in Brazil to fight for territorial rights and the rights to access natural resources (Schwartzman and Zimmerman, 2005). Nevertheless, despite success in a number of court cases, political illegality has often done little to deter the expansion of urban centres and farmland into Indigenous territories, as well as the profit brought by international mining companies. In one extreme case, the Munduruku tribe has been shown to have high levels of mercury poisoning as a result of gold production in the area (de Oliveira Santos et al., 2002). There have even been violent conflicts regarding land rights between the government and Indigenous populations, fuelled by the fact that integration of Indigenous peoples into wage labour sees them at the bottom end of one of the highest income inequalities in the world (Schwartzman and Zimmerman, 2005; Zhouri, 2010).

Involvement of South American Indigenous groups in conservation and anti-climate change campaigns, as well as their use of indigenous dress and rituals to make money from tourists and further their international cause, have led to them being seen as 'inauthentic'. Nevertheless, many tropical forest groups retain a local ecological knowledge of crucial importance to sustainable occupation in these habitats. For example, the Maya peoples who inhabit southern Mexico, Guatemala, Belize, El Salvador, and Honduras, use their shared cultural and historical knowledge to take an active part in the conservation of North and Central American tropical forests (Sanz, 2017a) and seek the benefits of tourism (e.g., Stronza, 2001; Coria and Calfucura, 2012; Sanz, 2017a, 2017b). While many of these groups are perhaps now loosely connected to the Classic Maya, following centuries of introduced disease, warfare, relocation, forced conversion, and intermarriage, they still share linguistic and artistic identities, including the production of brightly coloured yarn-based textiles (Grube, 2006; Comaroff and Comaroff, 2010). While one could cynically paint this simply as the economic 'selling' of an ethnicity, Maya groups are becoming increasingly involved in sustainable honey production in Mexico (Sanz, 2017a, 2017b). Moreover, Maya communities still practise the *milpa* form of cultivation, discussed in Chapter 6, that offers a sustainable, multi-crop approach to life in a dry tropical forest (Ford and Nigh, 2015). This makes them an ongoing source of contemporary, as well as past, human knowledge, as an active agricultural Indigenous group in a tropical forest setting, rather than 'pristine' foragers.

Indigenous tropical forest communities are crucial components of tropical forest prehistory, history, and ethnography and remain active adaptive agents in these settings today. They also have much to contribute to archaeology in tropical forests and, while their forms of knowledge have often been neglected in favour of European historical accounts, consultation with these groups is likely to provide a more complete understanding of the local archaeological landscape. For example, much was recently made regarding the recent 'discovery' of the City of the Monkey God in the Mosquitas region of Honduras mentioned in Chapter 1 (Preston, 2015). However, Christopher Begley (2017) notes how the Pech, an Indigenous group that live in this part of eastern Honduras, maintain that they always knew about this site, and its 'undiscovered' remains, and had intentionally left it alone. Similarly, Michael Heckenberger, who has been heavily involved in the realization of ancient urbanism in the Brazilian Amazon (Chapter 6), has always advocated the inclusion and consultation of Indigenous groups, such as the Xinguano peoples of the Upper Xingu Basin, in the identification of sites and interpretation of the findings (Fausto and Heckenberger, 2007; Heckenberger et al., 2007; Athayde et al., 2017). Moreover, Heckenberger has documented the importance of recognizing these groups as curators of the landscape who should be asked for permission prior to excavation and intrusive sampling (Heckenberger et al., 2007, 2008). Indeed, the practices of these groups open the door to imagining the possibilities of human tropical forest use highlighted throughout this book.

It is also worth noting here that ethnographic work among tropical forest communities has also completely challenged traditional approaches to social and cultural anthropology. For example, Eduardo Kohn has recently drawn upon four years of fieldwork among the Runa of Ecuador's Upper Amazon to explore the question: 'Can Forests Think?' (Kohn, 2013). By this, he follows earlier work by Deleuze and Guattari (1987) and Viveiros de Castro (1998) that urges anthropologists to acknowledge the social and symbolic agency of things that we, from a western viewpoint, would usually consider as being 'beyond the human'. Kohn uses the quote: 'Sleep face-up! If a jaguar comes he'll see you can look back at him and he won't bother you. If you sleep facedown he'll think you're *aicha* (prey) and he'll attack' (2013) from one of his informants to demonstrate that the Runa believe that jaguars, as well as other living things, can understand and act upon human signs and actions. As a result, any anthropology of these groups must also consider animals and plants as agentive and ontological equals within social systems, symbolic schemes, ritual, oral histories, and politics, rather than focusing simply on how humans categorize and manipulate the natural world. Given the western world's social, political, and moral interaction with domestic pets, farm animals, and even trees and forests, this may not be as 'other' as it might first appear. Nevertheless, the way the Runa think about their tropical forest world is offering a major challenge to traditionally entrenched western ideas of self, society, symbolism, and anthropological practice and writing.

NOT PRISTINE BUT THREATENED KNOWLEDGE

If existing Indigenous hunter-gatherer and agricultural tropical forest groups are no longer 'pristine' relics of past human adaptations in these environments, can their study and consultation still provide a sustainable adaptive counterpoint to industrial agriculture, mining, and urban expansion today? NGOs and anthropologists (Balée and Erickson, 2006) have continued to paint Indigenous groups as active conservationists. As defined by the *Oxford English Dictionary* (2017) conservation refers to the 'Prevention of wasteful use of a resource' or the 'Preservation, protection, or restoration of the natural environment and of wildlife'. As Hames (2007) notes, Smith and Wishnie (2000) take a more specific definition of conservation as 'actions that prevent or mitigate biodiversity loss and are designed to do so'. It is questionable to what extent small-scale foraging societies, with limited consumer demand or technology, truly 'design' conservation (Hames, 2007). However, sustainable hunting and gathering has been noted among many ethnographic groups. An analysis of the hunting yields of sixteen native groups and six Peruvian and Brazilian groups descended from European colonists in the Amazon demonstrated that the former had a far less negative impact on biodiversity, despite reliance on game for a larger fraction of their diet, due to focus on a wider variety of prey (Redford and Robinson, 1987). While some of this has been argued to be a result of a lack of extra-local or high population demand on the part of the Indigenous foragers and horticulturalists, the contacts noted for these groups above suggest that this cannot be the sole reason.

Van Vliet et al. (2009) have noted a similarly contrasting impact on local game populations by commercial 'bushmeat' hunters and Indigenous forest hunter-gatherers in the Central African rainforest. In part, this is related to the use of rifle technology and metal traps by the former who are focused on the fastest means of gaining meat at the least cost (Harrison, 2011). Indeed bushmeat hunters will focus on a particular portion of the forest and simply try to take as many animals as possible with limited focus on the age structure of their kills. By contrast, Indigenous groups seeking to subsist on forest game take a longer-term perspective, moving to different areas of the forest as populations dwindle in one region (Van Vliet et al., 2009). Therefore, although modern market hunting and indiscriminate shooting and snaring may be leading to 'empty forests' (Harrison, 2011), and the removal of often key species in seed dispersal and floristic diversity (Stoner et al., 2007; Wang et al., 2007), systematic and strategic hunting for subsistence, that follows fluctuations in prey availability and ecological characteristics, can be a highly sustainable source of protein in tropical rainforests (Van Vliet et al., 2009). Again, although the 'conservation' may not be intentional, and a longer-term perspective may simply reflect the demands for permanent occupation and a small-scale community, Indigenous groups can offer insights into the use of a local landscape as one that is 'lived in' rather than purely to be 'extracted from'.

This idea is further supported by ethnographic and archaeological evidence, mentioned in Chapters 2, 5, and 6, for Indigenous groups utilizing a huge variety of protein sources, including insects, molluscs, freshwater fish, and birds, that are rarely considered by other societies in these environments. It can also be seen in the use of different plant resources. The burning of grasslands and forest edges to inhibit tree invasion and enhance forage grasses attracting herbivores, as well as starch-rich seeds and 'honey harvesters' that facilitate continued hive access is well documented in different tropical forest fringe environments (Posey, 1998; Hill and Baird, 2003). Indeed, Indigenous tropical forest groups have been argued to enhance the environment in which they live, at least from the perspective of biodiversity (Balée and Erickson, 2006). Sustainable tree-cropping and the wide use and translocation of wild yams in different tropical forest regions also demonstrate a concern with the maintenance of access to reliable carbohydrate sources (Hong, 1987; Brosius, 1990, 1991; Dentan, 1991). Some studies have even documented cultural ideas of conservation of resources for the future among Indigenous groups in a non-tropical forest setting (Zavaleta, 1999). However, other studies, such as Smith's (2001) among the Machiguenga of the Amazon, fail to document a clear perception of resource depression or changes in soil nutrient quality through time, suggesting that 'conservation' is not an explicit concern (see also Vickers, 1995).

Yet, arguably a more important perspective of long-term human sustainability in tropical forest regions is available when we consider the connections and agency of Indigenous groups that have been discussed throughout this chapter. The debate should not necessarily focus on evidence as to whether Indigenous groups have a 'traditional' interest in conservation, but rather on the evidence that they have incorporated farming and trading demands into existing land-use practices. This moves us beyond the debate as to whether they are 'pristine', or whether their current 'traditional' foraging practices reflect prehistoric and sustainable strategies. Instead, these groups are key for consultation because they have the most experience of any societies in the world at negotiating the creation of new tropical forest adaptations in a globalized world market. Perhaps the best example of this would be the Maya of Mesoamerica. The Classic Maya themselves founded an urban society on the bringing together of key crops within a tropical forest setting. As discussed in Chapter 6, this was developed with a clear knowledge of local dynamics and environmental challenges in mind. The resulting *milpa* and 'forest garden' practices have been continued among Maya communities today, even as they are drawn into capitalist wage labour and globalized food access (Ford and Nigh, 2015). Furthermore, as mentioned above, Mayan honey production for commercial markets draws on a local knowledge of the forest ecosystem.

Similarly, the Mbuti of the Congo Basin, the Wanniya-laeto of Sri Lanka, and the Penan of Borneo have all adapted trade for crops and metal and plastic items into existing tropical forest hunting and gathering lifeways and strategies. Many individuals from these groups have also become engaged in capitalist wage labour but retain connections with their traditional group and traditional knowledge. In the case of the Wanniya-laeto, although the wider market of Sri Lanka now dominates their subsistence, they maintain their forest hunting strategies, as well as their cultural practices, in a museum that stores this information for future generations. The Raute of Nepal continue a mobile foraging strategy in the monsoonal forests of western Nepal, though their movement is now dictated by trips to local farming villages for rice and iron tools. Nevertheless, the majority of their protein still comes from the forest, with men and women of all ages participating in communal net hunts of macaque and langur monkeys that are shared across traditional kinship lines (Fortier, 2009) (Chapter 2). In Australia, the Jirrbal, as well as many other Indigenous groups, now live and eat in villages and towns alongside other Australians. However, they retain their oral traditions of past forest subsistence and ritual (Ferrier, 2015). Indigenous tropical forest groups are not perfect examples of 'isolated' hunter-gatherers. Instead, they exemplify the active meshing of new political, economic, and adaptive necessities within traditional landscapes and value systems.

This chapter has hopefully demonstrated that Indigenous tropical forest hunter-gatherers and agriculturalists are not 'pristine' or uncontacted, at least by other groups living in the same region, and have instead been agentively involved in increasingly global economic and political connections throughout the Holocene. In many cases, the existing perception of isolated, small-scale communities is a result of significant mortality or political marginalization and translocation during the last millennium. Yet, despite a move away from seeing Indigenous groups as timeless forest caretakers, they still undoubtedly possess an extensive understanding of their local environment and its biota (Gadgil et al., 1993; Hames, 2007). Furthermore, as is becoming increasingly apparent, these groups are also unparalleled in their knowledge of introducing new crops and economic structures into tropical forest environments. Therefore, while in some parts of the world there is a growing movement by governments to see Indigenous groups as 'false' traditional tropical forest tenants (Barume, 2000; Ocheje, 2007), they remain the best source of local knowledge relating to sustainable human–tropical forest symbiosis in the modern world. Not only do they demonstrate the possibilities of past human adaptations and connections within these environments, but they also offer the promise of consultation as to how the growing need for humans to occupy tropical forest landscapes but also retain forest cover and its crucial ecosystem services can be married together, including within new ontological perspectives of the world, in future conservation strategies.

8

The Tropical ‘Anthropocene’

A Modern Battleground or a Long-Term Framework?

> It was strangely like war. They attacked the forest as if it were an enemy to be pushed back from the beachheads, driven into the hills, broken into patches, and wiped out. Many operators thought they were not only making lumber but liberating the land from the trees . . .
>
> —Murray Morgan writing on the first logging of the US Olympic Peninsula, in *The Last Wilderness* (1955). Quoted by William Dietrich (1992: 21)

Although referencing temperate, rather than tropical, rainforest destruction in the United States of America the above passage highlights the shift in landscape valuation driven by modern demographic and economic pressures. Firstly, as a greater proportion of the world’s population shifts to the tropics over the course of the twenty-first century, more and more local smallholders will rely on tropical forests as a source of freshwater, agricultural land, and urban land, as well as timber, medicine, and food (Ghazoul and Sheil, 2010; The State of the Tropics Project, 2016). Furthermore, rather than solely being contexts for local subsistence and use, tropical forests are now also national and international ‘mines’ that provision high value wood, minerals, fuels, and land for multi-national businesses and markets. Notions that tropical forests should be removed, rather than managed or maintained, in order to increase local productivity and land value, have led to them becoming the most threatened terrestrial environments on the face of the Earth after the polar ice-caps. Certainly, the increasingly dramatic impacts these pressures are having upon them form part of broader discussions of a new, human-driven era of earth systems domination known as the ‘Anthropocene’ (Malhi et al., 2014). Disproportionate biodiversity, the regulatory role these habitats play in local and regional soil structure and chemistry, and their position within local, regional, and even global climate systems mean that human alterations to tropical forests, that have been argued to have changed in nature and scale since the European industrial revolution of the eighteenth century and the ‘Great Acceleration’ of the 1960s, have massive implications for the planet as a whole (Malhi et al., 2014; Malhi, 2017). As a result, tropical forests are a focal political, economic, and cultural battlefield between local populations reliant upon living *within* them, and business and governmental interests seeking to extract *from* them.

This chapter explores the tensions that exist in the human occupation and use of global tropical forest regions today, including the advance of urbanism and industrialization, exploitation of mineral, floral, and faunal resources by local groups and multi-national corporations, and their key position in discussions of anthropogenically induced climate change. I outline the key position of tropical forests in earth systems—including biological, chemical, and hydrological—as well as the main human-induced threats that have been argued to have initiated an 'Anthropocene' in tropical forests (Malhi et al., 2014). However, rather than seeing these threats as isolated to the industrial revolution or the Great Acceleration of technological developments since the 1960s, I seek to evaluate to what extent the 'Anthropocene' can be used as a framework to bring the archaeological evidence for human tropical forest modifications to bear on present-day interests, threats, and conservation solutions. Furthermore, I discuss the potential role of archaeology and palaeoenvironmental research in informing existing conservation methods and predicting their relative success and failure. Case studies document the ways in which information from the past is already being used in multidisciplinary conservation methods that take into account long-term histories of climatic and human impacts. This book has sought to highlight that humans have had a long history of tropical forest occupation and interference, not to justify unchecked modern alterations, but rather with the view that past examples of human sustainability can be used to inform existing practices and predictions for the future of these habitats.

TROPICAL FORESTS AND EARTH SYSTEMS

Tropical and subtropical forests lie at the heart of the global biosphere and at the crossroads of a number of earth systems and cultural frameworks. Their importance is highlighted in their provisioning, regulating, supporting, and cultural 'ecosystem services' (Millennium Ecosystem Assessment (MEA), 2005), that have been economically valued at double that of temperate forests (Costanza et al., 2014). Tropical forests are close to the equator, leading to greater light availability and a longer growing season. This, alongside the increased moisture and temperatures discussed in Chapter 2, mean that they contain the largest biodiversity of any terrestrial ecosystem. For example, a single tree in the Peruvian Amazon may be home to more ant species than the entire British Isles (Brandon, 2014), while 100 m^2 of Costa Rican lowland tropical forest is home to 233 vascular plant species (Whitmore et al., 1985). The variety of tropical forest formations adds to the diversity of these environments on a global scale, with different tree heights, root systems, transpiration rates, chemicals, flowering and fruiting seasons impacting the nature of animals, and other organisms, living within them (Chapter 2). In the case of lowland dry tropical forests, while absolute species-richness is lower than in lowland evergreen rainforests (Janzen, 1988), specialized plant and animal species have unique tolerances and capacities leaving them particularly vulnerable to environmental change (DRYFLOR et al., 2015). Given the disproportionate biodiversity housed by these habitats, even local and regional changes to their flora and fauna can leave massive imprints on the global biosphere.

The overall high diversity of plants and animals in the majority of tropical forests also has important local and regional nutritional implications for humans. While people may not get a large quantity of any one type of food at a given time, they often have access to small portions of a wide range of foods. This diversity of foods can be incredibly important in terms of micronutrients. For example, in Malawi, children living in places with higher forest cover, relative to open areas, have a more diverse diet, and are more likely to consume foods rich in vitamin A, with corresponding health benefits (Johnson et al., 2013). As we have seen, freshwater resources are also significant to humans in tropical forests. For example, in the Great Tonle Sap area of Cambodia nearly three-quarters of people depend on fish and inland aquatic animals for 80 per cent of their animal protein (Baran et al., 2007). Tropical forests also include rich coastal ecosystems and Malaysia's mangrove forests support over half of Malaysia's annual fish catch (Chong, 2007). Large biodiversity has also led to the evolution of a rich chemistry in tropical forest plants as a means of defence against pathogens and potential consumers (Ghazoul and Sheil, 2010). For example, *Catharanthus roseus*, the Madagascar periwinkle, has been used in the treatment of various forms of cancer, including leukaemia and Hodgkin's disease (Voeks, 2004) (Figure 8.1), and the chemical characterization of tropical forest plants has been an important focus of major drug companies (Mateo et al., 2001).

Tropical forests also support the maintenance of soil stability, nutrient cycling, and hydrological dynamics. Dense root systems anchor soils and act as breaks

Figure 8.1. Photograph of the Madagascan periwinkle (*Catharanthus Roseus*) that has been used in the treatment of various forms of cancer, including leukaemia and Hodgkin's disease (Voeks, 2004). Photograph created by Arts Vector and reproduced with permission from shutterstock.com.

against disturbance, as well as providing channels for the infiltration of water (Sidle et al., 2006; Ghazoul and Sheil, 2010). Forests also reduce the amount of rainfall that reaches the ground floor, maintaining soil stability (Greenway, 1987). As seen in Chapter 2, tropical forests are also essential in supporting nutrient cycling as moist, dark, high-oxygen environments provide the ideal home for fungi and bacteria to rapidly break down plant material into nutrient-rich humus available for new growth (van der Heijden et al., 2007; Powers et al., 2009). This process can be as rapid as 4–12 months in humid, organic-rich lowland rainforests (Whitmore, 1998). Healthy forests also play an important role in reducing the sediment content and pollutants of drinking water during the process of transpiration, and the filtering of overland water through roots, vegetation, leaf litter, and soil microbes that transforms and removes contaminants (Brauman et al., 2007; Acreman et al., 2012; Tanner et al., 2016). Seasonal tropical forests can even retain water and then release it during the dry season when it is most needed by plants and animals (Krishnaswamy 2013; Ogden et al. 2013). The removal of tropical forests can thus cause significant soil erosion, changes in local nutrient flows, and flooding with regional and continental signatures and effects (Ghazoul and Sheil, 2010; Larsen, 2017).

Tropical forests also regulate local and regional climate and are particularly efficient at the transfer of moisture from the biosphere back into the atmosphere due to their large leaf area, canopy layering, and density, as well as deep root systems (Lee et al., 2005). It has been found that 30–90 per cent of the annual rainfall in 138 different tropical forests is returned to the atmosphere as moisture (Kume et al., 2011). Furthermore, tropical trees engage in photosynthesis daily, resting at night rather than during the winter months, allowing them to convert solar energy and rainfall into moisture throughout the year (Sheil, 2003). This moisture increases the cloud cover that, in turn, reflects the sun's energy through albedo, and facilitates cooling (Boysen et al., 2014). The presence of many trees, of varying height, also facilitates the exchange of water, heat, and carbon dioxide with the atmosphere, and up to as much as two-thirds of the upper portion of canopy can be coupled to atmospheric changes in old growth rainforests (Seidler et al., 2013). Finally, it is also becoming increasingly apparent that pollen, bacteria, fungal spores, and volatile organic compounds distributed by tropical forests facilitate condensation and cloud formation and overall rainfall (Sheil, 2014). The forests lying between the Tropic of Cancer and Capricorn also store roughly one-quarter of the world's terrestrial carbon in their biomass (Union of Concerned Scientists, 2011; Brandon, 2014). This prominent role as a carbon sink has important implications for local and regional energetic transfers and lifecycles.

The role of tropical forests in climate regulation does not stop at the local or regional scale. Although debated, a growing body of research is highlighting that tropical forest modifications, such as deforestation, in one part of the world can affect climate and weather significant distances away through subtle atmospheric circulation and pressure patterns. For example, on a continental scale, the main source of rainfall in the Congo Basin is moisture evaporated over East Africa, while the Congo Basin is in turn responsible for a major proportion of the rainfall in the Sahel (van der Ent et al., 2010). Modelling studies have suggested that large-scale deforestation of the Amazon Basin could not only lead to increased heat, aridity, and reduced cloud cover in the region and across eastern South America, but also the redirection of storm tracks in the North Atlantic and Europe, cooling southern Europe and warming parts of Asia in the winter (Foley et al., 2007). Similarly, it has been suggested that rainfall in the Midwest of the United States could decline by *c.* 35 per cent should the entirety of the Central African tropical forests be cleared (Werth and Avissar, 2005). The global role of tropical forests in regulating climate is also highlighted by their position as the main terrestrial carbon sink. Given that the majority of this is stored in living organic biomass, disturbance and deforestation can lead to globally relevant emissions of CO_2 to the atmosphere (Anderson et al., 2011; Pan et al., 2011; Lewis et al., 2015).

Tropical forests also regulate the impacts of natural disasters on local and regional scales that may, in turn, be influenced by wider climate changes. Closed-canopy forests with high precipitation are more resistant to increased aridity and fire caused by droughts or lightning storms (Cochrane, 2003; Hoffmann et al., 2003; Brandon, 2014). Tropical forests will also reduce the frequency and extent of landslides (Guns and Vanacker, 2013), reduce the water-flow during times of flooding (Brauman et al., 2007), and, in the case of coastal forests and mangroves, even lessen the impacts of waves from cyclones and tsunamis (Baird and Kerr, 2008; Bayas et al., 2011; Yanagisawa et al., 2010). In the case of the 2004 Indian Ocean tsunami, for example, researchers have suggested that coastal forest plantations in Sri Lanka reduced the tsunami force, velocity, and depth relative to that seen in parts of Indonesia, with significant implications for human property and mortality (Samarakoon et al., 2013). Larsen (2017) also notes the important role forest cover plays in combating flooding linked to high rainfall in the Neotropics. Tropical forests, and the high biodiversity and clean water they contain, have also been argued to reduce human exposure to disease, as well as disease spread (Wood et al., 2014). While recent instances demonstrate that they can also be the source of severe diseases, such as Ebola, HIV/AIDS, and haemorrhagic fever, the spread of these diseases is in fact exacerbated by climate change and deforestation (Myers et al., 2013; Olivero et al., 2017; Faust et al., 2018).

Finally, while not a traditional earth system, tropical forests have ongoing cultural importance, including within indigenous rituals and ideologies (e.g. Balée, 1989, 2013; Hames, 2007; Ghazoul and Sheil, 2010; Ferrier, 2015). Forests are also becoming increasingly important symbols of health and well-being, as well as of environmental protection, in western societies. This also makes them significant sources of tourism revenue in the form of nature walks and the opportunity to meet endangered species for local populations and international visitors, alike (Cassells and Valentine, 1990; Tobias and Mendelsohn, 1991; Turton and Stork, 2006). This can have positive impacts on local populations in the form of new jobs, increased local income, as well as providing revenue and impetus for the preservation of tropical forests (Molnar et al., 2010). Nevertheless, perhaps the greatest cultural value of tropical forests is as living representations of past human practice and presence (Bonell and Bruijnzeel, 2005; Sanz, 2017a, 2017b). As the preceding chapters in this book have tried to demonstrate, tropical forests have been lived in by the earliest members of the hominin evolutionary branch, and also by some of the earliest members of our own species. From this point they have been used and altered for millennia, and have become home to a number of important archaeological sites. Tropical forests are not just crucial to earth systems, but are also a major part of human cultural heritage.

TROPICAL FORESTS IN THE 'ANTHROPOCENE'

As Yadvinder Malhi recently put it, the term 'Anthropocene' tries to capture 'that human activity is having a dominating presence on multiple aspects of the *natural world and the functioning of the Earth system*, and that this has consequences for how we view and interact with the natural world—and perceive our place in it' (2017: 78, my emphasis). This broad purview has been co-opted by different disciplines and put to varying uses. Literature and philosophy have taken it up as a new semantic context for human thinking, emotion, and art (Malhi, 2017). On the other hand, geologists, in the form of the Anthropocene Working Group of the International Union of Geological Sciences, have sought to formally define a new epoch with global stratigraphic identification that represents an empirical change in human impacts on the planet (Zalasiewicz et al., 2008, 2015). Ecology, climate science, and especially the social sciences, have sought more of a middle ground, using the term to think about changes in the scale of human impacts on the biosphere—including species distributions, extinctions, and environmental composition—and the Earth's climate, across timescales that stretch well beyond the industrial revolution (Ruddiman et al., 2015; Williams et al., 2015; Ellis et al., 2016). While the plethora of definitions, as well as political appropriation, have led some to question its utility as a term, its defining feature remains the highlighting of *global*, *multi-faceted*, and *pervasive* changes wrought by humans on a supposedly 'natural' landscape.

Given the importance of tropical forests to earth systems documented above, human threats to these environments today, in the form of extinction of indigenous plants and animals, introduction of exotic species, deforestation, landscape fragmentation, and changes in stand structure represent a major part of considerations of the Anthropocene in the mid-twentieth century (Malhi et al., 2014; Lewis et al., 2015). On a global scale, the removal of tropical forest is prominently caused by clearance for the cultivation of crops by small-scale farmers (Figure 8.2). Where regrowth occurs, as with shifting cultivation (Chapter 5), this is sustainable. However, large businesses and landowners seek more permanent landscape conversion (Corlett and Primack, 2011). For example, cattle-ranching is the largest driver of deforestation in every country in which the Amazon rainforest exists, accounting for 80 per cent of current deforestation. The Brazilian Amazon sustains more than 90 million head of cattle and is home to 550,000 km^2 of cattle ranches (Walker et al., 2000; Boucher et al., 2011; Butler, 2014). Low input costs and high consumer demand for low price Brazilian beef drives expansion and deforestation in aid of this industry. Poor yields require the firing of the landscape in the dry season to replenish soils and clear brush for grazing (Kaimowitz et al., 2004). The constant removal of forest pioneers and the loss of carbon from the soils is taking local areas beyond a threshold of possible reforestation, converting them into permanent grassland (Sierra et al., 2007; Lapola et al., 2010).

Tropical forest clearance for the growth of 'cash crops' represents a similar threat. For example, soybean cultivation is often cited as a major factor behind deforestation in the Amazon Basin (Fearnside, 2001) (Figure 8.3). Seeds from this plant (*Glycine mas* (L.) Merr.) provide high protein livestock feed, as well as a useful source of nutrition in seed or oil form for direct human consumption (Fearnside, 2001; Corlett and Primack, 2011). Brazil is the second largest producer

Figure 8.2. Satellite image of the Amazon Basin. At the bottom right and centre deforestation and cultivation is evident in the form of regular, rectangular plots. Red dots represent fires identified by MODIS and thick smoke can be seen in the bottom left of the image. Credit to Jacques Descloitres, MODIS Land Rapid Response Team, NASA/GSFC—The Amazon, Brazil. Reproduced with permission from wikimediacommons.org.

Figure 8.3. Photograph of an oil palm plantation and corresponding deforestation of the surrounding tropical evergreen rainforest in Southeast Asia. Photograph taken by Rich Carey and reproduced with permission from shutterstock.com.

of soybeans in the world after the United States of America and has 25 million hectares devoted to their cultivation (Figure 8.3). Although active clearance for soybean agriculture is actually focused on the *cerrado* savannas of Brazil (Fearnside, 2001), populations expelled from these landscapes move on to deforest other regions (Carvalho, 1999; Carvalho et al., 2002). Tree crops, such as the oil palm, seemingly retain a microclimate more similar to that of tropical forest but intensive management and simple structure makes them unattractive to native species (Sheldon et al., 2010). Moreover, prior to oil palm plantation, the peat swamps characteristic of the region are drained, leading to peat decomposition, and peat swamp forest decline, and the release of huge amounts of carbon dioxide into the atmosphere (Hooijer et al., 2010; Gaveau et al., 2014). In many cases drier, drained landscapes also lead to a frequency of peat and forest fires (Hooijer et al., 2010). Although prior to 2005 oil palm plantations were established on already-developed or burned land, there is a growing trend in Indonesia for forests to be permanently converted to plantations (Gaveau et al., 2016).

Logging is perhaps one of the more direct human threats to tropical forests worldwide (Figure 8.4). Contrary to popular belief, however, due to issues of accessibility, marketability, and value, much logging is actually 'selective' and focused on a few high-value species. Forestry, as a science, is about attempting to maintain a forest for constant use, rather than to destroy it. Nevertheless, it can be poorly implemented with unintended outcomes. In West Kalimantan, Borneo, while logging operations were recorded as removing only 15 mature dipterocarp trees per hectare, they only left 0.8 (or 3 per cent) trees in an undamaged state (Curran et al., 1999) (Figure 8.4). The repeated finding, cutting, and development of an infrastructure network to extract the logs can have massive, wide-reaching impacts on the canopy and soil of a local forest ecosystem. This is particularly the case when the construction of roads to improve large vehicle access leads to

Figure 8.4. Logging yard of tropical rainforest hardwood on the bank of the Mahakam River in Borneo, Indonesia. Photograph taken by Hilmawan Nurhatmadi and reproduced with permission from shutterstock.com.

repeated cycles of logging, and the arrival of landless farmers who clear the remaining land for agriculture (Corlett and Primack, 2011). Moreover, even where particularly valuable species are targeted, justifying a large investment in infrastructure, they are often the most endangered, such as Neotropical big-leaf mahogany (*Swietenia macriphylla*) (Grogan et al., 2010). On the other hand, however, there are also tropical forests that only exist today because they were originally protected for their timber value, such as the Sundarbans mangroves in Bangladesh (Biswas and Choudhury, 2007).

Tropical forest animal species are some of the most threatened in the world. Some of this is a result of the removal of habitats. However, hunting for subsistence and commercial 'bushmeat' hunting in particular has been argued to leave tropical regions such as Southeast Asia with 'empty forests' even as the trees remain standing (Harrison, 2011) (Figure 8.5). Detailed estimates at thirty Amazonian forest sites noted that tapirs were harvested at a five times greater rate than their maximum sustainable harvest within all hunting areas, while semi-arboreal and arboreal primates, as well as curassows were also over-hunted (Peres, 2001). Increases in human populations, forest fragmentation, the growing connection of Indigenous and small-scale farming populations to commercial markets, and a growing use of rifle technologies has increased this threat (Bennett and Robinson, 2000; Peres, 2001; Nasi et al., 2008). While Indigenous groups hunting for subsistence often maintain cultural restrictions, or hunt in ways that facilitate animal population regeneration, commercial hunters are more interested in maximizing

Figure 8.5. Eastern lesser bamboo lemurs (*Hapalemur griseus*) killed for bushmeat in northeastern Madagascar in 2010. Author of the photograph does not wish to be named for safety reasons. Photograph used with permission from wikimediacommons.org.

returns. For example, leg-hold cable snares that indiscriminately kill a variety of ground-dwelling species, of all ages, are commonly used in Africa due to their low cost (Fa and Brown, 2009). Animal parts have also become part of a huge trade, particularly in Asia, in traditional medicine ingredients (Corlett, 2007). In addition, tropical forest primates and reptiles are a significant part of a huge illegal trading network for pets (Broad et al., 2003; Duarte-Quiroga and Estrada, 2003; Lyons and Natusch, 2011).

Tropical forest fragmentation, and not just complete clearance, is also a major threat to plant and animal community structures. Small-scale clearance for cultivation and logging, as well as competing land ownership claims, often leaves behind mosaics of forest, grassland, and pioneer landscapes. Conservation studies have demonstrated that the remaining connections between habitats are more important than the absolute area of habitats for the persistence of a number of species. For example, many bird, mammal, and insect species specialized for life in a forest interior will not readily cross narrow open areas or a road (Laurance et al., 2009). The smaller the fragments, the faster the loss of biodiversity will be and studies of isolated Amazonian tropical forest fragments predict the loss of half of their forest-dependent birds within fifteen years (Ferraz et al., 2003). Such losses, in turn, affect the spread of plant seeds and their essential genetic diversity, making local forest trees more vulnerable to insect, fungal, and bacterial pests (Corlett and Primack, 2011). Forest fragmentation also results in more forest edges per area of forest, that influence forest temperature, humidity, and light levels as far in as 100 metres in Amazonian forest edges (Corlett and Primack, 2011). These new structures promote the spread and recurrence of forest fires (Harrison et al., 2009) and also promote the invasion of exotic plant species that suppress native pioneer species, such as *Piper aduncum* in New Guinea (Ghazoul and Sheil, 2010) and *Maesopsis eminii* in East Africa (Cordeiro et al., 2004).

The above threats operate on different scales and to different extents across the tropical forest regions of the world. However, there is one threat that affects all tropical forests: human-induced global climate change (Figure 8.6). It is estimated that with various scenarios of annual atmospheric carbon dioxide concentration increases the tropics will be at least 3–4°C warmer than they are today by 2100 (Meehl et al., 2007; Corlett and Primack, 2011). Although increasing temperatures beyond the tropics may offer new opportunities for tropical forest plant and animal communities, these taxa themselves have been shown to be vulnerable to the scale of temperature increases predicted. Ectotherms such as insects, amphibians, and reptiles, which rely on the environment to maintain their body heat, can only tolerate small variations in temperature (Deutsch et al., 2008; Huey et al., 2009). Even the mortality of endotherms, namely mammals, will be affected by things such as heatstroke, as has already been noted for flying foxes in tropical Australia (Welbergen et al., 2008). In montane tropical forest situations, with elevation gradients, it is likely that global temperature changes will generally lead to redistribution of current species communities. For example, moth species on Mount Kinabalu in Borneo have moved uphill at an average of *c.* 70 metres in the last 42 years (Chen et al., 2009). However, when warming is too fast, or a specialized animal species relies on a particular plant host with a given tolerance (Koh et al., 2004), species can suffer here as well.

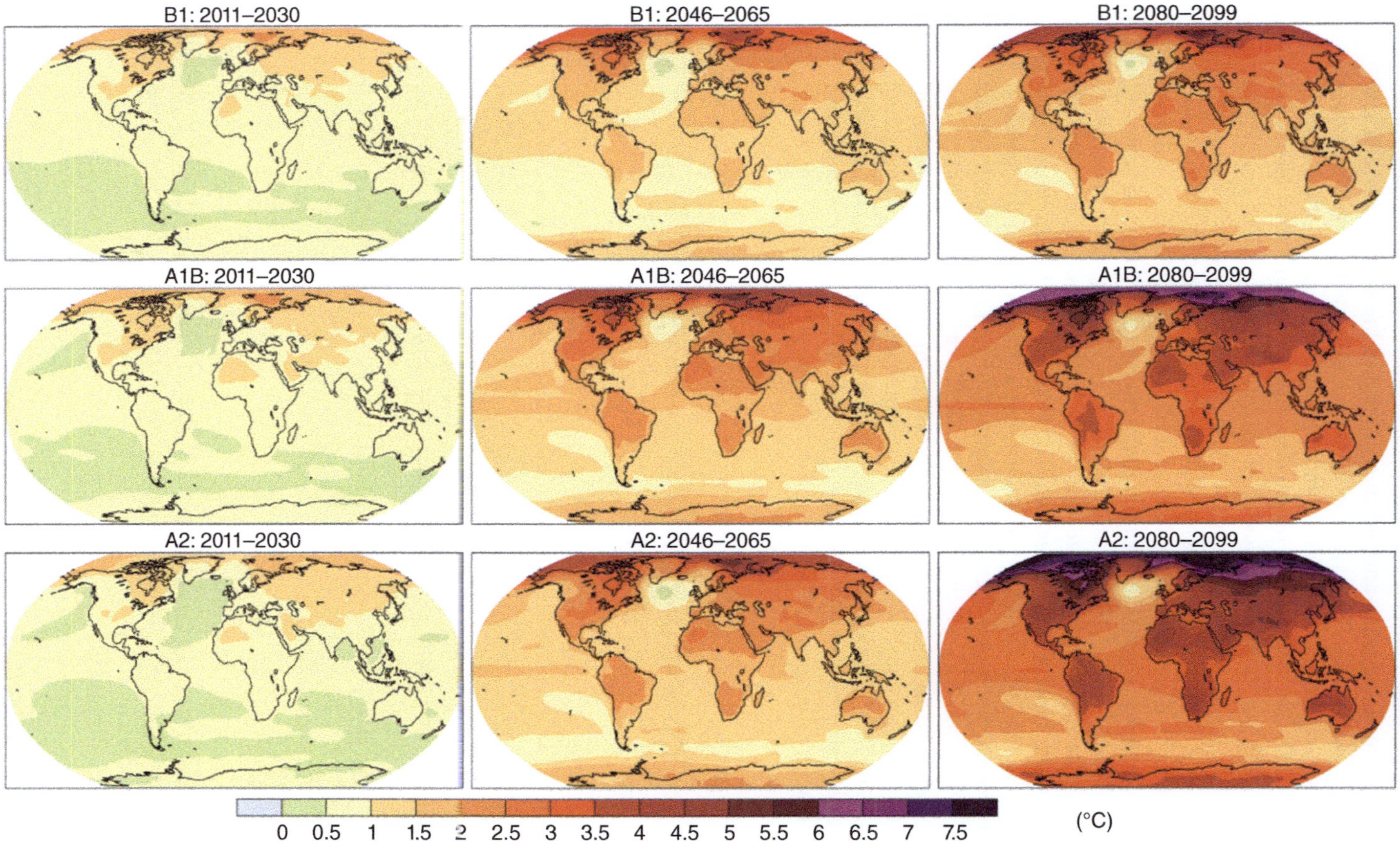

Figure 8.6. Map of multi-model mean of annual mean surface warming (surface air temperature change, °C) for different scenarios (B1—top; A1B—middle; A2—bottom) and three time periods (2011 to 2030—left; 2046 to 2065—middle; 2080 to 2099—right). In scenario B1 a temperature increase of *c.* 3°C is predicted across the tropics, rising to at least 5°C in scenarios A1B and A2. Reproduced with permission from Figure 10.8 in Meehl, G. A., Stocker, T. F., Collins, W. D., Friedlingstein, P., Gaye, A. T., Gregory, J. M., Kitoh, A., Knutti, R., Murphy, J. M., Noda, A., Raper, S. C. B., Watterson, I. G., Weaver, A. J., Zhao, Z.-C. 2007. Global climate projections. In S. Solomon, D. Qin, M. Manning, Z. Chen, M. Marquis, K. B. Avery, M. Tignor, H. L. Miller (eds.), *Climate Change 2007: The Physical Science Basis. Contribution of Working Group I to the Fourth Assessment Report of the Intergovernmental Panel on Climate Change*. Cambridge: Cambridge University Press.

Given the huge biodiversity of tropical forests, all of these threats have major implications for local, regional, and global ecology and life systems (Corlett and Primack, 2011; Malhi et al., 2014). Furthermore, the removal of dense tree cover results in reductions in local, regional, and perhaps even continental rainfall, alongside increases in precipitation seasonality, that have significant effects for hydrological, soil, and climate cycles on a variety of scales. The permanent conversion of tropical forest landscapes to mosaics or grassland also has major implications for the pace and extent of carbon dioxide release into the atmosphere, and subsequent global warming (Ghazoul and Sheil, 2010; Malhi et al., 2014). Particular concern has been voiced for the maintenance of dry tropical forests in Africa, South America, and Asia in this regard, given their greater accessibility and suitability for intensive corporate and urban land-use, and their lower, but more threatened, biodiversity (Ghazoul and Sheil, 2010; Malhi et al., 2014; DRYFLOR et al., 2015). Estimates of annual rainfall changes, and vegetation changes, have also indicated the susceptibility of lowland dry tropical forests in Africa and South Asia to climate change (Malhi and Wright, 2004). The presence of worsening, more frequent dry seasons in these environments, in turn, has impacts on local social stability and nutrient storage and recycling (Clark et al., 2010). Montane tropical forest habitats are also considered particularly vulnerable to changes in atmospheric carbon dioxide concentration and local temperature (Clark et al., 2010).

EXPANDING THE 'ANTHROPOCENE' IN TROPICAL FORESTS

As in the context of Indigenous peoples in tropical forests today (Chapter 7), conservation outlooks and reviews of the above threats tend to consider that mid-twentieth-century tropical forests were 'pristine' prior to the arrival of the modern forces of commercial mining and logging, urbanization, and the conversion of land to agricultural fields and pastures. Where human occupation and modification of tropical forests in the pre-industrial era have been considered, they have tended to be dismissed as purely 'local' relative to anthropogenic forces in the modern world (Ghazoul and Sheil, 2010; Malhi et al., 2014). Indeed, several widely used anthropogenic land cover change scenarios suggest that significant human colonization of tropical forest landscapes only began in the industrial era (Klein Goldewijk, 2001; Pongratz et al., 2008; Klein Goldewijk et al., 2011). Roberts et al. (2018a), however, recently used the KK10 scenario of anthropogenic land cover change to examine the potential influence humans may have had on tropical forest environments prior to the Industrial Revolution (taken here as AD 1750) in parts of the global dataset co-located with tropical forests (Figure 8.7). The results demonstrated peak tropical land-use in the Western Hemisphere prior to AD 1550. In the Eastern Hemisphere, while KK10 does suggest that peak land-use is reached prior to AD 1350 in some parts of southern China, the scenario generally implies that land-use in the Eastern Hemisphere largely followed a more linear pattern of land-use intensification over the pre-industrial Holocene.

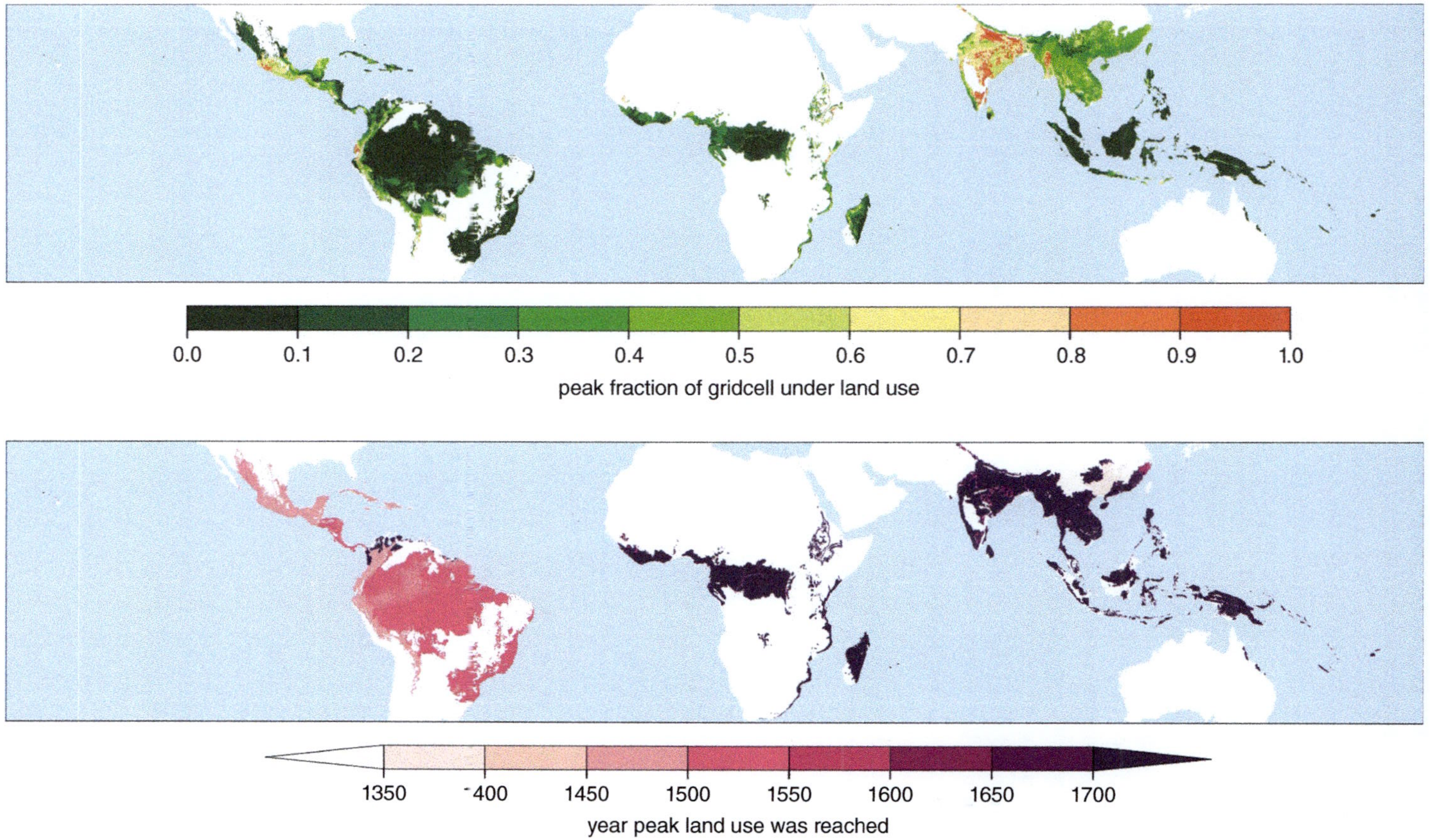

Figure 8.7. Top: Peak land-use in tropical forests prior to AD 1750 calculated using the KK10 model (Kaplan et al., 2011). The maximum magnitude of land-use as the fraction of each 5 arc-minute gridcell in the KK10 scenario is plotted for the regions defined as tropical forest according to the WWF terrestrial ecoregions map (Olson et al., 2001). Bottom: Timing of peak pre-industrial land-use. The year prior to AD 1750, when maximum land-use in tropical forests as depicted above in the KK10 scenario. Modified from Roberts et al. (2018a). With thanks to Jed Kaplan.

While this model emphasizes pre-industrial changes not often considered in conservation reviews and discussions of the 'Anthropocene' it almost certainly still underestimates them (Roberts et al., 2018a). The role of Late Pleistocene humans in megafaunal extinctions will continue to be debated (Chapter 4). However, the loss of tropical megafauna in South America, including large-bodied *Stegodon* spp. across Southeast Asia, Gomphotherioidea (gomphotheres) and Megalonychid (ground sloths) in the Neotropics, and Diprotodontidae in Australia (Corlett, 2013; Prado et al., 2015), would have had huge implications for local and regional species diversity, pollination, and carbon and vegetation dynamics (Doughty et al., 2015; Lewis and Maslin, 2015). In the case of South America, the loss of 83 per cent of megafauna by the Holocene suggests continental-scale implications for biodiversity (Lewis and Maslin, 2015). Impacts on small mammal taxa in the tropics remain even less well-understood, though the deliberate translocation of small mammals between Melanesian islands from 20 ka noted in Chapter 5 has had lasting effects on local ecological dynamics (Allen et al., 1989; Gosden and Robertson, 1991). The local and regional management of tropical forest mosaics (Haberle et al., 2010; Hunt et al., 2012) and translocation of edible tropical forest plants (Clement et al., 2015; ter Steege et al., 2013) in the Holocene had similar ecological legacies, though the exact time-depth of these practices remains to be definitively determined (Hunt et al., 2012; Roberts et al., 2017a).

The transition from the Terminal Pleistocene to the Holocene witnessed increasingly intensive human manipulation of plant and animal resources that resulted in genetic and phenotypic changes in various species as part of what has been termed the 'origins of agriculture' (Denham et al., 2003). Although tropical forests have often been neglected in these discussions, evidence for domestication, cultivation, and deliberate landscape modification had local, regional, and perhaps also continental effects on soil structure, hydrology, and biodiversity (Chapter 5) (Arroyo-Kalin, 2012; Bayon et al., 2012). Fuller et al. (2011a) have argued that the expansion of ungulate livestock into sub-Saharan Africa, monsoonal India, and central China, as well as the expansion of irrigated rice and water buffalo throughout Southeast Asia after 1,000 BC both led to significant atmospheric methane inputs. To what extent these changes in the tropics were the dominant force in the potentially anomalous increase in atmospheric CO_2 and methane concentrations at the start of the Holocene, to what extent this is associated with humans, and the degree to which changes would have had significant global ramifications, continues to be debated (Kaplan et al., 2011; Singarayer et al., 2011; Stocker et al., 2011; Ruddiman et al., 2015). Migrating farmers, bringing with them invasive commensal plant and animal species, also had significant impacts on biodiversity from the Middle Holocene (Boivin et al., 2016; Roberts et al., 2017a) (Chapter 5).

It is with the development of extensive 'low-density agrarian urbanism' in the Late pre-industrial Holocene that unequivocal anthropogenic influences on tropical forests, which may have led to continental and global feedbacks on climate, nutrient cycles, and biodiversity, appear to have occurred worldwide (Heckenberger et al., 2008; Fletcher, 2010, 2012; Lucero et al., 2015) (Chapter 6). In the case of the Greater Angkor region, the more than 3000 km^2 of engineered landscape that peaked from the ninth to the eleventh centuries AD remains the largest pre-industrial city on record (Fletcher, 2010). The deforestation and soil erosion that occurred as a result of these urban developments (Penny et al., 2006), as well as the vast network of agrarian settlements reaching out into the forests beyond them, almost certainly had regional, continental, and global implications for atmosphere, climate, and soil dynamics (Cook et al., 2012; Fischer-Kowalski et al., 2014; Lewis and Maslin, 2015). While the KK10 model documents that peak land-use in Amazonia occurred prior to AD 1750, in fitting with archaeological evidence for extensive 'Garden Cities', soil modifications, and extensive earthwork construction (Heckenberger et al., 2008; Heckenberger and Neves, 2009), entire regions such as Island Southeast Asia, and much of West and Central Africa, barely register landscape change in KK10, despite regional and even continental evidence for large-scale deforestation and soil erosion from 1,000 years ago (e.g., Penny et al., 2006, 2014; Bayon et al., 2012). Similarly extensive urbanism in South and Southeast Asia, documented in Chapter 5, undoubtedly led to periods of declines in occupation intensity or even complete abandonment that are not obviously apparent in the KK10 model. More sophisticated land-use models should improve this situation in future (Kay and Kaplan, 2015).

Of particular interest here is the fact that the KK10 model, as well as archaeological and palaeoenvironmental evidence for 'low-density agrarian urbanism' (Chapter 6), both confirm that dry tropical forests, notably those in West and Central Africa, across much of India, North and Central America, and the fringes of the Amazon Basin, were the most susceptible to human occupation and modification in the past as well as the present. This highlights that these forest formations were particularly vulnerable to human exploitation and modification, likely as a result of the fact that drier, more open conditions lend themselves to the expansion of agricultural field systems and the development of urban settlements focused on water control (Lucero et al., 2015). In turn, this supports conservation calls that these formations are particularly threatened by the expansion of agriculture, urbanism, hunting, and deforestation in the modern world (Malhi et al., 2014; DRYFLOR et al., 2015). The development of more detailed archaeological and palaeoenvironmental datasets on human land-use in tropical forests around the world is essential if we are to build up more detailed temporal comparisons of global anthropogenic environmental impacts. Acceptance that the 'Anthropocene' as a term is also relevant to pre-industrial human tropical forest impacts, and the recognition that pre-1750 AD impacts were quantitatively rather than qualitatively different from those that followed, further emphasizes the necessity of including the social sciences in discussions of the 'Anthropocene' (Ellis et al., 2016). It also opens up the possibility that archaeology and palaeoecology can be actively used to inform modern conservation discussions and policy (Roberts et al., 2018a).

CONFLICTING INTERESTS AND POSSIBLE SOLUTIONS

The archaeological record of urban forms in tropical forests tells us that a constant tension exists between the interests of small-scale users of tropical forest landscapes and political, economic, and religious power structures (Chapter 6). The same is also evident in the modern world and while the severity of threats, and the clear benefits of maintaining tropical forests, seem obvious to a casual observer, the conveyance of this information and ensuring beneficial conservation solutions for local Indigenous and smallholder groups is not always straightforward. Although the small-scale farmers within the extensive 'low density agrarian' urban forms encountered in Chapter 6 could continue farming and local adaptations as power structures rose and fell, the expansion of capitalism, national government policy, and often-uncontrolled population growth mean local populations often have limited power over the protection and use of tropical forests. For example, the countries of Africa that contain tropical forests have some of the highest population growth rates in the world (women bearing on average 4.5–6.5 children each—United Nations Population Division cited in Corlett and Primack, 2011), alongside some of the lowest per capita incomes (most people live on less than US $1 a day) (International Monetary Fund cited in Corlett and Primack, 2011). These growing, poor populations are heavily dependent upon subsistence agriculture, herding, and bushmeat. While often they utilize swidden systems that can be sustainable, they are forced to reduce fallow times in order to do so.

In Brazil, now an increasingly wealthy member of the BRIC economic nations with relatively low fertility and child mortality, the greatest threat to tropical forests comes from large businesses and government policy. The organization of cattle-ranching by large-scale landowners is now the biggest cause of deforestation in Brazil, where regional demand for meat makes pastureland more valuable than tropical forest (Nepstad et al., 2008; Barona et al., 2010; Nobre et al., 2016). Oil and gas exploration, alongside mineral mining interests, have led to the government considering 'selling' vast areas of tropical forest to international corporations (Finer and Orta-Martínez, 2010; BBC News, 2017). In Southeast Asia, the region facing the highest rates of forest loss and degradation anywhere in the tropical world (Corlett and Primack, 2011), it is the conversion of tropical forest into land for growing cash crops that is really changing the values placed on tropical forest ecosystem services (Laurance, 2013; van Straaten et al., 2015). In the case of oil palm, for example, huge Asian and international market demand has played a leading role in the replacement of lowland tropical forests by plantations in Island Southeast Asia (Hansen et al., 2009), often funded by huge multi-national corporations (Ghazoul and Sheil, 2010). The supremacy of global capitalism, growing land values, and increased incorporation into national education and wage labour systems, mean that small-scale farmers and Indigenous peoples are often left with little option but to convert, abandon, or sell their tropical forest land.

The archaeological and palaeoenvironmental record highlights that it is those groups with an intimate interest in the tropical forest landscape that will develop the most resilient adaptations. This is even clear on a national scale in the form of the recent success of Australia's tropical conservation policy. Since the arrival of European colonists in the eighteenth century, the tropical forests of Australia have been subject to clearance for mining, timber, and pasture by groups who felt they were 'liberating' the land for new, more economic uses (for example, see Ferrier, 2015). However, Australia now protects most of its remaining tropical forest. In particular, the Wet Tropics of the Queensland World Heritage Area, that stretches for 894,420 hectares and contains all of Australia's unique marsupial taxa (UNESCO, 2018), has proven to be a hugely successful tourist attraction, for Australians and international visitors, especially when combined with the Great Barrier Reef that sits just off the coast (Valentine and Hill, 2008) (Figure 8.8).

Figure 8.8. a) Kuranda scenic railway in northeastern Queensland, Australia. Photograph taken by Aboumael. Reproduced with permission from wikimediacommons.org. b) Indigenous performance at Kuranda in northeastern Queensland, Australia. Photograph taken by bgabel. Reproduced with permission from wikimediacommons.org.

High levels of individual earnings and good, widespread education relative to other tropical regions, as well as an ever-improving treatment of Indigenous groups, have led to a greater appreciation of the cultural, provisioning, and regulating services of Australia's tropical forests. While others have comprehensively reviewed the potential solutions to the threats discussed above (Ghazoul and Sheil, 2010), I wish to focus on how the reconciliation of national, small-scale landholder, and Indigenous interests in tropical forest protection often leads to the greatest results.

One very blatant way of conserving tropical forests around the globe has been the governmental and international declaration of a given area as 'protected'. The form of these 'protected areas' varies enormously, ranging in size from the 39,000 km^2 Tumucumaque National Park in Brazil to the 1.6 km^2 Bukit Timah Nature Reserve in Singapore (Corlett and Primack, 2011) (Figure 8.9). Reserves greater than 10,000 km^2 offer the best potential for the long-term conservation of entire tropical forest communities, including predators, large herbivores, and rare specialists (Laurance, 2005; DRYFLOR et al., 2015). However, even small reserves such as that at Bukit Timah protect unique hyper-diverse plant and invertebrate fauna, if not larger vertebrate animals (Corlett and Primack, 2011). Large-scale studies in the Neotropics suggest protected areas effectively slow deforestation (Brooks et al., 2009). Decisions may have to be made as to which formations or regions are most 'valuable'. Governments may prefer to protect a large area of cheap land versus a small one in a highly attractive region, and in many nations the majority of 'protected areas' lack economic value due to issues of accessibility. In terms of the forest types themselves, popular and governmental perceptions of

Figure 8.9. Bukit Timah, Singapore is one of the smallest government Protected Areas in tropical forests in the world at 1.6 km^2. Photograph of a hiking trail along old railway tracks within the reserve taken by N8Allen. Reproduced with permission from shutterstock.com.

Figure 8.10. Mountain gorillas (*Gorilla gberingei beringei*) in Bwindi National Park, Uganda. Photograph taken by Gudkov Andrey. Reproduced with permission from shutterstock.com.

natural value generally lead to lowland evergreen rainforests, and those forests with iconic vertebrates such as the orangutan (Ashton, 2010) or gorillas, being invested in most due to the more obvious publicity benefits of tourism (Figure 8.10). This is regularly to the detriment of the perhaps less-glamorous dry, coastal, peat swamp, or montane tropical forests (Ashton, 2010; DRYFLOR et al., 2016; Dittus, 2017) that are often home to more unique taxa and face the major pressures of human use and expansion, as well as climate change, more keenly.

It might seem that the wealth and resources available to national governments best enforce 'protected areas'. However, there are many types of 'protected areas' and questions arise as to what is permitted within each area and who should decide this. Government-instigated protected areas usually affect impoverished small-scale farmers or Indigenous groups the most, causing displacement and a lack of access to resources (Whitten and Balmford, 2006; Ghazoul and Sheil, 2010). This can cause resentment if the financial and ecological benefits of these protected areas remain unapparent on the local scale (Sodhi et al., 2010). In many cases, governments have even decided to evict Indigenous groups from forests that are crucial to their cultural and ecological identity in the name of nature conservation. For example, the Twa people were forcibly removed from Kahuzi-Biega National Park in eastern Congo by their government. This has led to pervasive malnutrition and disease among this population (Barume, 2000; Ocheje, 2007), meaning that conservation itself has threatened their cultural existence and way of life. In other cases, the removal of small-scale farmers from certain areas merely leads to migration and deforestation in others (Carvalho, 1999). In many poorer tropical nations, governments simply do not have the resources to transmit their power locally so more effective conservation and policy is actually likely only to be achieved through cooperation with local interest groups (Stevens, 1997; Sayer et al., 2013).

By contrast, as noted in Chapter 7, where Indigenous groups and local smallholders have been included in protected area management, the long-term effectiveness of these reserves has seen clear benefits (Stevens, 1997). For example, in the Brazilian Amazon, the 130,000 km^2 reserve of the Kayapó and Upper Xingu peoples is the largest protected tropical forest area in the world (Corlett and Primack, 2011). In the case of the Kayapó, it has been noted that Indigenous classification of botanical diversity and tropical forest formation gradients within this reserve prove essential in developing strategies for maintaining diversity (Posey, 1985). The Kayapó also appear to have knowledge of wildlife thresholds when hunting, providing an example of sustainable exploitation (Peres and Nascimento, 2006). Sometimes protection is often driven from the bottom up without any government involvement at all. Sheil et al. (2015) for example note that local monitoring by villages in the Mmberamo-Foja region of Papua (Indonesian New Guinea) deters unregulated exploitation. The Ijabait hereditary guardians live at strategic sites to protect, monitor, and control access to resource-rich lakes and tributaries along the Tariku River (Sheil et al., 2015). Such efforts present a cultural and ecological barrier to deforestation by small-scale farmers and large corporations, alike (Ricketts et al., 2010). Furthermore, the inclusion of populations *living in* tropical forests, in conservation strategies, is more likely to reconcile interests.

The rise of ecotourism—defined as 'responsible travel to natural areas that conserves the environment, *sustains the well-being of local people*, and involves interpretation and education' (TIES (The International Ecotourism Society), 2015, my emphasis)—provides a similar means of conveying ecosystem services to local and international consumers and policy makers simultaneously. The overall biodiversity of tropical forests makes them prime environments for the viewing of diverse plant and animal species. Costa Rica has proven particularly popular as an ecotourism destination for American and European tourists. For example, surveys undertaken at the Monteverde Cloud Forest Reserve demonstrate that the value placed by ecotourists from the United States on visiting Costa Rican tropical forests was $1,150 per visit in 1995 (Menkhaus and Lober, 1995). In the Ecuadorean Amazon, Indigenous groups combine with local NGOs and government initiatives to promote ecotourism to individual villages to provide additional wealth to these groups and also promote tropical forest preservation (Edwards et al., 1998; Zeppel, 2006). Similarly, gate fees from national parks in Uganda, such as Bwindi Impenetrable National Park and its gorillas, have generated employment and fed into community projects (Adams and Infield, 2003). Although spreading diseases to native fauna, noise pollution, littering, and the dubious benefits for poor local groups can lead to criticism (Butynski and Kalina, 1998; Southgate, 1998), where managed properly, ecotourism can provide a clear economic conservation incentive for local groups and governments (Ghazoul and Sheil, 2010).

Conveying a valuation of ecosystem services to regional and international (including governmental) consumers is also essential when attempting to stem the over-utilization of unsustainably sourced forest resources around the world. International markets drive over-exploitation in an era of mass consumption and

Figure 8.11. Coffee picker working in the Ciudad Barrios cooperative in El Salvador, certified by the Rainforest Alliance. Photograph taken by Robert Goodier. Reproduced with permission from wikimediacommons.org.

globalization, and must pay the cost for sustainable use rather than impoverished local groups that cannot afford it. This includes paying a premium on certified forest products, such as timber from the over 30 million hectares of tropical forest certified by the Forest Stewardship Council (FSC) (Forest500, 2018) and coffee from Rainforest Alliance Certified brands (Giovannucci and Ponte, 2005) (Figure 8.11). However, smallholders are often excluded from these schemes due to the high costs and technical and bureaucratic burdens of meeting certification standards (Gullison, 2003; Grieg-Gran et al., 2005). Similarly, a proliferation of certification can undermine credibility (Schulze et al., 2008). If certification is to be effective, domestic markets must also be encouraged to utilize these products, and become environmentally aware, as they offer a larger consumer base than export markets for many tropical producers (Ghazoul, 2001). For example, certification initiatives have proven particularly effective in Bolivia where local government laws, enforcement, tax benefits, and regulated extractions have subsidized and facilitated the use of certified forests by small-scale consumers and producers (Ebeling and Yasué, 2009).

Indeed, in order to maintain ecosystem services that will undoubtedly benefit their nations in the long term, local and international governments need to pay a cost for conservation rather than the smallholders least able to pay (Bawa et al., 2004; Sodhi et al., 2006; Ghazoul and Sheil, 2010). Timber and bushmeat exploitation remain crucial to the economies and subsistence of tropical nations, respectively. However, they must be regulated in some way if they are to continue to be sustainable. Reduced Impact Logging (RIL) initiatives seek to apply a series of principles in order to minimize the damage to trees, soils, and non-target plants and animals (Putz et al., 2008). The practices involved include the training of workers, protection of hillside forests, careful infrastructure planning, pre-impact inventories of species diversity, and extraction methods that avoid excessive soil or tree disturbance (Putz et al., 2008). These practices have been shown to have clear long-term economic, as well as 'ecosystem service', benefits to governments and local groups, though their benefits for long-term biodiversity recovery remain under study (Zarin et al., 2007). They can be expensive, however, and in order to regulate logging corporations and small-scale illegal logging state sponsors must be willing to make a short-term loss. A similar situation exists in the context of bushmeat exploitation. In Africa, where bushmeat is often a major source of protein for the local population, someone must be willing to provide alternative sources of protein if unsustainable bushmeat harvesting is to stop (Fa and Brown, 2009).

On a grander scale, the principle of subsidy is also behind the Reducing Emissions from Deforestation and forest Degradation (REDD+) initiative developed by Parties to the United Nations Framework Convention on Climate Change (UNFCCC) (unredd.net) (Ghazoul and Sheil, 2010). REDD+ places a financial value on the carbon stored in forests by offering incentives for developing countries to reduce emissions from forested lands, which currently make up 17 per cent of the global total, and invest in low-carbon approaches to sustainable development. Developing countries will receive results-based payments, subsidized by industrial nations, for results-based actions that include reducing logging impacts (e.g., through the RIL mentioned above), slowing deforestation, preventing fires, and restoring tropical forests to their 'native' form. The UN-REDD Programme currently supports sixty-four partner countries across Africa, Asia-Pacific, Latin America, and the Caribbean (UN-REDD Programme Collaborative Online Workshop, 2018), and represents an important mechanism for industrial countries to take on responsibility for the conservation of environments that have global economic and climatic significance (Figure 8.12). Although the economic benefits of this programme remain to be seen (Ghazoul and Sheil, 2010), this initiative not

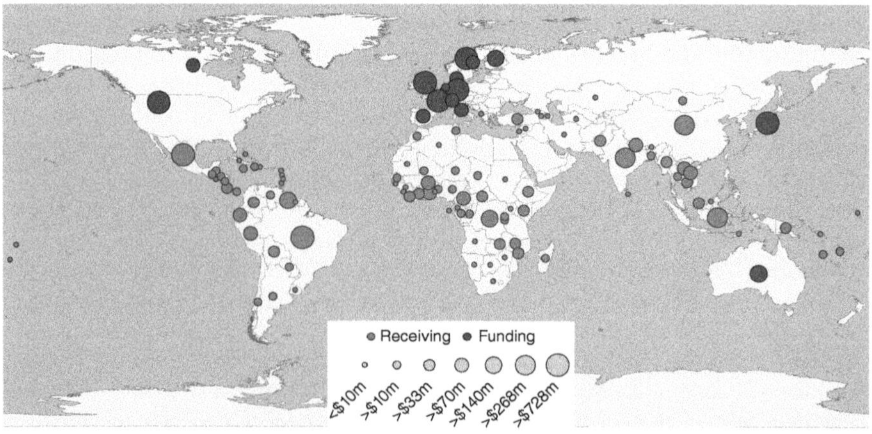

Figure 8.12. REDD+ funding and receiving amounts based on data reported by funding countries. Data from REDD+ database (FAO, 2018). Divisions based on natural jenks from ArcGIS. With thanks to Nils Vanwezer.

only rewards developing countries for meeting REDD+ targets but also helps them build the conservation capacities to meet these targets. It also ensures that the interests of national governments, business stakeholders, and Indigenous peoples are all consulted during the development and application of strategies in forest-dependent communities.

A LONG-TERM PERSPECTIVE: UNITING THE PAST AND THE FUTURE

Archaeology, history, and palaeoecology can do more than simply highlight the importance of identifying, and reconciling, tensions between small-scale groups living and using forests and broader political and economic powers in the resilience of tropical forest adaptations. The significant and ongoing role tropical forests have played in hominin and human adaptations, migrations, agriculture, and the development of complex settlement patterns and governance structures also highlights that these habitats preserve a significant proportion of national and international human cultural and ecological heritage. The recent discoveries made by remote sensing (including LiDAR), stripping away the forest to reveal kilometre upon kilometre of human land management and modification only highlight this further (Evans et al., 2013; Evans, 2016; Fisher et al., 2017). The growing archaeological evidence for tropical forest use and occupation, stretching back into the Pleistocene (Roberts et al., 2017a), as well as a potentially earlier beginning to the 'Anthropocene' in these habitats (Roberts et al., 2018a), means that archaeology can no longer be ignored by ecologists, conservationists, and government policy makers working in twenty-first century tropical forests (Bush and Silman, 2007; Ghazoul and Sheil, 2010; Levis et al., 2017; McMichael et al., 2017; Roberts et al., 2017a). Indeed, in specific examples, it is already beginning to make significant contributions to conservation and understanding of modern tropical forest structure, dynamics, and futures.

The active restoration of forest areas is a potential solution to tropical forest decline. This is most commonly done through planting tree seedlings in nurseries, though encouraging natural forest succession (Shono et al., 2007), planting live tree branches (Zahawi and Holl, 2009), and incorporating agricultural crops (Vieira et al., 2009) are other viable options. Regeneration should be planned and targeted on the regional level, for example in the rejoining of forest fragments, so it can do the most good to the forest community at large (Rodrigues et al., 2009). Successful animal reintroductions, although facing high levels of failure, can also make major contributions to ecosystem functioning as well as harnessing local pride and interest in the return of an important species (e.g., Corlett and Primack, 2011; Louys et al., 2014). Nevertheless, such a process can be costly, and there can be variable support. For example, while reintroduction of the tiger (*Panthera pardus*) into many tropical forest environments in India would yield international and national support, local populations will be concerned about growing potential conflicts with this species (Johnsingh and Madhusudan, 2009; Sankar et al., 2010). Similar concerns exist with regard to the reintroduction of the Sumatran orangutan (*Pongo abelii*) (Wilson et al., 2014). Furthermore, with both animals and plants there remains the question as to what a tropical forest should be 'restored' to. If there is no natural or 'pristine' forest state left, what do we protect?

Ecological restoration projects are drawing on the remains of plants and animals found at archaeological sites to guide species reintroductions. For example, on the Hawai'ian Island of Kaua'i a number of plant and animal extinctions occurred as a result of introduced domestic species, commensals such as rats, weeds, and diseases. Traditional conservation attempts to boost the population size of endangered plants in a given locale, particularly endemic wild flowering species, or to introduce species into new areas proved unsuccessful (Burney and Burney, 2007). By contrast, palaeoecological studies of botanical remains at archaeological sites in different biomes and at different altitudes enabled the past landscape of these species to be reconstructed. New conservation efforts, drawing upon this information, have achieved greater success in creating new populations of endangered species in regions where they were found in late prehistoric and early historical times (Burney and Burney, 2007). Animal reintroductions can also utilize archaeological and palaeoecological datasets. For example, it has been suggested that orangutans could be reintroduced into any large tropical forests in their previous Late Pleistocene Southeast Asian range not just those from which they have disappeared historically (Louys et al., 2014). Indeed, Louys et al. (2014) have used the Asian-Pacific fossil record to develop a tangible matrix of benefits, risks, and feasibility of the reintroduction of a variety of extirpated tropical species to their former ranges.

Archaeological and palaeoenvironmental datasets can also potentially provide insights into the reasons behind tropical floral and faunal extinctions, and thus be used to protect them. As noted above, human involvement in Late Pleistocene tropical forest megafaunal (animals >45 kg) extinctions remains hotly disputed, but more focused research and diverse methodologies may help to address this question in future. Increasingly resolved palaeoclimatic datasets may also enable conservationists to address when the thresholds of ecological preferences of extirpated or extinct species were crossed in the past (Louys et al., 2014; Wood et al., 2017). More detailed archaeological information can even provide insights into the interplay between animals and plants in forest persistence or decline. For example, while not in the tropics, analysis of a Holocene coprolite of the New Zealand kakapo (*Strigops habroptilus*) demonstrated abundant pollen from a currently threatened plant suggesting an important role for this bird in pollination (Wood et al., 2012). The benefits of understanding complex relationships such as these can be seen in the introduction of foreign, African spurred tortoises (*Centrocehlys sulcata*) as analogues of large grazing waterfowls on the island of Kauai, Hawai'i. While not 'natural', these tortoises suppress undergrowth, weed invasive species, increase soil nutrients, and aid seed germination, preventing the disappearance of endemic vegetation communities (Burney and Burney, 2016).

Focus on *how* humans have altered tropical forests for millennia, rather than *whether* they have, also provides conservation and ecological insights into the threats of different subsistence strategies and landscape management practices on these environments. For example, the archaeological record demonstrates that plant and animal management strategies developed *within* tropical forest settings, with long-term local knowledge, are largely sustainable. Arboriculture, mixed-cropping, and the translocation of tropical forest root crops and marsupials all appear to have been incredibly sustainable over the longer term (Chapter 5). By contrast, invasive management strategies, and particularly the introduction of foreign cropping systems and domesticated animals, have severe impacts on local soil stability, forest density, and biodiversity. The arrival of rice agriculture in the tropical forests of Southeast Asia and the expansion of millet into the Central African rainforests during the Late Holocene are both associated with large-scale forest clearance and soil erosion (Bellwood, 1993; Krigbaum, 2003; Bayon et al., 2012), while the land-use practices and species accompanying Polynesian settlers in the Pacific led to a decline in forest species diversity as well as mass avian extinctions (Steadman, 2006; Fall, 2010). These and other differences discussed in Chapter 5 should provide food for thought for corporations, governments, and smallholders seeking to transplant monoculture cropping and cattle-ranching into tropical forests.

The archaeological record can also be used to identify threats facing large human populations in tropical forests, as well as possible mitigation strategies for use in the future. For example, the extensive urban fringes around many ancient tropical forest urban centres discussed in Chapter 6 are attracting attention within present-day urban planning research in terms of their potential to solve contemporary challenges (e.g., Simon and Adam-Bradford, 2016). The role of this dispersed settlement pattern in aiding local agricultural adaptations to tropical forest settings, and in addressing the vulnerability of dense population centres to climate change and natural disasters such as mudslides and flooding, are of increasing interest, with archaeological data from tropical regions providing useful case studies of long-term dynamics (Simon and Adam-Bradford, 2016; Larsen, 2017). Also of interest in relation to dense populations in tropical forests are archaeologically discovered anthrosols, such as the fertile *terra preta* soils of the pre-Columbian Amazon. Research into these pre-Columbian 'soilscapes' has both encouraged the search for pantropical analogues (Sheil et al., 2012; Fraser et al., 2015) and inspired attempts to recreate similarly fertile soils, as a means of supporting increased agricultural production to sustain ever-growing human populations in these environments (Glaser and Birk, 2011). Such insights will prove valuable as global human demography increasingly converges on the tropics.

Further in-depth research and more diverse methodological approaches will only increase the cultural ecosystem service value of tropical forests and increasing numbers of world heritage sites are now being recommended from tropical forest habitats, ranging from early *Homo sapiens* cave sites in Sri Lanka (Roberts, 2017) to large-scale field systems in Bolivia (Rostain, 2017). UNESCO is now actively seeking to create joint world heritage sites of natural and cultural importance in tropical forest regions so that archaeological sites and their forest contexts are mutually protected within the framework of the UN 2030 Sustainable Development Programme (Sanz, 2017a, 2017b). Acknowledgement of the importance of protecting preserved knowledge of tropical forest adaptations has also reinforced the importance of incorporating Indigenous and traditional knowledge as part of the solution, and an integral part of conservation and policy consultation, rather than part of the problem of sustainable tropical forest development (Chapter 7)

(Sanz, 2017a, 2017b; Sheil, 2017). Moving forward, increasingly regular and intensive exchange between archaeologists, ecologists, anthropologists, biologists, geographers, as well as local Indigenous peoples and smallholders, is now being sponsored by international bodies such as UNESCO and the FAO (Bonell and Bruijnzeel, 2005; World Heritage Forests, 2011). This undoubtedly offers the best potential for uniting knowledge of human–tropical forest interactions from the past, present, and future in effective research and policy making.

This book has been an attempt to demonstrate that tropical forests have been a consistent part of the human journey. The archaeological research associated with this promotes recognition and conservation of tropical forests as 'novel ecosystems' (Hobbs et al., 2009; Clement et al., 2015; Roberts et al., 2017a), rather than primordial environments. From the evolution of our species we have occupied and manipulated tropical forest plants, animals, and soils, to survive in these habitats. Yet, this does not mean that we should consider modern logging, mining, commercial hunting, urban expansion, ranching, and the effects of human-induced climate change as business as usual. Instead, the archaeological record highlights exactly why these new developments are unsustainable. This temporal perspective shows that tropical forests represent crucial heritage capsules of human adaptation and innovation, but also demonstrates just how quickly our actions in the last century are tipping the balance of human impacts towards the permanent removal of tropical forests. Tropical forests are some of the oldest terrestrial ecosystems on the planet, existing for 100 million years (Couvreur et al., 2011). They were also likely the cradle for the emergence of those hominin ancestors *c.* 10–7 million years ago from which we later evolved. This makes the prospect of the potential end of our story with these environments all the more poignant and illustrative of the severity of human climatic and environmental tensions in the twenty-first century. It is to be hoped that growing interest in the human past within tropical forests can drive innovative solutions to ensure their persistence.

9

Forests of Plenty?

Comparisons and Conclusions

The clearest way into the Universe is through a forest wilderness.
—John Muir (Wolfe, 1938: 313)

The writings of John Muir were a key part of the early advocacy for the preservation of 'wilderness' areas in the United States. I hope to have demonstrated throughout this book that tropical forests are no longer a 'monstrous and boundless thicket, with no way of escape' (Chapter 1), and hold an important place in our biological and cultural history. Yet, while this certainly makes tropical forests key priorities for conservation, I also hope to have shown that they should not be seen as static or 'wild'. The development of new methodologies, theoretical realization that these environments are not 'primeval' or 'impenetrable', and a growth of research interest and funding are beginning to reimagine these environments as variable, dynamic, and important stages for human action. In this concluding chapter I argue that it is now time to move beyond dichotomies of whether tropical forests are 'pristine' or whether they have been constantly occupied and modified by humans through ongoing 'niche construction' (Roberts et al., 2017a). Instead, we are now able to begin to study, in detail, the sheer diversity of tropical forest 'prehistories', 'histories', and 'modernities' accessible to archaeologists, historians, palaeoecologists, and anthropologists. However, I want to go even further than the simple acknowledgement that a recent growth of data from, and interest in, tropical forests has given them greater visibility in discussions of human adaptations, past and present.

Instead, I wish to evaluate to what extent the information from these habitats can play a theoretical and methodological role in narratives of the human 'Universe' more broadly. To do this I return to a series of themes developed in Chapter 1 that pitch discussions of Miocene and Pleistocene tropical forest occupation, indigenous agricultural developments and external introductions, tropical forest urbanism, and ethnohistoric and ethnographic tropical forest foraging and farming against patterns of behaviour in other environments. Firstly, to what extent have adaptations to tropical forests been unique rather than just one part of broader human adaptive trends and capacities? Secondly, do human foraging, farming, and settlement in tropical forests face inevitable constraints, relative to more open or seasonal environments and is resilient occupation possible? It is clear that humans have influenced tropical forests, but how have these environments, in turn, altered the biological, cultural, economic, and political systems of our species? Finally, are tropical forests the ultimate marker for changes in human–environment relationships as we push our use of the Earth to the very limits of sustainability? Investigation of these themes ensures that tropical forests are truly brought in from the 'wilderness', not just in terms of available data, but also in terms of how we view ourselves and our ongoing interactions with the wider world.

TROPICAL FOREST PREHISTORIES, HISTORIES, AND MODERNITIES

Chapter 2 demonstrated that the term 'tropical forest' encompasses a huge variety of macro- and micro-scale habitat variability that, in turn, influences the animal and plant resources, hydrology, soil quality, and seasonality experienced by different human populations. Furthermore, growing palaeoenvironmental datasets across the tropical continents are revealing that tropical forests are temporally dynamic, with formations expanding and contracting at the expense of one another, or even disappearing entirely. In many cases the species composition, geomorphology, and extent of these environments have been significantly shaped, or 'constructed' by humans (Boivin et al., 2016; Roberts et al., 2017a). As a result, it is important that we not only acknowledge that a tropical forest prehistory or history for our species is possible (Mercader, 2002a, 2002b; Iriarte et al., 2007; Morrison, 2014; Roberts and Petraglia, 2015), but also that it is not singular. Instead, different foragers, farmers, and urban dwellers will have experienced very different conditions in these environments across space and time. Indeed, what has hopefully stood out from a number of chapters in this book is not only that there is now considerable evidence for the use of resources, cultivation and herding of plants and animals, and formation of complex settlements in tropical forests in the past, but that there is immense variability in how this came about in different parts of the world and even within the same region.

The evolution of the hominin clade through the Miocene, Pliocene, and into the Early Pleistocene has continued to be associated with a general movement away from tropical forest and towards the open 'savanna' (Dart, 1925; Plummer et al., 2009; Lee-Thorp et al., 2010; Domínguez-Rodrigo, 2014). Nevertheless, it is becoming increasingly apparent that hominin morphology, diet, and locomotion retained a complex relationship between forested and open areas during this time (Thorpe et al., 2007; Kivell et al., 2011; Roberts et al., 2016). Moreover, the emergence of C_4 biomass in local and regional environments varied in linearity and tempo between and within eastern and southern Africa throughout the Pliocene (Reed, 1997; Kingston and Harrison, 2007; Elton, 2008; Feakins et al., 2013; Robinson et al., 2017). Perhaps most significantly, however, is the fact that regardless of changing environments, hominin populations had behavioural and dietary agency. While Paranthropocines seemingly followed the trend of some other medium to large African mammalian taxa in specializing in C_4 resources use, other Pliocene and Pleistocene hominins, including members of the genus *Homo*, maintained more varied diets, likely including elements of woodland or forest (Sponheimer et al., 2013). This, and the current gap in the Miocene–Pleistocene fossil record from West and Central Africa, makes it likely that much more remains to be discovered regarding the involvement of tropical forests in hominin evolution during this time.

It is also likely that the maintenance of ecological and dietary mosaics, including tropical forests, assisted some of the earliest hominin forays beyond Africa during the Early and Middle Pleistocene (Chapter 3) (Roberts et al., 2016). In the case of our own species, its evolution and expansion within a context of climatic and environmental variability (Rabett, 2012; Brooks et al., 2018; Potts et al., 2018) necessitate systematic study of multiple tropical forest 'prehistories'. In Africa, current evidence suggests the ongoing use of a mixture of tropical forest–woodland–grassland mosaics (Taylor, 2011, 2016; Blome et al., 2012; Shipton et al., 2018), with tantalizing hints of tropical lowland evergreen rainforest occupation (Barham, 2001; Mercader, 2002a). Varied specialization in the hunting of primates in evergreen lowland rainforest, the deliberate maintenance of a diversity of evergreen rainforest and peat swamp forest formations and their resources, and the mobile use of montane rainforest tree crops and grassland mammals in South Asia, Island Southeast Asia, and Melanesia, respectively, (Chapter 4) further highlight how early members of our species occupied and manipulated tropical *forests* rather than forest. This diversity would be expanded even further when Late and Terminal Pleistocene occupation of temperate rainforests in Tasmania and South America are considered (Pike-Tay and Cosgrove, 2006; Piperno, 2011a; Dillehay et al., 2015, 2017).

Despite being neglected until relatively recently, tropical forests also evidently document rich prehistories of agricultural practice. Early Holocene cultivation of taro, banana, and sugarcane in montane Melanesia and, perhaps, also across Southeast Asia and even Australia, indicate new forms of biotic management without necessarily a full conversion to dense, sedentary populations (Golson, 1989; Denham et al., 2003, 2009; Denham, 2011; Golson et al., 2017; Roberts et al., 2017d). In contrast, a combination of non-domesticated tree crops, maintenance of proximity to, and even the corralling of, protein-rich freshwater resources, and small-scale, mobile clearing for manioc cultivation maintained a patchwork of agricultural practice that supported large, stable populations across significant portions of the Amazon Basin (Iriarte et al., 2007; Hemming, 2009; Arroyo-Kalin, 2012; Clement et al., 2015). A greater reliance on the tropical crop maize in the drier tropical forests of North and Central America, relative to that seen in Amazonia following its arrival there in the Middle Holocene (Hermenegildo et al., 2017), demonstrates different adaptive strategies. However, the 'forest gardening' of the Maya still involved persistent use of wild tropical forest tree crops and animals (Ford and Nigh, 2015; Lucero et al., 2015; Farahani et al., 2017). The ecological origins of the chicken, and perhaps also the water buffalo, as well as their subsequent expansion across a variety of environments will likely further expand the scope of tropical forest agricultural prehistories in future.

The existence and scale of prehistoric and historic emergence of urbanism in tropical forests continues to dominate public perceptions of tropical forest archaeological 'discovery' (Lucero et al., 2015; Chapter 1). The massive sprawls of Greater Angkor in Cambodia and Anuradhapura in Sri Lanka, and the growing recognition that lowland Classic Maya settlements document extensive settlement, have been used to highlight a trajectory towards 'low-density, agrarian-based urbanism' (Fletcher, 2009, 2012; Lucero et al., 2015). Nevertheless, low-density 'Garden City' settlements in Amazonia maintained a reliance on the wild tree crops and freshwater resources mentioned above, as well as 'agrarian' products (Heckenberger et al., 2008; Heckenberger and Neves, 2009). The capital of the Tu'i Tonga Empire in the insular tropical forests of the Pacific was more symbolic than 'agrarian', providing the ceremonial focal point for long-distance exchange networks that formed the basis for elite power and administration. Within the category of 'low-density, agrarian-based urbanism', while Greater Angkor and Anuradhapura developed centralized water 'tanks' or '*barays*' to combat seasonally dry climates, water management was dominated by small-scale communities in the Maya lowlands (Lucero et al., 2015). The continued growth of detailed remote sensing, historical ecology approaches, and palaeoenvironmental studies in these, and other, tropical forest regions in future will only further contribute to the diverse, contextually specific ways in which tropical forest 'urban' studies have challenged long-held preconceptions.

The 'pristine' and 'timeless' manner in which tropical forest groups have traditionally been ethnographically and ethnohistorically described has also necessitated theoretical overhaul (Bailey et al., 1989). However, here, multi-vocal *histories* are more appropriate than uniform narratives for a number of reasons. The arrival of European colonial powers, and their associated slavery, murder, and disease, led to massive mortality in the New World that has for a long time obscured evidence for vast, pre-Columbian anthropogenic landscapes and settlement networks (McMichael et al., 2017). By contrast, tropical forest groups in the Old World had faced economic and political marginalization and slavery, albeit on a different scale, long before European influence (Junker 2002a, 2002b; Mitchell, 2005; Morrison, 2014). Moreover, within both the Old and New World, different cultural groups did, and continue to, resist, manipulate, and adopt new political, economic, and cultural structures in the face of imperial expansion, new states, including modern nation-states, globalization, and the march of capitalism (Morrison, 2014; Ferrier, 2015; Roberts et al., 2018b). Perhaps most crucially, however, is the point that if a history is truly to be given back to Indigenous groups in tropical forests it is necessary to place their own historical methods, voices, and perspectives on the same level playing field as western history books, if not at the forefront of investigation and research (e.g., Ferrier, 2015).

Recognition of multifarious prehistories and histories also has considerable bearing on modern conservation strategies and priorities. Firstly, it offers the opportunity to scrutinize the particular changes in biodiversity, forest formation type, and landscape structure that have been wrought by human groups, and over how long a period. For example, the identification of extinct plant and animal taxa from a given region, and their specific tolerances, offers the potential for successful reintroduction when conditions change in a given locale or in a new, ecologically appropriate area (Burney and Burney, 2007; Louys et al., 2014). The knowledge of diverse human pasts in tropical forests can also facilitate the development of more successful approaches to increasing settlement and economic utilization of tropical forests in the modern world, as well as highlighting particular threats, such as natural disasters or soil erosion, in a particular region or tropical forest formation type (Larsen, 2017; McMichael et al., 2017; Roberts et al., 2017a). By bringing together prehistories, histories, and modernities in tropical forests, priorities and strategies of tropical forest conservation can start to be selected and refined on suitable local, regional, and global scales (Burney and Burney, 2007; Louys et al., 2014; Malhi et al., 2014; DRYFLOR et al., 2015; Roberts et al., 2018a).

UNIQUE ADAPTATIONS OR PART OF THE BIGGER HUMAN PICTURE?

Beyond compiling datasets from archaeology, palaeoenvironmental science, history, ethnography, and conservation to show a variety of past and present hominin tropical adaptations, this book has also sought to draw out some key ways in which tropical forests have uniquely informed broader theoretical and methodological approaches to the human story. On the most fundamental level, as we saw in Chapter 2, tropical forests are the oldest terrestrial ecosystems on the planet. As a result, they are the only environments to be present on the face of the Earth throughout the evolutionary emergence of our hominid and hominin ancestors from the Miocene onwards. While the intimacy of the relationship has changed, tropical forests, in some form, have been an ongoing part of hominin subsistence, locomotion, and migration in geological and palaeoanthropological time. For example, these environments shaped the taxonomic diversity and geographic extent of Miocene great apes across Eurasia and Africa, and climate-driven constrictions of these habitats likely played an important role in the current distributions of great ape taxa and the geographic emergence of the hominin clade in Africa towards the end of this geological epoch (Begun, 2010; Hamon et al., 2012). Furthermore, throughout the Pliocene, while these habitats were waning, hominin taxonomic variability may, in part, have been influenced by different degrees of reliance on tropical forest and grassland settings (Sponheimer et al., 2013).

However, the time-depth of their presence is not the only reason tropical forests are important elements of discussions of hominin biological and cultural change through time. Tropical forests account for a third of land-surface productivity and evapotranspiration, and are home to the highest proportion of the world's plant and animal species of any terrestrial environment (Malhi, 2012; Malhi et al., 2014). Yet they are also diverse, with different formations variably presenting complications of humidity or drought, abundant, large terrestrial protein mammalian resources or small arboreal mammals, a diversity of edible plants and fertile soils or dry, nutrient-poor soils (Chapter 2). As such, these environments offer the opportunity to look at the adaptive thresholds and flexibility of different hominin taxa entering them through time, acting as a microcosm of hominin adaptability to the Earth's environments more broadly (Roberts and Stewart, 2018). For example, although Early-Middle Pleistocene members of the genus *Homo* were found at latitudes of 40°N at Dmanisi, Georgia 1.8 Ma, in tropical Southeast Asia between 1.5 and 1.0 Ma (Larick et al., 2001; Zaim et al., 2011), and at the edge of the boreal zone at Pakefield in England 0.7 Ma (Parfitt et al., 2005, 2010), all palaeoenvironmental evidence points towards the use of fluctuating mosaics of forest and open, dry steppe. This remains the case for Middle Pleistocene *Homo floresiensis* in Flores, Indonesia, and *Homo neanderthalensis* across Eurasia (Roberts and Stewart, 2018).

By contrast, the emergence of our own species, *Homo sapiens*, as noted above, saw the specialized adaptation to a variety of different tropical forest types, ranging from thin-leaved, seasonally cold montane rainforest to low light, lowland, tropical evergreen rainforest. The flexibility of *H. sapiens* in this regard is perhaps also highlighted by the fact that it expanded in Southeast Asia alongside the modern 'Punung' rainforest fauna as earlier species of *Homo* disappeared (Roberts et al., 2016; Roberts and Amano, under review). Roberts and Stewart (2018) have argued that comparison of the specialized adaptations to extreme environments by *H. sapiens*, relative to the broadly uniform, mosaic preferences of early members of the genus *Homo*, is indicative of a new ecological capacity, and perhaps also niche, for our species. Tropical forests, alongside desertic, palaeoarctic, and high-altitude settings, have been an important part of the proposal of the ecological distinctiveness of our species. In turn, this encourages the use of new methodologies, including the stable isotope analysis of tooth enamel of different hominins in tropical environments (Roberts et al., 2016), and emphasizes the importance of presenting detailed palaeoenvironmental data alongside fossil finds (Roberts and Stewart, 2018). Moreover, if our species is indeed particularly distinctive ecologically, then this has major implications for climatic and environmental modelling of hominin expansions (Timmermann and Friedrich, 2016), and regions of priority for future archaeological and anthropological survey.

Tropical forests have also impacted theoretical, as well as methodological, approaches to the 'origins of agriculture' on a global scale. Traditional definitions have used the term 'agriculture' where domesticated plants and animals are the major subsistence source for a particular society (Price and Bar-Yosef, 2011). 'Agriculture' has been associated with sedentary societies, pottery production, and the emergence of inequality (Childe, 1936, 1950; Smith, 1999). Growing research in tropical forests, however, has highlighted the temporal depth of 'agricultural' activities such as cultivation and plant and animal translocation, as well as the management of forest mosaics using fire, that blurs the distinction between foraging and farming behaviours (Allen et al., 1989; Denham et al., 2003; Heckenberger et al., 2008; Denham, 2011; Clement et al., 2015; Hunt et al., 2012). Moreover, where plants have been domesticated in tropical forests, or domesticated plants introduced from outside, they are often incorporated as part of mobile use of the landscape or combined with the ongoing utilization of wild resources (Iriarte et al., 2007; Clement et al., 2015; Roberts et al., 2017d). A new focus on methodologically and theoretically appreciating the temporal span of farming *practice* and its biotic impacts, rather than a focus on a final, definitive 'agriculture', in part spurred on by growing research in the tropics, has left a lasting mark on approaches to studying new human ecological relationships with plants and animals during the Holocene on a global scale (Fuller et al., 2011a; Fuller and Hildebrand, 2013; Allaby et al., 2016).

A similar theoretical re-think of the term 'urbanism' in archaeological and historical research has been encouraged by research in tropical forests. Long-held ideas of urbanism in archaeology have focused on 'compact', often bounded, cities such as those found in Mesopotamia (4,000–3,100 BC), with demarcated relationships between the agricultural hinterland and the craft specialists and bureaucratic groups of the urban area (Adams, 1981; Postgate, 1992). Yet, research in Mesoamerica, Southeast Asia, and South Asia, reviewed in Chapter 6, has highlighted a very different urban trajectory towards 'low-density agrarian urbanism' (Fletcher, 2009, 2010, 2012). In this model, huge urban extents of population, such as the largest pre-industrial human population found anywhere in the world in the 1,000 km^2 of the Greater Angkor region of Cambodia in the late first millennium AD (Fletcher, 2012), incorporate agricultural land and hydrological management (Lucero et al., 2015). In many cases, sections of tropical forest were also managed for tree crops and wild fauna, with spotted deer being corralled in the case of the lowland Classic Maya (Ford and Nigh, 2015; Lucero et al., 2015). This research contests traditional western ideas of what an urban landscape should look like, and provides an important reference at a time when many modern industrial cities are also tending towards low-density agglomerations (Fletcher, 2012; Simon and Adam-Bradford, 2016).

Ethnographic, ethnohistorical, and archaeological reappraisal of Late Holocene populations in tropical forests also has an important message for how we consider and involve Indigenous groups more broadly in archaeology and anthropology (Morrison, 2014). These communities have been at the centre of the debate as to whether tropical forests were habitable with or without agricultural contacts (Bailey et al., 1989; Bailey and Headland, 1991; Hames, 2007). The growth of global trade networks, and later, the influence of colonial powers and their accompanying weapons, disease, and exploitation, have tended to be seen as things that happened to Indigenous tropical forest inhabitants in the Old and New World, alike. However, it is clear that in many cases, small-scale farmers and foragers demonstrated significant agency in the manipulation and varied adoption of broader economic, political, and cultural developments (Morrison and Junker, 2002; Mitchell, 2005; Fuller et al., 2011c; Morrison, 2014). Moreover, in many cases tropical forest subsistence was not something forced upon them or part of an innate isolation but was rather a usable form of resistance and identity (Morrison, 2014; Ferrier, 2015). In giving back a history and agency to Indigenous populations in tropical forests, growing ethnographic and archaeological collaborations with these groups are now trying to ensure that history, and even prehistory, is told by these groups in their own way and through their own cultural perspectives (Balée, 2013; Kohn, 2013; Ferrier, 2015; Sheil, 2017).

Archaeological and palaeoenvironmental research in tropical forests is also having a major effect on how ecologists and conservationists approach modern landscapes. In particular, while past human impacts on these environments have often been dismissed as qualitatively different to anthropogenic threats in the modern world (Malhi et al., 2014; Lewis et al., 2015), increasing evidence demonstrates that pre-industrial human populations stimulated continentally and globally relevant changes to these habitats in the Late Holocene (Roberts et al., 2018a). Moreover, these changes, and perhaps even early human influences, have had lasting impacts on species distributions, tropical forest formation extents, and landscape morphology that persist today (Roberts et al., 2017a). The argument for taking archaeological, historical, and palaeoecological evidence into account when studying modern impacts on tropical forests is an important part of a broader discussion relating to the concept of the 'Anthropocene'. While some argue that this term is only useful when applied as a formal, geological epoch in the modern world (Zalasiewicz et al., 2008, 2015), social scientists and palaeoecologists have argued that it is better used as a concept through which to study changes in the pervasiveness and timing of human impacts on earth systems (Ruddiman et al., 2015; Roberts et al., 2018a). This paves the way for a more integrated use of past, present, and future datasets when evaluating current and predicted human interactions with the planet's environments.

DESTINED FOR FAILURE?

Previous conservation and political assumptions that tropical forests were largely unoccupied and unaltered prior to the arrival of commercial mining companies, large-scale agricultural landowners, and industrial urban agglomerations from the eighteenth century onwards (Malhi et al., 2014; Lewis et al., 2015) gives them the guise of being innately 'delicate'. This, in turn, has fed into archaeological, historical, and ethnographic discussions and where sedentary tropical forest occupation and usage has been documented it was often deemed unsustainable (Meggers, 1954, 1971; Bailey et al., 1989; Beach, 1998; Beach et al., 2006; Lentz et al., 2014). The growing evidence for Late Pleistocene hunting and plant management, the emergence and expansion of agriculture in tropical forest settings, and the development of urban forms, as well as the huge diversity of tropical forest formations, allow us to take a more critical look into patterns of adaptive success in these environments. This highlights scenarios where human behaviours did cause lasting damage, leading to a deterioration of population size, political reconfiguration, or permanent landscape conversion. However, it also offers the opportunity to look at the contexts of resilient tropical adaptations of foragers, farmers, and urban dwellers, and potential solutions to climatic and environmental challenges facing the ever-expanding number of humans reliant on tropical forests in the modern world.

Arguments for the potential of Late Pleistocene *H. sapiens* to have had significant, long-term effects on tropical forests focus on its potential role in the extinction of megafauna (Lewis et al., 2015). Given the lack of previous evidence for hominin occupants in dense tropical forests, megafauna in these environments would have been potentially more vulnerable to human arrival and hunting of them less sustainable. For example, if humans had a hand in the demise of animals, such as gomphotheres and ground sloths in the Neotropics, and extinct marsupials (such as *Maokopia ronaldii* and *Thylogale hopeii*) in Melanesia, then they would have altered plant fertilization strategies, regional vegetation composition, and carbon cycles (Doughty et al., 2015; Lewis and Maslin, 2015). Nevertheless, their role remains debated and other instances of Late Pleistocene tropical forest occupation suggest human foraging could be remarkably resilient. For example, in Sri Lanka, humans living in evergreen lowland rainforests between 36 and 3 ka persistently specialized in the hunting of semi-arboreal and arboreal mammals, in these environments (Perera, 2010; Perera et al., 2011; Roberts, 2018). Similarly, at the Niah Caves, Borneo, deliberate vegetation burning to maintain mosaics of evergreen rainforest, peat swamp forest, and open grassland was practised from 45 ka into the Holocene (Hunt et al., 2012). Ethnographic examples of sustainable Indigenous hunting warn against assumptions of tropical forest mammal vulnerability (van Vliet et al., 2009).

Supposed nutrient-poor soils in tropical forests have been seen as a limiting factor for settled agricultural communities in these environments compared to temperate or more arid settings (Meggers, 1954; Beach et al., 2006). Indeed, traditional ideas of agriculture in Europe, northern America, and across much of Asia and Africa that focus on the herding of medium to large mammals and the cultivation of dry-adapted cereals seems improbable in many tropical forest settings and in the face of additional disease loads (Smith, 1999; Marshall and Hildebrand, 2002; Price and Bar-Yosef, 2011). Certainly, the introduction of external livestock or cereal-based cultivation systems into tropical forests, for example in the cases of the Bantu agricultural expansion through central, eastern, and southern Africa and the dispersal of rice farming throughout Southeast Asia, resulted in significant deforestation and soil erosion (Bellwood, 1993; Fuller et al., 2011a, 2011b; Bayon et al., 2012). Similarly, the transferal of pigs, and commensal taxa, onto Pacific islands previously untouched by humans had massive implications for seabird populations, nutrient cycles, and soil stability in the Late Holocene (Kirch, 2001, 2010; Steadman, 2006; Swift et al., 2017a, 2017b). It is also perhaps illustrative of the limitations of tropical forests for agriculture that significant modern staples—such as maize, water buffalo, and chickens—with tropical forest origins are most intensively kept or cultivated in drier, more open regions.

Nevertheless, a variety of agricultural strategies in tropical forests have also been developed and resiliently maintained. Such indigenous tropical forest agricultural systems make use of the unique ecological dynamics of these environments, focusing on tree crops, such as the banana, or starch-rich tubers such as yams, taro, and manioc (Iriarte et al., 2007; Denham, 2011). Furthermore, the growth of these domesticated resources often combines significant landscape alteration through burning or earthwork construction within systems of natural succession, and also frequently includes the combined use of domesticated tubers, cereals, or tree crops with the selective management of wild plant and animal resources (Golson, 1989; Denham et al., 2003; Hill and Baird, 2003; Iriarte et al., 2007; Denham, 2009, 2011; Clement et al., 2015). A focus away from traditional perceptions that 'agriculture' should involve intensive field systems and animal herding, leads to interesting comparisons in the origins, nature, and resilience of agricultural practice in different parts of the world and in different environments. For example, even following the formation of the Eurasian package of livestock and cereals in the Near East, periods of climatic change as well as environmental variability led to various combinations of herding, farming, fishing, and hunting in local subsistence strategies (Kujit and Goring-Morris, 2002; Yerkes et al., 2014).

Far from being destined to failure then, tropical forest agricultural practice actually highlights ways in which varied agricultural land-use can be sustainable, not just in these environments but also in other settings. It is not so important to look at basic assumptions of universal productivity and maximum calorific output, but rather to how different plants and animals were adapted to new human niches. This recognition has significant implications for agricultural sustainability in modern tropical landscapes. For example, the success of patchworks of indigenous crop management, tree crop usage, and the hunting of wild animals contrasts modern large-scale conversion of tropical land to monoculture or ranching, with consequences for biodiversity, soil stability, disease spread, and succession dynamics (Ghazoul and Sheil, 2010). Similarly, the introduction of cereals or livestock will not necessarily provide a more reliable source of nutrition than indigenous tubers or tree crops, as highlighted by the growing success of Neotropic starch crops, such as manioc, amongst smallholders and broader international initiatives in West and Central Africa (Cock, 1982; Polson and Spencer, 1991; Ghazoul and Sheil, 2010). The importance of transferring ecological knowledge, as well as the plants themselves, is evident in the colonial transfer of Senegambian African rice farmers as slaves to the Carolinas in order to improve yields in the New World (Mitchell, 2005). As a result, tropical forest modification, including food production strategies, forest burning, drainage system digging, and the construction of earthworks, should not necessarily be seen as inherently 'bad' when it is done with due knowledge of local plant and animal lifecycle dynamics.

The same point applies to the development of urban settlements and networks in tropical forests that are built upon these diverse systems of agricultural practice. Video games, films, and books continue to present tropical forests as settings for the search of the 'ruins' of ancient cities, with green overgrowth portraying the inevitable end of these societies. In archaeology, local, global, and popular studies have also often discussed urban settlements, such as those of the Classic Maya, as doomed to dangerous deforestation, soil erosion of already-fragile landscapes, and eventual collapse (Beach, 1998; Beach et al., 2002, 2006; Diamond, 2005). Even with a new 'low-density agrarian urbanism' trajectory established for prehistoric and historic tropical forest cities, the inherently extensive nature of such settlements has been seen to place an inescapable strain on the delicate balance between independent local agrarian activities and a centralized politics—particularly in the face of external pressures such as climate change (Fletcher, 2009, 2012; Buckley et al., 2010; Lentz et al., 2014; Lucero et al., 2015). Such developments appear to stand in stark contrast to the dense cities of craft and bureaucratic specialists, and focused agricultural hinterlands of agricultural field systems, that developed along fertile alluvial plains in Mesopotamia, the Indus region of South Asia, Bronze Age China, and Medieval Europe (Postgate, 1992; Cowgill, 2004; Pirenne, 2010; Xue and Sit, 2010). In the cases of Europe, China, and the Middle East, these urban networks, if not the hubs themselves, have remained in place for millennia (Fletcher, 2009, 2012).

Nevertheless, archaeological evidence highlights the potential longevity of, if not tropical forest settlements themselves, then resilient urban diaspora (Lucero et al., 2015). In cases such as that of Anuradhapura in Sri Lanka and the Late Holocene 'Garden Cities' of Amazonia, extensive human populations and urban networks survived for centuries, if not an entire millennium, longer than any industrial tropical forest city, and longer than many Medieval European centres to date (Roosevelt, 1999; Heckenberger and Neves, 2009; Fletcher, 2012). In some cases, longer-term studies of settlement patterns, such as in the Greater Angkor region of Cambodia and the Maya lowlands of Mesoamerica, highlight that a focus on spectacular periods of 'fluorescence' and 'collapse' has often missed the process and resilience of settlement patterns that persist through political crises

(Webster, 2002; Lucero et al., 2015). This growing interest in urban connectivity, as well as acceptance that the term 'urban' may define more diverse settlement patterns than we are frequently used to, also supports new approaches to classic urban examples in other environmental settings. For example, the 'Dark Ages' of much of northern and southern Europe are actually characterized by the reconfiguration of trading links and settlement patterns in different hubs rather than absolute collapse (Hamerow, 2007; Loveluck, 2013). Similarly, the coexistence of mobile pastoralists and political and bureaucratic 'cities of the dead' (Wengrow, 2006) during the Naqada I–III periods of ancient Egypt seems less aberrant as an 'urban' form when placed alongside the diverse, extensive agrarian settlements in tropical forests.

The documentation of resilient extensive settlement and bureaucratic networks in tropical forest settings, rather than inevitable ruin, enables a more detailed, contextual evaluation of urban experiences and challenges in these environments, and beyond. For example, 'low-density agrarian urbanism' is particularly well suited to the extensive, but sustainable occupation of seasonally dry tropical forest settings. Low-density settlement networks also seem to have offered an adaptive solution for communities with varying reliance on domesticated and wild resources in the grasslands of Mongolia (Roth et al., 2002), the grassland-woodland floodplain ecotones of the Inner Niger Delta, Mali, and the Middle Senegal River Valley (McIntosh and McIntosh, 1980; McIntosh, 1995, 1999; Connah, 2016), and the lowland evergreen rainforests of Amazonia (Heckenberger et al., 2008). However, due to the importance of the independent maintenance of dispersed agrarian holdings, herding, or wild resource use in each of these instances, such extensive, low-density trajectories seem to either discourage political centralization (e.g., McIntosh, 1995, 1999) or involve persistent internal strain between independent smallholders and centralized political forces that makes changing climatic and environmental situations more difficult to navigate (Fletcher, 2012; Lucero et al., 2015). Thus while such urban systems may seem attractive, they undoubtedly represent novel administrative challenges relative to more bounded, high-density settlement patterns.

The adoption of greater nuance in discussions of tropical forest forager, farmer, and urban resilience also re-centres the importance of the involvement of Indigenous and smallholder rites in conservation and policy discussions. The overall conservation capacities of Indigenous groups, for example, have fluctuated between absolute ideas of the pristine, curatorial 'Noble Savage' and the idea that these groups should be removed from protected areas for conservation to be effective (Balée, 1989, 1994; Stevens, 1997; Barume, 2000; Hames, 2007). However, both of these perspectives massively underestimate the agency and variability among these groups. Indigenous and smallholder bushmeat hunting can cause massive declines in biodiversity, particularly with the adoption of firearms (Fa and Brown, 2009), while smallholder land-use is one of the greatest causes of tropical forest deforestation worldwide (Corlett and Primack, 2011). Yet, it is also clear that some of the greatest conservation results occur when Indigenous and smallholder groups are included, or even initiate, local conservation efforts, and Indigenous hunting practices can be sustainable (Stevens, 1997; Peres and Nascimento, 2006; Ghazoul and Sheil, 2010; Sheil et al., 2015). What is evident, particularly when their history is re-evaluated (Chapter 8), is that

Indigenous tropical forest groups have the greatest experience in combining local ecological knowledge with new resources, power structures, and economic connections that will continue to characterize human interactions with tropical forest environments in the twenty-first century.

TROPICAL LEGACIES: CULTURAL, POLITICAL, SOCIAL, AND BIOLOGICAL OUTCOMES

Much of this book, and indeed this conclusion, has centred on how past and present human societies have occupied and modified tropical forests in multifarious ways. However, this leaves tropical forests as one of many of the world's environments, albeit a unique and previously under-appreciated one, that have acted as a theatre for human action and expression (Gamble, 1993, 2013; Iriarte et al., 2007; Rabett, 2012; Roberts and Stewart, 2018). To truly re-centre tropical forests in the human story, it is also important to evaluate and acknowledge the extent to which these environments have, in turn, actually had lasting impacts on our own species and its ancestors. 'Savanna' biomes have traditionally been forefronted in discussions of environmental drivers of hominin biological, ecological, and cultural evolution and variability from the Miocene through to the Pleistocene (de Menocal, 2004; Dennell and Roebroeks, 2005; Boivin et al., 2013; Domínguez-Rodrigo, 2014). Meanwhile, coastal settings have been emphasized as important to Middle and Late Pleistocene cognitive development and migration in our species (Mellars, 2006; Marean, 2016). In terms of agriculture and urbanism it is alluvial plains, often flanked by arid environments, which have been highlighted as significant (Hassan, 1988; Postgate, 1992; Wengrow, 2006). Tropical forests, by contrast, have remained conspicuously absent in contextual considerations of major human biological, economic, cultural, and political transitions.

As Chapter 3 demonstrated, tropical forests have clearly played a more complex and ongoing role in hominid and hominin evolution than has often been appreciated. Chapter 3 documents a variety of ways in which tropical forest settings continued to influence the morphology and diets of different hominin taxa (Sponheimer et al., 2013; Kimbel et al., 2014), and were likely part of the broader ecological mosaic of grassland–woodland–forest that accompanied the expansion of *Homo erectus* into tropical Southeast Asia during the Early and Middle Pleistocene (Marwick, 2009). In order to understand the technological and adaptive capacities argued to be characteristic of hominin evolution in the Early Pleistocene (Potts, 1998; Grove, 2011, 2015) it is therefore essential to study fluctuations in tropical forest presence and nature as well as simply the emergence of 'savanna' settings in Africa and beyond. It remains possible that tropical forest patches played a major role in variation in locomotion strategies (Green and Alemseged, 2012; Kimbel et al., 2014), overall size and body proportions (Argue et al., 2009, 2017; van den Bergh et al., 2016), and subsistence variability (Sponheimer et al., 2013) across Africa and Asia between the Miocene and Early Pleistocene. It is likely that these environments are a key part of the 'variability' that drove the emergence of the genus *Homo* and, later, our species (Potts, 2013; Potts et al., 2018).

Tropical forests have arguably left their mark on the biology of *H. sapiens*. Venkataraman et al. (2013a, 2013b) and Kraft et al. (2014), for example, have documented the specialized responses of human joint structures to the mechanical pressures of tree climbing amongst groups searching for honey and other resources in the forest canopy (Figure 9.1). The plastic and epigenetic responses of tropical forest forager biology to such challenges may not be uniform across our species but do highlight the influence of these habitats on the human body (Venkataraman et al., 2013a, 2013b). Similarly, the so-called 'pygmy', small-bodied phenotype evident in tropical forest forager groups in Central Africa, South America, and Southeast Asia, has been argued to be the product of convergent phenotypic adaptation to the pressures of high humidity, impoverished protein and carbohydrate access, and even disease loads common in dense, lowland evergreen rainforests (Verdu et al., 2009; Perry et al., 2014). It has also been argued that insular tropical forest environments led to the same morphological outcome in *H. floresiensis* on the island of Flores (Argue et al., 2009, 2017), as well as the hominin species represented at Callao Cave in the Philippines (Mijares et al., 2010). Although there is currently no evidence that these Pleistocene hominins occupied or utilized tropical rainforest habitats, the evolution of small stature in members of our species occupying tropical forests today hints at a lasting biological impact of these environments on *H. sapiens*, at least.

Although not widely appreciated, tropical forests have also had a lasting legacy on human economic structures. On the most basic level, a huge variety of domesticated fruits, vegetables, spices, cereals, condiments, and meats that can be found on supermarket and pharmacy shelves in the western world have their origin in tropical forests. Furthermore, global reliance on staples such as manioc, maize, and the water buffalo highlight the importance of these environments to human nutrition in many parts of the tropics, and beyond, today (Chapter 5). Perhaps more importantly, however, is the position of tropical forests in the development of incipient globalized trade networks between Eurasia, Africa, and Asia from the Bronze Age and Iron Age onwards (Fuller et al., 2011c; Boivin et al., 2012). The role of local tropical forest foragers and farmers in South and Southeast Asia in providing and moving spices and fruits that travelled across much of the globe should not be underestimated (Junker, 2002a, 2002b; Morrison, 2002a, 2002b, 2014). The connectivity of the wider world to the tropics was only to grow with the expansion of colonial powers, resulting in the trans-Atlantic transfer and planting of chocolate, coffee, tobacco, sugarcane, and rubber (Mintz, 1986; Mitchell, 2005; Hemming, 2009; Gallagher, 2016), and the pan-Eurasian movement of tea and fruits grown in Asian plantations (Chapter 7). The ongoing global significance of trade in these tropical resources, alongside the growth in luxury timber and cash crop markets, highlights the lasting effects tropical forests have had on our economic structures, and desires.

As well as economic impacts it should therefore be unsurprising that tropical forests have also left a cultural and political mark on human societies. 'Low-density agrarian urbanism' is a particularly emphatic example of this and it is no coincidence that this form of extensive settlement appears in the tropical forests of South Asia, Southeast Asia, and Mesoamerica (Fletcher, 2009, 2010, 2012; Lucero et al., 2015), and in a slightly different form in Amazonia and West and Central Africa (McIntosh, 1999; Heckenberger et al., 2008; Fletcher, 2012), throughout

Figure 9.1. A member of the Mbuti-Efe group climbs a tree to collect honey in the Ituri rainforest of the Democratic Republic of Congo. Photograph taken by, and reproduced with permission from, Bruno Zanzottera.

the Late Holocene. Where political and demographic factors bring together large numbers of people into a unified structure in tropical forests, the incorporation of agrarian land, independent landscape management, and even management of wild resources into extensive urban settlements is adaptive to a spectrum of tropical forest formations. This broader spread of agricultural land-use, including the accompanying soil impacts and deforestation, and reduced population per area, is encouraged by relatively widely-spaced resources, the potential for natural disasters, and, in the case of dry tropical forests, seasonal and longer-term fluctuations in rainfall (Lucero et al., 2015). While this urban trajectory inevitably places pressure on political hierarchies across such dispersed population (McIntosh, 1995, 1999; Roosevelt, 1999), it may well dominate the political and cultural landscapes of the increasingly occupied tropics in the future.

Tropical forests have also played a significant role in changing global political and cultural understandings of Indigenous groups and their rights. These environments have provided the backdrop to dichotomous anthropological characterization of Indigenous populations as primordial caretakers of the environment or as groups whose existence is only possible through attachment to local agricultural populations (Bailey et al., 1989; Bailey and Headland, 1991). As noted above, these positions have, respectively, similarly influenced political decisions to involve Indigenous groups, as well as local smallholders, in conservation initiatives (Hames, 2007), or to remove them completely from newly-formed 'Protected Areas' (Barume, 2000). However, ongoing ethnographic and ethnohistorical research in tropical forests is revealing that Indigenous groups should be seen as integral examples of not only how humans can sustainably survive in these environments, but also of how novel economies, resources, and growing population numbers can be incorporated into ecologically-tuned knowledge and subsistence systems. Some of the most successful conservation initiatives have placed local smallholders and Indigenous groups at the nexus of biotic conservation, new tourist structures, and sustainable cash crop and agricultural introductions (Ghazoul and Sheil, 2010; Sanz, 2017a, 2017b). The beneficial results of these policies, to cultures, political relationships, and tropical forest ecosystems alike, should act as a lesson for policy and conservation in other settings.

Finally, tropical forests, perhaps even more so than the polar ice-caps, have become the ultimate symbol of the global battle between consumer interests and environmental protection. The crucial importance of tropical forests in chemical, hydrological, and atmospheric earth systems, as well as the cultural heritage emphasized throughout this book (Roberts et al., 2018a; Chapters 2 and 8), mean that poorly informed alteration or wholesale conversion will have major implications, not just for local groups, but for humanity as a whole. As a result it is perhaps unsurprising that cooperatives such as the 'Rainforest Alliance' are at the heart of drives for global consumers to consider the sourcing and environmental implications of their products. Similarly, threats to wildlife, including the loss of habitat and illegal trading, as well as political initiatives against landscape modification undertaken by big oil and mining companies, more often than not involve tropical forest environments. Significant focus in terms of predicted temperature rises, and increasing unpredictability of rainfall, over the next century have also forefronted the effects on the increasingly populated tropics (Malhi and Phillips, 2005; Bermingham and Dick, 2014). While discussions of the impacts of logging, hunting, farming, and resource extraction can miss significant nuances (Chapter 8), tropical forests are forming a central part of a growing western popular consciousness that the relationship between our species and the planet is reaching a critical point.

DEFINING THRESHOLDS AND CROSSING A LINE?

I wish to end this book by arguing that tropical forests are not just a popular, but also an academic, marker for how we view the changing and ongoing interaction of our species with the planet. The term 'Anthropocene' was formulated as a way of defining a distinctive horizon of human influence of earth systems, contemporaneous with either the European industrial revolution of the eighteenth century or the Great Acceleration in the production of synthetic materials, mineral extraction, or nuclear fallout over recent decades (Zalasiewicz et al., 2008). Implicit in this is the consideration that prior to the 'modern' era, human impacts on the atmosphere, landscape, biota, hydrological cycles, and geological systems of the planet were of a different nature altogether (Chapter 8). Such a perspective was, and often still is, found in modern conservation work in environments across the planet that consider biotic diversity, species distributions, and vegetation formations as largely, if not completely, 'pristine' prior to the arrival of urbanization, industrial action, and global market forces (Whitmore, 1984; Pimm, 1991). It is also the prevailing perspective in much of the media that seeks to highlight the extent of modern ecological tensions and encourage political and economic action. In this scenario, archaeology and history, while potentially providing some anecdotal insights into environmental fragility and precursory human impacts, have little meaningful role in informing modern policy or conservation efforts.

Yet, as highlighted in Chapter 8, work undertaken in tropical forests reviewed throughout this book is encouraging prehistoric and historic human influences on different earth systems to be compared to those of the recent past. In this model, the 'Anthropocene' becomes a theoretical structure that facilitates evaluation as to *whether* and *how* archaeological and historical precedents were different from observed eighteenth- and twenty-first century anthropogenic impacts on certain environments and earth systems, rather than automatically discounting these datasets. It also enables a more flexible understanding of *when* human influences on different environments and earth systems began. For example, the disproportionate biodiversity in tropical forests means that it is possible that human deforestation and hunting pressures on 'keystone' taxa, perhaps as early as the Late Pleistocene, could have left an imprint on continental and global species diversity (Doughty et al., 2016). Similarly, the expansion of cereal agriculture and domesticated water buffalo herding within and beyond tropical ecosystems during the Middle Holocene could have been part of a postulated broader impact of agriculture on atmospheric, geological, and soil systems (Ruddiman, 2003; Fuller et al., 2011a; Bayon et al., 2012). Refining estimates of these human impacts on tropical forest, and other, ecosystems, as well as atmospheric and geomorphological systems, in different parts of the world will go a long way to establishing the longevity and continuum of wide-reaching human signatures within the planet's ecology and geology (Ellis et al., 2016).

On the basis of the currently available information it seems clear that human modifications of tropical forests began to have continentally and globally pervasive effects on earth systems from the Late Holocene onwards. Anthropogenically influenced deforestation and soil erosion begin to register on the scale of entire river catchments, including the Congo River in Africa and the Siem Reap River in Southeast Asia (Penny et al., 2006; Bayon et al., 2012; Garcin et al., 2018), while the expansion of agriculture into insular tropical forest settings in the Pacific and the Caribbean had significant results for local biodiversity (Steadman, 2006; Fitzpatrick and Keegan, 2007). Growing evidence for Late Holocene urbanism in North and Central America, South America, Southeast Asia, Africa, and the Pacific, albeit with an apparent tendency towards low-density settlement, would only have intensified anthropogenic landscape modifications and the impacts of agricultural practice in the tropics (Lucero et al., 2015). The atmospheric effects of expanding agricultural and urban populations—both in terms of CO_2 emissions and the loss of carbon from the terrestrial biosphere as well as methane emissions from the expansion of large domesticated ungulates in the tropics—remain debated (Ruddiman, 2003; Fuller et al., 2011a; Kaplan et al., 2011; Singarayer et al., 2011; Ruddiman et al., 2015). However, influences on tropical forest nutrient cycles, species, and climate, almost certainly began to occur worldwide by the Late pre-industrial Holocene. Given the chemical, atmospheric, and biological feedback of tropical forests with earth systems, this potentially had planet-wide ramifications.

When this pre-industrial alteration of tropical forests is recognized as quantitatively rather than qualitatively different to present-day human impacts, archaeological and palaeoecological datasets can begin to inform modern conservation (McMichael et al., 2017). For example, while its record of human tropical land-use remains incomplete, examination of pre-industrial land-use in the global KK10 model scenario of land cover change (Kaplan et al., 2011; Kaplan and Krumhardt, 2011) (Chapter 8) demonstrates an unbalanced late pre-industrial human impact on seasonally dry tropical forests, particularly across much of South Asia, North and Central America, Africa, and the Amazon Basin (Roberts et al., 2018a). This highlights that these forest formations were particularly vulnerable to human exploitation and modification, likely as a result of the fact that they are more hospitable for the expansion of agricultural field systems and more likely to stimulate large-scale agrarian urban settlements and administrations with a focus on water control (Lucero et al., 2015). Interestingly, these are also the tropical forests most under threat from the modern expansion of monoculture agriculture, urban centres, bushmeat hunting, deforestation for mining, land clearance, and cash crop production, and human-induced climate change (Malhi et al., 2008, 2014; DRYFLOR et al., 2015). While of a different scale in terms of quantitative effects, the placement of archaeological impacts on tropical forests on the same scale as modern anthropogenic alterations enables us to integrate archaeological, historical, and palaeoenvironmental datasets more effectively into conservation and environmental policy and predictions today.

Hopefully this book has shown that archaeological survey, excavation, and the application of archaeological science methods have the potential greatly to enrich our understanding of the human story. Palaeoanthropological, archaeological, and palaeoenvironmental study of much of West and Central Africa, the interior of much of Island and Mainland Southeast Asia, southern China, and the western Amazon Basin, has remained largely lacking to date, but promises to offer much to future understandings relating to the origins and expansion of our species and its hominin ancestors. Similarly, systematic study of West and Central Africa, southwestern India and Sri Lanka, and much of Southeast Asia will fill in blanks on the map in the context of the origins and expansion of agricultural practice, and even urban settlements, in tropical forests. Hopefully new generations of students will now see tropical forests as regions of exciting potential rather than hopeless, dangerous endeavours. Furthermore, this book will also hopefully have demonstrated the new insights a combination of ancient DNA analysis, stable isotope analysis, palaeoenvironmental coring, historical ecology, archaeobotany, and in-depth analysis of historical texts, as well as other methodologies, can provide in these, and more-explored, tropical regions in the pursuit of past human adaptations within, and modifications to, different tropical forests. The continued amassing of multi-proxy datasets on local, regional, and global scales will yield an increasingly clearer picture of our species' position within these dynamic environments across space and time.

However, given the overall importance of tropical forests to the functioning of the Earth's ecosystems, it is also important to build datasets and models that allow human tropical forest prehistories, histories, and modernities to be placed on the same scale. By using the 'Anthropocene' as a comparative concept, rather than a definitive epoch or stratigraphic layer, we can start to explore the continuum of anthropogenic tropical forest modifications on local, regional, and global scales. In this way we can begin to compare the effects of hunter-gatherer, agricultural, and urban occupation in different tropical forest types, and regions, to each other. Moreover, we can highlight significant changes in the scale of human effects on these environments, not by assuming *a priori* a post-colonial or post-industrial 'spike' but rather through the comparison of multiple environmental and archaeology datasets. In this way, archaeology, history, and palaeoecology can play more than a token role in modern conservation and planning agendas, providing detailed insights into past tropical forest vulnerabilities, changes in composition, and the resilience of different types of human subsistence, political organisation, and economic networks (Ellis et al., 2016). Similarly, the effects of human-induced climate change on tropical forest species compositions and extents today, can be compared to more constrained palaeoenvironmental datasets in different parts of the tropics in the past. Given that archaeology, history, and palaeoenvironmental science are the only disciplines that offer a long-term perspective on changes in these terrestrial habitats, modern policy-makers, funding-bodies, and conservation groups should support this goal, and these disciplines, in as many ways as they can.

I wish to finish this book with a passage from one of my favourite recent novels, *The Trees* by Ali Shaw (2016: 24–5):

> If this yew was as old as she said it was it would have been a sapling at the time of medieval kings. It would have grown fat in its infancy off soil stacked full by the bubonic plague. Troubled Catholics would have prayed around it, wondering what would become of their souls when their king rejected Rome. The crude props and

> theatre of the mummers would have played in its shadow, as might the fresh recited verses of a feted Elizabethan bard. It would have been marched past by the hard boots of Roundheads and the fine shoes of Cavaliers. Its leaves would have tried the first smog, sucked in the first fumes of the first factories. Young women would have sneaked into its quiet churchyard imploring heaven to keep safe their sweethearts, who ducked from the whistling artillery in the mud-sunk ditches of Europe. Time had trampled through this churchyard and left behind, as time always did, the sense that just because things had progressed, they were progressing towards a goal . . . Its bark was as grey and eroded as the graveyard's headstones, as if the aim of its long growth had been to turn itself into yet another monument to the dead.

This quotation encapsulates beautifully how western literature and culture frequently evoke the enduring witness of European temperate forest trees, including yews, oaks, and elm, in human history, as well as the poignant importance of their conservation. In writing this book I hope to have highlighted how tropical forests have stood watch over human history for even longer and on a more global scale. These forests have grown over a more extensive period than any other on the planet and framed the emergence of our hominin ancestors *c.* 8–5 Ma. They were one of the first environments our own species encountered in regions such as Southeast Asia and Melanesia from 45 ka, and one of the first settings to observe the development of agricultural practice in the Early Holocene. They have been home to vast pre-industrial urban settlements and formed a crucial part of early, incipient globalized trading networks. Tropical forests were onlookers to some of the worst crimes of European colonialism, but also to some of the first attempts to incorporate the interests, practice, and ontologies of Indigeneous peoples into archaeology, history, and anthropology. Modern tropical forest taxa, species distributions, and extents, as well as human biology, economy, culture, and political systems, are the legacy of this long interaction. We must hope that growing recognition of this longevity, and of the swathes of human prehistory and history held within tropical forests, will stimulate archaeology, history, anthropology, ecology, and palaeoenvironmental science to come together to ensure that modern economies, human activity, and conservation policies enable these environments to continue watching over human generations to come.

References

Abbas, B., Bintoro, M. H., Sudarsono, S., Surahman, M., Ehara, H. 2009. Genetic relationship of sago palm (*Metroxylon sagu* Rottb.) in Indonesia based on RAPD markers. *Biodiversitas* 10: 168–74.

Acabado, S. B. 2010. Landscapes and the archaeology of the Ifugao agricultural terraces: Establishing antiquity and social organisation. *Hukay: Journal for Archaeological Research in Asia and the Pacific* 15: 31–61.

Acreman, M., Albertengo, J., Amado, T., Amis, M., Anderson, A., Bacchur, I., Basch, G., Calegari, A., Chappell, N., Chettri, N., Coates, D., Cohen-Shacham, E., Corsi, S., Davidson, N., De Mello, C. R., Fleiner, R., Freidrich, T., Goddard, T., Gonzalez-Sanchez, E., Gregersen, H., Harwood, R., Hussein, M., Kassam, A., Kim, I.-J., Kim, K.-G., Laurent, F., Li, H., Lukacs, G., Mccartney, M., McInnes, R., Mello, I., Morales Sá, J. C., Moreno-Casasola, P., Nurbekov, A., Okruszko, T., Peiretti, R., Pretty, J., Raisch, R.,, Shahriar, M. W., Sharif, A., Shrestha, A. B., Sinun, W., Sturny, W., Thierfelder, C., Uphoff, N., Wani, S., Yakushina, E. 2012. *Report of the Work of the Expert Group on Maintaining the Ability of Biodiversity to Continue to Support the Water Cycle.* Hyderabad, India: Report to the Conference of the Parties to the Convention on Biological Diversity, Eleventh Meeting, 8–19 October 2012.

Adams, R. E. W., Jones, R. C. 1981. Spatial patterns and regional growth among Maya cities. *American Anthropologist* 46: 301–22.

Adams, R. McC. 1981. *Heartland of Cities.* Chicago: University of Chicago Press.

Adams, W. M., Infield, M. 2003. Who is on the Gorilla's payroll? Claims on tourist revenue from a Ugandan National Park. *World Development* 31: 177–90.

Aimi, M., Aziz, F. 1985. Vertebrate fossils from the Sangiran dome, Mojokerto, Trinil, and Sambungmacan areas. In N. Watanabe, D. Kadar (eds.), *Quaternary Geology of the Hominid Fossil Bearing Formations in Java.* Bandung, Indonesia: Geological Research and Development Centre Special Publication 4, pp. 155–97.

Algeo, T. J., Scheckler, S. E. 1998. Terrestrial–marine teleconnections in the Devonian: Links between the evolution of land plants, weathering processes and marine anoxic events. *Philosophical Transactions of the Royal Society B: Biological Sciences* 353: 13–130.

Allaby, M. 2010. *Oxford Dictionary of Ecology* (4th edition). Oxford: Oxford University Press.

Allaby, R. G., Kitchen, J., Fuller, D. Q. 2016. Surprisingly low limits of selection in plant domestication. *Evolutionary Bioinformatics* 2: 41–51.

Allaby, R. G., Stevens, C., Leilani, L., Osamu, M., Fuller, D. Q. 2017. Geographic mosaics and changing rates of cereal domestication. *Philosophical Transactions of the Royal Society B: Biological Sciences* 372. doi:10.1098/rstb.2016.0429.

Allen, J., Golson, J., Jones, R. (eds.). 1977. *Sunda and Sahul: Prehistoric Studies in Southeast Asia, Melanesia and Australia.* New York: Academic Press.

Allen, J., Gosden, C., White, J. P. 1989. Human Pleistocene adaptations in the tropical island Pacific: Recent evidence from New Ireland, a Greater Australian Outlier. *Antiquity* 63: 548–61.

Alpern, S. 1995. What Africans got for their slaves: A master list of European trade goods. *History in Africa* 22: 5–43.

Amblard, S., Pernès, J. 1989. The identification of cultivated pearl millet (*Pennisetum*) amongst plant impressions on pottery from Qued Chebbi (Dhar Oualata, Mauritia). *African Archaeological Review* 7: 117–26.

Ambrose, S. H. 1993. Isotopic analysis of paleodiets: Methodological and interpretive considerations. In M. K. Sandford (ed.), *Investigations of Ancient Human Tissue: Chemical Analyses in Anthropology*. New York: Gordon and Breach, pp. 59–130.

Anderson, D. D. 1997. Cave archaeology in Southeast Asia. *Geoarchaeology* 12: 607–38.

Anderson, J. M., Swift, M. J. 1983. Decomposition in tropical forests. In S. L. Sutton, T. C. Whitmore, A. C. Chawick (eds.), *Tropical Rain Forest: Ecology and Management*. Oxford: Blackwell Scientific Publications, pp. 287–309.

Anderson, R. G., Canadell, J. G., Randerson, J. T., Jackson, R. B., Hungate, B. A., Baldocchi, D. D., Ban-Weiss, G. A., Bonan, G. B., Caldeira, K., Cao, L., Diffenbaugh, N. S., Gurney, K. R., Kueppers, L. M., Law, B. E., Luyssaert, S., O'Halloran, T. L. 2011. Biophysical considerations in forestry for climate protection. *Frontiers in Ecology and the Environment* 9: 174–82.

Andrews, P., Martin, L. 1987. Cladistic relationships of extant and fossil hominoids. *Journal of Human Evolution* 16: 101–18.

Angel, S. 2012. *Planet of Cities*. Cambridge, MA: Lincoln Institute of Land Policy.

Angel, S., Sheppard, S. C., Civco, D. L., Buckleyt, R., Chabaeva, A., Gitlin, L., Kraley, A., Parent, J., Perlin, M. 2005. *The Dynamics of Global Urban Expansion*. Washington, DC: Transport and Urban Development Department, World Bank.

Aplin, K. P., Helgen, K. M. 2010. Quaternary murid rodents of Timor Part I: New material of *Coryphomys buehleri* Schaub, 1937, and description of a second species of the genus. *Bulletin of the American Museum of Natural History* 341: 1–80.

Argue, D., Morwood, M. J., Sutikna, T., Jatmiko, Saptomo, E. W. 2009. *Homo floresiensis*: A cladistic analysis. *Journal of Human Evolution* 57: 623–39.

Argue, D., Groves, C. P., Lee, M. S. Y., Jungers, W. L. 2017. The affinities of *Homo floresiensis* based on phylogenetic analyses of cranial, dental, and postcranial characters. *Journal of Human Evolution* 107: 107–33.

Arnason, U., Gullberg, A., Janke, A. 1998. Molecular timing of primate divergences as estimated by two nonprimate calibration points. *Journal of Molecular Evolution* 47: 718–27.

Arnau, G., Bhattacharjee, R., Sheela, M. N., Chaïr, H., Malapa, R., Lebot, V., Abraham, K., Perrier, X., Petro, D., Penet, L., Pavis, C. 2017. Understanding the genetic diversity and population structure of yam (*Dioscorea alata* L.) using microsatellite markers. *PLoS ONE* 12: e0174150.

Arroyo-Kalin, M. 2008. Steps towards an ecology of landscape: The pedo-stratigraphy of anthropogenic dark earths. In W. I. Woods, W. Teixera, J. Lehmann, C. Steiner, A. WinklerPrins (eds.), *Amazonian Dark Earths: Wim Sombroek's Vision*. Dordrecht: Springer Verlag, pp. 33–83.

Arroyo-Kalin, M. 2010. The Amazonian formative: Crop domestication and anthropogenic soils. *Diversity* 2: 473–504.

Arroyo-Kalin, M. 2012. Slash-burn-and-churn: Landscape history and crop cultivation in pre-Columbian Amazonia. *Quaternary International* 249: 4–18.

Arroyo-Kalin, M. 2014. Amazonian dark earths, geoarchaeology. In C. Smith (ed.), *Encyclopaedia of Global Archaeology*. New York: Springer, pp. 168–78.

Arroyo-Kalin, M. 2017. Las tierras antrópicas amazónicas: Algo más que un puñado de tierra. In S. Rostain, C. Jaimes (eds.), *Las Siete Maravillas de la Amazonía precolombina*. La Paz: 4-EIAA/BAS/Plural Publicaciones, pp. 99–118.

Asfaw, B., White, T., Lovejoy, O., Latimer, B., Simpson, S., Suwa, G. 1999. *Australopithecus garhi*: A new species of early hominid from Ethiopia. *Science* 284: 629–35.

Ashley, G. M., Bunn, H. T., Delaney, J. S., Barboni, D., Domínguez-Rodrigo, M., Mabulla, A. Z. P., Gurtov, A. N., Baluyot, R. D., Beverly, E. J., Baquedano, E. 2014. Paleoclimatic and paleoenvironmental framework of FLK North archaeological site, Olduvai Gorge, Tanzania. *Quaternary International* 322–3: 54–65.

Ashton, N., Lewis, S. G., De Groote, I., Duffy, S. M., Bates, M., Bates, R., Hoare, P., Lewis, M., Parfitt, S. A., Peglar, S., Williams, C., Stringer, C. 2014. Hominin footprints from Early Pleistocene deposits at Happisburgh, UK. *PLoS ONE* 9: e88329.

Ashton, P. S. 2010. Conservation of Borneo biodiversity: Do small lowland parks have a role, or are big inland sanctuaries sufficient? Brunei as an example. *Biodiversity and Conservation* 19: 343–56.

Ashton, P. S., Gunatilleke, C. V. S. 1987. New light on the plant geography of Ceylon I: Historical plant geography. *Journal of Biogeography* 14: 249–85.

Asouti, E., Fuller, D. Q. 2008. *Trees and Woodland in South India: An Archaeological Approach*. Walnut Creek, CA: Left Coast Press.

Athayde, S., Silva-Lugo, J., Schmink, M., Heckenberger, M. 2017. The same, but different: Indigenous knowledge retention, erosion, and innovation in the Brazilian Amazon. *Human Ecology* 45: 533–44.

Aung-Thwin, M. 1985. *Pagan: The Origins of Modern Burma*. Honolulu: University of Hawai'i Press.

Aung-Thwin, M. 1990. *Irrigation in the Heartland of Burma: Foundations of the Pre-Colonial Burmese State*. Occasional Paper 15. Monograph Series. Dekalb: Center for Southeast Asian Studies, Northern Illinois University.

Backwell, L., d'Errico, F., Wadley, L. 2008. Middle Stone Age bone tools from the Howiesons Poort layers, Sibudu Cave, South Africa. *Journal of Archaeological Science* 35: 1566–80.

Bacon, A.-M., Demeter, F., Duringer, P., Helm, C., Bano, M., The Long, V., Thi Kim Thuy, N., Antoine, P.-O., Thi Mai, B., Thi Mai Huong, N., Dodo, Y., Chabaux, F., Rihs, S. 2008. The Late Pleistocene Duoi U'Oi cave in northern Vietnam: Palaeontology, sedimentology, taphonomy and palaeoenvironments. *Quaternary Science Reviews* 27: 1627–54.

Bacus, E. A., Lucero, L. J. (eds.). 1999. *Complex Polities in the Ancient Tropical World*. Archaeological Papers of the American Anthropological Association 9. Arlington, VA: AAA.

Bae, C. J., Douka, K., Petraglia, M. D. 2017. On the origin of modern humans: Asian perspectives. *Science* 358: 9067.

Bahuchet, S., Guillaume, H. 1982. Aka-farmer relations in the northeast Congo Basin. In E. Leacock, R. Lee (eds.), *Politics and History in Band Societies*. Cambridge: Cambridge University Press, pp. 189–211.

Bahuchet, S., McKey, D., de Garine, I. 1991. Wild yams revisited: Is independence from agriculture possible for rain forest hunter-gatherers? *Human Ecology* 19: 213–42.

Bailey, J. 1863. An account of the wild tribes of the Veddahs of Ceylon, etc. *Transactions of the Ethnological Society* 2: 278–320.

Bailey, R. C., Headland, T. N. 1991. The tropical rain forest: Is it a productive environment for human foragers? *Human Ecology* 19: 261–85.

Bailey, R. C., Head, G., Jenike, M., Owen, B., Rechtman, R., Zechenter, E. 1989. Hunting and gathering in tropical rain forest: Is it possible? *American Anthropologist* 91: 59–82.

Bailey, R. C., Jenicke, M., Rechtman, R. 1991. Reply to Colinvaux and Bush. *American Anthropologist* 93: 160–2.

Baird, A. H., Kerr, A. M. 2008. Landscape analysis and tsunami damage in Aceh: Comment on Iverson & Prasad (2007). *Landscape Ecology* 23: 3–5.

Baker, L. 2017. What trees teach us about life and happiness. <http://www.bbc.com/culture/story/20170811-what-can-trees-teach-us-about-life-and-happiness>. Retrieved 24 March 2018.

Balée, W. 1988. Indigenous adaptation to Amazonian palm forests. *Principes* 32: 47–54.

Balée, W. 1989. The culture of Amazonian forests. In D. Posey, W. Balée (eds.), *Resource Management in Amazonian Indigenous and Folk Strategies*. New York: New York Botanical Garden, pp. 1–21.

Balée, W. 1992. People and fallow: A historical ecology of foraging in lowland South America. In K. Redford, C. Padoch (eds.), *Conservation of Neotropical Forests: Working from Traditional Resource Use*. New York: Columbia University Press, pp. 35–57.

Balée, W. 1994. *Footprints of the Forest: Ka'apar Ethnobotany—the Historical Ecology of Plant Utilization by an Amazonian People*. New York: Columbia University Press.

Balée, W. 2013. *Cultural Forests of the Amazon: A Historical Ecology of People and their Landscapes*. Tuscaloosa: University of Alabama Press.

Balée, W., Erickson, C. 2006. Time, complexity, and historical ecology. In W. Balée, C. Erickson (eds.), *Time and Complexity in Historical Ecology*. New York: Columbia University Press, pp. 1–17.

Bandaranayake, S. 2003. The pre-modern city in Sri Lanka: The 'first' and 'second' urbanisation. In P. J. J. Sinclair (ed.), *The Development of Urbanism from a Global Perspective*. Uppsala: Uppsala University. <http://www.arkeologi.uu.se/digitalAssets/9/9377_bandaranayake.pdf>.

BaoYin, Y., WeiWen, H., David, Z. 2007. New evidence for human occupation of the northern Tibetan Plateau, China during the Late Pleistocene. *Chinese Science Bulletin* 52: 2675–9.

Baran, E., Jantunen, T., Chong, C. K. 2007. Special review: Values of inland fisheries in the Mekong River Basin. In A. E. Neiland, C. Béné (eds.), *WorldFish Center Special Report on Tropical River Valuation*. Penang, Malaysia: WorldFish, pp. 4–58.

Barboni, D., Bonnefille, R., Prasad, S., Ramesh, B. R. 2003. Variation in the modern pollen from tropical evergreen forest and the monsoon seasonality gradient in SW India. *Journal of Vegetation Science* 14: 551–62.

Barboni, D., Ashley, G. M., Domínguez-Rodrigo, M., Bunn, H. T., Mabulla, A. Z. P., Baquedano, E. 2010. Phytoliths infer locally dense and heterogeneous paleovegetation at FLK North and surrounding localities during upper Bed I time, Olduvai Gorge, Tanzania. *Quaternary Research* 74: 344–54.

Barham, B., Coomes, O. 1994. Wild rubber: Industrial organisation and the microeconomics of extraction during the Amazon rubber boom (1860–1920). *Journal of Latin American Studies* 26: 37–72.

Barham, L. S. (ed.). 2000. *The Middle Stone Age of Zambia, South Central Africa*. Bristol: Western Academic and Specialist Press.

Barham, L. S. 2001. Central Africa and the emergence of regional identity in the Middle Pleistocene. In L. S. Barham, K. Robson-Brown (eds.), *Human Roots: Africa and Asia in the Middle Pleistocene*. Bristol: Western Academic and Specialist Press, pp. 65–80.

Barham, L. S. 2002. Backed tools in the Middle Pleistocene of Central Africa and their evolutionary significance. *Journal of Human Evolution* 43: 585–603.

Barham, L. S., Smart, P. L. 1996. An early date for the Middle Stone Age of central Zambia. *Journal of Human Evolution* 30: 287–90.

Barker, G. 2005. The archaeology of foraging and farming at Niah Cave, Sarawak. *Asian Perspectives* 44: 90–106.

Barker, G. (ed.). 2013. *Rainforest Foraging and Farming in Island Southeast Asia: The Archaeology of the Niah Caves, Sarawak, Volume 1*. Cambridge: McDonald Institute for Archaeological Research.

Barker, G., Farr, L. (eds.). 2016. *Archaeological Investigations in the Niah Caves, Sarawak: The Archaeology of the Niah Caves, Sarawak, Volume 2*. Cambridge: McDonald Institute for Archaeological Research.

Barker, G., Barton, H., Bird, M., Daly, P., Datan, I., Dykes, A., Farr, L., Gilbertson, D., Harrisson, B., Hunt, C., Higham, T., Kealhofer, L., Krigbaum, J., Lewis, H., McLaren, S., Paz, V., Pike, A., Piper, P., Pyatt, B, Rabett, R., Reynolds, T., Rose, J., Rushworth, G., Stephens, M., Stringer, C., Thompson, J., Turney, C. 2007. The 'human evolution' in lowland tropical Southeast Asia: The antiquity and behaviour of anatomically modern humans at Niah Cave (Sarawak, Borneo). *Journal of Human Evolution* 52: 243–61.

Barker, G., Hunt, C., Carlos, J. 2011. Transitions to farming in Island Southeast Asia: Archaeological, biomolecular and palaeoecological perspectives. In G. Barker, M. Janowski (eds.), *Why Cultivate? Anthropological and Archaeological Approaches to Foraging-Farming Transitions in Southeast Asia*. Cambridge: McDonald Institute for Archaeological Research, pp. 59–72.

Barker, J. S. F. 2014. Water buffalo: Domestication. In C. Smith (ed.), *Encyclopedia of Global Archaeology*. New York: Springer, pp. 7694–7.

Barona, E., Ramankutty, N., Hyman, G., Coomes, O. T. 2010. The role of pasture and soybean in deforestation of the Brazilian Amazon. *Environmental Research Letters* 5: 024002.

Barrau, J. 1955. *Subsistence Agriculture in Melanesia*. Noumea, New Caledonia: South Pacific Commission.

Barton, C. M., Riel-Salvatore, J., Anderies, J. M., Popescu, G. 2011. Modeling human ecodynamics and biocultural interactions in the late Pleistocene of western Eurasia. *Human Ecology* 39: 705–25.

Barton, H. 2005. The case for rainforest foragers: The starch record at Niah Cave, Sarawak. *Asian Perspectives* 44: 56–72.

Barton, H. 2016. Functional analysis of stone tools from the West Mouth. In G. Barker, L. Farr (eds.), *Archaeological Investigations in the Niah Caves, Sarawak: The Archaeology of the Niah Caves, Sarawak, Volume 2*. Cambridge: McDonald Institute for Archaeological Research, pp. 279–300.

Barton, H., Denham, T. 2011. Vegeculture and social life in island Southeast Asia. In G. Barker, M. Janowski (eds.), *Why Cultivate? Anthropological and Archaeological Approaches to Foraging–Farming Transitions in Southeast Asia*. Cambridge: McDonald Institute for Archaeological Research, pp. 17–25.

Barton, H., Paz, V. 2007. Subterranean diets in the tropical rain forests of Sarawak, Malaysia. In T. Denham, J. Iriarte, L. Vrydaghs (eds.), *Rethinking Agriculture: Archaeological and Ethnoarchaeological Perspectives*. Los Angeles: One World Archaeology, Left Coast Press, pp. 50–77.

Barton, H., Piper, P. J., Rabett, R., Reeds, I. 2009. Composite hunting technologies from the Terminal Pleistocene and Early Holocene, Niah Cave, Borneo. *Journal of Archaeological Science* 36: 1708–14.

Barton, H., Denham, T., Neumann, K., Arroyo-Kalin, M. 2012. Long-term perspectives on human occupation of tropical rainforests: An introductory overview. Guest editorial. *Quaternary International* 249: 1–3.

Barton, H., Paz, V., Carlos, A. J. 2016. Plant food remains from the Niah Caves: Macroscopic and microscopic approaches. In G. Barker, L. Farr (eds.), *Archaeological Investigations in the Niah Caves, Sarawak: The Archaeology of the Niah Caves, Sarawak, Volume 2*. Cambridge: McDonald Institute for Archaeological Research, pp. 455–68.

Barume, A. 2000. *Heading Towards Extinction? Indigenous Rights in Africa: The Case of the Twa of the Kahuzi-Biega National Park, Democratic Republic of Congo*. Copenhagen: International Work Group for Indigenous Affairs (IWGIA).

Bar-Yosef, O. 2011. Climatic fluctuations and early farming in West and East Asia. *Current Anthropology* 52: S175–S193.

Bar-Yosef, O., Belfer-Cohen, A. 1989. The origins of sedentism and farming communities in the Levant. *Journal of World Prehistory* 3: 447–98.

Bar-Yosef, O., Belfer-Cohen, A. 2001. From Africa to Eurasia: Early dispersals. *Quaternary International* 75: 19–28.

Bar-Yosef, O., Eren, M. I., Yuan, J., Cohen, D. J., Li, Y. 2012. Were bamboo tools made in prehistoric Southeast Asia? An experimental view from South China. *Quaternary International* 269: 9–21.

Basell, L. S. 2008. Middle Stone Age (MSA) site distributions in eastern Africa and their relationship to Quaternary environmental change, refugia and the evolution of *Homo sapiens*. *Quaternary Science Reviews* 27: 2484–98.

Bassani, E., Fagg, W. 1988. *Africa and the Renaissance: Art in Ivory*. London: Centre for African Art, Prestel-Verlag.

Basso, E. B. 1995. *The Last Cannibals: A South American Oral History*. Austin: University of Texas Press.

Bawa, K. S., Kress, W. J., Nadkarni, N. M., Lele, S., Raven, P. H., Janzen, D. H., Lugo, A. E., Ashton, P. S., Lovejoy, T. E. 2004. Tropical ecosystems into the 21st century. *Science* 306: 227–8.

Bayas, J. C. L., Marohn, C., Dercon, G., Dewi, S., Piepho, H. P., Joshi, L., van Noordwijk, M., Cadisch, G. 2011. Influence of coastal vegetation on the 2004 tsunami wave impact in West Aceh. *Proceedings of the National Academy of Sciences of the United States of America* 108: 18612–17.

Bayle des Hermens, R., Oslisly, R., Peyrot, B. 1987. Premières séries de Pierres Taillées du Paléolithique Inférieur découvertes au Gabon, Afrique Central. *L'Anthropologie* 91: 693–8.

Bayon, G., Dennielou, B., Etoubleau, J., Ponzevera, E., Toucanne, S. 2012. Intensifying weathering and land use in Iron Age Central Africa. *Science* 335: 1219–22.

Beach, T. 1998. Soil catenas, tropical deforestation, and ancient and contemporary soil erosion in the Peten, Guatemala. *Physical Geography* 19: 378–404.

Beach, T., Luzzadder-Beach, S., Dunning, N., Hageman, J., Lohse, J. 2002. Upland agriculture in the Maya Lowlands: Ancient conservation in Northwestern Belize. *Geography Review* 92: 372–97.

Beach, T., Dunning, N., Luzzadder-Beach, S., Cook, D. E., Lohse, J. 2006. Impacts of the ancient Maya on soils and soil erosion in the central Maya Lowlands. *Caterna* 65: 166–78.

Beauvilain, A. 2008. The contexts of discovery of *Australopithecuys bahrelghazali* (Abel) and of *Sahelanthropus tchadensis* (Toumaï): Unearthed, embedded in sandstone, or surface collected? *South African Journal of Science* 104: 165–8.

Beavitt, P. B., Kurui, E., Thompson, G. B. 1996. Confirmation of an early date for the presence of rice in Borneo: Preliminary evidence for possible Bidayuh/Aslian links. *Borneo Research Bulletin* 27: 29–37.

Begley, C. 2017. Ancient ruins keep being 'discovered': Were they ever lost? <https://aeon.co/ideas/ancient-ruins-keep-being-discovered-were-they-ever-lost>. Retrieved 24 March 2018.

Begun, D. R. 2004. *Sivapithecus* is east and *Dryopithecus* is west, and never the twain shall meet. *Anthropological Science* 113: 53–64.

Begun, D. R. 2010. Miocene hominids and the origins of the African apes and humans. *Annual Review of Anthropology* 39: 67–84.

Bellwood, P. 1985. *Prehistory of the Indo-Malaysian Archipelago*. New York: Academic Press.

Bellwood, P. 1993. Cultural and biological differentiation in Peninsular Malaysia: The last 10,000 years. *Asian Perspectives* 32: 37–60.

Bellwood, P. 1999. Archaeology of Southeast Asian hunters and gatherers. In R. B. Lee, R. Daly (eds.), *The Cambridge Encyclopedia of Hunters and Gatherers*. Cambridge: Cambridge University Press, pp. 284–8.

Bellwood, P. 2005. *First Farmers*. Oxford: Blackwell.

Bellwood, P. 2011. Holocene population history in the Pacific region as a model for worldwide food producer dispersals. *Current Anthropology* 52: S363–S378.

Bennett, B. C. 1992. Plants and people of the Amazonian rainforests: The role of ethnobotany in sustainable development. *BioScience* 42: 599–607.

Bennett, E. L., Robinson, J. G. 2000. *Hunting of Wildlife in Tropical Forests: Implications for Biodiversity and Forest Peoples*. Washington, DC: Global Environment Division, World Bank.

Benoit, J., Thackeray, F. J. 2017. A cladistics analysis of *Graecopithecus*. *South African Journal of Science* 113: 1–2.

Bermingham, E., Dick, C. W. 2014. Overview: The history and ecology of tropical rainforest communities. In E. Bermingham, C. W. Dick, C. Mortiz (eds.), *Tropical Rainforests: Past, Present and Future*. Chicago: University of Chicago Press, pp. 7–15.

Bermúdez de Castro, J. M., Arsuaga, J. L., Carbonell, E., Rosas, A., Martínez, I., Mosquera, M. 1997. A hominid from the Lower Pleistocene of Atapuerca, Spain: Possible ancestor to Neandertals and modern humans. *Science* 276: 1392–5.

Berna, F., Goldberg, P., Horwitz, L. K., Brink, J., Holt, S., Bamford, M., Chazan, M. 2012. Microstratigraphic evidence of *in situ* fire in the Acheulean strata of Wonderwerk Cave, Northern Cape province, South Africa. *Proceedings of the National Academy of Sciences of the United States of America* 109: E1215–E1220.

Bernard, A., Daux, A., Lécuyer, C., Brugal, J.-P., Genty, D., Wainer, K., Gardien, V., Fourel, F., Jaubert, J. 2009. Pleistocene seasonal temperature variations recorded in the $\delta^{18}O$ of *Bison priscus* teeth. *Earth and Planetary Science Letters* 283: 133–43.

Berndt, R. M., Berndt, C. H. 1993. *A World That Was*. Parkville: Melbourne Univeristy Press.

Bertaux, J., Sondag, F., Santos, R., Soubiès, F., Causse, C., Plagnes, V., Le Cornec, F., Seidel, A. 2002. Paleoclimatic record of speleothems in a tropical region: Study of laminated sequences from a Holocene stalagmite in Central-West Brazil. *Quaternary International* 89: 3–16.

Bettendorff, J. F. 1990 (1910). *Crônica dos Padres da Companhia de Jesus no Estado do Maranhão*. Belém: Fundação Cultural do Pará/Secretaria de Estado da Cultura.

Bettis (III), E. A., Milius, A. K., Carpenter, S. J., Larick, R., Zaim, Y., Rizal, Y., Ciochon, R. L., Tassier-Surine, S. A., Murray, D., Suminto Bronto, S. 2009. Way out of Africa: Early Pleistocene paleoenvironments inhabited by *Homo erectus* in Sangiran Java. *Journal of Human Evolution* 56: 11–24.

Bevan, B. W. 1989. *Geophysical Surveys at Three Sites Along the Lower Amazon River*. Pitman, NJ: Geosight.

Biesbrouck, K., Elders, S., Rossel, G. 1999. *Central African Hunter-Gatherers in a Multidisciplinary Perspective: Challenging Elusiveness*. Leiden: CNWS.

Birch, S. E. P., Miracle, P. T., Stevens, R. E., O'Connell, T. C. 2016. Late Pleistocene/Early Holocene migratory behavior of ungulates using isotopic analysis of tooth enamel and its effects on forager mobility. *PLoS ONE* 11: e0155714.

Bird, M. I., Taylor, D., Hunt, C. 2005. Palaeoenvironments of insular Southeast Asia during the last glacial period: A savanna corridor in Sundaland? *Quaternary Science Reviews* 24: 2228–42.

Bird, M. I., Hutley, L. B., Lawes, M. J., Lloyd, J., Luly, J. G., Ridd, P. V., Roberts, R. G., Ulm, S., Wurster, C. M. 2013. Humans, megafauna and environmental change in tropical Australia. *Journal of Quaternary Science* 28: 439–52.

Bird-David, N. 1992. Beyond the original affluent society: A culturalist reformulation. *Current Anthropology* 33: 25–47.

Bird-David, N. 1997. Economies: A cultural-economic perspective. *International Social Science Journal* 49: 463–75.

Biswas, S. R., Choudhury, J. K. 2007. Forests and forest management practices in Bangladesh: The question of sustainability. *International Forestry Review* 9: 627–40.

Blasco, F., Bellan, M. F., Aizpura, M. 1996. A vegetation map of tropical continental Asia at scale 1:5 million. *Journal of Vegetation Science* 7: 623–34.

Blench, R. M. 1996. Evidence for the inception of agriculture in the Nigeria–Cameroun borderland. In J. E. G. Sutton (ed.), *The Growth of Farming Communities in Africa from the Equator Southwards*. Nairobi: British Institute of East Africa, pp. 83–102.

Blench, R. M. 2007. Using linguistics to reconstruct African subsistence systems: Comparing crop names to trees and livestock. In T. P. Denham, J. Iriarte, L. Vrydaghs (eds.), *Rethinking Agriculture: Archaeological and Ethnoarchaeological Perspectives*. Walnut Creek, CA: Left Coast Press, pp. 408–38.

Blench, R. 2013. Was there once a zone of vegeculture linking Melanesia with northeast India? In G. R. Summerhayes, H. Bailey (eds.), *Pacific Archaeology: Documenting the Past 50,000 Years*. Papers from the 2011 Lapita Pacific Archaeology Conference. Otago: University of Otago Studies in Archaeology, no. 25 pp. 1–17.

Blinkhorn, J., Achyuthan, H., Petraglia, M., Ditchfield, P. 2013. Middle Palaeolithic occupation in the Thar Desert during the Upper Pleistocene. *Quaternary Science Reviews* 77: 233–8.

Blome, M. W., Cohen, A. S., Tryon, C. A., Brooks, A. S., Russell, J. 2012. The environmental context for the origins of modern human diversity: A synthesis of regional variability in African climate 150,000–30,000 years ago. *Journal of Human Evolution* 62: 563–92.

Blumenthal, S. A., Levin, N. E., Brown, F. H., Brugal, J.-P., Chritz, K. L., Harris, J. M., Jehle, G. E., Cerling, T. E. 2017. Aridity and hominin environments. *Proceedings of the National Academy of Sciences of the United States of America* 114: 7331–6.

Boas, F. 1911. *The Mind of Primitive Man*. New York: Macmillan.

Boas, F. 1940. *Race, Language, and Culture*. New York: Macmillan.

Bobe, R., Behrensmeyer, K., Chapman, R. E. 2002. Faunal change, environmental variability and late Pliocene hominin evolution. *Journal of Human Evolution* 42: 475–97.

Bocherens, H., Schrenk, F., Chaimanee, Y., Kullmer, O., Morike, D., Pushkina, D., Jaeger, J.-J. 2017. Flexibility of diet and habitat in Pleistocene South Asian mammals: Implications for the fate of the giant fossil ape *Gigantopithecus*. *Quaternary International* 434: 148–55.

Bocquet-Appel, J. P. 2011. When the world's population took off: The springboard of the Neolithic demographic transition. *Science* 333: 560–1.

Bodmer, R. E., Eisenberg, J. F., Redford, K. H. 1997. Hunting and the likelihood of extinction of Amazonian mammals. *Conservation Biology* 11: 460–6.

Boëda, E., Rocca, R., Da Costa, A., Fontugne, M., Hatte, C., Clemente-Conte, I., Santos, J. C., Lucas, L., Felice, G., Lourdeau, A., Villagran, X., Gluchy, M., Ramos, M. P., Viana, S., Lahaye, C., Guidon, N., Griggo, C., Pino, M., Pessis, A.-M., Borges, C., Gato, B. 2016. New data on a Pleistocene archaeological sequence in South America: Toca do Sítio do Meio, Piauí, Brazil. *PaleoAmerica* 2: 286–302.

Böhme, M., Spassov, N., Ebner, M., Geraads, D., Hristova, L., Kirscher, U., Kötter, S., Linnemann, U., Prieto, J., Roussiakis, S., Theodorou, G., Uhlig, G., Winklhofer, M. 2017. Messinian age and savannah environment of the possible hominin *Graecopithecus* from Europe. *PLoS ONE* 12: e0177347.

Boivin, N., Fuller, D. Q., Crowther, A. 2012. Old World globalization and the Columbian exchange: Comparison and contrast. *World Archaeology* 44: 452–68.

Boivin, N., Fuller, D. Q., Dennell, R., Allaby, R., Petraglia, M. D. 2013. Human dispersal across diverse environments of Asia during the Upper Pleistocene. *Quaternary International* 300: 32–47.

Boivin, N. L., Zeder, M. A., Fuller, D. Q., Crowther, A., Larson, G., Erlandson, J. M., Denham, T., Petraglia, M. D. 2016. Ecological consequences of human niche construction: Examining long-term anthropogenic shaping of global species distributions. *Proceedings of the National Academy of Sciences of the United States of America* 113: 6388–96.

Bonaccorso, E., Koch, I., Peterson, A. 2006. Pleistocene fragmentation of Amazon species' ranges. *Diversity and Distributions* 12: 157–64.

Bonell, M., Bruijnzeel, L. A. 2005. *Forests, Water and People in the Humid Tropics: Past, Present and Future Hydrological Research for Integrated Land and Water Management*. Cambridge: Cambridge University Press/UNESCO.

Bonnefille, R., Roeland, J. C., Guiot, J. 1990. Temperature and rainfall estimates for the past 40,000 years in equatorial Africa. *Nature* 246: 347–9.

Boomgaard, P. 1998. The VOC trade in forest products. In R. H. Grove, V. Damodaran, S. Sangwan (eds.), *Nature and the Orient: The Environmental History of South and Southeast Asia*. Delhi: Oxford University Press, pp. 375–95.

Borgerhoff Mulder, M., Coppolillo, P. 2005. *Conservation: Ecology, Economics, and Culture.* Princeton, NJ: Princeton University Press.

Bosso, C., Guber, D. 2006. Maintaining presence: Environmental advocacy and the permanent campaign. In N. Vig, M. Kraft (eds.), *Environmental Policy: New Directions for the Twenty First Century* (6th edition). Washington, DC: CQ Press, pp. 80–6.

Boucher, D., Elias, P., Lininger, K., May-Tobin, C., Roquemore, S., Saxon, E. 2011. *The Root of the Problem: What's Driving Tropical Deforestation Today?* Cambridge, MA: Union of Concerned Scientists.

Bourgeon, L., Burke, A., Higham, T. 2017. Earliest human presence in North America dated to the Last Glacial Maximum: New radiocarbon dates from Bluefish Caves, Canada. *PLoS ONE* 12: e0169486.

Bourgeron, P. S. 1983. Spatial aspects of vegetation structure. In F. B. Golley (ed.), *Tropical Rain Forest Ecosystems: Structure and Function.* Ecosystems of the World (14th edition). Amsterdam: Elsevier Scientific Publishing, pp. 29–47.

Bouzouggar, A., Barton, N., Vanhaeren, M., d'Errico, F., Collcutt, S., Higham, T., Hodge, E., Parfitt, S., Rhodes, E., Schwenninger, J.-L., Stringer, C., Turner, E., Ward, S., Moutmir, A., Stambouli, A. 2007. 82,000-year-old shell beads from North Africa and implications for the origins of modern human behavior. *Proceedings of the National Academy of Sciences of the United States of America* 104: 9964–9.

Bowman, D. M. J. S., Woinarski, J. C. Z. 1996. Australian monsoon rainforest mammals and possible implications for the conservation of tropical rainforest biodiversity. In D. S. Edwards, W. E. Booth, S. C. Choy (eds.), *Tropical Rainforest Research: Current Issues.* Monographiae Biologicae, vol. 74. Dordrecht: Springer, pp. 67–9.

Boysen, L. R., Brovkin, V., Arora, V. K., Cadule, P., de Noblet-Ducoudr, N., Kato, E., Pongratz, J., Gayler, V. 2014. Global and regional effects of land-use change on climate in 21st century simulations with interactive carbon cycle. *Earth System Dynamics Discussion* 5: 443–72.

Bramble, D. M., Lieberman, D. E. 2004. Endurance running and the evolution of *Homo. Nature* 432: 345–52.

Brand, W., Coplen, T. B., Vogl, J., Rosner, M., Prohaska, T. 2014. Assessment of international reference materials for isotope-ratio analysis (IUPAC Technical Report). *Pure and Applied Chemistry* 86: 425–67.

Brandon, K. 2014. *Ecosystem Services from Tropical Forests: Review of Current Science.* Washington, DC: Center for Global Development Working Paper 380.

Brandt, S., Hildebrand, E., Vogelsang, R., Wolfhagen, J., Wang, H. 2012. A new MIS 3 radiocarbon chronology for Mochena Borago Rockshelter, SW Ethiopia: Implications for the interpretation of Late Pleistocene chronostratigraphy and human behavior. *Journal of Archaeological Science: Reports* 11: 352–69.

Brauman, K. A., Daily, G. C., Duarte, T. K., Mooney, H. A. 2007. The nature and value of ecosystem services: An overview highlighting hydrologic services. *Annual Review of Environment and Resources* 32: 67–98.

Breeze, P. S., Drake, N. A., Groucutt, H. S., Parton, A., Jennings, R. P., White, T. S., Clark-Balzan, L., Shipton, C., Scerri, E. M. L., Stimpson, C. M., Crassard, R., Hilbert, Y., Alsharekh, A., Al-Omari, A., Petraglia, M. D. 2015. Remote sensing and GIS techniques for reconstructing Arabian paleohydrology and identifying archaeological sites. *Quaternary International* 382: 98–119.

British Broadcasting Corporation (BBC). 2007. *Bruce Parry's Amazon.* <http://www.bbc.co.uk/amazon/>. Retrieved 19 March 2018.

British Broadcasting Corporation (BBC). 2017. Brazil opens vast Amazon reserve to mining. <http://www.bbc.com/news/world-latin-america-41033228>. Retrieved 25 March 2018.

Broad, S., Mulliken, T., Roe, D. 2003. The nature and extent of legal and illegal trade in wildlife. In S. Oldfield (ed.), *The Trade in Wildlife: Regulation for Conservation.* London: Earthscan, pp. 3–22.

Bronk Ramsey, C., Dee, M., Lee, S., Nakagawa, T., Staff, R. A. 2010. Developments in the calibration and modeling of radiocarbon dates. *Radiocarbon* 52: 953–61.

Brooks, A. S., Yellen, J. E., Potts, R., Behrensmeyer, A. K., Deino, A. L., Leslie, D. E., Ambrose, S. H., Ferguson, J. R., d'Errico, F., Zipkin, A. M., Whittaker, S., Post, J., Veatch, E. G., Foecke, K., Clark, J. B. 2018. Long-distance stone transport and pigment use in the earliest Middle Stone Age. *Science* (15 March). doi:10.1126/science.aao2646.

Brooks, G. E. 2003. *Eurafricans in Western Africa: Commerce, Social Status, Gender, and Religious Observance from the Sixteenth to the Eighteenth Century*. Athens, OH: Ohio University Press.

Brooks, T. M., Wright, S. J., Sheil, D. 2009. Evaluating the success of conservation actions in safeguarding tropical forest biodiversity. *Conservation Biology* 23: 1448–57.

Brosius, J. P. 1986. River, forest and mountain: The Penan Gang landscape. *Sarawak Museum Journal* 36: 173–84.

Brosius, J. P. 1990. Penan hunter-gatherers of Sarawak, East Malaysia. *AnthroQuest* 42: 1–7.

Brosius, J. P. 1991. Foraging in tropical forests: The case of the Penan of Sarawak, East Malaysia (Borneo). *Human Ecology* 19: 123–50.

Brosius, J. P. 1999. Analyses and interventions. *Current Anthropology* 40: 277–309.

Brotton, J. 2014. *A History of the World in 12 Maps*. London: Penguin Books.

Brown, P., Sutikna, T., Morwood, M. J., Soejono, R. P., Jatmiko, Saptomo, E. W., Awe Due, R. 2004. A new small-bodied hominin from the late Pleistocene of Flores, Indonesia. *Nature* 431: 1055–61.

Brown, T. A., Jones, M. K., Powell, W., Allaby, R. G. 2009. The complex origins of domesticated crops in the Fertile Crescent. *Trends in Ecology & Evolution* 24: 103–9.

Bruenig, E. F. 1983. Vegetation structure and growth. In F. B. Golley (ed.), *Tropical Rain Forest Ecosystems: Structure and Function*. Amsterdam: Elsevier, pp. 49–73.

Bruenig, E. F. 1996. *Conservation and Management of Tropical Rainforests: An Integrated Approach to Sustainability*. Wallingford: CAB International.

Brumm, A., Aziz, F., van den Bergh, G. D., Morwood, M. J., Moore, M. W., Kurniawan, I., Hobbs, D. R., Fullagar, R. 2010. Early stone technology on Flores and its implications for *Homo floresiensis*. *Nature* 441: 624–8.

Brumm, A., van den Bergh, G. D., Storey, M., Kurniawan, I., Alloway, B. V., Setiawan, R., Setiyabudi, E., Grün, R., Moore, M. W., Yurnaldi, D., Puspaningrum, M. R., Wibowo, U. P., Insani, H., Sutisna, I., Westgate, J. A., Pearce, N. J. G., Duval, M., Meijer, H. J. M., Aziz, F., Sutikna, T., van der Kaars, S., Flude, S., Morwood, M. J. 2016. Age and context of the oldest known hominin fossils from Flores. *Nature* 534: 249–53.

Brunet, M., Beauvillain, A., Coppens, Y., Heintz, É., Moutaye, A. H. E., Pilbeam, D. 1995. The first australopithecine 2,500 kilometres west of the Rift Valley (Chad). *Nature* 378: 273–5.

Brunet, M., Guy, F., Pilbeam, D., Mackaye, H. T., Likius, A., Ahounta, D., Beauvilain, A., Blondel, C., Bocherens, H., Boisserie, J.-R., De Bonis, L., Coppens, Y., Dejax, J., Denys, C., Duringer, P., Eisenmann, V., Fanone, G., Fronty, P., Geraads, D., Lehmann, T., Lihoreau, F., Louchart, A., Mahamat, A., Merceron, G., Mouchelin, G., Otero, O., Campomanes, P. P., De Leon, M. P., Rage, J.-C., Sapanet, M., Schuster, M., Sudre, J., Tassy, P., Valentin, X., Vignaud, P., Viriot, L., Zazzo, A., Zollikofer, C. 2002. A new hominid from the Upper Miocene of Chad, Central Africa. *Nature* 418: 145–51.

Brunet, M., Guy, F., Pilbeam, D., Lieberman, D. E., Likius, A., Mackaye, H. T., de León, P., Zollikofer, C., Vignaud, P. 2005. New material of the earliest hominid from the Upper Miocene of Chad. *Nature* 434: 752–5.

Buchmann, N., Ehleringer, J. R. 1998. CO_2 concentration profiles, and carbon and oxygen isotopes in C_3 and C_4 crop canopies. *Agriculture and Forest Meteorology* 89: 45–58.

Buchmann, N., Guehl, J.-M., Barigah, T. S., Ehleringer, J. R. 1997. Interseasonal comparison of CO_2 concentrations, isotopic composition, and carbon dynamics in an Amazonian rainforest (French Guiana). *Oecologia* 110: 120–31.

Buckley, B. M., Anchukaitis, K. J., Penny, D., Fletcher, R., Cook, E. R., Sanod, M., Nam, L. C., Wichienkeeo, A., Minh, T. T., Hong, T. M. 2010. Climate as a contributing factor in the demise of Angkor, Cambodia. *Proceedings of the National Academy of Sciences of the United States of America* 107: 6748–52.

Buckley, B. M., Fletcher, R., Wang, S. S., Zottoli, B., Pottier, C. 2014. Monsoon extremes and society over the past millennium on mainland Southeast Asia. *Quaternary Science Reviews* 95: 1–19.

Bullock, S., Mooney, H., Medina, E. (eds.). 1995. *Seasonally Dry Tropical Forests*. Cambridge: Cambridge University Press.

Bulmer, S. 1966. The prehistory of the New Guinea Highlands: A discussion of archaeological field survey and excavations, 1959–1960. MA thesis, University of Auckland, Auckland.

Bulmer, S. 1977. Between the mountain and the plain: Prehistoric settlement and environment in the Kaironk Valley. In J. H. Winslow (ed.), *The Melanesian Environment*. Canberra: Australian National University Press, pp. 61–73.

Bulmer, S. 1991. Variation and change in stone tools in the highlands of Papua New Guinea: The witness of Wanelek. In A. Pawley (ed.), *Man and a Half: Essays in Pacific Anthropology and Ethnobiology in Honour of Ralph Bulmer*. Auckland: The Polynesian Society, pp. 470–8.

Bunker, S. G. 1984. Modes of extraction, unequal exchange, and the progressive underdevelopment of an extreme periphery: The Brazilian Amazon, 1600–1980. *American Journal of Sociology* 89: 1017–64.

Bunn, H. T. 2001. Hunting, power scavenging, and butchering by Hadza foragers and by Plio-Pleistocene *Homo*. In C. B. Stanford, H. T. Bunn (eds.), *Meat-Eating and Human Evolution*. Oxford: Oxford University Press, pp. 199–218.

Burkhill, I. H. 1966. *A Dictionary of the Economic Products of the Malay Peninsula*. Kuala Lumpur: Ministry of Agriculture and Cooperatives.

Burley, D. V., Barton, A., Dickinson, W. R., Connaughton, S. P., Taché, K. 2010. Nukuleka as a founder colony for west Polynesian settlement: New insights from recent excavations. *Journal of Pacific Archaeology* 1: 128–44.

Burney, D. A., Burney, L. P. 2007. Paleoecology and 'inter-situ' restoration on Kaua'i, Hawai'i. *Frontiers in Ecology and the Environment* 5: 483–90.

Burney, D. A., Burney, L. P. 2016. Monitoring results from a decade of native plant translocations at Makauwahi Cave Reserve, Kaua'i. *Plant Ecology* 217: 139–53.

Bush, M. B., Silman, M. R. 2007. Amazonian exploitation revisited: Ecological asymmetry and the policy pendulum. *Frontiers in Ecology and the Environment* 5: 457–65.

Bush, M. B., Correa-Metrio, A., van Woesik, R., Shadik, C. R., McMichael, C. N. H. 2017. Human disturbance amplifies Amazonian El Niño-Southern Oscillation signal. *Global Change Biology* 23: 3181–92.

Butler, R. 2014. Deforestation in the Amazon. <https://data.mongabay.com/brazil.html>. Retrieved 25 March 2018.

Butynski, T. M., Kalina, J. 1998. Gorilla tourism: A critical look. In E. J. Milner-Gulland, R. Mace (eds.), *Conservation of Biological Resources*. London: Wiley, pp. 294–313.

Calvin, M., Benson, A. A. 1948. The path of carbon in photosynthesis. *Science* 107: 476–80.

Campbell, B. 1983. *Human Ecology: The Story of Our Place in Nature from Prehistory to the Present*. New York: Aldine.

Campbell, D. G., Ford, A., Lowell, K., Walker, J., Lake, J. K., Ocampo-Raeder, C., Townesmith, A., Balick, M. 2006. The feral forests of the Eastern Petén. In C. Erickson, W. Balée (eds.), *Time and Complexity in the Neotropical Lowlands*. New York: Columbia University Press, pp. 21–55.

Campbell, M. C., Tishkoff, S. A. 2008. African genetic diversity: Implications for human demographic history, modern human origins, and complex disease mapping. *Annual Review of Genomics and Human Genetics* 9: 403–33.

Campbell, M. W., de Waal, F. B. M. 2014. Chimpanzees empathize with group mates and humans, but not with baboons or unfamiliar chimpanzees. *Proceedings of the Royal Society B: Biological Sciences* 281. doi:10.1098/rspb.2014.0013.

Cann, R. L., Stoneking, M., Wilson, A. C. 1987. Mitochondrial DNA and human evolution. *Nature* 325: 31–6.

Cannon, C. H., Morley, R. J., Bush, A. B. G. 2009. The current refugial rainforests of Sundaland are unrepresentative of their biogeographic past and highly vulnerable to disturbance. *Proceedings of the National Academy of Sciences of the United States of America* 106: 11188–93.

Cardich, A., Cardich, L., Hajduk, A. 1973. Secuencia arqueológica y cronología radiocarbónica de la cueva 3 de Los Toldos (Santa Cruz, Argentina). *Relaciones de la Sociedad Argentina de Antropología* 7: 85–123.

Carney, J. A. 2001. Rice in the Columbian exchange. *Journal of African History* 42: 377–96.

Carrasco, M. A., Barnosky, A. D., Graham, R. W. 2009. Quantifying the extent of North American mammal extinction relative to the pre-anthropogenic baseline. *PLoS ONE* 4: e8331.

Carson, J. F., Whitney, B. S., Mayle, F. E., Iriarte, J., Prümers, H., Soto, J. D., Watling, J. 2014. Environmental impact of geometric earthwork construction in pre-Columbian Amazonia. *Proceedings of the National Academy of Sciences of the United States of America* 111: 10497–502.

Carter, M. L., Bradbury, M. W. 2016. Oxygen isotope ratios in primate bone carbonate reflect amount of leaves and vertical stratification in the diet. *American Journal of Primatology* 78: 1086–97.

Carter, N. 2007. *The Politics of the Environment: Ideas, Activism, Policy* (2nd edition). Cambridge: Cambridge University Press.

Carvalho, G., Nepstad, D., McGrath, D., Vera-Diaz, M. D. C., Santilli, M., Barros, A. C. 2002. Frontier expansion in the Amazon: Balancing development and sustainability. *Environment* 44: 34–45.

Carvalho, R. 1999. *A Amazônia rumo ao círculo da soja. Amazônia Papers*, vol. I, 2. Programa Amazônia. São Paulo: Amigos da Terra.

Casey, J. 2013. The Stone to Metal Age in West Africa. In P. Mitchell, P. Lane (eds.), *The Oxford Handbook of African Archaeology*. Oxford: Oxford University Press, pp. 603–14.

Cassells, D. S., Valentine, P. S. 1990. Recreation management issues in tropical rainforest. James Cook University, Townsville: Proceedings of the Institute of Tropical Rainforest Studies Workshop no. 1.

Castillo, C. 2011. Rice in Thailand: The archaeobotanical contribution. *Rice* 4: 114–20.

Cavalli-Sforza, L. L. 1986. *African Pygmies*. New York: Academic Press.

Cerling, T. E., Harris, J. M. 1999. Carbon isotope fractionation between diet and bioapatite in ungulate mammals and implications for ecological and paleoecological studies. *Oecologia* 120: 347–63.

Cerling, T. E., Harris, J. M., MacFadden, B. J., Leakey, M. G., Quade, J., Eisenmann, V., Ehleringer, J. R. 1997. Global vegetation change through the Miocene/Pliocene boundary. *Nature* 389: 153–8.

Cerling, T. E., Hart, J. A., Hart, T. B. 2004. Isotope ecology in the Ituri forest. *Oecologia* 138: 5–12.

Cerling, T. E., Harris, J. M., Leakey, M. G., Passey, B. H., Levin, N. E. 2010. Stable carbon and oxygen isotopes in East African mammals: Modern and fossil. In L. Werdelin, W. J. Sanders (eds.), *Cenozoic Mammals of Africa*. London: University of California Press, pp. 941–52.

Cerling, T. E., Mbua, E., Kirera, F. M., Manthi, F. K., Grine, F. E., Leakey, M. G., Sponheimer, M., Uno, K. T. 2011. Diet of *Paranthropus boisei* in the early Pleistocene of East Africa.

Proceedings of the National Academy of Sciences of the United States of America 108: 9337–41.

Cerling, T. E., Manthi, F. K., Mbua, E. N., Leakey, L. N., Leakey, M. G., Leakey, R. E., Brown, F. H., Grine, F. E., Hart, J. A., Kaleme, P., Roche, H., Uno, K. T., Wood, B. A. 2013. Stable isotope-based diet reconstructions of Turkana Basin hominins. *Proceedings of the National Academy of Sciences of the United States of America* 110: 10501–6.

Chaimanee, Y., Suteethorn, V., Jintasakul, P., Vidthayanon, C., Marandat, B., Jaeger, J.-J. 2004. A new orang-utan relative from Late Miocene of Thailand. *Nature* 427: 439–41.

Chaïr, H., Traore, R. E., Duval, M. F., Rivallan, R., Mukherjee, A., Aboagye, L. M., Van Rensburg, W. J., Andrianavalona, V., Pinheiro de Carvalho, M. A. A., Saborio, F., Sri Prana, M., Komolong, B., Lawac, F., Lebot, V. 2016. Genetic diversification and dispersal of taro (*Colocasia esculenta* (L.) Schott). *PLoS ONE* 11: e0157712.

Champion, T. (ed.). 1989. *Centre and Periphery: Comparative Studies in Archaeology*. London: Unwin Hyman.

Channel 4. 2014. *The Island with Bear Grylls*. Shine Television, Bear Grylls Ventures.

Chase, A. F., Chase, D. Z., Weishampel, J. F., Drake, J. B., Shrestha, R. L., Slatton, K. C., Awe, J. J., Carter, W. E. 2011. Airborne LiDAR, archaeology, and the ancient Maya landscape at Caracol, Belize. *Journal of Archaeological Science* 38: 387–98.

Chase, D. Z., Chase, A. F. 2017. Caracol, Belize and changing perceptions of ancient Maya society. *Journal of Archaeological Research* 25: 185–249.

Chaubey, G., Endicott, P. 2013. The Andaman Islanders in a regional genetic context: Reexamining the evidence for an early peopling of the Archipelago from South Asia. *Human Biology* 85: 153–71.

Chen, F., Jiang, A., Ishwaran, N. 2014. Angkor site monitoring and evaluation by radar remote sensing. *Proceedings of SPIE, Volume 9260, Land Surface Remote Sensing* 2: 926018.

Chen, I. C., Shiu, H.-J., Benedick, S., Holloway, J. D., Chey, V. K., Barlow, H. S., Hill, J. K., Thomas, C. D. 2009. Elevation increases in moth assemblages over 42 years on a tropical mountain. *Proceedings of the National Academy of Sciences of the United States of America* 106: 1479–83.

Chernela, J. M. 1985. Indigenous fishing in the neotropics: The Turkanoan Uanano of the Blackwater Uaupés River Basin in Brazil and Colombia. *Interciencia* 10: 78–86.

Childe, V. G. 1936. *Man Makes Himself*. London: Watts.

Childe, V. G. 1950. The Urban Revolution. *Town Planning Review* 21: 3–17.

Chong, V. C. 2007. Mangroves and fisheries linkages: The Malaysian perspective. *Bulletin of Marine Science* 80: 755–72.

Christopher, E. 2006. *Slave Ship Sailors and Their Captive Cargoes, 1730–1807*. Cambridge: Cambridge University Press.

Ciochon, R. L. 2009. The mystery ape of Pleistocene Asia. *Nature* 459: 910–11.

Ciochon, R. L., Olsen, J., James, J. 1990. *Other Origins: The Search for the Giant Ape in Human Prehistory*. London: Bantam.

Cipriani, L. 1966. *The Andaman Islanders*. London: Weidenfeld & Nicolson.

Clark, D. B., Clark, D. A., Oberbauer, S. F. 2010. Annual wood production in a tropical rain forest in NE Costa Rica linked to climatic variation but not to increasing CO_2. *Global Change Biology* 16: 747–59.

Clark, G. R. 2016. Chiefly tombs, lineage history, and the ancient Tongan state. *Journal of Island and Coastal Archaeology* 11: 326–43.

Clark, G. R., Martinsson-Wallin, H. 2007. Monumental architecture in West Polynesia: Origins, chiefs and archaeological approaches. *Archaeology in Oceania* 42: 28–40.

Clark, G. R., Reepmeyer, C. 2014. Stone architecture, monumentality and the rise of the early Tongan chiefdom. *Antiquity* 88: 1244–60.

Clark, G. R., Reepmeyer, C., Melekiola, N., Woodhead, J., Dickinson, W. R., Martinsson-Wallin, H. 2014. Stone tools from the ancient Tongan state reveal prehistoric interaction centers in the Central Pacific. *Proceedings of the National Academy of Sciences of the United States of America* 111: 10491–6.

Clark, J. D. 1963. *Prehistoric Cultures of Northeast Angola and their Significance in Tropical Africa.* Lisbon: Companhis de Diamantes de Angola (DIAMANG).

Clark, J. D. 1966. *The Distribution of Prehistoric Culture in Angola.* Lisbon: Companhis de Diamantes de Angola (DIAMANG).

Clark, J. D. 1967. *The Atlas of African Prehistory.* Chicago: University of Chicago Press.

Clark, J. D. 1969. *Kalambo Falls Prehistoric Site, Volume I: The Geology, Palaeoecology and Detailed Stratigraphy of the Excavations.* Cambridge: Cambridge University Press.

Clark, J. D. 1971. Problems of archaeological nomenclature and definition in the Congo basin. *South African Archaeological Bulletin* 26: 67–78.

Clark, J. D. 1988. The Middle Stone Age of East Africa and the beginnings of regional identity. *Journal of World Prehistory* 2: 235–305.

Clark, J. D., Brown, K. S. 2001. The Twin Rivers Kopje, Zambia: Stratigraphy, fauna, and artifact assemblages from the 1954 and 1956 excavations. *Journal of Archaeological Science* 28: 305–30.

Clark, J. D., Beyene, Y., WoldeGabriel, G., Hart, W. K., Renne, P. R., Gilbert, H., Defleur, A., Suwa, G., Katoh, S., Ludwig, K. R., Boisserie, J.-R., Asfaw, B., White, T. D. 2003. Stratigraphic, chronological and behavioural contexts of Pleistocene *Homo sapiens* from Middle Awash, Ethiopia. *Nature* 423: 747–52.

Clark, P. U., Dyke, A. S., Shakun, J. D., Carlson, A. E., Clark, J., Wohlfarth, B., Mitrovica, J. X., Hostetler, S. W., McCabe, A. M. 2009. The Last Glacial Maximum. *Science* 325: 710–14.

Clarkson, C., Jones, S., Harris, C. 2012. Continuity and change in the lithic industries of the Jurreru Valley, India, before and after the Toba eruption. *Quaternary International* 258: 165–79.

Clarkson, C., Jacobs, Z., Marwick, B., Fullagar, R., Wallis, L., Smith, M., Roberts, R. G., Hayes, E., Lowe, K., Carah, X., Florin, S. A., McNeil, J., Cox, D., Arnold, L. J., Hua, Q., Huntley, J., Brand, H. E. A., Manne, T., Fairbairn, A., Shulmeister, J., Lyle, L., Salinas, M., Page, M., Connell, K., Park, G., Norman, K., Murphy, T., Pardoe, C. 2017. Human occupation of northern Australia by 65,000 years ago. *Nature* 547: 306–10.

Clemens, S. C., Prell, W. L., Murray, D. W., Shimmield, G., Weedon, G. 1991. Forcing mechanisms of the Indian Ocean monsoon. *Nature* 353: 720–5.

Clement, C. R. 1988. Domestication of the pejibaye palm (*Bactris gasipaes*): Past and present. In M. J. Balick (ed.), *The Palm—Tree of Life: Biology, Utilization and Conservation.* Bronx, NY: New York Botanical Garden, Advances in Economic Botany, Volume 6, pp. 155–74.

Clement, C. R. 1999. 1492 and the loss of crop genetic resources: I. Crop biogeography at contact. *Economic Botany* 53: 203–16.

Clement, C. R. 2006. Fruit trees and the transition to food production in Amazonia. In W. Balée, C. L. Erickson (eds.), *Time and Complexity in the Neotropical Lowlands: Studies in Historical Ecology.* New York: Columbia University Press, pp. 165–85.

Clement, C. R., Rival, L., Cole, D. M. 2009. Domestication of peach palm (*Bactris gasipaes Kunth*): The roles of human mobility and migration. In M. N. Alexiades (ed.), *Shifting Spaces, Changing Times: Mobility, Migration and Displacement in Indigenous Lowland South America.* Oxford: Berghahn Books, pp. 117–40.

Clement, C. R., de Cristo-Araújo, M., d'Eeckenbrugge, G. C., Pereira, A. A., Picanço-Rodrigues, D. 2010. Origin and domestication of native Amazonian crops. *Diversity* 2: 72–106.

Clement, C. R., Denevan, W. M., Heckenberger, M. J., Junqueira, A. B., Neves, E. G., Teixeira, W. G., Woods, W. I. 2015. The domestication of Amazonia before European

conquest. *Proceedings of the Royal Society B: Biological Sciences* 282. doi:10.1098/rspb.2015.0813.
Clist, B. 1993. Archaeological fieldwork and labwork in Gabon during 1992. *Nyame Akuma* 39: 26–32.
Clist, B. 1995. *Gabon 100 000 ans d'Histoire*. Saint Exupéry: Centre Culture Français.
Clutton-Brock, T. H., Harvey, T. 1977. Primate ecology and social organization. *Journal of Zoology* 183: 1–39.
Cochrane, M. A. 2003. Fire science for rainforests. *Nature* 421: 913–19.
Cock, J. H. 1982. Cassava: A basic energy source in the tropics. *Science* 218: 755–62.
Cockrill, W. R. (ed.). 1974. *The Husbandry and Health of the Domestic Buffalo*. Rome: Food and Agriculture Organization of the United Nations.
Cockrill, W. R. 1977. *The Water Buffalo*. Rome: Food and Agriculture Organization of the United Nations.
Coe, M. D. 1965. Tikal, Guatemala, and the emergent Maya civilization. *Science* 147: 1401–19.
Coe, M. D. 1967. *Tikal: A Handbook of the Ancient Maya Ruins*. Philadelphia: University Museum, University of Pennsylvania.
Coe, M. D. 1999. *The Maya*. London: Thames & Hudson.
Coe, M. D., Houston, S. 2015. *The Maya*. London: Thames & Hudson.
Cole, G. 1967. The Later Acheulian and Sangoan of southern Uganda. In W. Bishop, J. Clark (eds.), *Background to Evolution in Africa*. Chicago: University of Chicago Press, pp. 481–527.
Colinvaux, P. A. 2014. The Pleistocene vector of neotropical diversity. In E. Bermingham, C. W. Dick, C. Moritz (eds.), *Tropical Rainforests: Past, Present and Future*. Chicago: University of Chicago Press, pp. 78–106.
Colinvaux, P. A., Bush, M. B. 1991. The rain-forest ecosystem as a resource for hunting and gathering. *American Anthropologist* 93: 153–60.
Colunga-García Marín, P., Zizumbo-Villarreal, D. 2004. Domestication of plants in Maya Lowlands. *Economic Botany* 58: S101–S110.
Comaroff, J. L., Comaroff, J. 2010. *Ethnicity*. Chicago: University of Chicago Press.
Coningham, R. A. E. 1999. *Anuradhapura: The British-Sri Lankan Excavations at Anuradhapura Salgaha Watta 2*. British Archaeological Reports, International Series 824. Oxford: Archaeopress.
Coningham, R. A. E. 2006. *Anuradhapura: The British-Sri Lankan Excavations at Anuradhapura Salgaha Watta: Volume 2, The Artefacts*. Oxford: Archaeopress for the Society of South Asian Studies.
Coningham, R. A. E., Gunawardhana, P. 2013. *Anuradhapura: Volume 3, The Hinterland*. Oxford: Archaeopress for the Society for South Asian Studies.
Coningham, R. A. E., Gunawardhana, P., Manuel, M., Adikari, G., Katugampola, M., Young, R., Schmidt, A., Krishnan, K., Simpson, I. A., McDonnell, G., Batt, C. 2007. The state of theocracy: Defining an early medieval hinterland in Sri Lanka. *Antiquity* 81: 699–719.
Conklin, B. 1997. Body paint, feathers, and VCRs: Aesthetics and authenticity in Amazonian activism. *American Ethnologist* 24: 711–37.
Connah, G. 2016. *African Civilizations: An Archaeological Perspective* (3rd edition). Cambridge: Cambridge University Press.
Conrad, J. 1999. *Heart of Darkness* (2nd edition). Edited by D. C. R. A. Goonatilleke. Canada: Broadview Press Limited.
Cook, B. I., Anchukaitis, K. J., Kaplan, J. O., Puma, M. J., Kelley, M., Gueyffier, D. 2012. Pre-Columbian deforestation as an amplifier of drought in Mesoamerica. *Geophysical Research Letters* 39: L16706.
Coon, C., Bricker, H., Johnson, F., Lamberg-Karlovsky, C. 1968. *Yengema Cave Report*. Philadelphia: The University Museum, University of Pennsylvania.

Cooper, A., Turney, C., Hughen, K. A., Brook, B. W., McDonald, H. G., Bradshaw, C. J. A. 2015. Abrupt warming events drove Late Pleistocene Holarctic megafaunal turnover. *Science* 349: 602–6.

Cooper, Z. 2002. *Archaeology and History: Early Settlements in the Andaman Islands.* Oxford: Oxford University Press.

Coppens, Y. 1982. Qui fit quoi? Les plus anciennes industries préhistorique et leur artisans. *Bulletin de la Société Préhistorique Française* CRSM 6: 163–5.

Coppens, Y. 1991. L'évolution des hominidés, de leur locomotion et de leurs environnements. In Y. Coppen, B. Senut (eds.), *Origine(s) de la Bipédie chez les Hominidés.* Paris: Cahiers de Paleoanthropologie, CNRS, pp. 295–300.

Coppens, Y. 1994. East side story: The origins of mankind. *Scientific American* 270: 62–9.

Cordeiro, N. J., Patrick, D. A. G., Munisi, B., Gupta, V. 2004. Role of dispersal in the invasion of an exotic tree in an East African submontane forest. *Journal of Tropical Ecology* 20: 449–57.

Coria, J., Calfucura, E. 2012. Ecotourism and the development of indigenous communities: The good, the bad, and the ugly. *Ecological Economics* 73: 47–55.

Corlett, R. T. 2007. The impact of hunting on the mammalian fauna of tropical Asian forests. *Biotropica* 39: 292–303.

Corlett, R. T. 2013. The shifted baseline: Prehistoric defaunation in the tropics and its consequences for biodiversity conservation. *Biological Conservation* 163: 13–21.

Corlett, R. T., Primack, R. 2011. *Tropical Rain Forests: An Ecological and Biogeographical Comparison.* London: Wiley-Blackwell.

Cornelissen, E. 2002. Human responses to changing environments in Central Africa between 40,000 and 12,000 BP. *Journal of World Prehistory* 16: 197–235.

Cornelissen, E. 2013. Hunting and gathering in Africa's tropical forests at the end of the Pleistocene and in the Early Holocene. In P. Mitchell, P. Lane (eds.), *The Oxford Handbook of African Archaeology.* Oxford: Oxford University Press, pp. 403–18.

Corner, E. J. H. 1966. *The Natural History of Palms.* Berkeley: University of California Press.

Cosgrove, R., Field, J., Ferrier, Å. 2007. The archaeology of Australia's tropical rainforests. *Palaeogeography, Palaeoclimatology, Palaeoecology* 251: 150–73.

Costanza, R., de Groot, R., Sutton, P., van der Ploeg, S., Anderson, S. J., Kubiszewski, I., Farber, S., Kerry Turner, R. 2014. Changes in the global value of ecosystem services. *Global Environmental Change* 26: 152–8.

Coulibaly, S., Pasquet, R. S., Papa, R., Gepts, P. 2002. AFLP analysis of the phenetic organization and genetic diversity of *Vigna unguiculata* L. Walp. reveals extensive gene flow between wild and domesticated types. *Theoretical and Applied Genetics* 104: 358–66.

Couvreur, T. L. P., Forest, F., Baker, W. J. 2011. Origin and global diversification patterns of tropical rain forests: Inferences from a complete genus-level phylogeny of palms. *BMC Biology* 9: 44.

Cowgill, G. L. 2004. Origins and development of urbanism: Archaeological perspectives. *Annual Review of Anthropology* 33: 525–49.

Cowling, S. A., Maslin M. A., Sykes, M. T. 2001. Paleovegetation simulations of lowland Amazonia and implications for neotropical allopatry and speciation. *Quaternary Research* 55: 140–9.

Cowlishaw, G., Dunbar, R. I. M. 2000. *Primate Conservation Biology.* Chicago: University of Chicago Press.

Cranbrook, E. 2000. Northern Borneo environments of the past 40,000 years: Archaeological evidence. *Sarawak Museum Journal* 55: 61–109.

Crawford, H. E. W. 2004. *Sumer and the Sumerians* (2nd edition). Cambridge: Cambridge University Press.

Crevecoeur, I., Skinner, M. M., Bailey, S. E., Gunz, P., Bortoluzzi, S., Brooks, A. S., Burlet, C., Cornelissen, E., De Clerck, N., Maureille, B., Semal, P., Vanbrabant, Y., Wood, B. 2014. *First early hominin from Central Africa* (Ishango, Democratic Republic of Congo). PLoS ONE 9: e84652.

Crompton, R. H., Sellers, W. I., Thorpe, S. K. S. 2010. Arboreality, terrestriality and bipedalism. *Philosophical Transactions of the Royal Society B: Biological Sciences* 365: 3301–14.

Crosby, A. W. 1972. *The Columbian Exchange: Biological and Cultural Consequences of 1492*. Westport, Connecticut: Greenwood Press.

Cuarón, A. D. 2000. A global perspective on habitat disturbance and tropical rainforest mammals. *Conservation Biology* 14: 1574–9.

Curran, L. M., Caniago, I., Paoli, G. D., Astianti, D., Kusneti, M., Leighton, M., Nirarita, C. E., Haeruman, H. 1999. Impact of El Niño and logging on canopy tree recruitment in Borneo. *Science* 286: 2184–8.

Da Silva, M. N. F., Patton, J. L. 1993. Amazonian phylogeography: mtDNA sequence variation in arboreal echimyid rodents (*Caviomorpha*). *Molecular Phylogenetics and Evolution* 2: 243–55.

Daegling, D. J., Grine, F. 1994. Bamboo feeding, dental microwear, and diet of the Pleistocene ape *Gigantopithecus blacki*. *South African Journal of Science* 90: 527–32.

Daguan, Z. 2007. *A Record of Cambodia: The Land and its People* (edited and translated by P. Harris). Thailand: Silkworm Books.

Dai, A., Wigley, T. M. L. 2000. Global patterns of ENSO-induced precipitation. *Geophysical Research Letters* 27: 1283–6.

D'Andrea, A. C., Klee, M., Casey, J. 2001. Archaeological evidence for pearl millet (*Pennisetum glaucum*) in sub-Saharan West Africa. *Antiquity* 75: 341–8.

D'Andrea, A. C., Logan, A. L., Watson, D. J. 2006. Oil palm and prehistoric subsistence in tropical West Africa. *Journal of African Archaeology* 4: 195–222.

D'Andrea, A. C., Kahlheber, S., Logan, A. L., Watson, D. J. 2007. Early domesticated cowpea (*Vigna unguiculata*) from Central Ghana. *Antiquity* 81: 686–98.

Daniel, G. 1975. *150 Years of Archaeology*. London: Duckworth.

Dart, R. A. 1925. *Australopithecus africanus*: The man-ape of South Africa. *Nature* 115: 195–9.

Dart, R. A. 1926. Taungs and its significance. *Journal of Natural History* 26: 315–27.

Davis, C. C., Webb, C. O., Wurdack, K. J., Jaramillo, C. A., Donoghue, M. J. 2005. Explosive radiation of Malpighiales supports a mid-Cretaceous origin of modern tropical rain forests. *American Naturalist* 165: E36–E65.

Davy, J. 1821. *An Account of the Interior of Ceylon and of its Inhabitants with Travels in that Island*. London: Longman.

De Bonis, L., Bouvrain, G., Geraads, D., Koufos, G. 1990. New hominid skull material from the late Miocene of Macedonia in Northern Greece. *Nature* 345: 712–14.

De Heinzelin, J., Clark, J. D., White, T., Hart, W., Renne, P., Woldegabriel, G., Beyene, Y., Vrba, E. 1999. Environment and behaviour of 2.5-million-year-old Bouri hominids. *Science* 284: 625–9.

De la Penha, G. 1986. *O Museu Paraense Emilio Goeldi*. São Paulo: Museu Paraense Emilio Goeldi, Banco Safra SA, and CNPq.

De Luna, K. M. 2017. Conceptualizing vegetation in the Bantu Expansion: Reflections on linguistics in central African history. *Quaternary International* 448: 158–68.

De Maret, P. 2013. Archaeologies of the Bantu Expansion. In P. Mitchell, P. J. Lane (eds.), *The Oxford Handbook of African Archaeology*. Oxford: Oxford University Press, pp. 627–43.

De Oliveira Santos, E. C., de Jesus, I. M., Brabo, E. S., de M. Câmara, V., Loureiro, E. C. B., Mascarenhas, A. F., de F. Faial, K., da Silva, A. P., da Silva, R. U., Luiz, R. R., Higuchi, H. 2002. Exposure to mercury in the urban population of Rio Branco City, State of Acre, Brazil. *Bulletin of Environmental Contamination and Toxicology* 69: 314–19.

De Sy, V., Herold, M., Achard, F., Beuchle, R., Clevers, J. G. P. W., Lindquist, E., Verchot, L. V. 2015. Land use patterns and related carbon losses following deforestation in South America. *Environmental Research Letters* 10: 124004.

de Vos, J., Sondaar, P. Y., van den Bergh, G. D., Aziz, F. 1994. The *Homo* bearing deposits of Java and its ecological context. *Courier Forschungsinstitut Senckenberg* 171: 129–40.

DeCorse, C. 2001. Introduction. In C. DeCorse (ed.), *West Africa during the Atlantic Slave Trade: Archaeological Perspectives*. Leicester: Leicester University Press, pp. 1–13.

Deevey, E. S., Rice, D. S., Rice, P. M., Vaughan, H. H., Brenner, M., Flannery, M. S. 1979. Mayan urbanism: Impact on a tropical Karst environment. *Science* 206: 298–306.

DeFoliart, G. R. 1999. Insects as food: Why the western attitude is important. *Annual Review of Entomology* 44: 21–50.

Deleuze, G., Guattari, F. 1987. *A Thousand Plateaus: Capitalism and Schizophrenia*. London: Athlone Press.

Delfin, F., Salvador, J. M., Calacal, G. C., Perdigon, H. B., Tabbada, K. A., Villamor, L. P., Halos, S. C., Gunnarsdóttir, E., Myles, S., Hughes, D. A., Xu, S., Jin, L., Lao, O., Kayser, M., Hurles, M. E., Stoneking, M., De Ungria, M. C. A. 2011. The Y-chromosome landscape of the Philippines: Extensive heterogeneity and varying genetic affinities of Negrito and non-Negrito groups. *European Journal of Human Genetics* 19: 224–30.

Delfin, F., Min-Shan, K. A., Li, M., Gunnarsdóttir, E., Tabbada K. A., Salvador, J. M., Calacal, G. C., Sagum, M. S., Datar, F. A., Padilla, S. G., De Ungria, M. C., Stoneking, M. 2014. Complete mtDNA genomes of Filipino ethnolinguistic groups: A melting pot of recent and ancient lineages in the Asia-Pacific region. *European Journal of Human Genetics* 22: 228–37.

Demarest, A. 2004. *Ancient Maya: The Rise and Fall of a Rainforest Civilization*. Cambridge: Cambridge University Press.

DeMenocal, P. B. 2004. African climate change and faunal evolution during the Pliocene-Pleistocene. *Earth and Planetary Science Letters* 220: 3–24.

Demeter, F., Shackelford, L. L., Bacon, A.-M., Duringer, P., Westaway, K., Sayavongkhamdy, T., Braga, J., Sichanthongtip, P., Khamdalavong, P., Ponche, J.-L., Wang, H., Lundstrom, C., Patole-Edoumba, E., Karpoff, A.-M. 2012. Anatomically modern human in Southeast Asia (Laos) by 46 ka. *Proceedings of the National Academy of Sciences of the United States of America* 109: 14375–80.

Demeter, F., Shackelford, L., Westaway, K., Duringer, P., Bacon, A.-M., Ponche, J.-L., Wu, X., Sayavongkhamdy, T., Zhao, J.-X., Barnes, L., Boyon, M., Sichanthongtip, P., Sénégas, F., Karpoff, A.-M., Patole-Edoumba, E., Coppens, Y., Braga, J. 2015. Early modern humans and morphological variation in Southeast Asia: Fossil evidence from Tam Pa Ling, Laos. *PLoS ONE* 10: e0121193.

Denevan, W. 1966. *An Aboriginal Cultural Geography of the Llanos de Mojos de Bolivia*. Ibero-American, No. 48. Berkeley: University of California.

Denham, T. 2004. Early agriculture in the highlands of New Guinea: An assessment of Phase 1 at Kuk Swamp. In V. Attenbrow, R. Fullagar (eds.), *A Pacific Odyssey: Archaeology and Anthropology in the Western Pacific*. Papers in Honour of Jim Specht. Records of the Australian Museum Supplement 29. Sydney: Australian Museum, pp. 47–57.

Denham, T. 2008. Environmental archaeology: Interpreting practices-in-the-landscape through geoarchaeology. In B. David, J. Thomas (eds.), *Handbook of Landscape Archaeology*. Walnut Creek, CA: Left Coast Press, pp. 468–81.

Denham, T. 2009. A practice-centred method for charting the emergence and transformation of agriculture. *Current Anthropology* 50: 661–7.

Denham, T. 2011. Early agriculture and plant domestication in New Guinea and Island Southeast Asia. *Current Anthropology* 52: S379–S395.

Denham, T., Barton, H. 2006. The emergence of agriculture in New Guinea: Continuity from pre-existing foraging practices. In D. J. Kennett, B. Winterhalder (eds.), *Behavioral Ecology and the Transition to Agriculture*. Berkeley: University of California Press, pp. 237–64.

Denham, T., Haberle, S. G. 2008. Agricultural emergence and transformation in the Upper Wahgi Valley during the Holocene: Theory, method and practice. *The Holocene* 18: 499–514.

Denham, T., Haberle, S. G., Lentfer, C., Fullagar, R., Field, J., Therin, M., Porch, N., Winsborough, B. 2003. Origins of agriculture at Kuk Swamp in the Highlands of New Guinea. *Science* 301: 189–93.

Denham, T., Haberle, S. G., Lentfer, C. 2004. New evidence and interpretations for early agriculture in highland New Guinea. *Antiquity* 78: 839–57.

Denham, T., Haberle, S. G., Pierret, A. 2009. A multi-disciplinary method for the investigation of early agriculture: Learning lessons from Kuk. In S. Fairbairn, S. O'Connor, B. Marwick (eds.), *New Directions in Archaeological Science*. Canberra: Terra Australis 28, ANU E Press, pp. 139–54.

Denham, T., Golson, J., Hughes, P. 2017a. Phase 1: The case for 10,000-year-old agriculture at Kuk. In J. Golson, T. Denham, P. Hughes, P. Swadling, J. Muke (eds.), *Ten Thousand Years of Cultivation at Kuk Swamp in the Highlands of Papua New Guinea* (Terra Australis 46). Canberra: Australian National University Press, pp. 187–200.

Denham, T., Golson, J., Hughes, P. 2017b. Phase 2: Mounded cultivation during the Mid Holocene. In J. Golson, T. Denham, P. Hughes, P. Swadling, J. Muke (eds.), *Ten Thousand Years of Cultivation at Kuk Swamp in the Highlands of Papua New Guinea* (Terra Australis 46). Canberra: Australian National University Press, pp. 201–20.

Denham, T., Golson, J., Hughes, P. 2017c. Phase 3: The emergence of ditches. In J. Golson, T. Denham, P. Hughes, P. Swadling, J. Muke (eds.), *Ten Thousand Years of Cultivation at Kuk Swamp in the Highlands of Papua New Guinea* (Terra Australis 46). Canberra: Australian National University Press, pp. 221–38.

Dennell, R. 2016. Life without the Movius Line: The structure of the East and Southeast Asian Early Palaeolithic. *Quaternary International* 400: 14–22.

Dennell, R., Roebroeks, W. 2005. Out of Africa: An Asian perspective on early human dispersal from Africa. *Nature* 438: 1099–104.

Denslow, J. S. 1987. Tropical rainforest gaps and tree species diversity. *Annual Review of Ecology, Evolution, and Systematics* 18: 431–51.

Dentan, R. K. 1991. Potential food sources for foragers in Malaysian rainforest: Sago, yams, and lots of little things. *Bijdragen tot de Taal, Land-en Volkenkunde* 147: 420–44.

Deraniyagala, S. U. 1992. *The Prehistory of Sri Lanka: An Ecological Perspective* (2nd edition). Colombo: Department of Archaeological Survey.

Deraniyagala, S. U. 2007. The prehistory and protohistory of Sri Lanka. In P. L. Prematilleke, S. Bandaranayake, S. U. Deraniyagala, R. Silva (eds.), *The Art and Archaeology of Sri Lanka, 1*. Colombo: Central Cultural Fund, pp. 1–96.

d'Errico, F., Vanhaeren, M., Barton, N., Bouzouggar, A., Mienis, H., Richter, D., Hublin, J.-J., McPherron, S. P., Lozouet, P. 2009. Additional evidence on the use of personal ornaments in the Middle Paleolithic of North Africa. *Proceedings of the National Academy of Sciences of the United States of America* 106: 16051–6.

DeSouza, J. G., Schaan, D. P., Robinson, M., Barbosa, A. D., Aragão, L. E. O. C., Marimon Jr., B. H., Marimon, B. S., da Silva, I. B., Khan, S. S., Nakahara, F. R., Iriarte, J. 2018. Pre-Columbian earth-builders settled along the entire southern rim of the Amazon. *Nature Communications* 9: 1125.

Détroit, F., Dizon, E., Falguères, C., Hameau, S., Ronquillo, W., Sémah, F. 2004. Upper Pleistocene *Homo sapiens* from Tabon cave (Palawan, the Philippines): Description and dating of new discoveries. *Comptes Rendus Palevol* 3: 705–12.

Deutsch, C. A., Tewksbury, J. J., Huey, R. B., Sheldon, K. S., Ghalambor, C. K., Haak, D. C., Martin, P. R. 2008. Impacts of climate warming on terrestrial ectotherms across latitude. *Proceedings of the National Academy of Sciences of the United States of America* 105: 6668–72.

DeVore, I., Washburn, S. L. 1963. Baboon ecology and human evolution. In C. F. Howell, F. Bourlière (eds.), *African Ecology and Human Evolution*. Chicago: Aldine, pp. 335–67.

Dewar, G., Stewart, B. A. 2016. Paleoenvironments, sea levels and land use in Namaqualand, South Africa, during MIS 6–2. In S. C. Jones, B. A. Stewart (eds.), *Africa from MIS 6–2: Population Dynamics and Paleoenvironments*. Dordrecht: Springer, pp. 195–212.

Dexter, K. G., Smart, B., Baldauf, C., Baker, T. R., Bessike Balinga, M. P., Brienen, R. J. W., Fauset, S., Feldpausch, T. R., Ferreira-Da Silva, L., Ilunga Muledi, J., Lewis, S. L., Lopez-Gonzalez, G., Marimon-Junior, B. H., Marimon, B. S., Meerts, P., Page, N., Parthasarathy, N., Phillips, O. L., Sunderland, T. C. H., Theilade, I., Weintritt, J., Affum-Baffoe, K., Araujo, A., Arroyo, L., Begne, S. K., Carvalho-Das Neves, E., Collins, M., Cuni-Sanchez, A., Djuikouo, M. N. K., Elias, F., Foli, E. G., Jeffery, K. J., Killeen, T. J., Malhi, Y., Maracahipes, L., Mendoza, C., Monteagudo-Mendoza, A., Morandi, P., Oliveira-Dos Santos, C., Parada, A. G., Pardo, G., Peh, K. S.-H., Salomão, R. P., Silveira, M., Sinatora-Miranda, H., Slik, J. W. F., Sonke, B., Taedoumg, H. E., Toledo, M., Umetsu, R. K., Villaroel, R. G., Vos, V. A., White, L. J. T., Pennington, R. T. 2015. Floristics and biogeography of vegetation in seasonally dry tropical regions. *International Forestry Review* 17(S2): 10–32.

Dharmadasa, K. N. O., Samarasinghe, S. W. R. 1990. *The Vanishing Aborigines: Sri Lanka's Veddas in Transition*. New Delhi: Vikas Publishing House.

Diamond, J. 2005. *Collapse: How Societies Choose to Fail or Survive*. London: Penguin Books.

Dietrich, W. 1992. *The Final Forest: The Battle for the Last Great Trees of the Pacific Northwest*. London: Penguin Books.

Diffie, B. W., Winius, G. D. 1977. *Foundations of the Portuguese Empire, 1415–1580*. Minneapolis: University of Minnesota Press.

Dillehay, T. D. 1989. *Monte Verde: A Late Pleistocene Settlement in Chile. Volume I: The Paleo-environment and Site Context*. Washington, DC: Smithsonian Institution Press.

Dillehay, T. D. 2000. *The Settlement of the Americas: A New Prehistory*. New York: Basic Books.

Dillehay, T. D., Rossen, J., Andres, T. C., Williams, D. E. 2007. Preceramic adoption of peanut, squash, and cotton in northern Peru. *Science* 316: 1890–3.

Dillehay, T. D., Ramirez, C., Pino, M., Collins, M. B., Rossen, J., Pino-Navarro, J. D. 2008. Monte Verde: Seaweed, food, medicine, and the peopling of the South America. *Science* 320: 784–6.

Dillehay, T. D., Ocampo, C., Saavedra, J., Sawakuchi, A. O., Vega, R. M., Pino, M., Collins, M. B., Cummings, L. S., Arregui, I., Villagran, X. S., Hartmann, G. A., Mella, M., González, A., Dix, G. 2015. New archaeological evidence for an early human presence at Monte Verde, Chile. *PLoS ONE* 10: e0145471.

Dillehay, T. D., Goodbred, S., Pino, M., Vásquez Sánchez, V. F., Rosales Tham, T., Advosaio, J., Collins, M. B., Netherly, P. J., Hastorf, C. A., Chiou, K. L., Piperno, D., Rey, I., Velchoff, N. 2017. Simple technologies and diverse food strategies of the Late Pleistocene and Early Holocene at Huaca Prieta, Coastal Peru. *Science Advances* 3: e1602778.

DiMaggio, E. N., Campisano, C. J., Rowan, J., Dupont-Nivet, G., Deino, A. L., Bibi, F., Lewis, M. E., Souron, A., Garello, D., Werdelin, L., Reed, K. E., Arrowsmith, J. R. 2015. Late Pliocene fossiliferous sedimentary record and the environmental context of early *Homo* from Afar, Ethiopia. *Science* 347: 1355–9.

Dirks, P. H. G. M., Roberts, E. M., Hilbert-Wolf, H., Kramers, J. D., Hawks, J., Dosseto, A., Duval, M., Elliott, M., Evans, M., Grün, R., Hellstrom, J., Herries, A. I. R., Joannes-Boyau, R., Makhubela, T. V., Palczek, C. J., Robbins, J., Spandler, C., Wiersma, J., Woodhead, J., Berger, L. R. 2017. The age of *Homo naledi* and associated sediments in the Rising Star Cave, South Africa. eLife 2017: 6:e24231. doi:10.7554/eLife.24231.

Dirzo, R., Young, J. S., Mooney, H. A., Ceballos, G. (eds.). 2011. *Seasonally Dry Tropical Forests: Ecology and Conservation*. Washington, DC: Island Press.

Dittus, W. P. J. 2017. The biogeography and ecology of Sri Lankan mammals point to conservation priorities. *Ceylon Journal of Science* 46: 33–64.

Doebley, J. 2004. The genetics of maize evolution. *Annual Review of Genetics* 38: 37–59.

Domínguez-Rodrigo, M. 2014. Is the 'Savanna Hypothesis' a dead concept for explaining the emergence of the earliest hominins? *Current Anthropology* 55: 59–81.

Donkin, R. A. 1999. *Dragon's Brain Perfume: An Historical Geography of Camphor.* Leiden: Brill.

Doughty, C. E., Wolf, A., Morueta-Holme, N., Jorgensen, P. M., Sandel, B., Violle, C., Boyle, B., Kraft, N. J. B., Peet, R. K., Enquist, B. J., Svenning, J.-C., Blake, S., Galetti, M. 2016. Megafauna extinction, tree species range reduction, and carbon storage in Amazonian forests. *Ecography* 39: 194–203.

Douka, K. 2013. Exploring the great wilderness of prehistory: The chronology of the Middle to the Upper Paleolithic transition in the northern Levant. *Mitteilungen der Gesellschaft Urgeschicte* 22: 11–40.

Drake, N., Breeze, P. 2016. Climate change and modern human occupation of the Sahara from MIS 6–2. In S. Jones, B. Stewart (eds.), *Africa from MIS 6–2: Population Dynamics and Paleoenvironments.* Dordrecht: Springer, pp. 103–22.

Drake, N., Blench, R. M., Armitage, S. J., Bristow, C. S., White K. H. 2011. Ancient watercourses and biogeography of the Sahara explain the peopling of the desert. *Proceedings of the National Academy of Sciences of the United States of America* 108: 458–62.

DRYFLOR, Banda-R, K., Delgado-Salinas, A., Dexter, K. G., Linares-Palomino, R., Oliveira-Filho, A., Prado, D., Pullan, M., Quintana, C., Riina, R., Rodríguez, M., G.M., Weintritt, J., Acevedo-Rodríguez, P., Adarve, J., Álvarez, E., Aranguren B. A., Arteaga, J. C., Aymard, G., Castaño, A., Ceballos-Mago, N., Cogollo, A., Cuadros, H., Delgado, F., Devia, W., Dueñas, H., Fajardo, L., Fernández, A., Fernández, M. A., Franklin, J., Freid, E. H., Galetti, L. A., Gonto, R., González-M, R., Graveson, R., Helmer, E. H., Idárraga, A., López, R., Marcano-Vega, H., Martínez, O. G., Maturo, H. M., McDonald, M., McLaren, K., Melo, O., Mijares, F., Mohni, V., Molina, D., del Pilar Moreno, N., Nassar, J. M., Neves, D. M., Oakley, L. J., Oatham, M., Olvera-Luna, A. R., Pezzini, F. F., Dominguez, O. J. R., Rios, M. E., Rivera, O., Rodríguez, N., Rojas, A., Sarkinen, T., Sánchez, R., Smith, M., Vargas, C., Villanueva, B., Pennington, R. T. 2016. Plant diversity patterns in neotropical dry forests and their conservation implications. *Science* 353: 1383–7.

Duarte-Quiroga, A., Estrada, A. 2003. Primates as pets in Mexico City: An assessment of the species involved, source of origin, and general aspects of treatment. *American Journal of Primatology* 61: 53–60.

Dufour, D. L. 1987. Insects as food: A case study from the northwest Amazon. *American Anthropologist* 89: 383–97.

Dufour, D. L. 1990. Use of tropical rainforests by Native Amazonians. *BioScience* 40: 652–9.

Duin, R. S., Toinaike, K., Alupki, T., Opoya, A. 2013. Archaeology of engagement: Indigenous people, social memory, and making history in the Upper Maroni Basin (Northern Amazonia). *Current Anthropology* 56: 753–61.

Dunning, N. P., Beach, T. P., Luzzadder-Beach, S. 2012. Kax and kol: Collapse and resilience in lowland Maya civilization. *Proceedings of the National Academy of Sciences of the United States of America* 109: 3652–7.

Dupont, L., Jahns, S., Marret, F., Ning, S. 2000. Vegetation change in equatorial West Africa: Time slices for the last 150 ka. *Palaeogeography, Palaeoclimatology, Palaeocology* 155: 95–122.

Dutta, P. C. 1978. *The Great Andamanese: Past and Present.* Calcutta: Anthropological Survey of India.

Dwyer, P. D. 1983. Etolo hunting performance and energetics. *Human Ecology* 11: 145–74.

Dwyer, P. D., Minnegal, M. 1991. Hunting in lowland tropical rainforest: Towards a model of nonagricultural subsistence. *Human Ecology* 19: 187–212.

Ebeling, J., Yasué, M. 2009. The effectiveness of market-based conservation in the tropics: Forest certification in Ecuador and Bolivia. *Journal of Environmental Management* 90: 1145–53.

Edwards, S. N., McLaughlin, W. J., Ham, S. H. 1998. *Comparative Study of Ecotourism Policy in the Americas: Volume II—Latin America and the Caribbean*. Idaho: University of Idaho.

Ehleringer, J. R. and Osmond, C. B. 1991. Stable isotopes. In R. W. Pearcy, J. R. Ehleringer, H. A. Mooney, P. W. Rundel (eds.), *Plant Physiological Ecology*. London: Chapman and Hall, pp. 281–300.

Ellen, R. F. 1988. Foraging, starch extraction and the sedentary lifestyle in the lowland rainforest of central Seram. In J. Woodburn, T. Ingold, D. Riches (eds.), *History, Evolution and Social Change in Hunting and Gathering Societies*. London: Berg, pp. 117–34.

Ellingson, T. 2001. *The Myth of the Noble Savage*. Berkeley: University of California Press.

Ellis, E., Maslin, M., Boivin, N. L., Bauer, A. 2016. Involve social scientists in defining the Anthropocene. *Nature* 540: 192–3.

Elton, S. 2006. Forty years on and still going strong: The use of hominin-cercopithecid comparison in palaeoanthropology. *Journal of the Royal Anthropological Institute* 12: 19–38.

Elton, S. 2007. Environmental correlates of the cercopithecoid radiations. *Folia Primatologica* 74: 344–64.

Elton, S. 2008. The environmental context of human evolutionary history in Eurasia and Africa. *Journal of Anatomy* 212: 377–93.

Emmons, L. H. 1984. Geographic variation in densities and diversities of non-flying mammals in Amazonia. *Biotropica* 16: 210–22.

Endicott, K. 1999. The Batek of Peninsula Malaysia. In R. B. Lee, R. Daly (eds.), *The Cambridge Encyclopedia of Hunters and Gatherers*. Cambridge: Cambridge University Press, pp. 298–306.

Endicott, K. and Bellwood, P. 1991. The possibility of independent foraging in the rain forest of peninsular Malaysia. *Human Ecology* 19: 151–85.

Endicott, P., Thomas, M., Gilbert, P., Stringer, C., Lalueza-Fox, C., Willerslev, E., Hansen, A. J., Cooper, A. 2003. The genetic origins of the Andaman Islanders. *American Journal of Human Genetics* 72: 178–84.

Endicott, P., Ho, S. Y. W., Metspalu, M., Stringer, C. 2009. Evaluating the mitochondrial timescale of human evolution. *Trends in Ecology & Evolution* 24: 515–21.

Enzel, Y., Ely, L. L., Mishra, S., Ramesh, R., Amit, R., Lazar, B., Rajaguru, S. N., Baker, V. R., Sandler, A. 1999. High-resolution Holocene environmental changes in the Thar Desert, northwestern India. *Science* 284: 125–8.

Erickson, D. L., Smith, B. D., Clarke, A. C., Sandweiss, D. H., Tuross, N. 2005. An Asian origin for a 10,000-year-old domesticated plant in the Americas. *Proceedings of the National Academy of Sciences of the United States of America* 102: 18315–20.

Eriksson, J., Larson, G., Gunnarsson, U., Bed'hom, B., Tixier-Boichard, M., Strömstedt, L., Wright, D., Jungerius, A., Vereijken, A., Randi, E., Jensen, P., Andersson, L. 2008. Identification of the yellow skin gene reveals a hybrid origin of the domestic chicken. *PLoS Genetics* 4: e1000010.

Erlandson, J. M., Graham, M. H., Bourque, B. J., Corbett, D., Estes, J. A., Steneck, R. S. 2007. The kelp highway hypothesis: Marine ecology, the coastal migration theory, and the peopling of the Americas. *Journal of Island and Coastal Archaeology* 2: 161–74.

Estrada-Belli, F. 2011. *The First Maya Civilization: Ritual and Power Before the Classic Period*. London: Routledge.

Evans, D. 2016. Airborne laser scanning as a method for exploring long-term socio-ecological dynamics in Cambodia. *Journal of Archaeological Science* 74: 164–75.

Evans, D., Pottier, C., Fletcher, R., Hensley, S., Tapley, I., Milne, A., Barbetti, M. 2007. A comprehensive archaeological map of the world's largest preindustrial settlement complex at Angkor, Cambodia. *Proceedings of the National Academy of Sciences of the United States of America* 104: 14277–82.

Evans, D.., Fletcher, R. J., Pottier, C., Chevance, J.-B., Soutif, D., Tan, B. S., Im, S., Ea, D., Tin, T., Kim, S., Cromarty, C., De Greef, S., Hanus, K., Bâty, P., Kuszinger, R., Shimoda, I., Boornazian, G. 2013. Uncovering archaeological landscapes at Angkor using lidar. *Proceedings of the National Academy of Sciences of the United States of America* 110: 12595–600.

Fa, J. E., Brown, D. 2009. Impacts of hunting on mammals in African tropical moist forests: A review and synthesis. *Mammal Review* 39: 231–64.

Fagan, B. 2009. *The Great Warming: Climate Change and the Rise and Fall of Civilizations*. London: Bloomsbury.

Fage, J. D. 1980. Slaves and society in western Africa, *c.* 1445–*c.*1700. *Journal of African History* 21: 289–310.

Fairbairn, A. S., Hope, G. S., Summerhayes, G. R. 2006. Pleistocene occupation of New Guinea's highland and subalpine environments. *World Archaeology* 38: 371–86.

Falguères, C., Bahain, J.-J., Yokoyama, Y., Arsuaga, J. L., Bermudez de Castro, J. M., Carbonell, E., Bischoff, J. L., Dolo, J.-M. 1999. Earliest humans in Europe: The age of TD6 Gran Dolina, Atapuerca, Spain. *Journal of Human Evolution* 37: 343–52.

Fall, P. L. 2010. Pollen evidence for plant introductions in a Polynesian tropical island ecosystem, Kingdom of Tonga. In S. G. Haberle, J. Stevenson, M. Prebble (eds.), *Altered Ecologies: Fire, Climate and Human Influence on Terrestrial Landscapes*. Canberra: Australian National University Press, pp. 253–71.

FAO. 2018. Voluntary REDD+ Database (VRD). <http://www.fao.org/forestry/vrd/en/>. Retrieved 28 March 2018.

Farahani, A., Chiou, K. L., Harkey, A., Hastorf, C. A., Lentz, D. L., Sheets, P. 2017. Identifying 'plantscapes' at the Classic Maya village of Joya de Cerén, El Salvador. *Antiquity* 91: 980–97.

Farquhar, G. D., Ehleringer, J. R., Hubick, K. T. 1989. Carbon isotope discrimination and photosynthesis. *Annual Review of Plant Physiology and Plant Molecular Biology* 40: 503–37.

Fash, B. 2005. Iconographic evidence for water management and social organization at Copán. In E. W. Andres, W. L. Fash (eds.), *Copán: The History of an Ancient Maya Kingdom*. Santa Fe: School of American Research, pp. 103–38.

Faust, C. L., McCallum, H. I., Bloomfield, L. S. P., Gottdenker, N. L., Gillespie, T. R., Torney, C. J., Dobson, A. P., Plowright, R. K. 2018. Pathogen spillover during land conversion. *Ecology Letters* 21: 471–83.

Fausto, C., Heckenberger, M. 2007. *Time and Memory in Indigenous Amazonia: Anthropological Perspectives*. Gainesville: University Press of Florida.

Favreau, J. 2016. *The Jungle Book*. California: Walt Disney Pictures, Fairview Entertainment.

Feakins, S. J., deMenocal, P. B., Eglinton, T. I. 2005. Biomarker records of late Neogene changes in northeast African vegetation. *Geology* 33: 977–80.

Feakins, S. J., Levin, N. E., Liddy, H. M., Sieracki, A., Eglinton, T. I., Bonnefille, R. 2013. Northeast African vegetation change over 12 m.y. *Geology* 41: 295–8.

Feakins, S. J., Bentley, L. P., Salinas, N., Shenkin, A., Blonder, B., Goldsmith, G. R., Ponton, C., Arvin, L. J., Sin Wu, M., Peters, T., West, A. J., Martin, R. E., Enquist, B. J., Asner, G. P., Malhi, Y. 2016. Plant leaf wax biomarkers capture gradient in hydrogen isotopes in precipitation from the Andes and Amazon. *Geochimica et Cosmochimica Acta* 182: 155–72.

Fearnside, P. M. 2001. Soybean cultivation as a threat to the environment in Brazil. *Environmental Conservation* 28: 23–38.

Fedick, S. L. (ed.). 1996. *The Managed Mosaic: Ancient Maya Agriculture and Resource Use.* Salt Lake City: University of Utah Press.

Fedick, S. L., Ford, A. 1990. The prehistoric agricultural landscape of the central Maya lowlands: An examination of local variability in a regional context. *World Archaeology* 22: 18–33.

Feeley, K. J., Davies, S. J., Perez, R., Hubbell, S. P., Foster, R. B. 2011. Directional changes in the species composition of a tropical forest. *Ecology* 92: 871–82.

Fernando, P., Wikramanayakae, E. D., Janaka, H. K., Jayasinghe, L. K. A., Gunawardena, M., Kotagama, S. W., Weerakoon, D., Pastorini, J. 2008. Ranging behaviour of the Asian elephant in Sri Lanka. *Mammalian Biology—Zeitschrift für Sägetierkunde* 73: 2–13.

Ferraz, G., Russell, G. J., Stouffer, P. C., Bierregaard, R. O., Pimm, S. L., Lovejoy, T. E. 2003. Rates of species loss from Amazonian forest fragments. *Proceedings of the National Academy of Sciences of the United States of America* 100: 14069–73.

Ferrier, Å. 2015. *Journeys into the Rainforest: Archaeology of Culture Change and Continuity on the Evelyn Tableland, North Queensland.* Canberra: Australian National University.

Finer, M., Orta-Martínez, M. 2010. A second hydrocarbon boom threatens the Peruvian Amazon: Trends, projections, and policy implications. *Environmental Research Letters* 5: 014012.

Finlayson, C. 2014. *The Improbable Primate: How Water Shaped Human Evolution.* Oxford: Oxford University Press.

Firestone, R. B., West, A., Kennett, J. P., Becker, L., Bunch, T. E., Revay, Z. S., Schultz, P. H., Belgya, T., Kennett, D. J., Erlandson, J. M., Dickenson, O. J., Goodyear, A. C., Harris, R. S., Howard, G. A., Kloosterman, J. B., Lechler, P., Mayewski, P. A., Montgomery, J., Poreda, R., Darrah, T., Que Hee, S. S., Smith, A. R., Stich, A., Topping, W., Wittke, J. H., Wolbach, W. S. 2007. Evidence for an extraterrestrial impact 12,900 years ago that contributed to the megafaunal extinctions and the Younger Dryas cooling. *Proceedings of the National Academy of Sciences of the United States of America* 104: 16016–21.

Fischer-Kowalski, M., Krausmann, F., Pallua, I. 2014. A sociometabolic reading of the Anthropocene: Modes of subsistence, population size and human impact on Earth. *Anthropocene Review* 1: 8–33.

Fisher, C. T., Cohen, A. S., Fernández-Diaz, J. C., Leisz, S. J. 2017. The application of airborne mapping LiDAR for the documentation of ancient cities and regions in tropical regions. *Quaternary International* 448: 129–38.

Fitzpatrick, S. M. 2015. The Pre-Columbian Caribbean: Colonization, population dispersal, and island adaptations. *PaleoAmerica* 1: 305–31.

Fitzpatrick, S. M., Keegan, W. F. 2007. Human impacts and adaptations in the Caribbean islands: An historical ecology approach. *Earth and Environmental Science Transactions of the Royal Society of Edinburgh* 98: 29–45.

Fix, A. 2002. Foragers, farmers, and traders in the Malayan Peninsula: Origins of cultural and biological diversity. In K. Morrison, L. Junker (eds.), *Forager-Traders in South and Southeast Asia.* Cambridge: Cambridge University Press, pp. 185–202.

Flach, M. 1997. *Sago Palm Metroxylon sagu Rottb. Promoting the Conservation and Use of Underutilized and Neglected Crops.* 13. Rome: International Plant Genetic Resources Institute.

Flannery, T., White, J. P. 1991. Animal translocations: Zoogeography of New Ireland mammals. *National Geographic Research and Exploration* 7: 96–113.

Fleagle, J. G. 1983. Locomotor adaptations of Oligocene and Miocene hominoids and their phyletic implications. In R. L. Ciochon, R. S. Corruccini (eds.), *New Interpretations of Ape and Human Ancestry.* New York: Plenum, pp. 301–24.

Flenley, J., Bush, M. 2007. *Tropical Rainforest Responses to Climatic Change.* Dordrecht: Springer.

Fletcher, R. 1991. Very large mobile communities: Interaction stress and residential dispersal. In C. S. Gamble, W. A. Boismer (eds.), *Ethnoarchaeological Approaches to Mobile Campsites: Hunter-Gatherer and Pastoralist Case-Studies.* Ann Arbor: International Monographs in Prehistory. Ethnoarchaeological Series 1, pp. 395–420.

Fletcher, R. 1995. *The Limits of Settlement Growth: A Theoretical Outline.* Cambridge: Cambridge University Press.

Fletcher, R. 2004. Materiality, space, time and outcome. In J. L. Bintliff (ed.), *A Companion to Archaeology.* Oxford: Blackwell, pp. 110–40.

Fletcher, R. 2009. Low-density, agrarian-based urbanism: A comparative view. *Insights* (Institute of Advanced Study, Durham University) 2: 1–19.

Fletcher, R. 2010. Urban materialities: Meaning, magnitude, friction and outcomes. In D. Hicks, M. C. Beaudry (eds.), *The Oxford Handbook of Material Culture Studies.* Oxford: Oxford University Press, pp. 459–83.

Fletcher, R. 2012. Low-density, agrarian-based urbanism: Scale, power and ecology. In M. E. Smith (ed.), *The Comparative Archaeology of Complex Societies.* Cambridge: Cambridge University Press, pp. 285–320.

Fletcher, R., Penny, D., Evans, D., Pottier, C., Barbetti, M., Kummu, M., Lustig, T. and Authority for the Protection and Management of Angkor and the Region of Siem Reap (APSARA) Department of Monuments and Archaeology Team. 2008a. The water management network of Angkor, Cambodia. *Antiquity* 82: 658–70.

Fletcher, R., Pottier, C., Evans, D., Kummu, M. 2008b. The development of the water management system of Angkor: A provisional model. *Bulletin of the Indo-Pacific Prehistory Association* 28: 57–66.

Foley, J. A., Asner, G. P., Heil Costa, M., Coe, M. T., DeFries, R., Gibbs, H. K., Howard, E. A., Olson, S., Patz, J., Ramankutty, N., Snyder, P. 2007. Amazonia revealed: Forest degradation and loss of ecosystem goods and services in the Amazon Basin. *Frontiers in Ecology and the Environment* 5: 25–32.

Foley, R. A., Lewin, R. 2004. *Principles of Human Evolution* (2nd edition). Oxford: Wiley-Blackwell.

Ford, A. 1986. *Population Growth and Social Complexity: An Examination of Settlement and Environment in the Central Maya Lowlands.* Tempe: Arizona State University.

Ford, A., Nigh, R. 2015. *The Maya Forest Garden: Eight Millennia of Sustainable Cultivation of the Tropical Woodlands.* London: Routledge.

Forest500. 2018. Forest Stewardship Council. Global Canopy: Forest 500.org. <https://forest500.org/rankings/other-powerbrokers/forest-stewardship-council>. Retrieved 25 March 2018.

Fortier, J. 2009. *The Kings of the Forest: The Cultural Resilience of Himalayan Hunter-Gatherers.* Honolulu: University of Hawai'i Press.

Foster, L. 2002. *Handbook to Life in the Ancient Maya World.* New York: Oxford University Press.

Frankland, S. 1999. Turnbull's syndrome: Romantic fascination in the rainforest. In K. Biesbrouck, S. Elders, G. Rossel (eds.), *Central African Hunter-Gatherers in a Multidisciplinary Perspective: Challenging Elusiveness.* Leiden: CNWS, pp. 59–73.

Franklin, L. C. 2013. Corn. In A. F. Smith (ed.), *The Oxford Encyclopedia of Food and Drink in America* (2nd edition). Oxford: Oxford University Press, pp. 551–8.

Fraser, J. A., Frausin, V., Jarvis, A. 2015. An intergenerational transmission of sustainability? Ancestral habitus and food production in a traditional agro-ecosystem of the Upper Guinea Forest, West Africa. *Global Environmental Change* 31: 226–38.

Freeland, T., Heung, B., Burley, D. V., Clark, G., Knudby, A. 2016. Automated feature extraction for prospection and analysis of monumental earthworks from aerial LiDAR in the Kingdom of Tonga. *Journal of Archaeological Science* 69: 64–74.

Freire, G. N. 2007. Indigenous shifting cultivation and the New Amazonia: A Piaroa example of economic articulation. *Human Ecology* 35: 681–96.

Frere, S. S. 1987. *Britannia: A History of Roman Britain* (3rd revised edition). London: Routledge & Kegan Paul.

Friesem, D. E., Lavi, N. 2017. Foragers, tropical forests and the formation of archaeological evidences: An ethnoarchaeological view from South India. *Quaternary International* 448: 117–28.

Friesem, D. E., Lavi, N., Madella, M., Boaretto, E., Ajithparsad, P., French, C. 2017. The formation of fire residues associated with hunter-gatherers in humid tropical environments: A geo-ethnoarchaeological perspective. *Quaternary Science Reviews* 171: 85–99.

Fryd, V. G. 1995. Rereading the Indian in Benjamin West's 'Death of General Wolfe'. *American Art* 9: 72–85.

Fu, Q., Li, H., Moorjani, P., Jay, F., Slepchenko, S. M., Bondarev, A. A., Johnson, P. L. F., Aximu-Petri, A., Prüfer, K., de Filippo, C., Meyer, M., Zwyns, N., Salazar-García, D. C., Kuzmin, Y. V., Keates, S. G., Kosintsev, P. A., Razhev, D. I., Richards, M. P., Peristov, N. V., Lachmann, M., Douka, K., Higham, T. F. G., Slatkin, M., Hublin, J.-J., Reich, D., Kelso, J., Viola, T. B., Pääbo, S. 2014. Genome sequence of a 45,000-year-old modern human from western Siberia. *Nature* 514: 445–9.

Fullagar, R., Field, J., Denham, T. P., Lentfer, C. 2006. Early and mid Holocene tool-use and processing of taro (*Colocasia esculenta*), yam (*Dioscorea* sp.) and other plants at Kuk Swamp in the highlands of Papua New Guinea. *Journal of Archaeological Science* 33: 595–614.

Fuller, D. Q. 2002. Fifty years of archaeobotanical studies in India: Laying a solid foundation. In S. Settar, R. Korisettar (eds.), *Indian Archaeology in Retrospect. Archaeology and Interactive Disciplines*, vol. III. Delhi: Manohar, pp. 247–363.

Fuller, D. Q. 2003. An agricultural perspective on Dravidian historical linguistics: Archaeological crop packages, livestock and Dravidian crop vocabulary. In P. Bellwood, C. Renfrew (eds.), *Examining the Farming/Language Dispersal Hypothesis*. Cambridge: McDonald Institute for Archaeological Research, pp. 191–213.

Fuller, D. Q. 2006a. Agricultural origins and frontiers in South Asia: A working synthesis. *Journal of World Prehistory* 20: 1–86.

Fuller, D. Q. 2006b. Silence before sedentism and the advent of cash-crops: A status report on early agriculture in South Asia from plant domestication to the development of political economies (with an excursus on the problem of semantic shift among millets and rice). In T. Osada (ed.), *Proceedings of the Pre-Symposium of RIHN and 7th ESCA Harvard-Kyoto Roundtable*. Kyoto: Research Institute for Humanity and Nature, pp. 175–213.

Fuller, D. Q. 2007a. Contrasting patterns in crop domestication and domestication rates: Recent archaeobotanical insights from the Old World. *Annals of Botany* 100: 903–24.

Fuller, D. Q. 2007b. Non-human genetics, agricultural origins and historical linguistics in South Asia. In M. D. Petraglia, B. Allchin (eds.), *The Evolution and History of Human Populations in South Asia*. Dordrecht: Springer, pp. 393–443.

Fuller, D. Q. 2011. Finding plant domestication in the Indian subcontinent. *Current Anthropology* 52: S347–S362.

Fuller, D. Q., Hildebrand, E. 2013. Domesticating plants in Africa. In P. Mitchell, P. Lane (eds.), *The Oxford Handbook of African Archaeology*. Oxford: Oxford University Press, pp. 507–25.

Fuller, D., Lucas, L. 2017. Adapting crops, landscapes, and food choices: Patterns in the dispersal of domesticated plants across Eurasia. In N. Boivin, R. Crassard, M. Petraglia (eds.), *Human Dispersal and Species Movement from Prehistory to the Present*. Cambridge: Cambridge University Press, pp. 304–31.

Fuller, D. Q., Madella, M. 2009. Banana cultivation in South Asia and East Asia: A review of the evidence from archaeology and linguistics. *Ethnobotany Research & Applications* 7: 333–51.

Fuller, D. Q., Allaby, R. G., Stevens, C. 2010. Domestication as innovation: The entanglement of techniques, technology and chance in the domestication of cereal crops. *World Archaeology* 42: 13–28.

Fuller, D. Q., van Etten, J., Manning, K., Castillo, C., Kingwell-Banham, E., Weisskopf, A., Qin, L., Satao, Y.-I., Himans, R. J. 2011a. The contribution of rice agriculture and livestock pastoralism to prehistoric methane levels. *The Holocene* 21: 743–59.

Fuller, D. Q., Willcox, G., Allaby, R. G. 2011b. Cultivation and domestication had multiple origins: Arguments against the core area hypothesis for the origins of agriculture in the Near East. *World Archaeology* 43: 628–52.

Fuller, D. Q., Boivin, N., Hoogervorst, T., Allaby, R. 2011c. Across the Indian Ocean: The prehistoric movement of plants and animals. *Antiquity* 85: 544–58.

Fuller, D. Q., Denham, T., Arroyo-Kalin, M., Lucas, L., Stevens, C. J., Qin, L., Allaby, R. G., Purugganan, M. D. 2014. Convergent evolution and parallelism in plant domestication revealed by an expanding archaeological record. *Proceedings of the National Academy of Sciences of the United States of America* 111: 6147–52.

Fumihito, A., Miyake, T., Sumi, S., Takada, M., Ohno, S., Kondo. N. 1994. One subspecies of the red junglefowl (Gallus gallus gallus) suffices as the matriarchic ancestor of all domestic breeds. *Proceedings of the National Academy of Sciences of the United States of America* 91: 12505–9.

Fuss, J., Spassov, N., Begun, D. R., Böhme, M. 2017. Potential hominin affinities of *Graecopithecus* from the Late Miocene of Europe. *PLoS ONE* 12: e0177127.

Gabunia, L., Vekua, A., Lordkipanidze, D., Swisher III, C. C., Ferring, R., Justus, A., Nioradze, M., Tvalchrelidze, M., Antón, S. C., Bosinski, G., Jöris, O., de Limley, M.-A., Majsuradze, G., Mouskhelishvili, A. 2000. Earliest Pleistocene hominid cranial remains from Dmanisi, Republic of Georgia: Taxonomy, geological setting, and age. *Science* 288: 1019–25.

Gadgil, M., Berkes, F., Folke, C. 1993. Indigenous knowledge for biodiversity conservation. *AMBIO* 22: 151–6.

Gaffney, D., Ford, A., Summerhayes, G. 2015. Crossing the Pleistocene–Holocene transition in the New Guinea Highlands: Evidence from the lithic assemblage of Kiowa rockshelter. *Journal of Anthropological Archaeology* 39: 223–46.

Gallagher, D. 2016. American plants in sub-Saharan Africa: A review of the archaeological evidence. *Azania: Archaeological Research in Africa* 51: 24–61.

Galluzzi, G., Dufour, D., Thomas, E., van Zonneveled, M., Salamanca, A. F. E., Toro, A., Rivera, A., Duque, H. S., Baron, H. S., Gallego, G., Scheldeman, X., Mejia, A. G. 2015. An integrated hypothesis on the domestication of *Bactris gasipaes*. *PLoS ONE* 10: e0144644.

Gamble, C. 1993. *Timewalkers: The Prehistory of Global Colonization*. Stroud: Alan Sutton.

Gamble, C. 1997. Handaxes and Palaeolithic individuals. In N. Ashton, F. Healey, P. Pettitt (eds.), *Stone Age Archaeology*. Oxford: Oxbow Books, Monograph 102, pp. 105–9.

Gamble, C. 2013. *Settling the Earth: The Archaeology of Deep Human History*. Cambridge: Cambridge University Press.

Gangwar, A. K., Ramakrishnan, P. S. 1990. Ethnobiological notes on some tribes of Arunachal Pradesh, Northeastern India. *Economic Botany* 44: 94–105.

Garcia, T., Féraud, G., Falguères, C., de Lumley, H., Perrenoud, C., Lordkipanidze, D. 2010. Earliest human remains in Eurasia: New $^{40}Ar/^{39}Ar$ dating of the Dmanisi hominid-bearing levels, Georgia. *Quaternary Geochronology* 5: 443–51.

Garcin, Y., Vincens, A., Williamson, D., Buchet, G., Guiot, J. 2007. Abrupt resumption of the African Monsoon at the Younger Dryas–Holocene climatic transition. *Quaternary Science Reviews* 26: 690–704.

Garcin, Y., Deschamps, P., Menot, G., de Saulieu, G., Schefuss, E., Sebag, D., Dupont, L. M., Oslisly, R., Brademann, B., Mbusnum, K. G., Onana, J.-M., Ako, A. A., Epp, L. S., Tjallingii, R., Strecker, M. R., Brauer, A., Sachse, D. 2018. Early anthropogenic impact on Western Central African rainforests 2,600 y ago. *Proceedings of the National Academy of Sciences of the United States of America*. doi:http://doi.org/10.1073/pnas.1715336115.

Gathorne-Hardy, F. J., Syaukani, Davies, R. G., Eggleton, P., Jones, D. T. 2002. Quaternary rainforest refugia in south-east Asia: Using termites (Isoptera) as indicators. *Biological Journal of the Linnean Society* 75: 453–66.

Gaveau, D. L. A., Salim, M. A., Hergoualc'h, K., Locatelli, B., Sloan, S., Wooster, M., Marlier, M. E., Molidena, E., Yaen, H., DeFries, R., Verchot, D. L., Murdiyarso, D., Nasi, R., Holmgren, P., Sheil, D. 2014. Major atmospheric emissions from peat fires in Southeast Asia during non-drought years: Evidence from the 2013 Sumatran fires. *Scientific Reports* 4: 6112.

Gaveau, D. L., Sheil, D., Husnayaen, M. A. S., Arjasakusuma, S., Ancrenaz, M., Pacheco, P., Meijaard, E. 2016. Rapid conversions and avoided deforestation: Examining four decades of industrial plantation expansion in Borneo. *Scientific Reports* 6: 32017.

Geist, H. J., Lambin, E. F. 2002. Proximate causes and underlying driving forces of tropical deforestation. *BioScience* 52: 143–50.

Ghazoul, J. 2001. Barriers to biodiversity conservation in forest certification. *Conservation Biology* 15: 315–17.

Ghazoul, J. 2015. *Forests: A Very Short Introduction*. Oxford: Oxford University Press.

Ghazoul, J. 2016. *Dipterocarp Biology, Ecology, and Conservation*. Oxford: Oxford University Press.

Ghazoul, J., Sheil, D. 2010. *Tropical Rain Forest Ecology, Diversity, and Conservation*. Oxford: Oxford University Press.

Gifford-Gonzales, D. 2000. Animal disease challenges to the emergence of pastoralism in Sub-Saharan Africa. *African Archaeological Review* 17: 95–139.

Gifford-Gonzales, D. 2015. 'Animal disease challenges' fifteen years later: The hypothesis in light of new data. *Quaternary International* 436: 283–93.

Gill, R. 2000. *The Great Maya Droughts*. Albuquerque: University of New Mexico.

Gilliland, K., Simpson, I. A., Adderley, W. P., Burbidge, C. I., Cresswell, A. J., Sanderson, D. C. W., Coningham, R. A. E., Manuel, M., Strickland, K., Gunawardhana, P., Adikari, G. 2013. The dry tank development and disuse of water management infrastructure in the Anuradhapura hinterland, Sri Lanka. *Journal of Archaeological Science* 40: 1012–28.

Giovannucci, D., Ponte, S. 2005. The collective formulation and effectiveness of public & private sustainability standards. *Food Policy Journal* 30: 284–301.

Giovas, C. M., LeFebvre, M. J., Fitzpatrick, S. M. 2012. New records for prehistoric introduction of Neotropical mammals to the West Indies: Evidence from Carriacou, Lesser Antilles. *Journal of Biogeography* 39: 476–87.

Glaser, B., Birk, J. J. 2011. State of the scientific knowledge on properties and genesis of anthropogenic dark earths in Central Amazonia (*terra preta de Índio*). *Geochimica et Cosmochimica Acta* 82: 39–51.

Goebel, T., Waters, M. R., O'Rourke, D. H. 2008. The Late Pleistocene dispersal of modern humans in the Americas. *Science* 319: 1497–502.

Goldsmith, F. B. (ed.). 1998. *Tropical Rain Forest: A Wider Perspective*. London: Chapman and Hall.

Golson, J. 1977. No room at the top: Agricultural intensification in the New Guinea Highlands. In J. Allen, J. Golson, R. Jones (eds.), *Sunda and Sahul: Prehistoric Studies in Southeast Asia, Melanesia and Australia*. London: Academic Press, pp. 601–38.

Golson, J. 1989. The origins and development of New Guinea agriculture. In D. R. Harris, G. C. Hillman (eds.), *Foraging and Farming: The Evolution of Plant Exploitation*. London: Unwin Hyman, pp. 109–36.

Golson, J. 1991. Bulmer Phase II: Early agriculture in the New Guinea highlands. In A. Pawley (ed.), *Man and a Half: Essays in Pacific Anthropology and Ethnobiology in Honour of Ralph Bulmer*. Auckland: The Polynesian Society, pp. 484–91.

Golson, J. 1996. New Guinea: The making of a prehistory. In M. Julien, M. Orliac, C. Orliac (eds.), *Mémoire de Pierre, mémoire d'homme: tradition et archéologie en Océanie*. Paris: Publications de la Sorbonne, pp. 153–79.

Golson, J., Hughes, P. J. 1980. The appearance of plant and animal domestication in New Guinea. *Journal de la Société des Océanistes* 36: 294–303.

Golson, J., Denham, T., Hughes, P., Swadling, P., Muke, J. 2017. *Ten Thousand Years of Cultivation at Kuk Swamp in the Highlands of Papua New Guinea* (Terra Australis 46). Canberra: Australian National University Press.

Gomes, D. 2001. Santarém: Symbolism and power in the tropical forest. In C. McEwan, C. Barreto, E. G. Neves (eds.), *Unknown Amazon: Culture in Nature in Ancient Brazil*. London: British Museum Press, pp. 134–55.

Gond, V., Fayolle, A., Pennec, A., Cornu, G., Mayaux, P., Camberlin, P., Doumenge, C., Fauvet, N., Gourlet-Fleury, S. 2013. Vegetation structure and greenness in Central Africa from Modis multi-temporal data. *Philosophical Transactions of the Royal Society B: Biological Sciences* 368: 20120309.

Gongora, J., Rawlence, N. J., Mobegi, V. A., Jianlin, H., Alcalde, J. A., Matus, J. T., Hanotte, O., Moran, C., Austin, J. J., Ulm, S., Anderson, A. J., Larson, G., Cooper, A. 2008. Indo-European and Asian origins for Chilean and Pacific chickens revealed by mtDNA. *Proceedings of the National Academy of Sciences of the United States of America* 105: 10308–13.

Goodall, J. 1986. *The Chimpanzees of Gombe: Patterns of Behaviour*. Cambridge, MA: Harvard University Press.

Goody, J. 1971. *Technology, Tradition and the State in Africa*. London: Oxford University Press.

Goren-Inbar, N., Feibel, C. S., Verosub, K. L., Melamed, Y., Kislev, M. E., Tchernov, E., Saragusti, I. 2000. Pleistocene milestones on the Out-of-Africa corridor at Gesher Benot Ya'aqov, Israel. *Science* 289: 944–7.

Gosden, C. 1995. Arboriculture and agriculture in coastal Papua New Guinea. In J. Allen, J. F. O'Connell (eds.), *Transitions: Pleistocene to Holocene in Australia and Papua New Guinea. Antiquity* 69, Special Number 265: 807–17.

Gosden, C. 2010. When humans arrived in the New Guinea Highlands. *Science* 330: 41–2.

Gosden, C., Robertson, N. 1991. Models for Matenkupkum: Interpreting a late Pleistocene site from Southern New Ireland, Papua New Guinea. In J. Allen, C. Gosden (eds.), *Report of the Lapita Homeland Project*. Occasional Papers in Prehistory 20. Canberra: Department of Prehistory. Research School of Pacific Studies, The Australian National University, pp. 20–91.

Gottman, J. 1961. *Megalopolis: The Urbanized Northeastern Seaboard of the United States*. New York: Twentieth Century Fund.

Gradstein, F. M., Ogg, J. G., Smith, A. G. (eds.). 2004. *A Geologic Time Scale 2004*. Cambridge: Cambridge University Press.

Graham, E. 2011. *Maya Christians and their Churches in Sixteenth-Century Belize*. Gainesville: University Press of Florida.

Grainger, A. 1996a. The degradation of tropical rain forest in Southeast Asia: Taxonomy and appraisal. In M. E. Eden, J. T. Parry (eds.), *Land Degradation in the Tropics*. London: Mansell Publications, pp. 61–75.

Grainger, A. 1996b. Forest environments. In W. Adams, A. Goudie, A. Orme (eds.), *The Physical Geography of Africa*. Oxford: Oxford University Press. pp. 173–95.

Grann, D. 2009. *The Lost City of Z: A Tale of Deadly Obsession in the Amazon*. New York: Vintage Books.

Gray, J. 2016. *The Lost City of Z*. California: Plan B Entertainment.

Green, D. J., Alemseged, Z. 2012. *Australopithecus afarensis* scapular ontogeny, function, and the role of climbing in human evolution. *Science* 338: 514–17.

Green, R. C., Leach, H. M. 1989. New information for the Ferry Berth site, Mulifanua, Western Samoa. *Journal of the Polynesian Society* 98: 319–30.

Green, R. E. J., Krause, J., Briggs, A. W., Maricic, T., Stenzel, U., Kircher, M., Patterson, N., Li, H., Zhai, W., Fritz, M. H.-Y., Hansen, N. F., Durand, E. Y., Malaspinas, A.-S., Jensen, J. D., Marques-Bonet, T., Alkan, C., Prüfer, K., Meyer, M., Burbano, H. A., Good, J. M., Schultz, R., Aximu-Petri, A., Butthof, A., Höber, B., Höffner, B., Siegemund, M., Weihmann, A., Nusbaum, C., Lander, E. S., Russ, C., Novod, N., Affourtit, J., Egholm, M., Verna, C., Rudan, P., Brajkovic, D., Kucan, Z., Gušic, I., Doronichev, V. B., Golovanova, L. V., Lalueza-Fox, C., de la Rasilla, M., Fortea, J., Rosas, A., Schmitz, R. W., Johnson, P. L. F., Eichler, E. E., Falush, D., Birney, E., Mullikin, J. C., Slatkin, M., Nielsen, R., Kelso, J., Lachmann, M., Reich, D., Pääbo, S. 2010. A draft sequence of the Neandertal genome. *Nature* 328: 710–22.

Greenway, D. R. 1987. Vegetation and slope stability. In M. G. Anderson, K. S. Richards (eds.), *Slope Stability: Geotechnical Engineering and Geomorphology*. New York: Wiley, pp. 187–230.

Gregory, L. C., Meert, J. G., Bingen, B., Pandit, M. K., Torsvik, T. H. 2009. Paleomagnetism and geochronology of the Malani Igneous Suite, northwest India: Implications for the configuration of Rodinia and assembly of Gondwana. *Precambrian Research* 170: 13–26.

Grieg-Gran, M., Porras, L. T., Wunder, S. 2005. How can market mechanisms for forest environmental services help the poor? Preliminary lessons from Latin America. *World Development* 33: 1511–27.

Griffin, P. B., Headland, T. N. 1985. A selected bibliography of the Agta Negritos of eastern Luzon, Philippines. In P. B. Griffin, A. Estioko-Griffin (eds.), *The Agta of Northeastern Luzon: 25 Recent Studies*. Cebu City: San Carlos Publications, pp. 166–75.

Grobman, A., Bonavia, D., Dillehay, T. D., Piperno, D. R., Iriarte, J., Holst, I. 2012. Preceramic maize from Paredones and Huaca Prieta, Peru. *Proceedings of the National Academy of Sciences of the United States of America* 109: 1755–9.

Grogan, J., Blundell, A. G., Landis, R. M., Youatt, A., Gullison, R. E., Martinez, M., Kómetter, R., Lentini, M., Rice, R. E. 2010. Over-harvesting driven by consumer demand leads to population decline: Big-leaf mahogany in South America. *Conservation Letters* 3: 12–20.

Grollemund, R., Branford, S., Bosten, K., Meade, A., Venditti, C., Pagel, M. 2015. Bantu expansion shows that habitat alters the route and pace of human dispersals. *Proceedings of the National Academy of Sciences of the United States of America* 112: 13296–301.

Groube, L. 1989. The taming of the rain forests: A model for Late Pleistocene forest exploitation in New Guinea. In D. R. Harris, G. C. H. Hillman (eds.), *Foraging and Farming: The Evolution of Plant Exploitation*. London: Unwin Hyman, pp. 292–304.

Groucutt, H. S., Petraglia, M. D. 2012. The prehistory of the Arabian Peninsula: Deserts, dispersals, and demography. *Evolutionary Anthropology* 21: 113–25.

Groucutt, H. S., Petraglia, M. D., Bailey, G., Scerri, E. M. L., Parton, A., Clark-Balzan, L., Jennings, R. P., Lewis, L., Blinkhorn, J., Drake, N. A., Breeze, P. S., Inglis, R. H., Devès, M. H., Meredith-Williams, M., Boivin, N., Thomas, M. H., Scally, A. 2015. Rethinking the dispersal of *Homo sapiens* out of Africa. *Evolutionary Anthropology* 24: 149–64.

Groucutt, H. S., Grün, R., Zalmout, I. A. S., Drake, N. A., Armitage, S. J., Candy, I., Clark-Wilson, R., Louys, J., Breeze, P. S., Duval, M., Buck, L. T., Kivell, T. L., Pomeroy, E., Stephens, N. B., Stock, J. T., Stewart, M., Price, G. J., Kinsley, L., Wai Sung, W., Alsharekh, A., Al-Omari, A., Zahir, M., Memesh, A. M., Abdulshakoor, A. J., Al-Masari, A. M., Bahameem, A. A., Al Murayyi, K. M. S., Zahrani, B., Scerri, E. L. M., Petraglia, M. D. 2018. *Homo sapiens* in Arabia by 85,000 years ago. *Nature Ecology and Evolution*. doi:10.1038/s41559-018-0518-2.

Grove, M. 2011. Change and variability in Plio-Pleistocene climates: Modelling the hominin response. *Journal of Archaeological Science* 38: 3038–47.

Grove, M. 2015. Palaeoclimates, plasticity, and the early dispersal of *Homo sapiens*. *Quaternary International* 369: 17–37.

Grove, M., Lamb, H., Roberts, H., Davies, S., Marshall, M., Bates, R., Huws, D. 2015. Climatic variability, plasticity, and dispersal: A case study from Lake Tana, Ethiopia. *Journal of Human Evolution* 87: 32–47.

Grube, N. 2006. Maya today: From Indios deprived of rights to the Maya movement. In N. Grube, E. Eggebrecht, M. Seidel (eds.), *Maya: Divine Kings of the Rain Forest*. Cologne: Könemann Press, pp. 417–25.

Grün, R., Stringer, C. B. 2000. Tabun revisited: Revised ESR chronology and new ESR and U-series analyses of dental material from Tabun C1. *Journal of Human Evolution* 39: 601–12.

Grün, R., Brink, J. S., Spooner, N. A., Taylor, L., Stringer, C. B., Franciscus, R. G., Murray, A. S. 1996. Direct dating of Florisbad hominid. *Nature* 382: 500–1.

Guha, R. 1999. *Environment and Ethnicity in India: 1200–1991*. Cambridge: Cambridge University Press.

Guidon, N., Arnaud, B. 1991. The chronology of the New World: Two faces of one reality. *World Archaeology* 23:167–78.

Guillén, A. K. Z., Barrett, G. M., Takenaka, O. 2005. Genetic diversity among African great apes based on mitochondrial DNA sequences. *Biodiversity Conservation* 14: 2221–33.

Gullison, R. E. 2003. Does forest certification conserve biodiversity? *Oryx* 37: 153–65.

Guns, M., Vanacker, V. 2013. Forest cover change trajectories and their impact on landslide occurrence in the tropical Andes. *Environmental Earth Sciences* 70: 2941–52.

Guy, F., Lieberman, D. E., Pilbeam, D., Ponce de León, M., Likius, A., Mackaye, H. T., Vignaud, P., Zollikofer, C., Brunet, M. 2005. Morphological affinities of the Sahelanthropus tchadensis (Late Miocene hominid from Chad) cranium. *Proceedings of the National Academy of Sciences of the United States of America* 102: 18836–41.

Haberle, S. G. 1994. Anthropogenic indicators in pollen diagrams: Problems and prospects for late Quaternary palynology in New Guinea. In J. G. Hather (ed.), *Tropical Archaeobotany: Applications and New Developments*. London: Routledge, pp. 172–201.

Haberle, S. G. 1995. Identification of cultivated Pandanus and Colocasia in pollen records and the implications for the study of early agriculture in New Guinea. *Vegetation History and Archaeobotany* 4: 195–210.

Haberle, S. G. 1998. Late Quaternary vegetation change in the Tari Basin, Papua New Guinea. *Palaeogeography, Palaeoclimatology, Palaeoecology* 137: 1–24.

Haberle, S. G. 2003. The emergence of an agricultural landscape in the highlands of New Guinea. *Archaeology in Oceania* 38: 149–58.

Haberle, S. G. 2005. A 23,000-yr pollen record from Lake Euramoo, Wet Tropics of NE Queensland, Australia. *Quaternary Research* 64: 343–56.

Haberle, S. G., Lusty, C. A. 2000. Can climate influence cultural development? A view through time. *Environment and History* 6: 349–69.

Haberle, S. G., Hope, G. S., DeFretes, Y. 1991. Environmental change in the Baliem Valley, montane Irian Jaya, Republic of Indonesia. *Journal of Biogeography* 18: 25–40.

Haberle, S. G., Hope, G. S., van der Kaars, W. A. 2001. Biomass burning in Indonesia and Papua New Guinea: Natural and human induced fire events in the fossil record. *Palaeogeography, Palaeoclimatology, Palaeoecology* 171: 259–68.

Haberle, S. G., Rule, S., Roberts, P., Heijnis, H., Jacobsen, G., Turney, C., Cosgrove, R., Ferrier, A., Moss, P., Mooney, S., Kershaw, P. 2010. Paleofire in the wet tropics of northeast Queensland, Australia. *PAGES News* 18: 78–80.

Haberle, S. G., Lentfer, C., O'Donnell, S., Denham, T. 2012. The palaeoenvironments of Kuk Swamp from the beginnings of agriculture in the highlands of Papua New Guinea. *Quaternary International* 249: 129–39.

Haberle, S. G., Lentfer, C., Denham, T. 2017. Palaeoecology. In J. Golson, T. Denham, P. Hughes, P. Swadling, J. Muke (eds.), *Ten Thousand Years of Cultivation at Kuk Swamp in the Highlands of Papua New Guinea* (Terra Australis 46). Canberra: Australian National University Press, pp. 145–62.

Haffer, J. 1969. Speciation in Amazonian forest birds. *Science* 165: 131–7.

Haile-Selassie, Y. 2001. Late Miocene hominids from the Middle Awash, Ethiopia. *Nature* 412: 178–81.

Haile-Selassie, Y., Suwa, G., White, T. D. 2004. Late Miocene teeth from Middle Awash, Ethiopia, and early hominid dental evolution. *Science* 303: 1503–5.

Hall, K. R. 1985. *Maritime Trade and State Development in Early Southeast Asia*. Honolulu: University of Hawai'i Press.

Hamerow, H. 2007. Agrarian production and the emporia of mid Saxon England, ca. AD 650–850. In J. Henning (ed.), *Post-Roman Towns, Trade and Settlement in Europe and Byzantium: The Heirs of the Roman West*. Vol. 1. Berlin: Walter de Gruyter, pp. 219–32.

Hames, R. 2007. The ecologically noble savage debate. *Annual Review of Anthropology* 36: 177–90.

Hamilton, A. C. 1972. The interpretation of pollen diagrams from highland Uganda. *Palaeoecology of Africa* 7: 45–9.

Hamilton, A. C., Karamura, D., Kakudidi, E. 2016. History and conservation of wild and cultivated plant diversity in Uganda: Forest species and banana varieties as case studies. *Plant Diversity* 38: 23–44.

Hamon, N., Spulchre, P., Donnadieu, Y., Henrot, A.-J., François, L., Jaeger, J.-J., Ramstein, G. 2012. Growth of subtropical forests in Miocene Europe: The roles of carbon dioxide and Antarctic ice volume. *Geology* 40: 567–70.

Hansen, M. C., Stehman, S. V., Potapov, P. V., Arunarwati, B., Stolle, F., Pittmann, K. 2009. Quantifying changes in rates of forest clearing in Indonesia from 1990 to 2005 using remotely sensed data sets. *Environmental Research Letters* 4: 034001.

Harcourt, C. S., Sayer, J. A. (eds.). 1996. *The Conservation Atlas of Tropical Forests: The Americas*. New York: Simon & Schuster.

Harcourt-Smith, W. E. H., Aiello, L. C. 2004. Fossils, feet and the evolution of human bipedal locomotion. *Journal of Anatomy* 204: 403–16.

Hare, V. J., Loftus, E., Jeffrey, A., Bronk Ramsey, C. 2018. Atmospheric CO_2 effect on stable carbon isotope composition of terrestrial fossil archives. *Nature Communications*. doi:10.1038/s41467-017-02691-x.

Harmand, S., Lewis, J. E., Feibel, C. S., Lepre, C. J., Prat, S., Lenoble, A., Boës, X., Quinn, R. L., Brenet, M., Arroyo, A., Taylor, N., Clément, S., Daver, G., Brugal, J.-P., Leakey, L., Mortlock, R. A., Wright, J. D., Lokorodi, S., Kirwa, C., Kent, D. V., Roche, H. 2015. 3.3-million-year-old stone tools from Lomekwi 3, West Turkana, Kenya. *Nature* 521: 310–15.

Harris, D. R. 1972. The origins of agriculture in the tropics. *American Scientist* 60: 180–93.

Harris, D. R. 1989. An evolutionary continuum of people–plant interaction. In D. R. Harris, G. C. Hillman (eds.), *Foraging and Farming: The Evolution of Plant Exploitation*. London: Unwin Hyman, pp. 11–26.

Harrison, M. E., Page, S. E., Limin, S. H. 2009. The global impact of Indonesian forest fires. *Biologist* 56: 156–63.

Harrison, P. D., Turner, B. L. II. 1978. *Pre-Hispanic Maya Agriculture*. Albuquerque: University of New Mexico Press.

Harrison, R. D. 2011. Emptying the forest: Hunting and the extirpation of wildlife from tropical nature reserves. *BioScience* 61: 919–24.

Harrison, T., Xueping, J., Su, D. 2002. On the systematic status of the late Neogene hominoids from Yunnan Province, China. *Journal of Human Evolution* 43: 207–27.

Hart, T. B., Hart, J. A. 1986. The ecological basis of hunter-gatherer subsistence in African rain forests. *Human Ecology* 14: 29–55.

Harvati, K., Stringer, C., Grun, R., Aubert, M., Allsworth-Jones, P., Folorunso, C. A. 2011. The Late Stone Age calvaria from Iwo Eleru, Nigeria: Morphology and chronology. *PLoS ONE* 6: e24024.

Harvey, E. L., Fuller, D. Q., Pal, J. N., Gupta, M. C. 2005. Early agriculture of Neolithic Vindyhas (North-Central India). In U. Franke-Vogt, J. Weisshaar (eds.), *South Asian Archaeology 2003. Proceedings of the European Association for South Asian Archaeology Conference, Bonn, Germany, 7th–11th July 2003*. Aachen: Linden Soft, pp. 329–34.

Hassan, F. 1988. The Predynastic of Egypt. *Journal of World Prehistory* 2: 135–86.

Hatch, M., Slack, C., Johnson, H. 1967. Further studies on a new pathway of photosynthesis carbon dioxide fixation in sugarcane and its occurrence in other species. *Biochemical Journal* 102: 417–22.

Hättenschwiler, S., Coq, S., Barantal, S., Tanya Handa, I. 2011. Leaf traits and decomposition in tropical rainforests: Revisiting some commonly held views and towards a new hypothesis. *New Phytologist* 189: 950–65.

Hawks, J. 2009. Sahelanthropus: 'The femur of Toumaï?' Johnhawks.net. <http://johnhawks.net/weblog/fossils/sahelanthropus/femur-toumai-bergeret-recherche-2009.html>. Retrieved 20 March 2018.

Head, L. 1996. Rethinking the prehistory of hunter-gatherers, fire and vegetation change in northern Australia. *The Holocene* 6: 481–7.

Headland, T. N. 1987. The wild yam question: How well could independent hunter-gatherers live in a tropical rain forest ecosystem? *Human Ecology* 15: 463–91.

Headland, T. N. 1988. Ecosystemic change in a Philippine tropical rain forest and its effect on a Negrito foraging society. *Tropical Ecology* 29: 121–35.

Headland, T. N. 1993. The Tasaday 'Cave People'. <http://www-01.sil.org/~headlandt/caveppl.htm>. Retrieved 19 March 2018.

Headland, T. N., Reid, L. A. 1989. Hunter-gatherers and their neighbors from prehistory to the present. *Current Anthropology* 30: 43–66.

Heaney, L. R. 1991. A synopsis of climatic and vegetational change in Southeast Asia. *Climatic Change* 19: 53–61.

Hecht, S. B. 1992. Logics of livestock and deforestation: The case of Amazonia. In T. Downing, S. Hecht, H. Pearson, C. Garcia Downing (eds.), *Development or Destruction: The Conversion of Tropical Forest to Pasture in Latin America*. Boulder, CO: Westview Press, pp. 7–25.

Hecht, S. B. 2003. Indigenous soil management and the creation of Amazonian dark earths: Implications of Kayapó practices. In J. Lehmann, D. C. Kern, B. Glaser, W. I. Woods (eds.), *Amazonian Dark Earths: Origin, Properties, Management*. Dordrecht: Springer, pp. 355–72.

Hecht, S. B., Norgaard, R. B., Possio, G. 1988. The economics of cattle ranching in eastern Amazonia. *Interciencia* 13: 233–40.

Heckenberger, M. J. 1996. *War and Peace in the Shadow of Empire: Sociopolitical Change in the Upper Xingu of Southeastern Amazonia, AD 1400–2000*. Pittsburgh: University of Pittsburgh Press.

Heckenberger, M. J. 2013. Tropical garden cities: Archaeology and memory in the southern Amazon. *Cadernos do CEOM—Patrimônio, Memória e Identidade* 26: 186–207.

Heckenberger, M. J., Neves, E. G. 2009. Amazonian archaeology. *Annual Review of Anthropology* 38: 251–66.

Heckenberger, M. J., Kuikuro, A., Tabata Kuikuro, U. T., Russell, J. C., Schmidt, M., Fausto, C., Franchetto, B. 2003. Amazonia 1492: Pristine forest or cultural parkland? *Science* 301: 1710–14.

Heckenberger, M. J., Russell, J. C., Toney, J. R., Schmidt, M. J. 2007. The legacy of cultural landscapes in the Brazilian Amazon: Implications for biodiversity. *Philosophical Transactions of the Royal Society B: Biological Sciences* 362: 197–208.

Heckenberger, M. J., Russell, J. C., Fausto, C., Toney, J. R., Schmidt, M. J., Pereira, E., Franchetto, B., Kuikuro, A. 2008. Pre-Columbian urbanism, anthropogenic landscapes, and the future of the Amazon. *Science* 321: 1214–17.

Hemley, R. 2003. *Invented Eden: The Elusive, Disputed History of the Tasaday.* New York: Farrar, Straus & Giroux.

Hemming, J. 1995. *Red Gold: The Conquest of the Brazilian Indians.* London: Papermac.

Hemming, J. 2009. *Tree of Rivers: The Story of the Amazon.* London: Thames & Hudson.

Hemming, J. 2017. The Lost City of Z is a very long way from a true story—and I should know. <https://www.spectator.co.uk/2017/04/the-lost-city-of-z-is-a-very-long-way-from-a-true-story-and-i-should-know/>. Retrieved 23 March 2018.

Henneberg, M., Eckhardt, R. B., Chavanaves, S., Hsü, K. J. 2014. Evolved developmental homeostasis disturbed in LB1 from Flores, Indonesia, denotes Down syndrome and not diagnostic traits of the invalid species *Homo floresiensis. Proceedings of the National Academy of Sciences of the United States of America* 111: 11967–72.

Henry, A. G., Ungar, P. S., Passey, B. H., Sponheimer, M., Rossouw, L., Bamford, M., Sandberg, P., de Ruiter, D. J., Berger, L. 2012. The diet of *Australopithecus sediba. Nature* 487: 90–3.

Henshilwood, C. S. 2008. Winds of change: Palaeoenvironments, material culture and human behaviour in the Late Pleistocene (~77 ka–48 ka ago) in the western Cape Province, South Africa. *South African Archaeological Society* 10: 35–51.

Henshilwood, C. S., d'Errico, F. E., Marean, C. W., Milo, R. G., Yates, R. 2001. An early bone tool industry from the Middle Stone Age at Blombos Cave, South Africa: Implications for the origins of modern human behaviour, symbolism and language. *Journal of Human Evolution* 41: 631–78.

Henshilwood, C. S., d'Errico, F., Yates, R., Jacobs, Z., Tribolo, C., Duller, G. A. T., Mercier, N., Sealy, J. C., Valladas, H., Watts, I., Wintle, A. G. 2002. Emergence of modern human behavior: Middle Stone Age engravings from South Africa. *Science* 295: 1278–80.

Henshilwood, C. S., d'Errico, F., van Nierkerk, K. L., Coquinot, Y., Jacobs, Z., Lauritzen, S.-E., Menu, M., Garcia-Moreno, R. 2011. A 100,000-year-old ochre-processing workshop at Blombos Cave, South Africa. *Science* 334: 219–22.

Hermenegildo, T., O'Connell, T. C., Guapindaia, V. L. C., Neves, E. G. 2017. New evidence for subsistence strategies of late pre-colonial societies of the mouth of the Amazon based on carbon and nitrogen isotopic data. *Quaternary International* 448: 139–49.

Hernández, E. X., Baltazar, E. B., Tacher, S. L. 1995. *La milpa en Yucatán, un sistema de producción agŕicola tradicional,* 2 vols. Mexico City: Colegio de Postgraduados.

Hernández, M. I. R. 2017. Paisajes productivos sostenibles en México: de la idea a la realidada. In N. Sanz (ed.), *Tropical Forest Conservation: Long-Term Processesof Human Evolution, Cultural Adaptations and Consumption Patterns.* Mexico City: UNESCO, pp. 320–35.

Herries, A. I. R., Hopley, P., Adams, J., Curnoe, D., Maslin, M. 2010. Geochronology and palaeoenvironments of the Southern African hominin-bearing localities: A reply to Wrangham et al., 2009: 'Shallow-water habitats as sources of fallback foods for hominins'. *American Journal of Physical Anthropologists* 143: 640–6.

Hershkovitz, I., Marder, O., Ayalon, A., Bar-Matthews, M., Yasur, G., Boaretto, E., Caracuta, V., Alex, B., Frumkin, A., Goder-Goldberger, M., Gunz, P., Holloway, R. L., Latimer, B., Lavi, R., Matthews, A., Slon, V., Bar-Yosef Mayer, D., Berna, F., Bar-Oz, G., Yeshurun, R., May, H., Hans, M. G., Weber, G. W., Barzilai, O. 2015. Levantine cranium from Manot Cave (Israel) foreshadows the first European modern humans. *Nature* 520: 216–19.

Hershkovitz, I., Weber, G. W., Quam, R., Duval, M., Grün, R., Kinsley, L., Ayalon, A., Bar-Matthews, M., Valladas, H., Mercier, N., Arsuaga, J. L., Martinón-Torres, M., Bermúdez de Castro, J. M., Fornai, C., Martin-Frances, L., Sarig, R., May, H., Krenn, V. A., Slon, V., Rodríguez, L., García, R., Lorenzo, C., Carretero, J. M., Frumkin, A., Shahack-Gross, R., Bar-Yosef Mayer, D. E., Cui, Y., Wu, X., Peled, N., Groman-Yaroslavski, I., Wessbrod, L., Yeshurun, R., Tsatskin, A., Zaidner, Y., Weinstein-Evron, M. 2018. The earliest modern humans outside Africa. *Science* 359: 456–9.

Hesse, H. 1984. *Bäume: Betrachtungen und Gedichte*. Berlin: Insel Verlag.

Heusser, C. J., Heusser, L. E., Lowell, T. V. 1999. Paleoecology of the Southern Chilean Lake District-Isla Grande de Chiloé during Middle-Late Llanquihue glaciation and deglaciation. *Geografiska Annaler. Series A, Physical Geography.* 81: 231–84.

Hewlett, B. S. 1996. Cultural diversity among African pygmies. In S. Kent (ed.), *Cultural Diversity among Twentieth Century Foragers: An African Perspective*. Cambridge: Cambridge University Press, pp. 215–44.

Hewlett, B. S. (ed.). 2014. *Hunter-Gatherers of the Congo Basin: Cultures, Histories, and Biology of African Pygmies*. London: Transaction Publishers.

Higham, C. 2001. *The Civilization of Angkor*. London: Weidenfeld & Nicolson.

Higham, C. 2002. *Early Cultures of Mainland Southeast Asia*. Bangkok: River Books.

Higham, T. F. G., Barton, H., Turney, C. S. M., Barker, G., Bronk Ramsey, C., Brock, F. 2009. Radiocarbon dating of charcoal from tropical sequences: Results from the Niah Great Cave, Sarawak, and their broader implications. *Journal of Quaternary Science* 24: 189–97.

Hilbert, D. W., Ostendorf, B., Hopkins, M. S. 2001. Sensitivity of tropical forests to climate change in the humid tropics of north Queensland. *Austral Ecology* 26: 590–603.

Hilbert, L., Neves, E. G., Pugliese, F., Whitney, B. S., Shock, M., Veasey, E., Zimpel, C. A., Iriarte, J. 2017. Evidence for mid-Holocene rice domestication in the Americas. *Nature Ecology and Evolution.* doi:10.1038/s41559-017-0322-4.

Hill, R., Baird, A. 2003. Kuku-Yalanji rainforest Aboriginal people and carbohydrate resource management in the Wet Tropics of Queensland, Australia. *Human Ecology* 31: 27–52.

Hill, R., Smyth, D., Shipton, H., Fischer, P. 2001. Cattle, mining or fire? The historical cause of recent contractions of open forest in the wet tropics of Queensland through invasion by rainforest. *Pacific Conservation Biology* 7: 187–94.

Hilton, A. 1985. *The Kingdom of Kongo*. Oxford: Oxford University Press.

Hobbs, R. J., Higgs, E., Harris, J. A. 2009. Novel ecosystems: Implications for conservation and restoration. *Trends in Ecology and Evolution* 24: 599–605.

Hoefs, J. 1997. *Stable Isotope Geochemistry* (4th edition). Berlin: Springer-Verlag.

Hoffmann, W. A., Orthen, B., do Nascimento, P. K. V. 2003. Comparative fire ecology of tropical savanna and forest trees. *Functional Ecology* 17: 720–6.

Holdridge, L. R. 1947. Determination of world plant formations from simple climatic data. *Science* 105: 367–8.

Holdridge, L. R. 1967. *Life Zone Ecology*. San Jose, Costa Rica: Tropical Science Center.

Holdridge, L. R., Grenke, W. C., Hatheway, W. H., Liang, T., Tosi Jr, J. 1971. *Forest Environments in Tropical Life Zones: A Pilot Study*. New York: Pergamon Press.

Holen, S. R., Deméré, T. A., Fisher, D. C., Fullagar, R., Paces, J. B., Jefferson, G. T., Beeton, J. M., Cerutti, R. A., Rountrey, A. N., Vescera, L., Holen, K. A. 2017. A 130,000-year-old archaeological site in southern California, USA. *Nature* 544: 479–83.

Hommon, R. J. 1986. Social evolution in Hawai'i. In P. Kirch (ed.), *Island Societies: Archaeological Approaches to Evolution and Transformation*. Cambridge: Cambridge University Press, pp. 55–68.

Hong, E. 1987. *Natives of Sarawak: Survival in Borneo's Vanishing Forests*. Malaysia: Institut Masyarakat.

Hooghiemstra, H., Van der Hammen, T. 1998. Neogene and Quaternary development of the neotropical rain forest: The forest refugia hypothesis, and a literature overview. *Earth-Science Reviews* 44: 147–83.

Hooijer, A., Page, S., Canadell, J. G., Silvius, M., Kwadijk, J., Wösten, H., Jauhiainen, J. 2010. Current and future CO_2 emissions from drained peatlands in Southeast Asia. *Biogeosciences* 7: 1505–14.

Hope, G. S. 2009. Environmental change and fire in the Owen Stanley Ranges, Papua New Guinea. *Quaternary Science Reviews* 28: 2261–76.

Hope, G. S., Golson, J. 1995. Late Quaternary change in the mountains of New Guinea. *Antiquity* 69: 818–30.

Hope, G. S., Kershaw, A., van der Kaars, W., Xiangjun, S., Liew, P.-M., Heusser, L. E., Takahara, H., McGlone, M., Miyoshi, N., Moss, P. T. 2004. History of vegetation and habitat change in the Austral-Asian region. *Quaternary International* 118–19: 103–26.

Houghton, R. A., Byers, B., Nassikas, A. A. 2015. A role for tropical forests in stabilizing atmospheric CO_2. *Nature Climate Change* 5: 1022–3.

Howell, F. C., Bourlière, F. 1963. *African Ecology and Human Evolution*. Chicago: Aldine.

Hublin, J.-J., Ben-Ncer, A., Bailey, S. E., Freidline, S. E., Neubauer, S., Skinner, M. M., Bergmann, I., Le Cabec, A., Benazzi, S., Harvati, K., Gunz, P. 2017. New fossils from Jebel Irhoud, Morocco and the pan-African origin of *Homo sapiens*. *Nature* 546: 289–92.

Hudson, B. 2001. The origins of Bagan: New dates and old inhabitants. *Asian Perspectives* 40: 48–74.

Hudson, B. 2004. The origins of Bagan: The archaeological landscape of Upper Burma to AD 1300. PhD thesis, University of Sydney, Sydney.

Huey, R. B., Deutsch, C. A., Tewksbury, J. J., Vitt, L. J., Hertz, P. E., Álvarez Pérez, H. J., Garland Jr, T. 2009. Why tropical forest lizards are vulnerable to climate warming. *Proceedings of the Royal Society B: Biological Sciences* 276: 1939–48.

Huffman, T. N. 2009. Mapungubwe and Great Zimbabwe: The origin and spread of social complexity in southern Africa. *Journal of Anthropological Archaeology* 28: 37–54.

Hung, H.-C., Iizuka, Y., Bellwood, P., Nguyen, K. D., Bellina, B., Silapanth, P., Dizon, E., Santiago, R., Datan, I., Manton, J. H. 2007. Ancient jades map 3,000 years of prehistoric exchange in Southeast Asia. *Proceedings of the National Academy of Sciences of the United States of America* 104: 19745–50.

Hunn, E. 1982. Mobility as a factor limiting resource use in the Columbian Plateau of North America. In N. Williams, E. Hunn (eds.), *Resource Managers: North American and Australian Foragers*. Boulder, CO: Westview Press, pp. 17–43.

Hunt, C. O., Barker, G. W. 2014. Missing links, cultural modernity and the dead: Anatomically modern humans in the great cave of Niah (Sarawk, Borneo). In R. Dennell, M. Porr (eds.), *Southern Asia, Australia, and the Search for Human Origins*. Cambridge: Cambridge University Press, pp. 90–107.

Hunt, C. O., Premathilake, P. 2012. Early Holocene vegetation, human activity and climate from Sarawak, Malaysian Borneo. *Quaternary International* 249: 105–19.

Hunt, C. O., Rabett, R. J. 2014. Holocene landscape intervention and plant food production strategies in island and mainland Southeast Asia. *Journal of Archaeological Science* 51: 22–33.

Hunt, C. O., Gilbertson, D. D., Rushworth, G. 2012. A 50,000-year record of late Pleistocene tropical vegetation and human impact in lowland Borneo. *Quaternary Science Reviews* 37: 61–80.

Hunt, C. O., Kealhoffer, L., Premathilake, R., Rushworth, G., Gilbertson, D., Jones, S., Barker, G. 2016. Palynology, phytoliths and diatoms in and adjacent to the Great Cave: Stratigraphic and taphonomic studies of Late Quaternary vegetation history. In G. Barker, L. Farr (eds.), 2016. *Archaeological Investigations in the Niah Caves, Sarawak*.

The Archaeology of the Niah Caves, Sarawak, Volume 2. Cambridge: McDonald Institute for Archaeological Research, pp. 149–76.

Hutterer, K. L. 1974. The evolution of Philippine lowland societies. *Mankind* 9: 287–99.

Hutterer, K. L. 1976. An evolutionary approach to the Southeast Asian cultural sequence. *Current Anthropology* 17: 221–42.

Hutterer, K. L. 1983. The natural and cultural history of Southeast Asian agriculture. *Anthropos* 78: 169–212.

Ichikawa, M. 1983. An examination of the hunting-dependent life of the Mbuti Pygmies, Eastern Zaire. *African Study Monographs (Kyoto)* 4: 55–76.

Ichikawa, M. 1996. The co-existence of man and nature in the Central African rainforest. In R. F. Ellen, K. Fukui (eds.), *Redefining Nature*. Oxford: Berg, pp. 467–92.

Ichikawa, M. 1999. Interactive process of man and nature in the Ituri forest of the Democratic Republic of Congo: An approach from historical ecology. In K. Biesbruck, S. Elder, G. Rossel (eds.), *Challenging Elusiveness: Central African Hunter-Gatherers in a Multidisciplinary Perspective*. Research School for Asian, African, and Amerindian Studies (CNWS). Netherlands: Universiteit Leiden, pp. 141–52.

Ichita, S., Takeshi, N., Takeshi, T. 1999. Preliminary report on the survey of Sambor Prei Kuk Monuments: A study on Khmer architecture in Pre-Angkor Period. *(I) Summaries of Technical Papers of Annual Meeting Architectural Institute of Japan. F-2. History and Theory of Architecture*: 187–8.

Ingicco, T., van den Bergh, G. D., Jago-on, C., Bahain, J.-J., Chacón, M. G., Amano, N., Forestier, H., King, C., Manalo, K., Nomade, S., Pereira, A., Reyes, M. C., Sémah, A.-M., Shao, Q., Voinchet, P., Falguères, C., Albers, P. C. H., Lising, M., Lyras, G., Yurnaldi, D., Rochette, P., Bautista, A., de Vos, J. 2018. Earliest known hominin activity in the Philippines by 709 thousand years ago. *Nature* 557: 233–7.

International Chicken Polymorphism Map Consortium. 2004. A genetic variation map for chicken with 2.8 million single-nucleotide polymorphisms. *Nature* 432: 717–22.

Iriarte, J., Denham, T., Vrydaghs, L. 2007. *Rethinking Agriculture: Archaeological and Ethnoarchaeological Perspectives*. Walnut Creek, CA: Left Coast Press.

Iriarte, J., Power, M. J., Rostain, S., Mayle, F. E., Jones, H., Watling, J., Whitney, B. S., McKey, D. B. 2012. Fire-free land use in pre-1492 Amazonian savannas. *Proceedings of the National Academy of Sciences of the United States of America* 109: 6473–8.

Isaac, G. L. 1976. Stages of cultural elaboration in the Pleistocene: Possible archaeological indicators of the development of language capabilities. In S. R. Harnad, H. D. Steklis, J. Beckman Lancaster (eds.), *Origins and Evolution of Languages and Speech*. New York: Annals of the New York Academy of Sciences 280, pp. 276–88.

Ishizawa, Y., Kono, Y. 1989. *Study on Pagan: Research Report*. California: Institute of Asian Cultures, Sophia University.

Iten, O. 1986. Die Tasaday: Ein Philippinischer Steinzeitschwindel. *Neue Zurcher Zeitung, Zurich* 12: 77–89.

ITV. 2002. *I'm a Celebrity . . . Get Me Out of Here!* LWT, ITV Studios.

Jablonski, N. G. 2004. The evolution of human skin and skin color. *Annual Review of Anthropology* 33: 585–623.

Jackson, P. C., Meinzer, F. C., Goldstein, G., Holbrook, N. M., Cavelir, J., Rada, F. 1993. Environmental and physiological influences on carbon isotope composition of gap and understory plants in a lowland tropical forest. In J. R. Ehleringer, A. E. Hall, G. D. Farquhar (eds.), *Stable Isotopes and Plant Carbon-Water Relations*. San Diego, CA: Academic Press, pp. 131–40.

Jacobs, B. 2004. Palaeobotanical studies from tropical Africa: Relevance to the evolution of forest, woodland and savannah biomes. *Philosophical Transactions of the Royal Society B: Biological Sciences* 359: 1573–83.

Jahns, S., Huls, M., Sarnthein, M. 1998. Vegetation and climate history of west equatorial Africa based on a marine pollen record off Liberia (site GIK 16776) covering the last 400,000 years. *Review of Palaeobotany and Palynology* 102: 277–88.

Janzen, D. H. 1988. Tropical dry forests: The most endangered major tropical ecosystem. In E. O. Wilson (ed.), *Biodiversity*. Washington, DC: National Academy Press, pp. 130–7.

Jaramillo, C. A., Rueda, M. J., Mora, G. 2006. Cenozoic plant diversity in the Neotropics. *Science* 311: 1893–6.

Joglekar, P. P. 2004. *Fauna from Lahuradewa: Preliminary Report.* Lucknow: National Seminar on the Archaeology of the Gange Plain, Joint Annual Conference of the Indian Archaeological Society, Indian Society of Prehistoric and Quaternary Studies, Indian History and Culture Society, December 2004.

Johanson, D. C., Maitland, A. E. 1981. *Lucy: The Beginning of Humankind.* St Albans: Granada.

Johns, R. J. 1986. The instability of the tropical ecosystem in New Guinea. *Blumea* 31: 341–71.

Johnsingh, A. J. T., Madhusudan, M. D. 2009. Tiger reintroduction in India: Conservation tool or costly dream? In M. W. Hayward, M. J. Somers (eds.), *Reintroduction of Top-Order Predators.* Chichester: Wiley-Blackwell, pp. 146–63.

Johnson, A., Earle, T. 1987. *The Evolution of Human Societies: From Forager Group to Agrarian State.* Stanford, CA: Stanford University Press.

Johnson, K. B., Jacob, A., Brown, M. E. 2013. Forest cover associated with improved child health and nutrition: Evidence from the Malawi Demographic and Health Survey and satellite data. *Global Health: Science and Practice* 1: 237–48.

Johnson, K. R., Ellis, B. 2002. A tropical rainforest in Colorado 1.4 million years after the Cretaceous–Tertiary boundary. *Science* 296: 2379–83.

Jolly, C. J. 1970. The seed-eaters: A new model of hominid differentiation based on a baboon analogy. *Man* 5: 5–26.

Jolly, D., Haxeltine, A. 1997. Effect of low glacial atmospheric CO_2 on tropical African montane vegetation. *Science* 276: 786–8.

Jones, M. 2007. *Feast: Why Humans Share Food.* Oxford: Oxford University Press.

Jones, R., Meehan, B. 1989. Plant foods of the Gidjingal: Ethnographic and archaeological perspectives from northern Australia on tuber and seed exploitation. In D. R. Harris, G. C. Hillman (eds.), *Foraging and Farming: The Evolution of Plant Exploitation.* London: Unwin Hyman, pp. 120–35.

Joordens, J. C. A., d'Errico, F., Wesselingh, F. P., Munro, S., de Vos, J., Walling, J., Ankjærgaard, C., Reimann, T., Wijbrans, J. R., Kuiper, K. F., Mücher, H. J., Coqueugniot, H., Prié, V., Joosten, I., van Os, B., Schulp, A. S., Panuel, M., van der Haas, V., Lustenhouwer, W., Reijmer, J. J. G., Roebroeks, W. 2015. *Homo erectus* at Trinil on Java used shells for tool production and engraving. *Nature* 518: 228–31.

Joshi, M., Singh, P. P. 2003. Tropical deforestation and forest degradation: A case study from India. XII World Forestry Congress, Québec City, Canada. <http://www.fao.org/docrep/article/wfc/xii/0250-a1.htm>.

Junker, L. L. 1990. The organization of intra-regional and long-distance trade in prehispanic Philippine complex societies. *Asian Perspectives* 29: 167–209.

Junker, L. L. 1999. *Raiding, Trading, and Feasting: The Political Economy of Philippine Chiefdoms.* Honolulu: University of Hawai'i Press.

Junker, L. L. 2002a. Economic specialization and inter-ethnic trade between foragers and farmers in the prehispanic Philippines. In K. Morrison, L. Junker (eds.), *Forager-Traders in South and Southeast Asia.* Cambridge: Cambridge University Press, pp. 203–41.

Junker, L. L. 2002b. Introduction. In K. Morrison, L. Junker (eds.), *Forager-Traders in South and Southeast Asia.* Cambridge: Cambridge University Press, pp. 131–66.

Kaeppler, A. L. 1978. Melody, drone, and decoration: Underlying structures and surface manifestations in Tonga art and society. In M. Greenhalgh, V. Megaw (eds.), *Art in Society: Studies in Styles, Culture and Aesthetics*. London: Duckworth, pp. 261–74.

Kaimowitz, D., Mertens, B., Wunder, S., Pacheco, P. 2004. *Hamburger Connection Fuels Amazon Destruction: Cattle Ranching and Deforestation in Brazil's Amazon*. Bogor, Indonesia: Center for International Forestry Research.

Kajale, M. D. 1989. Mesolithic exploitation of wild plants in Sri Lanka: Archaeobotanical study at the cave site of Beli-Lana. In D. R. Harris, G. C. Hillman (eds.), *Foraging and Farming: The Evolution of Plant Exploitation*. London: Unwin Hyman, pp. 269–81.

Kaplan, J. O., Krumhardt, K. M. 2011. The KK10 anthropogenic land cover change scenario. doi:/10.1594/PANGAEA.871369.

Kaplan, J. O., Krumhardt, K. M., Ellis, E. C., Ruddiman, W. F., Lemmen, C., Goldewijk, K. K. 2011. Holocene carbon emissions as a result of anthropogenic land cover change. *Holocene* 21: 775–91.

Kato, S. 1997. *The Ancient City of Kharakhorum*. Beijing: UNESCO/New World Press.

Katoh, S., Beyene, Y., Itaya, T., Hyodo, H., Hyodo, M. Yagi, K., Gouzu, C., WoldeGabriel, G., Hart, W. K., Ambrose, S. H., Nakaya, H., Bernor, R. L., Boisserie, J.-R., Bibi, F., Saegusa, H., Sasaki, T., Sano, K., Asfaw, B., Suwa, G. 2016. New geological and palaeontological age constrain for the gorilla–human lineage split. *Nature* 530: 215–18.

Kay, A. U., Kaplan, J. O. 2015. Human subsistence and land use in sub-Saharan Africa, 1000 BC to AD 1500: A review, quantification, and classification. *Anthropocene* 9: 14–32.

Kelley, J. 2002. The hominoid radiation in Asia. In W. Hartwig (ed.), *The Primate Fossil Record*. Cambridge: Cambridge University Press, pp. 369–84.

Kelly, K. 1997. The archaeology of African-European interaction: Investigating the social role of trade, traders and the use of space in the seventeenth and eighteenth century *Hueda* Kingdom, Republic of Benin. *World Archaeology* 28: 351–69.

Kemp, B. J. 1972. Temple and town in Nubia. In P. J. Ucko, R. Tringham, G. W. Dimbleby (eds.), *Man, Settlement, Urbanism*. London: University of London, Institute of Archaeology, pp. 657–80.

Kennedy, K. A. R. 2000. *God-Apes and Fossil Men: Palaeoanthropology of South Asia*. Ann Arbor: University of Michigan Press.

Kershaw, A. P. 1985. An extended late Quaternary vegetation record from north-eastern Queensland and its implications for the seasonal tropics of Australia. *Proceedings of the Ecological Society of Australia* 13: 179–89.

Kershaw, A. P. 1986. Climatic change and Aboriginal burning in north-east Australia during the last two glacial/interglacial cycles. *Nature* 322: 47–9.

Kershaw, A. P., Bretherton, S. C., van der Kaars, S. 2007. A complete pollen record of the last 230 ka from Lynch's Crater, north-eastern Australia. *Palaeogeography, Palaeoclimatology, Palaeoecology* 251: 23–45.

Kienast, M., Steinke, S., Stattegger, K., Calvert, S. E. 2001. Synchronous tropical South China Sea SST change and Greenland warming during deglaciation. *Science* 291: 2132–4.

Kikkawa, J., Dwyer, P. D. 1992. Use of scattered resources in rain forest of humid tropical lowlands. *Biotropica* 24: 293–308.

Kimbel, W. H., Suwa, G., Asfaw, B., Rak, Y., White, T. D. 2014. *Ardipithecus ramidus* and the evolution of the human cranial base. *Proceedings of the National Academy of Sciences of the United States of America* 111: 948–53.

Kingston, J. D., Harrison, T. 2007. Isotopic dietary reconstructions of Pliocene herbivores at Laetoli: Implications for early hominin paleoecology. *Palaeogeography, Palaeoclimatology, Palaeoecology* 243: 272–306.

Kirch, P. V. 1984. *The Evolution of the Polynesian Chiefdoms*. Cambridge: Cambridge University Press.

Kirch, P. V. 1990. The evolution of socio-political complexity in prehistoric Hawaii: An assessment of the archaeological evidence. *Journal of World Prehistory* 4: 311–45.

Kirch, P. V. 1997. *The Lapita Peoples: Ancestors of the Oceanic World.* London: Wiley-Blackwell.

Kirch, P. V. 2000. *On the Road of the Winds: An Archaeological History of the Pacific Islands before European Contact.* Berkeley: University of California Press.

Kirch, P. V. 2001. Pigs, humans, and tropic competition on small Oceania islands. In A. Anderson, T. Murray (eds.), *Australian Archaeologist: Collected Papers in Honour of Jim Allen.* Canberra: Australian National University Press, pp. 427–39.

Kirch, P. V. 2010. Peopling of the Pacific: A holistic anthropological perspective. *Annual Review of Anthropology* 39: 131–48.

Kirch, P. V. 2012. *A Shark Going Inland Is My Chief: The Island Civilization of Ancient Hawaii.* Berkeley: University of California Press.

Kirch, P. V., Ellison, J. C. 1994. Paleoenvironmental evidence for human colonization of remote Oceanic islands. *Antiquity* 68: 310–21.

Kivell, T. L., Schmitt, D. 2009. Independent evolution of knuckle-walking in African apes shows that humans did not evolve from a knuckle-walking ancestor. *Proceedings of the National Academy of Sciences of the United States of America* 106: 14241–6.

Kivell, T. L., Kibii, J. M., Churchill, S. E., Schmid, P., Berger, L. R. 2011. *Australopithecus sediba* hand demonstrates mosaic evolution of locomotor and manipulative abilities. *Science* 333: 1411–17.

Kjær, A., Barford, A. S., Asmussen, C. B., Seberg, O. 2004. Investigation of genetic and morphological variation in the sago palm (*Metroxylon sagu*; Arecaceae) in Papua New Guinea. *Annals of Botany* 94: 109–17.

Klein, R. 1999. *The Human Career: Human Biological and Cultural Origins.* Chicago: University of Chicago Press.

Klein Goldewijk, K. 2001. Estimating global land use change over the past 300 years: The HYDE database. *Global Biogeochemical Cycles* 15: 417–33.

Klein Goldewijk, K., Beusen, A., Van Drecht, G., de Vos, M. 2011. The HYDE 3.1 spatially explicit database of human-induced global land-use change over the past 12,000 years. *Global Ecology and Biogeography* 20: 73–86.

Klinge, H., Medina, E., Herrera, R. 1977. Studies on the ecology of Amazon caatinga forest in southern Venezuela. *Acta Científica Venezolana* 28: 270–6.

Knott, C. D. 1998. Changes in orangutan caloric intake, energy balance, and ketones in response to fluctuating fruit availability. *International Journal of Primatology* 19: 1061–79.

Knox, R. 1981 (1681). *A Historical Relation of Ceylon.* Colombo: Tisara Prakasakayo.

Koch, P. L. 1998. Isotopic reconstruction of past continental environments. *Annual Review of Earth and Planetary Sciences* 26: 573–613.

Kochert, G., Stalker, H. T., Gimenes, M., Galgaro, L., Moore, K. 1996. RFLP and cytogenetic evidence for the progenitor species of allotetraploid cultivated peanut, *Arachis hypogaea* (Leguminosae). *American Journal of Botany* 83: 1282–91.

Koh, L. P., Dunn, R. R., Sodhi, N. S., Colwell, R. K., Proctor, H. C., Smith, V. S. 2004. Species coextinctions and the biodiversity crisis. *Science* 305: 1632–4.

Kohn, E. 2013. *How Forests Think: Toward an Anthropology Beyond the Human.* Berkeley: University of California Press.

Kohn, M. J., Schoeninger, M. J., Valley, J. W. 1996. Herbivore tooth oxygen isotope compositions: Effects of diet and physiology. *Geochimica et Cosmochimica Acta* 60: 3889–96.

Kolb, M. J. 1993. *Na Wahi Pana O Hāmoa: The Renowned Places of Hāmoa. A Historical and Archaeological Survey of a Windward East Maui Community.* Hāna District, Maui: Report for the Native Hawaiian Culture and Arts Program.

Kolb, M. J. 1994. Ritual activity and chiefly economy at an upland religious site on Maui, Hawai'i. *Journal of Field Archaeology* 21: 417–36.

Kolb, M. J. 1999. Staple finance, ritual pig sacrifice, and ideological power in ancient Hawai'i. In E. A. Bacus, L. J. Lucero, J. Allen (eds.), *Complex Polities in the Ancient Tropical World*. Arlington: American Anthropological Association, pp. 89–107.

Kolb, M., Cordy, J. R., Conte, P. J. 1997. *The Communities of Keokea and Waiohuli: Archaeological Research in Kula, Maui*. Report. Maui: State Historical Preservation Division.

Kourampas, N., Simpson, I. A., Perera, N., Deraniyagala, S. U., Wijeyapala, W. H. 2009. Rockshelter sedimentation in a dynamic tropical landscape: Late Pleistocene–Early Holocene archaeological deposits in Kitulgala Beli-lena, Southwestern Sri Lanka. *Geoarchaeology: An International Journal* 24: 677–714.

Koutavas, A., Lynch-Stieglitz, J., Marchitto, T. M., Sachs, J. P. 2002. El Niño-like pattern in ice age tropical Pacific sea surface temperature. *Science* 297: 226–30.

Kraft, T. S., Venkataraman, V. V., Dominy, N. J. 2014. A natural history of human tree climbing. *Journal of Human Evolution* 71: 105–18.

Krigbaum, J. 2003. Neolithic subsistence patterns in northern Borneo reconstructed with stable carbon isotopes of enamel. *Journal of Anthropological Archaeology* 22: 292–304.

Krigbaum, J. 2005. Reconstructing human subsistence in the West Mouth (Niah Cave, Sarawak) burial series using stable isotopes of carbon. *Asian Perspectives* 44: 73–89.

Krigbaum, J., Berger, M. H., Daegling, D. J., McGraw, W. S. 2013. Stable isotope canopy effects for sympatric monkeys at Taï Forest, Côte d'Ivoire. *Biology Letters* 9: 20130466.

Krishnaswamy, J., Bonell, M., Venkatesh, B., Purandara, B. K., Lele, S., Kiran, M. C., Reddy, V., Badiger, S., Rakesh, K. N. 2013. The groundwater recharge response and hydrologic services of tropical humid forest ecosystems to use and reforestation: Support for the 'infiltration-evapotranspiration trade-off hypothesis'. *Journal of Hydrology* 498: 191–209.

Krützen, M., Willems, E. P., van Schaik, C. P. 2011. Culture and geographic variation in orangutan behavior. *Current Biology* 21: 1808–12.

Kuchikura, Y. 1988. Efficiency and focus of blowpipe hunting among Semaq Beri hunter-gatherers of Peninsular Malaysia. *Human Ecology* 16: 271–305.

Kuhlwilm, M., Gronau, I., Hubisz, M. J., de Filippo, C., Prado-Martinez, J., Kircher, M., Fu, Q., Burbano, H. A., Lalueza-Fox, C., de la Rasilla, M., Rosas, A., Rudan, P., Brajkovic, D., Kucan, Z., Gusic, I., Marques-Bonet, T., Andres, A. M., Viola, B., Paabo, S., Meyer, M., Siepel, A., Castellano, S. 2016. Ancient gene flow from early modern humans into Eastern Neanderthals. *Nature* 530: 429–33.

Kujit, I., Goring-Morris, N. 2002. Foraging, farming, and social complexity in the pre-pottery Neolithic of the southern Levant: A review and synthesis. *Journal of World Prehistory* 16: 361–440.

Kull, C., Imhof, S., Grosjean, M., Zech, R., Veit, H. 2008. Late Pleistocene glaciations in the Central Andes: Temperature versus humidity control—A case study from the eastern Bolivian Andes (17°S) and regional synthesis. *Global and Planetary Change* 60: 148–64.

Kumaran, N. K. P., Padmalal, D., Nair, M. K., Limaye, R. B., Guileria, J. S., Srivastava, R., Shukla, A. 2014. Vegetation response and landscape dynamics of Indian Summer Monsoon variations during Holocene: An eco-geomorphological appraisal of tropical evergreen forest subfossil logs. *PLoS ONE* 9: e93596.

Kume, T., Tanaka, N., Kuraji, K., Komatsu, H., Yoshiuji, N., Saitoh, T. M., Suzuki, M., Kumagair, T. 2011.Ten-year evapotranspiration estimates in a Bornean tropical rainforest. *Agricultural and Forest Meteorology* 15: 1183–92.

Kunimatsu, Y., Nakatsukasa, M., Sawada, Y., Sakai, T., Hyodo, M., Hyodo, H., Itaya, T., Nakaya, H., Saegusa, H., Mazurier, A., Saneyoshi, M., Tsujikawa, H., Yamamoto, A., Mbua, E. 2007. A new Late Miocene great ape from Kenya and its implications for the origins of African great apes and humans. *Proceedings of the National Academy of Sciences of the United States of America* 104: 19220–5.

Kuper, A. 2003. The return of the native. *Current Anthropology* 44: 389–402.

Lambeck, K., Yokoyama, Y., Purcell, T. 2002. Into and out of the Last Glacial Maximum: Sea-level change during Oxygen Isotope Stages 3 and 2. *Quaternary Science Reviews* 21: 343–60.

Lang, R., LeFurgy, J. 2003. Edgeless cities: Examining the non-centered metropolis. *Housing Policy Debate* 14: 427–60.

Langergraber, K. E., Prufer, K., Rowney, C., Boesch, C., Crockford, C., Fawcett, K., Inoue, E., Inoue-Muruyama, M., Mitani, J. C., Muller, M. N., Robbins, M. M., Schubert, G., Stoinski, T. S., Viola, B., Watts, D., Wittig, R. M., Wrangham, R. W., Zuberbühler, K., Pääbo, S., Vigilant, L. 2012. Generation times in wild chimpanzees and gorillas suggest earlier divergence times in great ape and human evolution. *Proceedings of the National Academy of Sciences of the United States of America* 109: 15716–21.

Langlie, B. S., Mueller, N. G., Spengler, R. N., Fritz, G. J. 2014. Agricultural origins from the ground up: Archaeological approaches to plant domestication. *American Journal of Botany* 101: 1601–17.

Lansing, J. S. 2007 (1991). *Priests and Programmers: Technologies of Power in the Engineered Landscape of Bali.* Princeton and Oxford: Princeton University Press.

Lapola, D. M., Schaldach, R., Alcamo, J., Bondeau, A., Koch, J., Koelking, C. Priess, J. A. 2010. Indirect land-use changes can overcome carbon savings from biofuels in Brazil. *Proceedings of the National Academy of Sciences of the United States of America* 107: 3388–93.

Larick, R., Ciochon, R. L., Zaim, Y., Sudijono, Suminto, Rizal, Y., Aziz, F., Reagan, M., Heizler, M. 2001. Early Pleistocene $^{40}Ar/^{39}Ar$ ages for Bapang formation hominins, Central Java, Indonesia. *Proceedings of the National Academy of Sciences of the United States of America* 98: 4866–71.

Larsen, M. C. 2017. Contemporary human uses of tropical forested watersheds and riparian corridors: Ecosystem services and hazard mitigation, with examples from Panama, Puerto Rico, and Venezuela. *Quaternary International* 448: 190–200.

Larsen, M. C., Wieczorek, G. F. 2006. Geomorphic effects of large debris flows and flash floods, northern Venezuela, 1999. In E. Latrubesse (ed.), *Tropical Geomorphology with Special Reference to South America*. Berlin: Zeitschrift für Geomorphologie Supplementband 145, pp. 147–75.

Larson, G. 2011. Genetics and domestication: Important questions for new answers. *Current Anthropology* 52: S485–S495.

Larson, G., Cucchi, T., Fujita, M., Matisoo-Smith, E., Robins, J., Anderson, A., Rolett, B., Spriggs, M., Dolman, G., Kim, T.-H., Thury, N. T. D., Randi, E., Doherty, M., Awe Due, R., Bollt, R., Djubiantono, T., Griffin, B., Intoh, M., Keane, E., Kirch, P., Li, K.-T., Morwood, M., Pedriña, L. M., Piper, P. J., Rabett, R. J., Shooter, P., Van den Bergh, G., West, E., Wickler, S., Yuan, J., Cooper, A., Dobney, K. 2007. Phylogeny and ancient DNA of Sus provides insights into Neolithic expansion in island Southeast Asia and Oceania. *Proceedings of the National Academy of Sciences of the United States of America* 104: 4834–9.

Larson, G., Liu, R., Zhao, X., Yuan, J., Fuller, D., Barton, L., Dobney, K., Fan, Q., Gu, Z., Liu, X.-H., Luo, Y., Lv, P., Andersson, L., Li, N. 2010. Patterns of East Asian pig domestication, migration, and turnover revealed by modern and ancient DNA. *Proceedings of the National Academy of Sciences of the United States of America* 107: 7686–91.

Larson, G., Piperno, D. R., Allaby, R. G., Purugganan, M. D., Andersson, L., Arroyo-Kalin, M., Barton, L., Vigueira, C. C., Denham, T., Dobney, K., Doust, A. N., Gepts, P., Thomas, Gilbert, M. T. P., Gremillion, K. J., Lucas, L., Lukens, L., Marshall, F. B., Olsen, K. M., Pires, J. C., Richerson, P. J., Rubio de Casas, R., Sanjur, O. I., Thomas, M. G., Fuller, D. Q. 2014. Current perspectives and the future of domestication studies. *Proceedings of the National Academy of Sciences of the United States of America* 111: 6139–46.

Latimer, B., Lovejoy, C. O. 1989. The calcaneus of *Australoptihecus afarensis* and its implications for the evolution of bipedality. *American Journal of Physical Anthropology* 78: 369–86.

Laurance, W. F. 2005. When bigger is better: The need for Amazonian mega-reserves. *Trends in Ecology & Evolution* 20: 645–8.

Laurance, W. F. 2013. Emerging threats to tropical forests. In M. Lowman, S. Devy, T. Ganesh (eds.), *Treetops at Risk: Challenges of Global Canopy Ecology and Conservation*. Dordrecht: Springer, pp. 71–9.

Laurance, W. F., Cochrane, M. A., Bergen, S., Fearnside, P. M., Dolmônica, P., Barber, C., D'Angelo, S., Fernandes, T. 2001. The future of the Brazilian Amazon. *Science* 291: 438–9.

Laurance, W. F., Goosem, M., Laurance, S. G. W. 2009. Impacts of roads and linear clearings on tropical forests. *Trends in Ecology & Evolution* 24: 659–69.

Leacock, E., Lee, R. B. 1982. *Politics and History in Band Societies*. New York: Cambridge University Press.

Leakey, L. S. B. 1959. A new fossil skull from Olduvai. *Nature* 184: 491–3.

Leakey, M. G., Feibel, C. S., McDougall, I., Walker, A. 1995a. New four-million-year-old hominid species from Kanapoi and Allia Bay, Kenya. *Nature* 376: 565–71.

Leakey, M. G., Ungar, P. S., Walker, A. 1995b. A new genus of large primate from the Late Oligocene of Lothidok, Turkana District, Kenya. *Journal of Human Evolution* 28: 519–31.

Leakey, M. G., Feibel, C. S., McDougall, I., Ward, C., Walker, A. 1998. New specimens and confirmation of an early age for *Australopithecus anamensis*. *Nature* 393: 62–6.

Leakey, R. E., Leakey, M. G., Walker, A. 1988. Morphology of *Afropithecus turkanensis* from Kenya. *American Journal of Physical Anthropology* 76: 289–307.

Leavesley, M. G. 2002. Buang Merabek: Early evidence for human occupation in the Bismarck Archipelago, Papua New Guinea. *Australian Archaeology* 54: 55–7.

Leavesley, M. G. 2004. Trees to the sky: Prehistoric hunting in New Ireland, Papua New Guinea. PhD thesis, Australian National University, Canberra.

Leavesley, M. G. 2005. Prehistoric hunting strategies in New Ireland, Papua New Guinea: The evidence of the cuscus (*Phalanger orientalis*) remains from Buang Merabak cave. *Asian Perspectives* 44: 207–18.

Leavesley, M. G. 2006. Late Pleistocene complexities in the Bismarck Archipelago. In I. Lilley (ed.), *Archaeology of Oceania: Australia and the Pacific Islands*. Carlton: Blackwell, pp. 189–204.

Leavesley, M. G. 2007. A shark-tooth ornament from Pleistocene Sahul. *Antiquity* 81: 308–15.

Lebatard, A.-E., Bourlès, D. L., Duringer, P., Jolivet, M., Braucher, R., Carcaillet, J., Schuster, M., Arnaud, N., Monié, P., Lihoreau, F., MacKaye, H. T., Vignaud, P., Brunet, M. 2008. Cosmogenic nuclide dating of *Sahelanthropus tchadensis* and *Australopithecus bahrelghazali*: Mio-Pliocene hominids from Chad. *Proceedings of the National Academy of Sciences of the United States of America* 105: 3226–31.

Lebot, V., Trilles, B., Noyer, J. L., Modesto, J. 1998. Genetic relationships between *Dioscorea alata* L. cultivars. *Genetic Resources and Crop Evolution* 45: 499–509.

Lebot, V., Prana, M. S., Kreike, N., van Heck, H., Pardales, J., Okpul, T., Gendua, T., Thongjiem, M., Hue, H., Viet, N., Yap, T. C. 2004. Characterisation of taro (*Colocasia esculenta* (L.) Schott) genetic resources in Southeast Asia and Oceania. *Genetic Resources and Crop Evolution* 51: 381–92.

Lee, J.-E., Oliveira, R. S., Dawson, T. E., Fung, I. 2005. Root functioning modifies seasonal climate. *Proceedings of the National Academy of Sciences of the United States of America* 102: 17576–81.

Lee, R., DeVore, I. 1968. *Man the Hunter*. Chicago: Aldine.

Lee-Thorp, J. A., van der Merwe, N. J. 1987. Carbon isotope analysis of fossil bone apatite. *South African Journal of Science* 83: 712–15.

Lee-Thorp, J. A., van der Merwe, N. J., Brain, C. K. 1989a. Isotopic evidence for dietary differences between two extinct baboon species from Swartkrans (South Africa). *Journal of Human Evolution* 18: 183–90.

Lee-Thorp, J. A., Sealy, J. C., van der Merwe, N. J. 1989b. Stable carbon isotope ratio differences between bone collagen and bone apatite, and their relationship to diet. *Journal of Archaeological Science* 16: 585–99.

Lee-Thorp, J. A., van der Merwe, N. J., Brain, C. K. 1994. Diet of *Australopithecus robustus* at Swartrans deduced from stable carbon isotope ratios. *Journal of Human Evolution* 27: 361–72.

Lee-Thorp, J. A., Thackeray, J. F., van der Merwe, N. J. 2000. The hunters and the hunted revisited. *Journal of Human Evolution* 39: 565–76.

Lee-Thorp, J. A., Sponheimer, M., van der Merwe, N. J. 2003. What do stable isotopes tell us about hominid dietary and ecological niches in the Pliocene? *International Journal of Osteoarchaeology* 13: 104–13.

Lee-Thorp, J. A., Sponheimer, M., Passey, B. H., de Ruiter, D. J., Cerling, T. E. 2010. Stable isotopes in fossil hominin tooth enamel suggest a fundamental dietary shift in the Pliocene. *Philosophical Transactions of the Royal Society B: Biological Sciences* 365: 3389–96.

Lee-Thorp, J. A., Likius, A., Mackaye, H. T., Vignaud, P., Sponheimer, M., Brunet, M. 2012. Isotopic evidence for an early shift to C4 resources by Pliocene hominins in Chad. *Proceedings of the National Academy of Sciences of the United States of America* 109: 20369–72.

Lentfer, C. 2009. Tracing domestication and cultivation of bananas from phytoliths: An update from Papua New Guinea. *Ethnobotany Research and Applications* 7: 247–70.

Lentfer, C., Denham, T. 2017. The archaeobotany of Kuk. In J. Golson, T. Denham, P. Hughes, P. Swadling, J. Muke (eds.), *Ten Thousand Years of Cultivation at Kuk Swamp in the Highlands of Papua New Guinea* (Terra Australis 46). Canberra: Australian National University Press, pp. 163–86.

Lentz, D. L., Dunning, N. P., Scarborough, V. L., Magee, K., Thompson, K. M., Weave, E., Carr, C., Terry, R. E., Islebe, G., Tankersley, K. B., Graziosa Sierra, L., Jones, J. G., Buttles, P., Valdez, F., Ramos Hernandez, C. 2014. Farms, forests and the edge of sustainability at the ancient Maya city of Tikal. *Proceedings of the National Academy of Sciences of the United States of America* 111: 18513–18.

Lentz, D. L., Magee, K., Weaver, E., Jones, J., Tankersley, K. B., Hood, A., Islebe, G., Ramos, C., Dunning, N. 2015. Agroforestry and agricultural practices of the ancient Maya at Tikal. In D. Lentz, N. Dunning, V. Scarborough (eds.), *Tikal: Palaeoecology of an Ancient Maya City*. Cambridge: Cambridge University Press, pp. 152–85.

L'environnement des hominidés au Plio-Pléistocène. 1985. Paris: Fondation, Signer-Polignac, Masson.

Levin, N. E., Cerling, T. E., Passey, B. H., Harris, J. M., Ehleringer, J. R. 2006. A stable isotope aridity index for terrestrial environments. *Proceedings of the National Academy of Sciences of the United States of America* 103: 11201–5.

Levin, N. E., Simpson, S. W., Quade, J., Cerling, T. E., Frost, S. R. 2008. Herbivore enamel carbon isotopic composition and the environmental context of *Ardipithecus* at Gona, Ethiopia. In J. Quade, J. G. Wynn (eds.), *The Geology of Early Humans in the Horn of Africa*. Boulder, CO: Geological Society of America Special Paper 446, pp. 215–34.

Levin, N. E., Haile-Selassie, Y., Frost, S. R., Saylor, B. Z. 2015. Dietary change among hominins and cercopithecids in Ethiopia during the early Pliocene. *Proceedings of the National Academy of Sciences of the United States of America* 112: 12304–9.

Levis, C., Figueira de Souza, P., Schietti, J., Emilio, T., da Veiga Pinto, J. L. P., Clement, C. R., Costa, F. R.C. 2012. Historical human footprint on modern tree species composition in the Purus-Madeira interfluve, central Amazonia. *PLoS ONE* 7: e48559.

Levis, C., Costa, F. R. C., Bongers, F., Peña-Claros, M., Clement, C. R., Junqueira, A. B., Neves, E. G., Tamanaha, E. K., Figueiredo, F. O. G., Salomão, R. P., Castilho, C. V., Magnusson, W. E., Phillips, O. L., Guevara, J. E., Sabatier, D., Molino, J.-F., Cardenas López, D., Mendoza, A. M., Pitman, N. C. A., Duque, A., Núñez Vargas, P., Zartman, C. E., Vasquez, R., Andrade, A., Camargo, J. L., Feldpausch, T. R., Laurance, S. G. W., Laurance, W. F., Killeen, T. J., Mendonça Nascimento, H. E., Montero, J. C., Mostacedo, B., Amaral, I. L., Guimaraes Vieira, I. C., Brienen, R., Castellanos, H., Terborgh, J., de Jesus Veiga Carim, M., da Silva Guimares, J. R., de Souza Coelho, L., de Almeida Matos, F. D., Wittmann, F., Mogollón, H. F., Damasco, G., Dávila, N., García-Villacorta, R., Coronado, E. N. H., Emilio, T., de Andrade Lima Filho, D., Schietti, J., Souza, P., Targhetta, N., Comiskey, J. A., Marimon, B. S., Marimon Jr B.-H., Neill, D., Alonso, A., Arroyo, L., Carvalho, F. A., de Souza, F. C., Dallmeier, F., Pansonato, M. P., Duivenvoorden, J. F., Fine, P. V. A., Stevenson, P. R., Araujo-Murakami, A., Aymard, G. A., Baraloto, C., do Amaral, D. D., Engel, J., Henkel, T. W., Maas, P., Petronelli, P., Cardenas Revilla, J. D., Stropp, J., Daly, D., Gribel, R., Rios Paredes, M., Silveira, M., Thomas-Caesar, R., Baker, T. R., da Silva, N. F., Ferreira, L. V., Peres, C. A., Silman, M. R., Cerón, C., Valverde, F. C., Di Fiore, A., Jimenez, E. M., Penuela Mora, M. C., Toledo, M., Barbosa, E. M., de Matos Bonates, L. C., Arboleda, N. C., de Sousa Farias, E., Fuentes, A., Guillaumet, J.-L., Moller Jorgensen, P., Malhi, Y., de Andrade Miranda, I. P., Phillips, J. F., Prieto, A., Rudas, A., Ruschel, A. R., Silva, N., von Hildebrand, P., Vos, V. A., Zent, E. L., Zent, S., Cintra, B. B. L., Nascimento, M. T., Oliveira, A. A., Ramirez-Angulo, H., Ramos, J. F., Rivas, G., Schöngart, J., Sierra, R., Tirado, M., van der Heijden, G., Torre, E. V., Wang, O., Young, K. R., Baider, C., Cano, A., Farfan-Rios, W., Ferreira, C., Hoffman, B., Mendoza, C., Mesones, I., Torres-Lezama, A., Medina, M. N. U., van Andel, T. R., Villarroel, D., Zagt, R., Alexiades, M. N., Balslev, H., Garcia-Cabreara, K., Gonzales, T., Hernandez, L., Huamantupa-Chuquimaco, I., Manzatto, A. G., Milliken, W., Cuenca, W. P., Pansini, S., Pauletto, D., Arevalo, F. R., Costa Reis, N. F., Sampaio, A. F., Giraldo, L. E. U., Valderrama Sandoval, E. H., Valenzuela Gamarra, L., Vela, C. I. A., ter Steege, H. 2017. Persistent effects of pre-Columbian plant domestication on Amazonian forest composition. *Science* 355: 925–31.

Lévi-Strauss, C. 1963. *Totemism*. London: Merlin Press.

Lévi-Strauss, C. 1966. *The Savage Mind*. Chicago: University of Chicago Press.

Lewis, L., Perera, N., Petraglia, M. 2014. First technological comparison of Southern African Howiesons Poort and South Asian microlithic industries: An exploration of inter-regional variability in microlithic assemblages. *Quaternary International* 350: 7–25.

Lewis, M. E., Werdelin, L. 2007. Patterns of change in the Plio-Pleistocene carnivorans of eastern Africa: Implications for hominin evolution. In R. Bobe, Z. Alemseged, A. K. Behrensmeyer (eds.), *Hominin Environments in the East African Pliocene: An Assessment of the Faunal Evidence*. Dordrecht: Springer, pp. 77–106.

Lewis, M. E., Werdelin, L. 2010. Carnivoran dispersal Out of Africa during the Early Pleistocene: Relevance for hominins. In J. G. Fleagle, J. J. Shea, F. E. Grine, A. L. Baden, R. E. Leakey (eds.), *Out of Africa I: The First Hominin Colonization of Eurasia*. Dordrecht: Springer, pp. 13–26.

Lewis, S. L., Maslin, M. A. 2015. Defining the Anthropocene. *Nature* 519: 171–80.

Lewis, S. L., Edwards, D. P., Galbraith, D. 2015. Increasing human dominance of tropical forests. *Science* 349: 827–32.

Lieberman, D. E., McBratney, B. M., Krovitz, G. 2002. The evolution and development of cranial form in *Homo sapiens*. *Proceedings of the National Academy of Sciences of the United States of America* 99: 1134–9.

Lieberman, V. B. 2003. *Strange Parallels: Southeast Asia in Global Context, c. 800–1830, Volume 1: Integration on the Mainland*. Cambridge: Cambridge University Press.

Lindsay, L. A. 2008. *Captives as Commodities: The Transatlantic Slave Trade*. London: Prentice Hall.

Lioubine, V., Guede, F. 2000. *The Paleolithic of Republic Côte d'Ivoire (West Africa)*. In Russian with an English summary. St. Petersburg: Russian Academy of Sciences.

Liu, W., Jin, C.-Z., Zhang, Y.-Q., Cai, Y.-J., Xing, S., Wu, X.-J., Cheng, H., Edwards, R. L., Pan, W.-S., Qin, D.-G., An, Z.-S., Trinkaus, E., Wu, X.-Z. 2010. Human remains from Zhirendong, South China, and modern human emergence in East Asia. *Proceedings of the National Academy of Sciences of the United States of America* 107: 19201–6.

Liu, W., Schepartz, L. A., Xing, S., Miller-Antonio, S., Wu, X., Trinkaus, E., Martinón-Torres, M. 2013. Late Middle Pleistocene hominin teeth from Panxian Dadong, South China. *Journal of Human Evolution* 64: 337–55.

Liu, W., Martinón-Torres, M., Cai, Y.-J., Xing, S., Tong, H.-W., Pei, S.-W., Sier, M. J., Wu, X.-H., Edwards, R. L., Cheng, H., Li, Y.-Y., Yang, X.-X., Bermúdez de Casto, J. M., Wu, X.-J. 2015. The earliest unequivocally modern humans in southern China. *Nature* 526: 696–700.

Liu, Y.-P., Wu, G.-S., Yao, Y.-G., Miao, Y.-W., Luikart, G., Baig, M., Beja-Pereira, A., Ding, Z.-L., Palanichamy, M. G., Zhang, Y.-P. 2006. Multiple maternal origins of chickens: Out of the Asian jungles. *Molecular Phylogenetics and Evolution* 38: 12–19.

Livingstone Smith, A., Cornelissen, E., de Francquen, C., Nikis, N., Mees, F., Mukendi, J. T., Beeckman, H., Bourland, N., Hubau, W. 2017. Forests and rivers: The archaeology of the north eastern Congo. *Quaternary International* 448: 95–116.

Lombard, M., Phillipson, L. 2010. Indications of bow and stone-tipped arrow use 64000 years ago in KwaZulu-Natal, South Africa. *Antiquity* 84: 635–48.

Lombardo, U., Prümers, H. 2010. Pre-Columbian human occupation patterns in the eastern plains of the Llanos de Moxos, Bolivian Amazonia. *Journal of Archaeological Science* 37: 1875–85.

Lombardo, U., Denier, S., May, J.-H., Rodrigues, L., Veit, H. 2013. Human–environment interactions in pre-Columbian Amazonia: The case of the Llanos de Moxos, Bolivia. *Quaternary International* 312: 109–19.

Loudon, J. E., Sandberg, P. A., Wrangham, R. W., Fahey, B., Sponheimer, M. 2016. The stable isotope ecology of Pan in Uganda and beyond. *American Journal of Primatology* 78: 1070–85.

Lourandos, H. 1997. *Continent of Hunter-Gatherers: New Perspectives in Australian Prehistory*. Cambridge: Cambridge University Press.

Louys, J., Curnoe, D., Tong, H. 2007. Characteristics of Pleistocene megafauna extinctions in Southeast Asia. *Palaeogeography, Palaeoclimatology, Palaeoecology* 243: 152–73.

Louys, J., Corlett, R. T., Price, G. J., Hawkins, S., Piper, P. J. 2014. Rewilding the tropics, and other conservation translocations strategies in the tropical Asia-Pacific region. *Ecology and Evolution* 4: 4380–98.

Love, M. 2007. Recent research in the southern highlands and Pacific coast of Mesoamerica. *Journal of Archaeological Research* 15: 275–328.

Lovejoy, C. O. 1981. The origin of man. *Science* 211: 341–50.

Lovejoy, C. O., Suwa, G., Spurlock, L., Asfaw, B., White, T. D. 2009. The pelvis and femur of *Ardipithecus ramidus*: The emergence of upright walking. *Science* 326: 71–71e6.

Loveluck, C. 2013. *Northwest Europe in the Early Middle Ages, c. AD 600–1150: A Comparative History*. Cambridge: Cambridge University Press.

Lu, F. 2010. Patterns of indigenous resilience in the Amazon: A case study of Huaorani hunting in Ecuador. *Journal of Ecological Anthropology* 14: 5–21.

Lucero, L. J. 1994. Household and community integration among hinterland elites and commoners: Maya residential ceramic assemblages of the Belize River area. PhD thesis, Ann Arbor, UCLA, University Microfilms.

Lucero, L. J. 1999. Water control and Maya politics in the southern Maya lowlands. In E. A. Bacus, L. J. Lucero (eds.), *Complex Polities in the Ancient Tropical World* (Archeological Papers of the American Anthropological Association number 9). Arlington: American Anthropological Association, pp. 34–49.

Lucero, L. J. 2002. The collapse of the Classic Maya: A case for the role of water control. *American Anthropologist* 104: 814–26.

Lucero, L. J. 2006. *Water and Ritual: The Rise and Fall of Classic Maya Rulers*. Austin: University of Texas Press.

Lucero, L. J., Gunn, J. D., Scarborough, V. L. 2011. Climate change and Classic Maya water management. *Water* 3: 479–94.

Lucero, L. J., Fletcher, R., Coningham, R. 2015. From 'collapse' to urban diaspora: The transformation of low-density, dispersed agrarian urbanism. *Antiquity* 89: 1139–54.

Luiselli, L. 2003. Comparative abundance and population structure of sympatric Afrotropical tortoises in six rainforest areas: The differential effects of 'traditional veneration' and of 'subsistence hunting' by local people. *Acta Oecologica* 24: 157–63.

Luomala, K. 1960. The native fog in the Polynesian system of values. In S. Diamond (ed.), *Culture in History: Essays in Honor of Paul Radin*. New York: Columbia University Press, pp. 190–240.

Lyons, J. A., Natusch, D. J. D. 2011. Wildlife laundering through breeding farms: Illegal harvest, population declines and a means of regulating the trade of green pythons (Morelia viridis) from Indonesia. *Biological Conservation* 144: 3073–81.

Macaulay, V., Hill, C., Achilli, A., Rengo, C., Clarke, D., Meehan, W., Blackburn, J., Semino, O., Scozzari, R., Cruciani, F., Taha, A., Shaari, N. K., Raja, J. M., Ismail, P., Zainuddin, Z., Goodwin, W., Bulbeck, D., Bandelt, H.-J., Oppenheimer, S., Torroni, A., Richards, M. 2005. Single, rapid coastal settlement of Asia revealed by analysis of complete mitochondrial genomes. *Science* 308: 1034–6.

MacGregor, R. 1941. The domestic buffalo. *Vet Record* 53: 443–50.

Macho, G. A., Lee-Thorp, J. A. 2014. Niche partitioning in sympatric *Gorilla* and *Pan* from Cameroon: Implications for life history strategies and for reconstructing the evolution of hominin life history. *PLoS ONE* 9: e102794.

Mackay, A. L. 1991. *A Dictionary of Scientific Quotations*. Boca Raton, FL: CRC Press.

MacKinnon, K., Hatta, G., Halim, H., Mangalik, A. 1997. *The Ecology of Kalimantan*. Oxford: Oxford University Press.

McBrearty, S. 1988. The Sangoan-Lupemban and Middle Stone Age sequence at the Muguruk site, Western Kenya. *World Archaeology* 19: 379–420.

McBrearty, S., Brooks, A. S. 2000. The revolution that wasn't: A new interpretation of the origin of modern human behavior. *Journal of Human Evolution* 39: 453–563.

McBrearty, S., Jablonski, N. G. 2005. First fossil chimpanzee. *Nature* 437: 105–8.

McCarroll, D., Loader, N. J. 2006. Isotopes in tree rings. In M. Leng (ed.), *Isotopes in Palaeoenvironmental Research*. Dordrecht: Springer, pp. 67–116.

McDougall, I., Brown, F. H., Fleagle, J. G. 2005. Stratigraphic placement and age of modern humans from Kibish, Ethiopia. *Nature* 433: 733–6.

McEachern, S. 2001. State formation and enslavement in northern Cameroon and northeastern Nigeria. In C. De Corse (ed.), *West Africa during the slave Trade: Archaeological and Historical Perspectives*. Leicester: Leicester University Press, pp. 131–51.

McElroy, A., Townsend, P. K. 1985. *Medical Anthropology in Ecological Perspective*. Boulder, CO: Westview Press.

McIntosh, S. K. (ed.). 1995. *Excavations at Jenné-Jeno, Hambarketolo, and Kaniana (Inland Niger Delta, Mali): The 1981 Season*. Berkeley: University of California Press.

McIntosh, S. K. 1999. Pathways to complexity: An African perspective. In S. K. McIntosh (ed.), *Beyond Chiefdoms: Pathways to Complexity in Africa*. Cambridge: Cambridge University Press, pp. 1–30.

McIntosh, S. K., McIntosh, R. J. 1980. *Prehistoric Investigations in the Region of Jenne, Mali*. Cambridge Monographs in African Archaeology, No. 2. Oxford: BAR.

McKey, D. 1979. The distribution of secondary compounds within plants. In G. A. Rosenthal, D. H. Jansen (eds.), *Herbivores: Their Interaction with Secondary Plant Metabolites*. New York: Academic Press, pp. 55–133.

McKnight, T. L., Hess, D. 2000. *Climate Zones and Types. Physical Geography: A Landscape Appreciation*. Upper Saddle River, NJ: Prentice Hall.

McMichael, C. N. H., Bush, M. B., Piperno, D. R., Silman, M. R., Zimmerman, A. R., Anderson, C. 2012. Spatial and temporal scales of pre-Columbian disturbance associated with western Amazonian lakes. *The Holocene* 22: 131–41.

McMichael, C. N. H., Matthew-Bird, F., Farfan-Rios, W., Feeley, K. J. 2017. Ancient human disturbances may be skewing our understanding of Amazonian forests. *Proceedings of the National Academy of Sciences of the United States of America* 114: 522–7.

McNulty, K. P., Begun, D. R., Kelley, J., Manthi, F. K., Mbua, E. N. 2015. A systematic revision of Proconsul with the description of a new genus of early Miocene hominoid. *Journal of Human Evolution* 84: 42–61.

Maezumi, S. Y., Alves, D., Robinson, M., de Souza, J. G., Levis, C., Barnett, R. L., de Oliveira, E. A., Urrego, D., Schaan, D., Iriarte, J. 2018. The legacy of 4,500 years of polyculture agroforestry in the eastern Amazon. *Nature Plants*. doi:10.1038/s41477-018-0205-y.

Magee, K., Weaver, E., Jones, J., Tankersley, K. B., Hood, A., Islebe, G., Ramos, C., Dunning, N. 2015. Agroforestry and agricultural practices of the ancient Maya at Tikal. In D. Lentz, N. Dunning, V. Scarborough (eds.), *Tikal: Paleoecology of an Ancient Maya City*. Cambridge: Cambridge University Press, pp. 152–85.

Magill, C. R., Ashley, G. M., Freeman, K. H. 2013. Ecosystem variability and early human habitats in eastern Africa. *Proceedings of the National Academy of Sciences of the United States of America* 110: 1167–74.

Malapa, R., Arnau, G., Noyer, J. L., Lebot, V. 2005. Genetic diversity of the greater yam (*Dioscorea alata* L.) and relatedness to *D. nummularia* Lam. and *D. transversa* Br. as revealed with AFLP markers. *Genetic Resources and Crop Evolution* 52: 919–29.

Maley, J. 1996. The African rainforest, main characteristics of changes in vegetation and climate from the Upper Cretaceous to the Quaternary. In I. J. Alexander, M. D. Swaine, R. Watling (eds.), *Essays on the Ecology of the Guinea-Congo Rainforest. Proceedings of the Royal Society of Edinburgh B Series* 104: 31–73.

Maley, J. 2002. A catastrophic destruction of African forests about 2500 years ago still exerts a major influence on present vegetation formations. *IDS Bulletin: Institute of Development Studies Journal* 33: 13–30.

Maley, J., Brenac, P. 1998. Vegetation dynamics, palaeoenvironments, and climate changes in the forest of Western Cameroon during the last 28,000 years B.P. *Review of Palaeobotany and Palynology* 99: 157–87.

Malhi, Y. 2012. The productivity, metabolism and carbon cycle of tropical forest vegetation. *Journal of Ecology* 100: 65–75.

Malhi, Y. 2017. The concept of the Anthropocene. *Annual Review of Environment and Resources* 42: 77–104.

Malhi, Y., Phillips, O. (eds.). 2005. *Tropical Forests and Global Atmospheric Change*. Oxford: Oxford University Press.

Malhi, Y., Wright, J. 2004. Spatial patterns and recent trends in the climate of tropical rainforest regions. *Philosophical Transactions of the Royal Society B: Biological Sciences* 359: 311–29.

Malhi, Y., Timmons Roberts, J., Bettis, R. A., Killeen, T. J., Li, W., Nobre, C. A. 2008. Climate change, deforestation, and the fate of the Amazon. *Science* 319: 169–72.

Malhi, Y., Adu-Bredu, S., Asare, R. A., Lewis, S. L., Mayaux, P. 2013. African rainforests: Past, present and future. *Philosophical Transactions of the Royal Society B: Biological Sciences* 368: 20120312.

Malhi, Y., Gardner, T. A., Goldsmith, G. R., Silman, M. R., Zelazowski, P. 2014. Tropical forests in the Anthropocene. *Annual Review of Environment and Resources* 39: 125–59.

Man, E. H., Temple, R. C. 1880. Note on two maps of the Andaman Islands. *Journal of the Royal Geographic Society* 50: 255–9.

Mann, C. C. 2008. Ancient earthmovers of the Amazon. *Science* 321: 1148–52.
Mann, G. V., Roels, O. A., Price, D. J., Merrill, J. M. 1962. Cardiovascular disease in African pygmies: A survey of the health status, serum lipids and diet of pygmies in Congo. *Journal of Chronic Diseases* 15: 341–71.
Manning, K. 2008. Mobility amongst LSA Sahelian pastoral groups: A view from the Lower Tilemsi Valley, Eastern Mali. *Archaeological Review from Cambridge* 23: 125–45.
Manning, K. 2010. A developmental history of West African agriculture. In P. Allsworth Jones (ed.), *West African Archaeology: New Developments, New Perspectives*. BAR International Series. Oxford: Archaeopress.
Manning, K., Pelling, R., Higham, T., Schwenninger, J.-L., Fuller, D. 2011. 4500-year old domesticated pearl millet (*Pennisetum glaucum*) from the Tilemsi Valley, Mali: New insights into an alternative cereal domestication pathway in Africa. *Journal of Archaeological Science* 38: 312–22.
Marcus, J. 1993. Ancient Maya political organization. In J. A. Sabloff, J. S. Henderson (eds.), *Lowland Maya Civilization in the Eighth Century AD*. Washington, DC: Dumbarton Oaks Research Library and Collection, pp. 111–83.
Marean, C. W. 2016. The transition to foraging for dense and predictable resources and its impact on the evolution of modern humans. *Philosophical Transactions of the Royal Society B: Biological Sciences* 317: 20150239.
Marean, C. W., Bar-Matthews, M., Bernatchez, J., Fisher, E., Goldberg, P., Herries, A. I. R., Jacobs, Z., Jerardino, A., Karkanas, P., Minichillo, T., Nilssen, P. J., Thompson, E., Watts, I., Williams, H. M. 2007. Early human use of marine resources and pigment in South Africa during the Middle Pleistocene. *Nature* 449: 905–8.
Marshall, F., Hildebrand, E. 2002. Cattle before crops: The beginnings of food production in Africa. *Journal of World Prehistory* 16: 99–143.
Martin, S., Grube, N. 2008. *Chronicle of the Maya Kings and Queens: Deciphering the Dynasties of the Ancient Maya*. London: Thames & Hudson.
Marwick, B. 2009. Biogeography of Middle Pleistocene hominins in mainland Southeast Asia: A review of current evidence. *Quaternary International* 202: 51–8.
Marwick, B., Clarkson, C., O'Connor, S., Collins, S. 2016. Early modern human lithic technology from Jerimalai, East Timor. *Journal of Human Evolution* 101: 45–64.
Maslin, M. A., Trauth, M. H. 2009. Plio-Pleistocene East African pulsed climate variability and its influence on early human evolution. In F. E. Grine, R. E. Leakey, J. G. Fleagle (eds.), *The First Humans: Origins of the Genus Homo*. Dordrecht: Springer, pp. 151–8.
Maslin, M. A., Brierley, C. M., Milner, A. M., Shultz, S., Trauth, M. H., Wilson, K. E. 2014. East African climate pulses and early human evolution. *Quaternary Science Reviews* 101: 1–17.
Masson, M. A. 2012. Maya collapse cycles. *Proceedings of the National Academy of Sciences of the United States of America* 109: 18237–8.
Masson, M. A., Freidel, D. A. 2012. An argument for Classic-era Maya market exchange. *Journal of Anthropological Archaeology* 31: 455–84.
Mateo, N., Nader, W., Tamayo, G. 2001. Bioprospecting. In S. A. Levin (ed.), *Encyclopedia of Biodiversity*. Vol. 1. San Diego, CA: Academic Press, pp. 471–88.
Matheny, R. T. 1986. Early states in the Maya lowlands during the late Preclassic period: Edzna and El Mirador. In E. P. Benson (ed.), *City-States of the Maya: Art and Architecture*. Denver, CO: Rocky Mountain Institute for Pre-Columbian Studies.
Matsuoka, Y., Vigouroux, Y., Goodman, M. M., Sanchez, J., Buckler, E., Doebley, J. 2002. A single domestication for maize shown by multilocus microsatellite genotyping. *Proceedings of the National Academy of Sciences of the United States of America* 99: 6080–4.
Matthews, R. J. 1992. Defining the style of the period: Jemdet Nasr 1926–28. *Iraq* 54: 1–34.
Maxwell, J. J., Howarth, J. D., Vandergoes, M. J., Jacobsen, G. E., Barber, I. G. 2016. Timing and importance of arboriculture and agroforestry in a temperate East Polynesia society, the Moriori, Rekohu (Chatham Island). *Quaternary Science Reviews* 149: 306–25.

Mayaux, P., Bartholomé, E., Fritz, S., Belward, A. 2004. A new land-cover map of Africa for the year 2000. *Journal of Biogeography* 31: 861–77.

Mayle, F. E., Beerling, D. J., Gosling, W. D., Bush, M. B. 2004. Responses of Amazonian ecosystems to climatic and atmospheric carbon dioxide changes since the Last Glacial Maximum. *Philosophical Transactions of the Royal Society B: Biological Sciences* 359. doi:10.1098/rstb.2003.1434.

Medina, J. T. 1934. *The Discovery of the Amazon According to the Account of Friar Gaspar de Carvajal and Other Documents.* Translated and edited by B. T. Lee, H. C. Heaton. New York: American Geographical Society Special Publication 17.

Medina-Elizalde, M., Rohling, E. J. 2012. Collapse of Classic Maya civilization related to modest reduction in precipitation. *Science* 335: 956–9.

Medina-Elizalde, M., Burns, S. J., Lea, D. W., Asmerom, Y., von Gunten, L., Polyak, V., Vuille, M., Karmalkar, A. 2010. High resolution stalagmite climate record from the Yucatán Peninsula spanning the Maya Terminal Classic period. *Earth and Planetary Science Letters* 298: 255–62.

Medway, L. 1972. Phenology of a tropical rainforest in Malaya. *Biological Journal of the Linnean Society* 4: 117–46.

Meehl, G. A., Stocker, T. F., Collins, W. D., Friedlingstein, P., Gaye, A. T., Gregory, J. M., Kitoh, A., Knutti, R., Murphy, J. M., Noda, A., Raper, S. C. B., Watterson, I. G., Weaver, A. J., Zhao, Z.-C. 2007. Global climate projections. In S. Solomon, D. Qin, M. Manning, Z. Chen, M. Marquis, K. B. Avery, M. Tignor, H. L. Miller (eds.), *Climate Change 2007: The Physical Science Basis. Contribution of Working Group I to the Fourth Assessment Report of the Intergovernmental Panel on Climate Change.* Cambridge: Cambridge University Press, pp. 749–845.

Meggers, B. J. 1954. Environmental limitation on the development of culture. *American Anthropologist* 56: 801–24.

Meggers, B. J. 1971. *Amazonia: Man and Culture in a Counterfeit Paradise.* Chicago: Harlan Davidson.

Meggers, B. J. 1977. Vegetational fluctuation and prehistoric cultural adaptations in Amazonia: Some tentative correlations. *World Archaeology* 8: 287–303.

Meggers, B. J. 1987. The early history of man in Amazonia. In T. Whitmore, G. Prance (eds.), *Biogeography and Quaternary History in Tropical America.* Oxford: Clarendon Press, pp. 151–74.

Meggers, B. J. 2003. Natural versus anthropogenic sources of Amazonian biodiversity: The continuing quest for El Dorado. In G. A. Bradshaw, P. A. Marquet (eds.), *How Landscapes Change.* Dordrecht: Springer, pp. 89–107.

Meggers, B. J., Evans, C. 1957. *Archeological Investigations at the Mouth of the Amazon.* Washington, DC: US Government Printing Office.

Meggers, B. J., Miller, E. T. 2002. Hunter-gatherers in Amazonia during the Pleistocene–Holocene transition. In J. Mercader (ed.), *Under the Canopy: The Archaeology of Tropical Rain Forests.* Piscataway, NJ: Rutgers University Press, pp. 291–316.

Meggers, B. J., Ayensu, E. S., Duckworth, W. D., Association for Tropical Biology. 1973. *Tropical Forest Ecosystems in Africa and South America: A Comparative Review.* Washington, DC: Smithsonian Institution Press.

Meijer, H. J. M., Awe Due, R. 2010. A new species of giant marabou stork (Aves: Ciconiiformes) from the Pleistocene of Liang Bua, Flores (Indonesia). *Zoological Journal of the Linnean Society* 160: 707–24.

Mellars, P. 2005. The impossible coincidence: A single-species model for the origins of modern human behavior in Europe. *Evolutionary Anthropology* 14: 12–27.

Mellars, P. 2006. Why did modern human populations disperse from Africa ca. 60,000 years ago? A new model. *Proceedings of the National Academy of Sciences of the United States of America* 103: 9381–6.

Mellars, P. A., Goric, K. C., Carr, M., Soares, P. A., Richards, M. B. 2013. Genetic and archaeological perspectives on the initial Modern Human colonization of Southern Asia. *Proceedings of the National Academy of Sciences of the United States of America* 110: 10699–704.

Meltzer, D. J., Adovasio, J. M., Dillehay, T. D. 1994. On a Pleistocene human occupation at Pedra Furada, Brazil. *Antiquity* 68: 695–714.

Meltzer, D. J., Grayson, D. K., Ardila, G., Barker, A. W., Dincauze, D. F., Haynes, C. V., Mena, F., Nunez, L., Stanford, D. J. 1997. On the Pleistocene antiquity of Monte Verde, Southern Chile. *American Antiquity* 62: 659–63.

Menkhaus, S., Lober, D. J. 1995. International ecotourism and the valuation of tropical rainforests in Costa Rica. *Journal of Environmental Management* 47: 1–10.

Mercader, J. 2002a. Forest people: The role of African rainforests in human evolution and dispersal. *Evolutionary Anthropology* 11: 117–24.

Mercader, J. (ed.). 2002b. *Under the Canopy: The Archaeology of Tropical Rain Forests.* Piscataway, NJ: Rutgers University Press.

Mercader, J., Martí, R. 1999. Archaeology in the tropical forest of Banyang-Mbo, SW Cameroon. *Nyame Akuma* 52: 17–24.

Mercader, J., Martí, R. 2002. The hunter-gatherer occupation of Atlantic central Africa: New evidence from Equatorial Guinea and Cameroon. In J. Mercader (ed.), *Under the Canopy: The Archaeology of Tropical Rainforests.* Piscataway, NJ: Rutgers University Press, pp. 64–92.

Mercader, J., Runge, F., Vrydaghs, L., Doutrelepont, H., Corneille, E., Juan-Tresseras, J. 2000. Phytoliths from archaeological sites in the tropical forest of Ituri, Democratic Republic of Congo. *Quaternary Research* 54: 102–12.

Mercader, J., Panger, M., Boesch, C. 2002. Excavation of a chimpanzee stone tool site in the African rainforest. *Science* 296: 1452–5.

Mercader, J., Martí, R., González, I., Sánchez, A., García, P. 2003. Archaeological site formation in tropical forests: Insights from the Ituri rock shelters, Congo. *Journal of Archaeological Science* 30: 45–65.

Mercader, J., Barton, H., Gillespie, J., Harris, J., Kuhn, S., Tyler, R., Boesch, C. 2007. 4,300-year-old chimpanzee sites and the origins of percussive stone technology. *Proceedings of the National Academy of Sciences of the United States of America* 104: 3043–8.

Mercader, J., Bennett, T., Esselmont, C., Simpson, S., Walde, D. 2013. Phytoliths from Midle Stone Age habitats in the Mozambican Rift (105–29 ka). *Journal of Human Evolution* 64: 328–36.

Meyer, M., Kircher, M., Gansauge, M.-T., Li, H., Racimo, F., Mallick, S., Schraiber, J. G., Jay, F., Prüfer, K., de Filippo, C., Sudmant, P. H., Alkan, C., Fu, Q., Do, R., Rohland, N., Tandon, A., Siebauer, M., Green, R. E., Bryc, K., Briggs, A. W., Stenzel, U., Dabney, J., Shendure, J., Kitzman, J., Hammer, M. F., Shunkov, M. V., Derevianko, A. P., Patterson, N., Andrés, A. M., Eichler, E. E., Slatkin, M., Reich, D., Kelso, J., Pääbo, S. 2012. A high-coverage genome sequence from an archaic Denisovan individual. *Science* 338: 222–6.

Meyer, M., Arsuaga, J. L., de Filippo, C., Nagel, S., Aximu-Petri, A., Nickel, B., Martinez, I., Gracia, A., Bermudez de Castro, J. M., Carbonell, E., Viola, B., Kelso, J., Prufer, K., Pääbo, S. 2016. Nuclear DNA sequences from the Middle Pleistocene Sima de los Huesos hominins. *Nature* 531: 504–7.

Michel, V., Valladas, H., Shen, G., Wang, W., Zhao, J.-X., Shen, C.-C., Valensi, P., Bae, C. J. 2016. The earliest modern *Homo sapiens* in China? *Journal of Human Evolution* 101: 101–4.

Mijares, A. S. B., Detroit, F., Piper, P., Grün, R., Bellwood, P., Aubert, M., Champion, G., Cuevas, N., De Leon, A., Dizon, E. 2010. New evidence for a 67,000-year-old human presence at Callao Cave, Luzon, Philippines. *Journal of Human Evolution* 59: 123–32.

Miksic, J. N. 1990. *Borobudur: Golden Tales of the Buddhas.* Boston: Shambala.

Miksic, J. N. 1999. Water, urbanization, and disease in ancient Indonesia. In E. A. Bacus, L. J. Lucero (eds.), *Complex Polities in the Ancient Tropical World.* Archeological Papers of the American Anthropological Association No. 9. Arlington: American Anthropological Association, pp. 167–84.

Millennium Ecosystem Assessment (MEA). 2005. *Ecosystems and Human Well-Being: Current State and Trends.* Washington, DC: Island Press.

Miller, M. E. 1999. *Maya Art and Architecture.* London: Thames & Hudson.

Miner, E. R. (ed.). 1972. *John Dryden.* London: Bell.

Mintz, S. W. 1986. *Sweetness and Power: The Place of Sugar in Modern History.* London: Penguin Books.

Mitchell, P. 2005. *African Connections: Archaeological Perspectives on Africa and the Wider World.* Lanham, MD: AltaMira Press.

Mitchell, P. 2015. *Horse Nations: The Worldwide Impact of the Horse of Indigenous Societies Post-1492.* Oxford: Oxford University Press.

Mitchell, P. 2017. Disease: A hitherto unexplored constraint on the spread of dogs (*Canis lupis familiaris*) in pre-Columbian South America. *Journal of World Prehistory* 30: 301–49.

Mittermeier, R. A. 1987. Effects of hunting on rain forest primates. In C. Marsh, R. Mittermeier (eds.), *Primate Conservation in the Tropical Rain Forest.* New York: Alan R. Liss, pp. 109–48.

Molnar, A., France, M., Purdy, L., Karver, J. 2010. *Community-Based Forest Management: The Extent and Potential Scope of Community and Smallholder Forest Management and Enterprises.* Washington, DC: Rights and Resources Initiative.

Montané, J. 1968. Paleoindian remains from laguna Tagua-tagua Central Chile. *Science* 161: 1137–8.

Moreno-Mayar, V. J., Potter, B. A., Vinner, L., Steinrücken, M., Rasmussen, S., Terhorst, J., Kamm, J. A., Albrechtsen, A., Malaspinas, A.-S., Sikora, M., Reuther, J. D., Irish, J. D., Malhi, R. S., Orlando, L., Song, Y. S., Nielsen, R., Meltzer, D. J., Willerslev, E. 2018. Terminal Pleistocene Alaskan genome reveals first founding population of Native Americans. *Nature* 553: 203–8.

Morison, S. E. 1993. *The European Discovery of America, Vol. 2: The Southern Voyages 1492–1616.* New York: Oxford University Press.

Moritz, C., Dick, C. W., Bermingham, E. 2014. From the past to the future: Evolution, ecology, and conservation of tropical rainforests. In E. Bermingham, C. W. Dick, C. Moritz (eds.), *Tropical Rainforests: Past, Present, and Future.* Chicago: University of Chicago Press, pp. 1–6.

Morley, R. J. 2000. *Origin and Evolution of Tropical Rain Forests.* Chichester: John Wiley.

Morley, R. J. 2003. Interplate dispersal paths for megathermal Angiosperms. *Perspectives in Plant Ecology Evolution and Systematics* 6: 5–20.

Morris, B. 1977. Tappers, trappers, and the Hill Pandaram (South India). *Anthropos* 72: 225–41.

Morris, B. 1982. *Hill Traders: A Socioeconomic Study of the Hill Pandaram.* London: Athlone Press.

Morrison, K. 1997. Commerce and culture in South Asia: Perspectives from archaeology and history. *Annual Review of Anthropology* 26: 87–108.

Morrison, K. 2002a. Historicizing adaptation, adapting to history: Forager-traders in South and Southeast Asia. In K. Morrison, L. Junker (eds.), *Forager-Traders in South and Southeast Asia.* Cambridge: Cambridge University Press, pp. 1–20.

Morrison, K. 2002b. Introduction. In K. Morrison, L. Junker (eds.), *Forager-Traders in South and Southeast Asia.* Cambridge: Cambridge University Press, pp. 21–40.

Morrison, K. 2014. Human–forest relationships and the erasure of history. In S. B. Hecht, K. D. Morrison, C. Padoch (eds.), *The Social Lives of Forests: Past, Present, and Future of Woodland Resurgence.* Chicago: University of Chicago Press, pp. 143–7.

Morrison, K. 2018a (forthcoming). Provincializing the Anthropocene: Eurocentrism in the earth system. In M. Rangarajan (ed.), *Nature in History*. Oxford: Oxford University Press.

Morrison, K. 2018b (forthcoming). Discourses of the remnant: People, forests, and sacred groves of Southern India. In A. Baviskar (ed.), *Nature Today: Studies of Ecology and Environment*. Delhi: Oxford University Press.

Morrison, K., Junker, L. (eds.). 2002. *Forager-Traders in South and Southeast Asia*. Cambridge: Cambridge University Press.

Morwood, M. J., Brown, P., Jatmiko, Sutikna, T., Saptomo, E. W., Westaway, K. E., Due, R. A., Roberts, R. G., Maeda, T., Wasisto, S., Djubiantono, T. 2005. Further evidence for small-bodied hominins from the Late Pleistocene of Flores, Indonesia. *Nature* 437: 1012–17.

Moss, P. T., Kershaw, A. P. 2007. A late Quaternary marine palynological record (oxygen isotope stages 1 to 7) for the humid tropics of northeastern Australia based on ODP site 820. *Palaeogeography, Palaeoclimatology, Palaeoecology* 251: 4–22.

Mountain, M.-J. 1991. Highland New Guinea hunter-gatherers: The evidence of Nombe Rockshelter, Simbu, with emphasis on the Pleistocene, 2 vols. PhD thesis, Australian National University, Canberra.

Movius, H. L. 1948. The Lower Palaeolithic cultures of Southern and Eastern Asia. *Transactions of the American Philosophical Society* 38: 329–420.

Moy, C. M., Seltzer, G. O., Rodbell, D. T., Anderson, D. M. 2002. Variability of El Niño/ Southern Oscillation activity at millennial timescales during the Holocene epoch. *Nature* 420: 162–5.

Moyà-Solà, S., Alba, D. M., Almécija, S., Casanovas-Vilar, I., Köhler, M., De Esteban-Trivigno, S., Robles, J. M., Galindo, J., Fortuny, J. 2009. A unique Middle Miocene European hominoid and the origins of the great ape and human clade. *Proceedings of the National Academy of Sciences of the United States of America* 106: 9601–6.

Mudar, K. M., Anderson, D. 2007. New evidence for Southeast Asian Pleistocene foraging economies: Faunal remains from the early levels of Lang Rongrien rockshelter, Krabi, Thailand. *Asian Perspectives* 46: 298–334.

Murakami, H. 2002. *After the Quake* (translated by Jay Rubin). New York: Knopf.

Murphy, P. G., Lugo, A. E. 1986. Ecology of tropical dry forest. *Annual Review of Ecology and Systematics* 17: 67–88.

Murray, M. A., Fuller, D. Q., Cappeza, C. 2007. Crop production on the Senegal River in the early first millennium AD: Preliminary archaeobotanical results from Cubalel. In R. T. J. Cappers (ed.), *Fields of Change: Progress in African Archaeobotany*. Groningen: Barkhuis, pp. 63–70.

Murty, M. L. K. 1978. Symbiosis and traditional behaviour in subsistence economies of the Kunchapuri Yeruklas of South India: A predictive model. *Puratattva* 10: 50–61.

Myers, S. S., Gaffikin, L., Golden, C. D., Ostfeld, R. S., Redford, K. H., Ricketts, T. H., Turner, W. R., Osofsky, S. A. 2013. Human health impacts of ecosystem alteration. *Proceedings of the National Academy of Sciences of the United States of America* 110: 18753–60.

Nance, J. 1975. *The Gentle Tasaday: A Stone Age People in the Philippine Rain Forest*. New York: Harcourt Brace Jovanovich.

Nasi, R., Brown, D., Wilkie, D., Bennett, E., Tutin, C., Van Tol, G., Christophersen, T. 2008. *Conservation and Use of Wildlife-Based Resources: The Bushmeat Crisis*. Bogor, Indonesia: Secretariat of the Convention on Biological Diversity and Center for International Forestry Research (CIFOR).

Nations, J. D., Nigh, R. B. 1980. The evolutionary potential of Lacandon Maya sustained yield tropical forest agriculture. *Journal of Anthropological Research* 36: 1–30.

Naughty Dog, Bend Studio, Bluepoint Games, Playspree. 2016. *Uncharted 4: A Thief's End*. Tokyo: Sony Interactive Entertainment.

Neich, R. 2006. Pacific voyaging after the exploration period. In K. R. Howe (ed.), *Vaka Moana, Voyages of the Ancestors: The Discovery and Settlement of the Pacific*. Auckland: David Bateman, pp. 198–245.

Nelson, S. 2003. *The Extinction of* Sivapithecus*: Faunal and Environmental Changes in the Siwaliks of Pakistan*. American School of Prehistoric Research Monographs, volume 1. Boston: Brill Academic Publishers.

Nengo, I., Tafforeau, P., Gilbert, C. C., Fleagle, J. G., Miller, E. R., Feibel, C., Fox, D. L., Feinberg, J., Pugh, K. D., Berruyer, C., Mana, S., Engle, Z., Spoor, F. 2017. New infant cranium from the African Miocene sheds light on ape evolution. *Nature* 548: 169–74.

Nepstad, D. C., Stickler, C. M., Soares-Filho, B., Merry, F. 2008. Interactions among Amazon land use, forests and climate: Prospects for a near-term forest tipping point. *Philosophical Transactions of the Royal Society B: Biological Sciences* 363: 1737–46.

Neumann, K., Hildebrand, E. 2009. Early bananas in Africa: The state of the art. *Ethnobotany Research and Applications* 7: 353–62.

Neumann, K., Bostoen, K., Höhn, A., Kahlheber, S., Ngomanda, A., Tchiengué, B. 2011. First farmers in the Central African rainforest: A view from southern Cameroon. *Quaternary International* 249: 53–62.

Neves, E. G. 2013. Was agriculture a key productive activity in pre-colonial Amazonia? The stable productive basis for social equality in the Central Amazon. In E. S. Brondizio, E. F. Moran (eds.), *Human–Environment Interactions: Current and Future Directions*. Dordrecht: Springer, pp. 371–88.

Neves, E. G., Bartone, R. N., Petersen, J. B., Heckenberger, M. J. 2001. The timing of Terra Preta formation in the central Amazon: New data from three sites. In B. Glaser, W. I. Woods (eds.), *Amazonian Dark Earths: Explorations in Space and Time*. Dordrecht: Springer, pp. 125–34.

Neves, E. G., Petersen, J. B., Bartone, R. N., Da Silva, C. A. 2003. Historical and socio-cultural origins of Amazonian dark earths. In J. Lehmann, D. C. Kern, B. Glaser, W. I. Woods (eds.), *Amazonian Dark Earths: Origin, Properties, Management*. Dordrecht: Kluwer Academic Publishers, pp. 29–49.

Neves, W. A., Hubbe, M. 2005. Cranial morphology of early Americans from Lagoa Santa, Brazil: Implications for the settlement of the New World. *Proceedings of the National Academy of Sciences of the United States of America* 102: 18309–14.

Ngomanda, A., Cheptow-Lusty, A., Makaya, M., Favier, C., Schevin, P., Maley, J., Fontugne, M., Oslisly, R., Jolly, D. 2009. Western equatorial African forest-savanna mosaics: A legacy of late Holocene climatic change? *Climate of the Past* 5: 647–59.

Nicholl, C. 1995. *The Creature in the Map: A Journey to El Dorado*. New York: William Morrow and Company.

Nicholson, C. F., Blake, R. W., Lee, D. R. 1995. Livestock, deforestation, and policy making: Intensification of cattle production systems in Central America revisited. *Journal of Dairy Science* 78: 719–34.

Nigh, R. 2008. Trees, fire and farmers: Making woods and soil in the Maya forest. *Journal of Ethnobiology* 28: 231–43.

Nimuendaju, C. U. 1949. Os Tapajo. *Boletim do Museu Paraense Emilio Goeldi* 10: 93–108.

Nissen, H. J. 1988. *The Early History of the Ancient Near East, 9000–2000* BC. Chicago: University of Chicago Press.

Nobre, C. A., Sampaio, G., Borma, L. S., Castilla-Rubio, J. C., Silva, J. S., Cardoso, M. 2016. Land-use and climate change risks in the Amazon and the need of a novel sustainable development paradigm. *Proceedings of the National Academy of Sciences of the United States of America* 113: 10759–68.

Nogueira-Filho, S. L. G., Nogueira, S. S. C., Fragoso, J. M. V. 2009. Ecological impacts of feral pigs in the Hawaiian Islands. *Biodiversity and Conservation* 18: 3677–83.

Nolan, C. 2016. *Interstellar*. Hollywood, California: Paramount Pictures, Warner Bros. Pictures.

Northrup, D. 2010. *The Atlantic Slave Trade* (3rd edition). Boston, MA: Houghton Mifflin.

Noss, A. J. 1997. The economic importance of communal net hunting among the BaAka of the Central African Republic. *Human Ecology* 25: 71–89.

Noss, A. J. 1998. The impacts of cable snare hunting on wildlife populations in the forests of the Central African Republic. *Conservation Biology* 12: 390–8.

Nunn, N., Qian, N. 2010. The Columbian exchange: A History of disease, food, and ideas. *Journal of Economic Perspectives* 24: 163–88.

O'Brien, E. 1981. The projectile capabilities of an Acheulian handaxe from Olorgesailie. *Current Anthropology* 22: 76–9.

O'Connell, J. F., Allen, J. 2012. The restaurant at the end of the universe: Modeling the colonization of Sahul. *Australian Archaeology* 74: 5–31.

O'Connor, S., Ono, R., Clarkson, C. 2011. Pelagic fishing at 42,000 years before the present and the maritime skills of modern humans. *Science* 334: 1117–21.

O'Connor, S., Robertson, G., Aplin, K. P. 2014. Are osseous artefacts a window to perishable material culture? Implications of an unusually complex bone tool from the Late Pleistocene of East Timor. *Journal of Human Evolution* 67: 108–19.

O'Reilly, D. J. W. 2007. *Early Civilizations of Southeast Asia*. New York: AltaMira Press.

Ocheje, P. D. 2007. 'In the public interest': Forced evictions, land rights and human development in Africa. *Journal of African Law* 51: 173–214.

Ochoa, G. A. 2012. Ice age hunter-gatherers and the colonization of Mesoamerica. In D. L. Nichols, C. A. Pool (eds.), *The Oxford Handbook of Mesoamerican Archaeology*. Oxford: Oxford University Press, pp. 129–40.

Ochsenius, C., Gruhn, R. (eds.). 1979. *Taima-taima: A Late Paleo-Indian Kill Site in Northernmost South America. Final Reports of the 1976 Excavations*. Federal Republic of Germany: CIPICS/South American Quaternary Documentation Programme.

Ockwell, D., Lovett, J. C. 2005. Fire assisted pastoralism vs. sustainable forestry: The implications of missing markets for carbon in determining optimal land use in the wet-dry tropics of Australia. *Journal of Environmental Management* 75: 1–9.

Odum, H. T., Pigeon, R. F. (eds.). 1970. *A Tropical Rain Forest*. Tennessee: US Atomic Energy Commission.

Ogden, F. L., Crouch, T. D., Stallard, R. F., Hall, J. S. 2013. Effect of land cover and use on dry season river runoff, runoff efficiency, and peak storm runoff in the seasonal tropics of central Panama. *Water Resources Research* 49: 8443–62.

Oliver, J. R., Alexander, C. S. 2003. Ocupaciones humanas del Plesitoceno terminal en el Occidente de Venezuela. *Maguare* 17: 83–246.

Olivero, J., Fa, J. E., Real, R., Farfán, M. A., Márquez, A. L., Vargas, J. M., Gonzalez, J. P., Cunningham, A. A., Nasi, R. 2017. Mammalian biogeography and the Ebola virus in Africa. *Mammal Review* 47: 24–37.

Olsen, K. M. 2002. Population history of *Manihot esculenta* (Euphorbiaceae) inferred from nuclear DNA sequences. *Molecular Ecology* 11: 901–11.

Olsen, K. M., Schaal, B. A. 1999. Evidence on the origin of cassava: Phylogeography of Manihot esculenta. *Proceedings of the National Academy of Sciences of the United States of America* 96: 5586–91.

Olsen, K. M., Schaal, B. A. 2006. DNA sequence data and inference on cassava's origin of domestication. In M. A. Zeder, D. G. Bradley, E. Emshwiller, B. D. Smith (eds.), *Documenting Domestication: New Genetic and Archaeological Paradigms*. Berkeley: University of California Press, pp. 123–33.

Olson, D. M., Dinerstein, E., Wikramanyake, E. D., Burgess, N. D., Powell, G. V. N., Underwood, E. C., D'Amico, J. A., Itoua, I., Strand, H. E., Morrison, J. C., Loucks,

C. J., Allnutt, T., Ricketts, T. H., Kura, Y., Lamoreux, J. F., Wettengel, W. W., Hedao, P., Kassem, K. R. 2001. Terrestrial ecoregions of the world: A new map of life on Earth. *Bioscience* 51: 933–8.

Osborne, P. L. 2000. *Tropical Ecosystems and Ecological Concepts*. Cambridge: Cambridge University Press.

Pagani, L., Lawson, D. J., Jagoda, E., Mörseburg, A., Eriksson, A., Mitt, M., Clemente, F., Hudjashov, G., DeGiorgio, M., Saag, L., Wall, J. D., Cardona, A., Mägi, R., Wilson Sayres, M. A., Kaewert, S., Inchley, C., Scheib, C. L., Jarve, M., Karmin, M., Jacobs, G. S., Antao, T., Iliescu, F. M., Kushniarevich, A., Qyub, Q., Tyler-Smith, C., Xue, Y., Yunusbayev, B., Tambets, K., Basu Mallick, C., Saag, L., Pocheshkhova, E., Andriadze, G., Muller, C., Westaway, M. C., Lambert, D. M., Zoraqi, G., Turdikulova, S., Dalimova, D., Sabitov, Z., Sultana, G. N. N., Lachance, J., Tishkoff, S., Momynaliev, K., Isakova, J., Damba, L. D., Gubina, M., Nymadawa, P., Evseeva, I., Atramentova, L., Utevska, O., Ricaut, F.-X., Brucato, N., Sudoyo, H., Letellier, T., Cox, M. P., Barashkov, N. A., Škaro, V., Mulahasanovic, L., Primorac, D., Sahakyan, H., Mormina, M., Eichstaedt, C. A., Lichman, D. V., Abdullah, S., Chaubey, G., Wee, J. T. S., Mihailov, E., Karunas, A., Litvinov, S., Khusainova, R., Ekomasova, N., Akhmetova, V., Khidiyatova, I., Marjanović, D., Yepiskoposyan, L., Behar, D. M., Balanovska, E., Metspau, A., Derenko, M., Malyarchuk, B., Voevoda, M., Fedorova, S. A., Osipova, L. P., Lahr, M. M., Gerbault, P., Leavesley, M., Migliano, A. B., Petraglia, M., Balanovksy, O., Khsnutdinova, E. K., Metspalu, E., Thomas, M. G., Manica, A., Nielsen, R., Villems, R., Willerslev, E., Kivisild, T., Metspalu, M. 2016. Genomic analyses inform on migration events during the peopling of Eurasia. *Nature* 538: 238–42.

Pagán-Jiménez, J. R., Rodríguez-Ramos, R., Reid, B. A., van den Bel, M., Hofman, C. L. 2015. Early dispersals of maize and other food plants into the southern Caribbean and northeastern South America. *Quaternary Science Reviews* 123: 231–46.

Page, S. E., Rieley, J. O., Wust, R. A. J. 2006. Lowland tropical peatland of Southeast Asia. In I. P. Martini, A. Martínez-Cortizas, W. Chesworth (eds.), *Peatlands: Evolution and Records of Environmental and Climate Changes*. Amsterdam: Elsevier, pp. 145–72.

Paine, O. C. C., Koppa, A., Henry, A., Leichliter, J. N., Codron, D., Codron, J., Lambert, J. E., Sponheimer, M. 2018. Grass leaves as potential hominin dietary resources. *Journal of Human Evolution* 117: 44–52.

Pan, Y., Birdsey, R. A., Fang, J., Houghton, R., Kauppi, P. E., Kurz, W. A., Phillips, O. L., Shvidenko, A., Lewis, S. L., Canadell, J. G., Ciais, P., Jackson, R. B., Pacala, S. W., McGuire, A. D., Piao, S., Rautiainen, A., Sitch, S., Hayes, D. 2011. A large and persistent carbon sink in the world's forests. *Science* 333: 988–93.

Panger, M. A, Brooks, A. S., Richmond, B. G., Wood, B. 2002. Older than the Oldowan? Rethinking the emergence of hominin tool use. *Evolutionary Anthropology* 11: 235–45.

Papworth S., Milner-Gulland, E. J., Slocombe, K. 2013. Hunted woolly monkeys (*Lagothrix poeppigii*) show threat-sensitive responses to human presence. *PLoS ONE* 8: e62000.

Parfitt, S. A., Barendregt, R. W., Breda, M., Candy, I., Collins, M. J., Coope, G. R., Durbidge, P., Field, M. H., Lee, J. R., Lister, A. M., Mutch, R., Penkman, K. E. H., Preece, R. C., Rose, J., Stringer, C. B., Symmons, R., Whittaker, J. E., Wymer, J. J., Stuart, A. J. 2005. The earliest record of human activity in northern Europe. *Nature* 438: 1008–12.

Parfitt, S. A., Ashton, N. M., Lewis, S. G., Abel, R. L., Coope, G. R., Field, M. H., Gale, R., Hoare, P. G., Larkin, N. R., Lewis, M. D., Karloukovski, V., Maher, B. A., Peglar, S. M., Preece, R. C., Whittaker, J. E., Stringer, C. B. 2010. Early Pleistocene human occupation at the edge of the boreal zone in northwest Europe. *Nature* 466: 229–33.

Parker, G. G. 1995. Structure and microclimate of forest canopies. In M. D. Lowman, N. M. Nadkarni (eds.), *Forest Canopies*. San Diego: Academic Press, pp. 73–106.

Parker, H. 1909. *Ancient Ceylon: An Account of the Aborigines and of Part of the Early Civilization*. London: Luzac.

Parr, L. A., Cohen, M., de Waal, F. B. M. 2005. The influence of social context on the use of blended and graded facial displays in chimpanzees (*Pan troglodytes*). *International Journal of Primatology* 26: 73–103.

Parton, P., Clark, G., Reepmeyer, C., Burley, D. 2018. The field of war: LiDAR identification of earthwork defences on Tongatapu Island, Kingdom of Tonga. *Journal of Pacific Archaeology* 8: 11–24.

Pasveer, J. M. 2004. *The Djef Hunters: 26,000 Years of Rainforest Exploitation on the Bird's Head of Papua, Indonesia*. Leiden: Balkema.

Patin, E., Laval, G., Barreiro, L. B., Salas, A., Semino, O., Santachiara-Benerecetti, S., Kidd, K. K., Kidd, J. R., Van der Veen, L., Hombert, J.-M., Gessain, A., Froment, A., Bahuchet, S., Heyer, E., Quintana-Murci, L. 2009. Inferring the demographic history of African farmers and pygmy hunter-gatherers using a multilocus resequencing data set. *PLoS Genetics* 5: e1000448.

Patterson, N., Richter, D. J., Gnerre, S., Lander, E. S., Reich, D. 2006. Genetic evidence for complex speciation of humans and chimpanzees. *Nature* 441: 1103–8.

Patton, J. L., da Silva, M. N. F. 2014. The history of Amazonian mammals: Mechanisms and timing of diversification. In E. Bermingham, C. W. Dick, C. Mortiz (eds.), *Tropical Rainforests: Past, Present and Future*. Chicago: University of Chicago Press, pp. 107–26.

Pavlides, C. 2004. From Misisil Cave to Eliva hamlet: Rediscovering the Pleistocene in interior West New Britain. In V. Auenbrow, R. Fullager (eds.), *A Pacific Odyssey: Archaeology and Anthropology in the Western Pacific. Papers in Honour of Jim Specht*. Records of the Australian Museum, Supplement 29. Sydney: Australian Museum, pp. 97–108.

Pavlides, C., Gosden, C. 1994. 35,000-year-old sites in the rainforests of West New Britain, Papua New Guinea. *Antiquity* 68: 604–10.

Pawlik, A. F., Piper, P. J., Mijares, A. S. B. 2014a. Modern humans in the Philippines: Colonization, subsistence and new insights into behavioural complexity. In R. Dennell, M. Poor (eds.), *Southern Asia, Australia and the Search for Human Origins*. Cambridge: Cambridge University Press, pp. 135–47.

Pawlik, A. F., Piper, P. J., Faylona, M. G. P. G., Padilla Jr, S. G., Carlos, J., Mijares, A. S. B., Vallejo, B., Reyes, M., Amano, N., Ingicco, T., Porr, M. 2014b. Adaptation and foraging from the Terminal Pleistocene to the Early Holocene: Excavation at Bubog on Ilin Island, Philippines. *Journal of Field Archaeology* 39: 230–47.

Paz, V. J. 2002. Island Southeast Asia: Spread or friction zone? In P. Bellwood, C. Renfrew (eds.), *Examining the Farming/Language Dispersal Hypothesis*. Cambridge: McDonald Institute for Archaeological Research, pp. 275–85.

Paz, V. J. 2005. Rock shelters, caves, and archaeobotany in island Southeast Asia. *Asian Perspectives* 44: 107–18.

Penny, D., Pottier, C., Fletcher, R., Barbetti, M., Fink, D., Hua, Q. 2006. Vegetation and land-use at Angkor, Cambodia: A dated pollen sequence from the Bakong Temple moat. *Antiquity* 80: 599–614.

Penny, D., Chevance, J.-B., Tang, D., De Greef, S. 2014. The environmental impact of Cambodia's ancient city of Mahendraparvata (Phnom Kulen). *PLoS ONE* 9: e84252.

Perera, N. 2010. *Prehistoric Sri Lanka: Late Pleistocene Rockshelters and an Open Air Site*. Oxford: Archaeopress.

Perera, N., Kourampas, N., Simpson, I. A., Deraniyagala, S. U., Bulbeck, D., Kamminga, J., Perera, J., Fuller, D. Q., Szabo, K., Oliveira, N. V. 2011. People of the ancient rainforest: Late Pleistocene foragers at the Batadomba-lena rockshelter, Sri Lanka. *Journal of Human Evolution* 61: 254–69.

Perera, N., Roberts, P., Petraglia, M. 2016. Bone technology in South Asia from Late Pleistocene rockshelter deposits in Sri Lanka. In M. C. Langley (ed.), *Osseous Projectile Weaponry: Towards an Understanding of Pleistocene Cultural Variability*. VERT Series. New York: Springer Verlag, pp. 173–88.

Peres, C. A. 2001. Synergistic effects of subsistence hunting and habitat fragmentation on Amazonian forest vertebrates. *Conservation Biology* 15: 1490–505.

Peres, C. A., Dolman, P. M. 2000. Density compensation in Neotropical primate communities: Evidence from 56 hunted and nonhunted Amazonian forests of varying productivity. *Oecologia* 122: 175–89.

Peres, C. A., Nascimento, H. S. 2006. Impact of game hunting by the Kayapó of southeastern Amazonia: Implications for wildlife conservation in tropical forest indigenous reserves. *Biodiversity and Conservation* 15: 2627–53.

Pérez, P. L. 1928. Los Aetas y Ilongotes de Filipinas. *Archivo Ibero Americano* 15: 71–106.

Perrier, X., Bakry, F., Carreel, F., Jenny, C., Horry, J.-P., Lebot, V., Hippolyte, I. 2009. Combining biological approaches to shed light on the evolution of edible bananas. *Ethnobotany Research and Applications* 7: 199–216.

Perrier, X., De Langhe, E., Donohue, M., Lentfer, C., Vrydaghs, L., Bakry, F., Carreel, F., Hippolyte, I., Horry, J.-P., Jenny, C., Lebot, V., Risterucci, A.-M., Tomekpe, K., Doutrelepont, H., Ball, T., Manwaring, J., de Maret, P., Denham, T. 2011. Multidisicplinary perspectives on banana (Musa spp.) domestication. *Proceedings of the National Academy of Sciences of the United States of America* 108: 11311–18.

Perry, G. H., Dominy, N. J. 2009. Evolution of the human pygmy phenotype. *Trends in Ecology & Evolution* 24: 218–25.

Perry, G. H., Verdu, P. 2017. Genomic perspectives on the history and evolutionary ecology of tropical rainforest occupation by humans. *Quaternary International* 448: 150–7.

Perry, G. H., Foll, M., Grenier, J.-C., Patin, E., Nedelec, Y., Pacis, A., Barakatt, M., Gravel, S., Zhou, X., Nsobya, S. L., Excoffier, L., Quintana-Murci, L., Dominy, N. J., Barreiro, L. B. 2014. Adaptive, convergent origins of the pygmy phenotype in African rainforest hunter-gatherers. *Proceedings of the National Academy of Sciences of the United States of America* 111: E3596–E3603.

Perry, L., Dickau, R., Zarrillo, S., Holst, I., Pearsall, D. M., Piperno, D. R., Berman, M. J., Cooke, R. G., Rademaker, K., Ranere, A. J., Raymond, J. S., Sandweiss, D. H., Scaramelli, F., Tarble, K., Zeidler, J. A. 2007. Starch fossils and the domestication and dispersal of chili peppers (Capsicum spp. L.) in the Americas. *Science* 315: 986–8.

Peterhans, J. C. K., Wrangham, R. W., Carter, M. L., Hauser, M. D. 1993. A contribution to tropical rain forest taphonomy: Retrieval and documentation of chimpanzee remains from Kibale Forest, Uganda. *Journal of Human Evolution* 25: 485–514.

Petraglia, M. D., Haslam, M., Fuller, D. Q., Boivin, N., Clarkson, C. 2010. Out of Africa: New hypotheses and evidence for the dispersal of *Homo sapiens* along the Indian Ocean rim. *Annals of Human Biology* 37: 288–311.

Pickersgill, B. 1976. Pineapple. In N. W.Simmonds (ed.), *Evolution of Crop Plants*. London: Longman, pp. 14–18.

Pickford, M. 2012. Orrorin and the African ape/hominid dichotomy. In S. C. Reynolds, A. Gallagher (eds.), *African Genesis: Perspectives on Hominin Evolution*. Cambridge: Cambridge University Press, pp. 99–119.

Pickford, M., Senut, B., Gommery, D., Treil, J. 2002. Bipedalism in Orrorin tugenensis revealed by its femora. *Comptes Rendus Palevol* 1: 191–203.

Pike, A. 2016. Uranium-series dating of the Niah 'Deep Skull'. In G. Barker, L. Farr (eds.), *Archaeological Investigations in the Niah Caves, Sarawak. The Archaeology of the Niah Caves, Sarawak, Volume 2*. Cambridge: McDonald Institute for Archaeological Research, pp. 233–4.

Pike-Tay, A., Cosgrove, R. 2006. Mousterians of the Antipodes? Exploring modern human behaviour in Late Pleistocene Tasmania. In J. M. Maillo, E. Baquedano (eds.), *Miscelánea en Homenaje a Victoria Cabrera*. Zona Arqueológica, ser. 7. Madrid: Museo Regional, pp. 336–47.

Pikirayi, I. 2001. *The Zimbabwe Culture*. Walnut Creek, CA: AltaMira Press.

Pilbeam, D. R., Young, N. M. 2001. *Sivapithecus* and hominoid evolution: Some brief comments. In L. de Bonis, G. Koufos, P. Andrews (eds.), *Hominoid Evolution and Climatic Change in Europe, Volume Two: Phylogeny of the Neogene Hominoid Primates of Eurasia*. Cambridge: Cambridge University Press, pp. 349–64.

Pimm, S. L. 1991. *The Balance of Nature: Ecological Issues in the Conservation of Species and Communities*. Chicago: University of Chicago Press.

Piper, P. J., Rabett, R. J. 2009. Disentangling the Harrisson archive to interpret the spatial and temporal distribution of vertebrate remains at Niah Caves, Borneo. *International Journal of Osteoarchaeology* 19: 464–75.

Piper, P. J., Rabett, R. J. 2014. Late Pleistocene subsistence strategies in island Southeast Asia and their implications for understanding the development of modern human behaviour. In R. Dennell, M. Poor (eds.), *Southern Asia, Australia and the Search for Human Origins*. Cambridge: Cambridge University Press, pp. 118–34.

Piper, P. J., Hung, H.-C., Campos, F. Z., Bellwood, P., Santiago, R. 2009. A 4000-year-old introduction of domestic pigs into the Philippine Archipelago: Implications for understanding routes of human migration through island Southeast Asia and Wallcea. *Antiquity* 83: 687–95.

Piperno, D. R. 2011a. Prehistoric human occupation and impacts on Neotropical forest landscapes during the Late Pleistocene and Early/Middle Holocene. In M. Bush, J. Flenley, W. Gosling (eds.), *Tropical Rainforest Responses to Climatic Change*. Berlin: Springer, pp. 185–212.

Piperno, D. R. 2011b. The origins of plant cultivation and domestication in the New World tropics: Patterns, process, and new developments. *Current Anthropology* 52(S4): S453–S470.

Piperno, D. R., Dillehay, T. D. 2008. Starch grains on human teeth reveal early broad crop diet in northern Peru. *Proceedings of the National Academy of Sciences of the United States of America* 105: 19622–7.

Piperno, D. R., Flannery, K. V. 2001. The earliest archaeological maize (Zea mays L.) from Highland Mexico: New accelerator mass spectrometry dates and their implications. *Proceedings of the National Academy of Sciences of the United States of America* 98: 2101–3.

Piperno, D. R., Stothert, K. E. 2003. Phytolith evidence for Early Holocene Cucurbita domestication in southwest Ecuador. *Science* 299: 1054–7.

Piperno, D. R., Ranere, A. J., Holst, I., Iriarte, J., Dickau, R. 2009. Starch grain and phytolith evidence for early ninth millennium BP maize from the Central Balsas River Valley, Mexico. *Proceedings of the National Academy of Sciences of the United States of America* 106: 5019–24.

Piperno, D. R., McMichael, C., Bush, M. B. 2017. Further evidence for localized, short-term anthropogenic forest alterations across pre-Columbian Amazonia. *Proceedings of the National Academy of Sciences of the United States of America* 114: E4118–E4119.

Pirenne, H. 2010. *Medieval Cities: Their Origins and the Revival of Trade* (translated by F. D. Halsey). Princeton, NJ: Princeton University Press.

Pitulko, V. V., Tikhonov, A. N., Pavlova, E. Y., Nikolskiy, P. A., Kuper, K. E., Polozov, R. N. 2016. Early human presence in the Arctic: Evidence from 45,000-year-old mammoth remains. *Science* 351: 260–3.

Plummer, T. W., Ditchfield, P. W., Bishop, L. C., Kingston, J. D., Ferraro, J. V., Braun, D. R., Hertel, F., Potts, R. 2009. Oldest evidence of tool-making hominins in a grassland-dominated ecosystem. *PLoS ONE* 4: e7199.

Plumptre, A. J., Davenport, T. R. B., Behangana, M., Kityo, R., Eilu, G., Ssegawa, P., Ewango, C., Meirte, D., Kahindo, C., Herremans, M., Peterhans, J. K., Pilgrim, J. D., Wilson, M., Languy, M., Moyer, D. 2007. The biodiversity of the Albertine Rift. *Biological Conservation* 134: 178–94.

Polley, H. W., Johnson, H. B., Marino, B. D., Mayeux, H. S. 1993. Increases in C_3 plant water-use efficiency and biomass over glacial to present CO_2 concentrations. *Nature* 361: 61–4.

Polson, R. A., Spencer, D. S. C. 1991. The technology adoption process in subsistence agriculture: The case of cassava in southwestern Nigeria. *Agricultural Systems* 36: 65–78.

Pongratz, J., Reick, C., Raddatz, T., Claussen, M. 2008. A reconstruction of global agricultural areas and land cover for the last millennium. *Global Biogeochemical Cycles* 22. doi:10.1029/2007GB003153.

Pookajorn, S. 1996. Human activities and environmental changes during the late Pleistocene to middle Holocene in southern Thailand and Southeast Asia. In L. G. Straus, B. V. Eriksen, M. Erlandson, D. R. Yesner (eds.), *Humans and the End of the Ice Age: The Archaeology of the Pleistocene–Holocene Transition*. New York: Plenum Press, pp. 201–13.

Posey, D. A. 1978. Ethnoentomological survey of Amerind groups in lowland Latin America. *The Florida Entomologist* 61: 225–9.

Posey, D. A. 1984. A preliminary report on diversified management of tropical forest by the Kayapó Indians of the Brazilian Amazon. *Advances in Economic Botany* 1: 112–26.

Posey, D. A. 1985. Indigenous management of tropical forest ecosystems: The case of the Kayapó Indians of the Brazilian Amazon. *Agroforestry Systems* 3: 139–58.

Posey, D. A. 1998. Diachronic ecotones and anthropogenic landscapes in Amazonia: Contesting the consciousness of conservation. In W. Balée (ed.), *Advances in Historical Ecology*. New York: Columbia University Press, pp. 104–18.

Posey, D. A., Balée, W. L. (eds.). 1989. *Resource Management in Amazonia: Indigenous and Folk Strategies*. New York: New York Botanical Garden.

Possehl, G. L., Rissman, P. 1992. The chronology of prehistoric India from earliest times to the Iron Age. In R. W. Ehrich (ed.), *Chronologies in Old World Archaeology*. Chicago: University of Chicago Press, pp. 465–90.

Postgate, J. N. 1992. *Early Mesopotamia: Society and Economy at the Dawn of History*. London and New York: Routledge.

Posth, C., Wiβing, C., Kitagawa, K., Pagani, L., van Holstein, L., Racimo, F., Wehrberger, K., Conard, N. J., Joachim Kind, C., Bocherens, H., Krause, J. 2017. Deeply divergent archaic mitochondrial genome provides lower time boundary for African gene flow into Neanderthals. *Nature Communications* 16046. doi:10.1038/ncomms16046.

Posth, C., Nagele, K., Colleran, H., Valentin, F., Bedford, S., Kami, K. W., Shing, R., Buckley, H., Kinaston, R., Walworth, M., Clark, G. R., Reepmeyer, C., Flexner, J., Maric, T., Moser, J., Gresky, J., Kiko, L., Robson, K. J., Auckland, K., Oppenheimer, S. J., Hill, A. V. S., Mentzer, A. J., Zech, J., Petchey, F., Roberts, P., Jeong, C., Gray, R. D., Krause, J., Powell, A. 2018. Language continuity despite population replacement in remote Oceania. *Nature Ecology and Evolution* 2: 731–40.

Pottier, C. 2000. A la recherche de Goloupura. Melanges du Centenaire de l'EFEO. *Bulletin de l'École Française d'Extrême-Orient* 87: 79–107.

Pottier, C. 2006. Early urban settlements in Angkor. In *Reassessing East Asia in the Light of Urban and Architectural History*. Kyoto: International Conference on East Asian Architectural Culture, pp. 133–40.

Potts, R. 1994. Variables versus models of early Pleistocene hominid land-use. *Journal of Human Evolution* 27: 7–24.

Potts, R. 1998. Environmental hypotheses of hominin evolution. *Yearbook of Physical Anthropology* 107: 93–137.

Potts, R. 2013. Hominin evolution in settings of strong environmental variability. *Quaternary Science Reviews* 73: 1–13.

Potts, R., Behrensmeyer, A. K., Faith, J. T., Tryon, C. A., Brooks, A. S., Yellen, J. E., Deino, A. L., Kinyanjui, R., Clark, J. B., Haradon, C., Levin, N. E., Meijer, H. J. M., Veatch, E. G.,

Owen, R. B., Renaut, R. B. 2018. Environmental dynamics during the onset of the Middle Stone Age in eastern Africa. *Science*. doi:10.1126/science.aao2200.

Powers, J. S., Montgomery, R. A., Adair, E. C., Brearley, F. Q., DeWalt, S. J., Castanho, C. T., Chave, J., Deinert, E., Ganzhorn, J. U., Gilbert, M. E., González-Iturbe, J. A., Bunyavejchewin, S., Grau, H. R., Harms, K. E., Hiremath, A., Iriarte-Vivar, S., Manzane, E., de Oliveira, A. A., Poorter, L., Ramanamanjato, J.-B., Salk, C., Varela, A., Weiblen, G. D., Lerdau, M. T. 2009. Decomposition in tropical forests: A pan-tropical study of the effects of litter type, litter placement and mesofaunal exclusion across a precipitation gradient. *Journal of Ecology* 97: 801–11.

Powis, T. G., Cyphers, A., Gaikwad, N. W., Grivetti, L., Cheong, K. 2011. Cacao use and the San Lorenzo Olmec. *Proceedings of the National Academy of Sciences of the United States of America* 108: 8595–600.

Poznik, G. D., Henn, B. M., Yee, M.-C., Sliwerska, E., Euskirchen, G. M., Lin, A. A., Snyder, M., Quintana-Murci, L., Kidd, J. M., Underhill, P. A., Bustamante, C. D. 2013. Sequencing Y chromosomes resolves discrepancy in time to common ancestor of males versus females. *Science* 341: 562–5.

Prado, J. L., Martinez-Maza, C., Alberdi, M. T. 2015. Megafauna extinction in South America: A new chronology for the Argentine Pampas. *Palaeogeography, Palaeoclimatology, Palaeoecology* 425: 41–9.

Prado-Martinez, J., Sudmant, P. H., Kidd, J. M., Li, H., Kelley, J. L., Lorente-Galdos, B., Veeramah, K. R., Woerner, A. E., O'Connor, T. D., Santpere, G., Cagan, A., Theunert, C., Casals, F., Laayouni, H., Munch, K., Hobolth, A., Halager, A. E., Malig, M., Hernandez-Rodriguez, J., Hernando-Herraez, I., Prüfer, K., Pybus, M., Johnstone, L., Lachmann, M., Alkan, C., Twigg, D., Petit, N., Baker, C., Hormozdiari, F., Fernandez-Callejo, M., Dabad, M., Wilson, M. L., Stevison, L., Camprubí, C., Carvalho, T., Ruiz-Herrera, A., Vives, L., Mele, M., Abello, T., Kondova, I., Bontrop, R. E., Pusey, A., Lankester, F., Kiyang, J. A., Bergl, R. A., Lonsdorf, E., Myers, S., Ventura, M., Gagneux, P., Comas, D., Siegismund, H., Blanc, J., Agueda-Calpena, L., Gut, M., Fulton, L., Tishkoff, S. A., Mullikin, J. C., Wilson, R. K., Gut, I. G., Gonder, M. K., Ryder, O. A., Hahn, B. H., Navarro, A., Akey, J. M., Bertranpetit, J., Reich, D., Mailund, T., Schierup, M. H., Hvilsom, C., Andrés, A. M., Wall, J. D., Bustamante, C. D., Hammer, M. F., Eichler, E. E., Marques-Bonet, T. 2013. Great ape genetic diversity and population history. *Nature* 499: 471–5.

Prasad, B. V. R., Ricker, C. E., Watkins, W. S., Atkins, W. S., Dixon, M. E., Rao, B. B., Naidu, J. M., Jorde, L. B., Bamshad, M. 2001. Mitochondrial DNA variation in Nicobarese Islanders. *Human Biology* 73: 715–25.

Premathilake, R. 2012. Human used upper montane ecosystem in the Horton Plains, Central Sri Lanka: A link to late glacial–early Holocene climate and environmental changes. *Quaternary Science Reviews* 50: 23–42.

Premathilake, R. 2015. Investigating the precursors and appearance of banana and rice cultivation in Sri Lanka: With the background of long-term climate and environmental changes. *Association for Environmental Archaeology Newsletter* 128: 2–5.

Premathilake, R., Risberg, J. 2003. Late Quaternary history of the Horton Plains, central Sri Lanka. *Quaternary Science Reviews* 22: 1525–41.

Preston, D. 2015. Exclusive: Lost city discovered in the Honduran rain forest. *National Geographic*. <https://news.nationalgeographic.com/2015/03/150302-honduras-lost-city-monkey-god-maya-ancient-archaeology/>.

Price, T. D., Bar-Yosef, O. 2011. The origins of agriculture: New data, new ideas. *Current Anthropology* 52: S163–S174.

Proctor, J. 1999. Heath forests and acid soils. *Botanical Journal of Scotland* 51: 1–14.

ProWildlife. 2007. *Going to Pot: The Neotropical Bushmeat Crisis and its Impact on Primate Populations*. Munich: ProWildlife.

Pruetz, J. D., Bertolani, P. 2007. Savanna chimpanzees, *Pan troglodytes verus*, hunt with tools. *Current Biology* 17: 412–17.

Pruetz, J. D., Bertolani, P. 2009. Chimpanzee (*Pan troglodytes verus*) behavioral responses to stresses associated with living in a savanna-mosaic environment: Implications for hominin adaptations to open habitats. *Paleoanthropology* 2009: 252–62.

Pruetz, J. D., Bertolani, P., Ontl, K. B., Lindshield, S., Shelley, M., Wessling, E. G. 2015. New evidence on the tool-assisted hunting exhibited by chimpanzees (Pan troglodytes verus) in a savannah habitat at Fongoli, Senegal. *Proceedings of the Royal Society B: Biological Sciences*. doi:10.1098/rsos.140507.

Prüfer, K., Munch, K., Hellmann, I., Akagi, K., Miller, J. R., Walenz, B., Koren, S., Sutton, G., Kodira, C., Winer, R., Knight, J. R., Mullikin, J. C., Meader, S. J., Ponting, C. P., Lunter, G., Higashino, S., Hobolth, A., Dutheil, J., Karakoc, E., Alkan, C., Sajjadian, S., Catacchio, C. R., Ventura, M., Marques-Bonet, T., Eichler, E. E., Andre, C., Atencia, R., Mugisha, L., Junhold, J., Patterson, N., Siebauer, M., Good, J. M., Fischer, A., PTak, S. E., Lachmann, M., Symer, D. E., Mailund, T., Schierup, M. H., Andrés, A. M., Kelso, J., Pääbo, S. 2012. The bonobo genome compared with the chimpanzee and human genomes. *Nature* 486: 527–31.

Prüfer, K., Racimo, F., Patterson, N., Jay, F., Sankararaman, S., Sawyer, S., Heinze, A., Renaud, G., Sudmant, P. H., de Filippo, C., Li, H., Mallick, S., Dannemann, M., Fu, Q., Kircher, M., Kuhlwilm, M., Lachmann, M., Meyer, M., Ongyerth, M., Siebauer, M., Theunert, C., Tandon, A., Moorjani, P., Pickrell, J., Mullikin, J. C., Vohr, S. H., Green, R. E., Hellmann, I., Johnson, P. L., Blanche, H., Cann, H., Kitzman, J. O., Shendure, J., Eichler, E. E., Lein, E. S., Bakken, T. E., Golovanova, L. V., Doronichev, V. B., Shunkov, M. V., Derevianko, A. P., Viola, B., Slatkin, M., Reich, D., Kelso, J., Pääbo, S. 2014. The complete genome sequence of a Neanderthal from the Altai Mountains. *Nature* 505: 43–9.

Prüfer, K., Filippo, C. D., Grote, S., Mafessoni, F., Korlevic, P., Hajdinjak, M., Vernot, B., Skov, L., Hsieh, P., Peyrégne, S., Reher, D., Hopfe, C., Nagel, S., Maricic, T., Fu, Q., Theunert, C., Rogers, R., Skoglund, P., Chintalapati, M., Dannemann, M., Nelson, B. J., Key, F. M., Rudan, P., Kucan, Ž., Gušic, I., Golovanova, L. V., Doronichev, V. B., Patterson, N., Reich, D., Eichler, E. E., Slatkin, M., Schierup, M. H., Andrés, A. M., Kelso, J., Meyer, M., Pääbo, S. 2017. A high-coverage Neandertal genome from Vindija Cave in Croatia. *Science* 358: 655–8.

Putz, F. E., Sist, P., Fredericksen, T. S., Dykstra, D. 2008. Reduced-impact logging: Challenges and opportunities. *Forest Ecology and Management* 256: 1427–33.

Quade, J., Cerling, T. E. 1995. Expansion of C4 grasses in the late Miocene of northern Pakistan: Evidence from stable isotopes in paleosols. *Palaeogeography, Palaeoclimatology, Palaeoecology* 115: 91–116.

Quade, J., Cerling, T. E., Andrews, P., Alpagut, B. 1995. Paleodietary reconstruction of Miocene faunas from Paşalar, Turkey using stable carbon and oxygen isotopes of fossil tooth enamel. *Journal of Human Evolution* 28: 373–84.

Quesada, C. A., Lloyd, J., Anderson, L. O., Fyllas, N. M., Schwarz, M., Czimczik, C. I. 2011. Soils of Amazonia with particular reference to the RAINFOR sites. *Biogeosciences* 8: 1415–40.

Quinn, R. L., Lepre, C. J., Feibel, C. S., Wright, J. D., Mortlock, R. A., Harmand, S., Brugal, J.-P., Roche, H. 2013. Pedogenic carbonate stable isotopic evidence for wooded habitat preference of early Pleistocene tool makers in the Turkana Basin. *Journal of Human Evolution* 65: 65–78.

Rabett, R. J. 2012. *Human Adaptation in the Asian Palaeolithic*. Cambridge: Cambridge University Press.

Rabett, R. J. 2016. Bone and tusk tools from the West Mouth and Lobang Hangus. In G. Barker, L. Farr (eds.), *Archaeological Investigations in the Niah Caves, Sarawak. The Archaeology of the Niah Caves, Sarawak, Volume 2*. Cambridge: McDonald Institute for Archaeological Research, pp. 301–24.

Rabett, R. J., Barker, G. 2007. Through the looking glass: New evidence on the presence and behaviour of Late Pleistocene humans at Niah Cave, Sarawak, Borneo. In P. Mellars, K. Boyle, O. Bar-Yosef, C. Stringer (eds.), *Rethinking the Human Revolution*. Cambridge: McDonald Institute for Archaeological Research, pp. 383–94.

Rabett, R. J., Piper, P. J. 2012. The emergence of bone technologies at the end of the Pleistocene in Southeast Asia: Regional and evolutionary implications. *Cambridge Archaeological Journal* 22: 37–56.

Rabett, R. J., Piper, P. J., Barker, G. 2006. Bones from Hell: Preliminary results of new work on the Harrisson faunal assemblage from the deepest part of Niah Cave, Sarawak. In E. A. Bacus, I. C. Glover, V. C. Pigott (eds.), *Uncovering Southeast Asia's Past: Selected Papers from the 10th International Conference of European Association of Southeast Asian Archaeologists*. Singapore: NUS Press, pp. 46–59.

Rabett, R. J., Barker, G., Hunt, C., Naruse, T., Piper, P., Raddatz, E., Reynolds, T., Nguyen, V. S., Stimpson, C., Szabo, K., Nguyen, C. T., Wilson, J. 2009. The Tràng An project: Late-to-post-Pleistocene settlement of the lower Song Hong Valley, North Vietnam. *Journal of the Royal Asiatic Society of Great Britain and Ireland* 19: 83–109.

Rabett, R. J., Ludgate, N., Stimpson, C., Hill, E., Hunt, C., Ceron, J., Farr, L., Morley, M., Reynolds, T., Zukswert, H., Simpson, D., Nyiri, B., Verhoeven, M., Appleby, J., Meneely, J., Phan, L., Dong, N.-N., Lloyd-Smith, L., Hawkes, J., Blyth, A., Cao Tấn, N. 2017. Tropical limestone forest resilience and late Pleistocene foraging during MIS-2 in the Tràng An massif, Vietnam. *Quaternary International* 448: 62–81.

Radcliffe-Brown, A. R. 1932. *The Andaman Islanders*. Cambridge: Cambridge University Press.

Raichlan, D. A., Gordon, A. D., Harcourt-Smith, W. E. H., Foster, A. D., Randall Haas, W. Jr. 2010. Laetoli footprints preserve earliest direct evience of human-like bipedal biomechanics. *PLoS ONE* 5: e9768.

Ranasinghe, R., Tennekoon, K. H., Karunanayake, E. H., Lembring, M., Allen, M. 2015. A study of genetic polymorphisms in mitochondrial DNA hypervariable regions I and II of the five major ethnic groups and Vedda population in Sri Lanka. *Legal Medicine* 17: 539–46.

Ranaweera, L., Kaewsutthi, S., Win Tun, A., Boonyarit, H., Poolsuwan, S., Lertrit, P. 2014. Mitochondrial DNA history of Sri Lankan ethnic people: Their relations within the island and with the Indian subcontinental populations. *Journal of Human Genetics* 59: 28–36.

Ranere, A. J., Piperno, D. R., Holst, I., Dickau, R., Iriarte, J. 2009. Preceramic human occupation of the Central Balsas Valley, Mexico: Cultural context of early domesticated maize and squash. *Proceedings of the National Academy of Sciences of the United States of America* 106: 5014–18.

Rao, V. K. 2013. The terracotta plaques of Pagan: Indian influence and Burmese innovations. *Ancient Asia* 4: 7.

Rathje, W. L. 1971. The origin and development of lowland Classic Maya civilization. *American Antiquity* 36: 275–85.

Redford, K. H. 1990. The ecologically noble savage. *Orion Nature Quaterly* 9: 25–9.

Redford, K. H., Robinson, J. 1987. The game of choice: Patterns of Indian and colonist hunting in the Neotropics. *American Anthropologist* 89: 650–66.

Reece, R. 2002. The end of the city in Roman Britain. In J. Rich (ed.), *The City in Late Antiquity*. London: Routledge, pp. 136–44.

Reed, K. E. 1997. Early hominid evolution and ecological change through the African Plio-Pleistocene. *Journal of Human Evolution* 32: 289–322.

Reed, K. E. 2008. Paleoecological patterns at the Hadar hominin site, Afar Regional State, Ethiopia. *Journal of Human Evolution* 54: 743–68.

Refisch, J., Koné, I. 2005. Impact of commercial hunting on monkey populations in the Taï region, Côte d'Ivoire. *Biotropica* 37: 136–44.

Reich, D., Thangaraj, K., Patterson, N., Price, A. L., Singh, L. 2009. Reconstructing Indian population history. *Nature* 461: 489–94.

Reich, D., Patterson, N., Kircher, M., Delfin, F., Nandineni, M. R., Pugach, I., Min-Shan Ko, A., Ko, Y.-C., Jinam, T. A., Phipps, M. E., Saitou, N., Wollstein, A., Kayser, M., Paabo, S., Stoneking, M. 2011. Denisova admixture and the first modern human dispersals into Southeast Asia and Oceania. *American Journal of Human Genetics* 89: 516–28.

Reid, L. A. 1987. The early switch hypothesis: Linguistic evidence for contact between negritos and Austronesians. *Man and Culture in Oceania* 3: 41–59.

Reid, L. A. 1992. The Tasaday language: A key to Tasaday prehistory. In T. N. Headland (ed.), *The Tasaday Controversy: Assessing the Evidence.* American Anthropological Association Scholarly Series 28. Washington, DC: AAA, pp. 180–93.

Rerkasem, K., Lawrence, D., Padoch, C., Schmidt-Vogt, D., Ziegler, A. D., Bruun, T. B. 2009. Consequences of swidden transitions for crop and fallow biodiversity in Southeast Asia. *Human Ecology* 37: 347–60.

Reynolds, T. E. G. 1990. The Hoabinhian: A review. In G. L. Barnes (ed.), *Bibliographic Review of Far Eastern Archaeology.* Oxford: Oxbow Books, pp. 1–30.

Rice, D. S., Rice, P. M. 1984. Lessons from the Maya. *Latin American Antiquity* 19: 7–34.

Richerson, P. J., Boyd, P., Bettinger, R. L. 2001. Was agriculture impossible during the Pleistocene but mandatory during the Holocene? A climate change hypothesis. *American Antiquity* 66: 387–411.

Richmond, B. G., Begun, D. R., Strait, D. S. 2001. Origin of human bipedalism: The knuckle-walking hypothesis revisited. *American Journal of Physical Anthropology Supplement* 33: 70–105.

Richmond, B. G., Jungers, W. L. 2008. *Orrorin tugenensis* femoral morphology and the evolution of hominin bipedalism. *Science* 319: 1662–5.

Richter, D., Grün, R., Joannes-Boyau, R., Steele, T. E., Amani, F., Rué, M., Fernandes, P., Raynal, J.-P., Geraads, D., Ben-Ncer, A., Hublin, J.-J., McPherron, S. P. 2017. The age of the hominin fossils from Jebel Irhoud, Morocco, and the origins of the Middle Stone Age. *Nature* 546: 293–6.

Ricketts, T. H., Soares-Filho, B., da Fonseca, G. A. B., Nepstad, D., Pfaff, A., Petsonk, A., Anderson, A., Boucher, D., Cattaneo, A., Conte, M., Creighton, K., Linden, L., Maretti, C., Moutinho, P., Ullman, R., Victurine, R. 2010. Indigenous lands, protected areas, and slowing climate change. *PLoS Biology* 8: e1000331.

Rink, W. J., Schepartz, L. A., Miller-Antonio, S., Huang, W. W., Hou, Y., Bakken, D., Richter, D., Jones, H. L. 2003. Electron spin resonance (ESR) dating of mammalian tooth enamel at Panxian Dadong Cave, Guizhou, China. In C. Shen, S. Keates (eds.), *Current Research in Chinese Pleistocene Archaeology.* Oxford: BAR International Series 117. Oxford: Archaeopress, pp. 111–18.

Riswan, S., Kartawinata, K. 1991. Species strategy in early stage of secondary succession associated with soil properties status in a lowland mixed dipterocarp forest and kerangas forest in East Kalimantan. *Tropics* 1: 13–34.

Roberts, P. 2017. Early human adaptation to Late Pleistocene-Holocene rainforests in South Asia. In N. Sanz (ed.), *Tropical Forest Conservation: Long-Term Processesof Human Evolution, Cultural Adaptations and Consumption Patterns.* Mexico City: UNESCO, pp. 28–43.

Roberts, P. 2018. *Late Pleistocene Tropical Rainforest Forager Sustainability and Resilience.* In N. Sanz (ed.), Exploring Frameworks for Tropical Forest Conservation: Integrating Natural and Cultural Diversity for Sustainability. A Global Perspective. Mexico City: UNESCO, pp. 116–135.

Roberts, P., Amano, N. Under review. Plastic pioneers: Hominin biogeography across the Movius Line during the Late Pleistocene. In S. Kadowaki, A. Noguchi (eds.), *Across the Movius Line.* The Science of Human History in Asia and the Pacific Series. Dordrecht: Springer.

Roberts, P., Petraglia, M. D. 2015. Pleistocene rainforests: Barriers or attractive environments for early human foragers? *World Archaeology.* doi:10.1080/00438243.2015.1073119.

Roberts, P., Stewart, B. 2018. Defining a new niche for our species: The Late Pleistocene 'generalist specialist'. *Nature Human Behaviour.* doi:10.1038/s41562-018-0394-4.

Roberts, P., Lee-Thorp, J. A., Mitchell, P. J., Arthur, C. 2013. Stable carbon isotopic evidence for climate change across the late Pleistocene to early Holocene from Lesotho, southern Africa. *Journal of Quaternary Science* 28: 360–9.

Roberts, P., Perera, N., Wedage, O., Deraniyagala, S., Perera, J., Eregama, S., Gledhill, A., Petraglia, M. D., Lee-Thorp, J. A. 2015a. Direct evidence for human reliance on rainforest resources in late Pleistocene Sri Lanka. *Science* 347: 1246–9.

Roberts, P., Boivin, N., Petraglia, M. D. 2015b. The Sri Lankan 'Microlithic' tradition c. 38,000 to 3000 years ago: Tropical technologies and adaptations of *Homo sapiens* at the southern edge of Asia. *Journal of World Prehistory* 29: 69–112.

Roberts, P., Boivin, N., Lee-Thorp, J., Petraglia, M., Stock, J. 2016. Tropical forests and the genus *Homo*. *Evolutionary Anthropology* 25: 306–17.

Roberts, P., Hunt, C., Arroyo-Kalin, M., Evans, D., Boivin, N. 2017a. The deep human prehistory of global tropical forests and its relevance for modern conservation. *Nature Plants* 3: 17093.

Roberts, P., Perera, N., Wedage, O., Deraniyagala, S., Perera, J., Eregama, S., Gledhill, A., Petraglia, M. D., Lee-Thorp, J. A. 2017b. Fruits of the forest: Human stable isotope ecology and rainforest adaptations in Late Pleistocene and Holocene (~ 36 to 3 ka) Sri Lanka. *Journal of Human Evolution* 106: 102–18.

Roberts, P., Blumenthal, S. A., Dittus, W., Wedage, O., Lee-Thorp, J. A. 2017c. Stable carbon, oxygen, and nitrogen isotope analysis of plants from a South Asian tropical forest: Implications for primatology. *American Journal of Primatology*. doi:10.1002/ajp.22656.

Roberts, P., Gaffney, D., Lee-Thorp, J., Summerhayes, G. 2017d. Persistent tropical foraging in the Highlands of Terminal Pleistocene-Holocene New Guinea. *Nature Ecology and Evolution* 1: 0044. doi:10.1038/s41559-016-0044.

Roberts, P., Boivin, N., Kaplan, J. O. 2018a. Finding the Anthropocene in tropical forests. *Anthropocene* 23: 5–16.

Roberts, P., Gillingwater, T. H., Mirazon Lahr, M., Lee-Thorp, J., MacCallum, M., Petraglia, M., Wedage, O., Heenbanda, U., Wanniya-Laeto, U. 2018b. Historical tropical forest reliance amongst the Wanniyalaeto (Vedda) of Sri Lanka: An isotopic perspective. *Human Ecology*. https://doi.org/10.1007/s10745-018-9997-7.

Robinson, J. G., Bennett, E. L. 2000. *Hunting for Sustainability in Tropical Forests*. New York: Columbia University Press.

Robinson, J. G., Redford, K. H. 1986. Body size, diet, and population density of Neotropical forest mammals. *American Naturalist* 128: 665–80.

Robinson, J. R., Rowan, J., Campisano, J., Wynn, J. G., Reed, K. E. 2017. Late Pliocene environmental change during the transition from *Australopithecus* to *Homo*. *Nature Ecology and Evolution* 1: 0159.

Robinson, M. M., Dowsett, H. J., Chandler, M. A. 2008. Pliocene role in assessing future climate impacts. *EoS, Transactions, American Geophysical Union* 89: 501–2.

Rodman, P. S., McHenry, H. M. 1980. Bioenergetics and the origin of hominid bipedalism. *American Journal of Physical Anthropology* 52: 103–6.

Rodrigues, R. R., Lima, R. A. F., Gandolfi, S., Nave, A. G. 2009. On the restoration of high diversity forests: 30 years of experience in the Brazilian Atlantic Forest. *Biological Conservation* 142: 1242–51.

Rodríguez-Trejo, D. A., Martínez-Hernández, P. A., Ortiz-Contla, H., Chavarría-Sánchez, M. R. 2011. The present status of fire ecology, traditional use of fire, and fire management in Mexico and Central America. *Fire Ecology* 7: 40–56.

Roe, P. G. 1982. *The Cosmic Zygote: Cosmology in the Amazon Basin*. New Brunswick: Rutgers University Press.

Rogers, J. D., Ulambayar, E., Gallon, M. 2005. Urban centres and the emergence of empires in Eastern Inner Asia. *Antiquity* 79: 801–18.

Roosevelt, A. C. 1991. *Moundbuilders of the Amazon: Geophysical Archaeology in Marajó Island.* San Diego: Academic Press.

Roosevelt, A. C. 1993. The rise and fall of the Amazon chiefdoms. *L'Homme* 33: 255–83.

Roosevelt, A. C. 1994. *Amazonian Indians from Prehistory to the Present: Anthropological Perspectives.* Tuscon: University of Arizona Press.

Roosevelt, A. C. 1995. Early pottery in the Amazon: Twenty years of scholarly obscurity. In W. K. Barnet, J. W. Hoopes (eds.), *The Emergence of Pottery: Technology and Innovation in Ancient Societies.* Washington, DC: Smithsonian Institution Press, pp. 115–31.

Roosevelt, A. C. 1999. The development of prehistoric complex societies: Amazonia, a tropical forest. In E. A. Bacus, L. J. Lucero, J. Allen (eds.), *Complex Polities in the Ancient Tropical World.* Arlington: American Anthropological Association, pp. 13–34.

Roosevelt, A. C. 2000. The Lower Amazon: A dynamic human habitat. In D. L. Lentz (ed.), *Imperfect Balance: Landscape Transformations in the Precolumbian Americas.* New York: Columbia University Press, pp. 455–92.

Roosevelt, A. C. 2006. Geophysical archaeology in the lower Amazon: A research strategy. In J. Wiseman, F. El-Baz (eds.), *Remote Sensing in Archaeology: Interdisciplinary Contributions to Archaeology.* New York: Springer, pp. 443–76.

Roosevelt, A. C. 2013. Prehistory of Amazonia. In C. Renfrew, P. Bahn (eds.), *The Cambridge World Prehistory.* Cambridge: Cambridge University Press, pp. 1175–99.

Roosevelt, A. C., Douglas, J., Brown, L. 2002. The migrations and adaptations of the first Americans: Clovis and pre-Clovis viewed from South America. In N. G. Jablonski (ed.), *The First Americans: The Pleistocene Colonization of the New World.* San Francisco: California Academy of Sciences, pp. 159–235.

Roscoe, P. 2002. The hunters and gatherers of New Guinea. *Current Anthropology* 43: 153–62.

Roseberry, W. 1989. *Economic Anthropology.* Stanford, CA: Stanford University Press.

Ross, N. J. 2011. Modern tree species composition reflects ancient Maya 'forest gardens' in northwest Belize. *Ecological Applications* 21: 75–84.

Rossetti, D. D. F., Toledo, P. M. D., Moraes-Santos, H. M., Santos Jr, A. E. de, 2004. Reconstructing habitats in central Amazonia using megafauna, sedimentology, radiocarbon, and isotope analyses. *Quaternary Research* 61: 289–300.

Rostain, S. 2008. Agricultural earthworks on the French Guiana coast. In H. Silverman, W. Isbell (eds.), *Handbook of South American Archaeology.* Dordrecht: Springer, pp. 279–302.

Rostain, S. 2014. *Islands in the Rainforest: Landscape Management in Pre-Columbian Amazonia.* London: Routledge.

Rostain, S. 2017. Sitios monumentales precolombinos en Amazonia. In N. Sanz (ed.), *Tropical Forest Conservation: Long-Term Processesof Human Evolution, Cultural Adaptations and Consumption Patterns.* Mexico City: UNESCO, pp. 44–65.

Roth, H., Erdenebat, U., Pohl, E., Nagel, E. 2002. *Qara Qrum-city (Mongolia) I: Preliminary Report of the Excavations 2000/2001.* Bonn: Bonn Contributions to Asian Archaeology.

Rothhammer, F., Dillehay, T. D. 2009. The Late Pleistocene colonization of South America: An interdisciplinary perspective. *Annals of Human Genetics* 73: 540–9.

Rousseau, J. J. 1992 (1754). *Discourse on the Origin and Basis of Inequality Among Men* (translated by D. A. Cress). Indianapolis: Hackett.

Rowland, M. J. 2004. Return of the 'noble savage': Misrepresenting the past, present and future. *Australian Aboriginal Studies* 2: 2–14.

Rozzi, R., Winkler, D. E., De Vos, J., Schulz, E., Palombo, M. R. 2013. The enigmatic bovid Duboisia santeng (Dubois, 1891) from the Early-Middle Pleistocene of Java: A multiproxy approach to its paleoecology. *Palaeogeography, Palaeoclimatology, Palaeoecology* 377: 73–85.

Ruddiman, W. F. 2003. The anthropogenic greenhouse era began thousands of years ago. *Climatic Change* 61: 261–93.

Ruddiman, W. F., Ellis, E. C., Kaplan, J. O., Fuller, D. Q. 2015. Geology: Defining the epoch we live in. *Science* 348: 38–9.

Ruddle, K., Johnson, D., Townsend, P. K., Rees, J. D. 1978. *Palm Sago: A Tropical Starch from Marginal Lands.* Honolulu: University Press of Hawai'i.

Ruff, C. 2009. Relative limb strength and locomotion in *Homo habilis. American Journal of Physical Anthropology* 138: 90–100.

Sabloff, J. A. 1990. *The New Archaeology and the Ancient Maya.* New York: Freeman.

Sabloff, J. A. 2007. It depends on how you look at things: New perspectives on the post-Classic period in the northern Maya lowlands. *Proceedings of the American Philosophical Society* 151: 11–25.

Sahlins, M. 1968. Notes on the original affluent society. In R. B. Lee, I. De Vore (eds.), *Man the Hunter.* New York: Aldine, pp. 85–9.

Sahlins, M. 1972. *Stone Age Economics.* New York: Walter de Gruyter.

Said, E. 1978. *Orientalism.* New York: Pantheon.

Salazar, E. 1998. De vuelta al Sangay: Investigaciones arqueologicas en el alto Upano, Amazonia, Ecuatoriana. *Bulletin de l'Institute Français des Études Andines* 27: 213–40.

Samarakoon, M. B., Tanaka, N., Iimura, K. 2013. Improvement of effectiveness of existing *Casuarina Equisetifolia* forests in mitigating tsunami damage. *Journal of Environmental Management* 114: 105–14.

Samper Carro, S. C., O'Connor, S., Louys, J., Hawkins, S., Mahirta, M. 2016. Human maritime subsistence strategies in the Lesser Sunda Islands during the terminal Pleistocene-early Holocene: New evidence from Alor, Indonesia. *Quaternary International* 416: 64–79.

Sankar, K., Qureshi, Q., Nigam, P., Malik, P. K., Sinha, R. N., Mehrotra, R. N., Gopal, R., Bhattacharjee, S., Mondal, K., Gupta, S. 2010. Monitoring of reintroduced tigers in Sariska Tiger Reserve, Western India: Preliminary findings on home range, prey selection and food habits. *Tropical Conservation Science* 3: 301–18.

Sanz, N. 2017a. Perspectives from local and indigenous producers in Mexico. In N. Sanz (ed.), *Tropical Forest Conservation: Long-Term Processesof Human Evolution, Cultural Adaptations and Consumption Patterns.* Mexico City: UNESCO, pp. 296–9.

Sanz, N. 2017b. The way forward for more sustainable natural and cultural diversity all over the world. In N. Sanz (ed.), *Tropical Forest Conservation: Long-Term Processes of Human Evolution, Cultural Adaptations and Consumption Patterns.* Mexico City: UNESCO, pp. 337–51.

Sarasin, P., Sarasin, F. 1893. *Ergebnisse Naturwissenschaftlicher Forschungen auf Ceylon: die Weddas von Ceylon und die sie umgebenden Völkerschaften, 4–6.* Wiesbaden: C. W. Kreidel.

Sarasin, P., Sarasin, F. 1908. *Ergebnisse Naturwissenschaftlicher Forschungen auf Ceylon, 4: die Steinzeit auf Ceylon.* Wiesbaden: C. W. Kreidel.

Saraswat, K. S. 1992. Archaeobotanical remains in ancient cultural and socio-economical dynamics of the Indian subcontinent. *Palaeobotanist* 40: 514–45.

Saraswat, K. S. 2004. Plant economy of early farming communities. In B. P. Singh (ed.), *Early Farming Communities of the Kaimur (Excavations at Senuwar).* Jaipur: Publication Scheme, pp. 416–535.

Saturno, W. A., Stuart, D., Beltrán, B. 2006. Early Maya writing at San Bartolo, Guatemala. *Science* 311: 1281–3.

Savage-Rumbaugh, E. S., Lewin, R. 1996. *Kanzi: The Ape at the Brink of the Human Mind.* London: Wiley.

Sawada, Y., Pickford, M., Senut, B., Itaya, T., Hyodo, M., Miura, T., Kashine, C., Chujo, T., Fujii, H. 2002. The age of Orrorin tugenensis, an early hominid from the Tugen Hills, Kenuya. *Comptes Rendus Palevol* 1: 293–303.

Sayer, J. A., Harcourt, C. S., Collins, N. M. (eds.). 1992. *The Conservation Atlas of Tropical Forests: Africa.* Basingstoke: Macmillan.

Sayer, J. A., Sunderland, T., Ghazoul, J., Pfund, J.-L., Sheil, D., Meijaard, E., Venter, M., Boedhihartono, A. K., Day, M., Garcia, C. 2013. Ten principles for a landscape approach to reconciling agriculture, conservation, and other competing land uses. *Proceedings of the National Academy of Sciences of the United States of America* 110: 8349–56.

Scally, A., Durbin, R. 2012. Revising the human mutation rate: Implications for understanding human evolution. *Nature Reviews Genetics* 13: 745–53.

Scarborough, V. L. 1993. Water management in the southern Maya lowlands: An accretive model for the engineered landscape. *Research in Economic Anthropology* 7: 17–69.

Scarborough, V. L., Gallopin, G. G. 1991. A water storage adaptation in the Maya lowlands. *Science* 251: 658–62.

Scarborough, V. L., Lucero, L. J. 2010. The non-hierarchical development of complexity in the semitropics: Water and cooperation. *Water History* 2: 185–205.

Scarborough, V. L., Dunning, N. P., Tankersley, K. B., Carr, C., Weaver, E., Grazioso, L., Lane, B., Jones, J.G., Buttles, P., Valdez, F., Lentz, D. L. 2012. Water and sustainable land use at the ancient tropical city of Tikal, Guatemala. *Proceedings of the National Academy of Sciences of the United States of America* 109: 12408–13.

Scerri, E. 2017. The North African Middle Stone Age and its place in recent human evolution. *Evolutionary Anthropology* 26: 119–35.

Schaan, D., Pärssinen, M., Saunaloma, S., Ranzi, A., Bueno, M., Barbosa, A. 2013. New radiometric dates for pre-Columbian (2000–700 BP) earthworks in western Amazonia, Brazil. *Journal of Field Archaeology* 37: 132–42.

Schadeberg, T. 1999. Batwa: The Bantu name for invisible people. In K. Briesbrouck, S. Elders, G. Rossel, R. te Leiden, C. voor Niet-Westerse Studies (eds.), *Central African Hunter-Gatherers in a Multidisciplinary Perspective: Challenging Elusiveness.* Leiden: CNWS, Universiteit Leiden, pp. 21–40.

Schepartz, L. A., Miller-Antonio, S. 2004. Asia and the Middle Pleistocene in global perspective. *Asian Perspectives* 43: 187–90.

Schepartz, L. A., Miller-Antonio, S., Bakken, D. A. 2000. Upland resources and the early Palaeolithic occupation of Southern China, Vietnam, Laos, Thailand and Burma. *World Archaeology* 32: 1–13.

Scherf, B. D. 2000. *World Watch List for Domestic Animal Diversity* (3rd edition). Rome: Food and Agriculture Organization of the United Nations.

Schick, K. D., Toth, N., Garufi, G. 1999. Continuing investigations into the stone tool-making and tool-using capabilities of a Bonobo (*Pan paniscus*). *Journal of Archaeological Science* 26: 821–32.

Schiffels, S., Durbin, R. 2014. Inferring human population size and separation history from multiple genome sequences. *Nature Genetics* 46: 919–25.

Schmidt, M. J., Py-Daniel, A. R., de Paula Moraes, C., Valle, R. B. M., Caromano, C. F., Texeira, W. G., Barbosa, C. A., Fonseca, J. A., Magalhães, M. P., do Carmo Santos, D. S., da Silva e Silva, R., Guapindaia, V. L., Moraes, B., Lima, H. P., Neves, E. G., Heckenberger, M. J. 2014. Dark earths and the human built landscape in Amazonia: A widespread pattern of anthrosol formation. *Journal of Archaeological Science* 42: 152–65.

Schoeninger, M. J., Reeser, H., Hallin, K. 2003. Paleoenvironment of *Australopithecus anamensis* at Allia Bay, East Turkana, Kenya: Evidence from mammalian herbivore enamel stable isotopes. *Journal of Anthropological Archaeology* 22: 200–7.

Schulze, M. D., Whitacre, D. F. 1999. A classification and ordination of the tree community of Tikal National Park, Petén, Guatemala. *Bulletin of the Florida Museum of Natural History* 41: 169–297.

Schulze, M. D., Grogan, J., Vidal, E. 2008. Forest certification in Amazonia: Standards matter. *Oryx* 42: 229–39.

Schwartz, J. H., Tattersall, I. 2002. *The Human Fossil Record, Vol. 1: Terminology and Cranial Morphology of Genus Homo (Europe).* New York: Wiley/Liss.

Schwartzman, S., Zimmerman, B. 2005. Conservation alliances with indigenous peoples of the Amazon. *Conservation Biology* 19: 721–7.

Scientific American. 2009. Measuring the daily destruction of the world's rainforests. Scientificamerican.com. <https://www.scientificamerican.com/article/earth-talks-daily-destruction/>. Retrieved 19 March 2018.

Scott, G. R., Gilbert, L. 2009. The oldest handaxes in Europe. *Nature* 461: 82–5.

Seelenfreund, D., Clarke, A. C., Oyanedel, N., Piña, R., Lobos, S., Matisoo-Smith, E., Seelenfreund, A. 2010. Paper mulberry (*Broussonetia papyrifera*) as a commensal model for human mobility in Oceania: Anthropological, botanical and genetic considerations. *New Zealand Journal of Botany* 48: 231–47.

Seelenfreund, D., Piña, R., Ho, K.-Y., Lobos, S., Moncada, X., Seelenfreund, A. 2011. Molecular analysis of Broussonetia papyrifera (L.) Vent. (Magnoliophyta: Urticales) from the Pacific, based on ribosomal sequences of nuclear DNA. *New Zealand Journal of Botany* 49: 413–20.

Seidler, R., Bawa, K. S., Lowman, M., Nadkarni, N. M. 2013. Forest canopies as Earth's support systems: Priorities for research and conservation. In M. Lowman, S. Devy, T. Ganesh (eds.), *Treetops at Risk: Challenges of Global Canopy Ecology and Conservation*. Dordrecht: Springer, pp. 55–70.

Seligmann, C. G., Seligmann, B. Z. 1911. *The Veddas*. Cambridge: Cambridge University Press.

Sémah, A.-M., Sémah, F. 2012. The rain forest in Java through the Quaternary and its relationships with humans (adaptation, exploitation and impact on the forest). *Quaternary International* 249: 120–8.

Sémah, F., Sémah, A.-M., Simanjuntak, T. 2002. More than a million years of human occupation in insular Southeast Asia: The early archaeology of eastern and central Java. In J. Mercader (ed.), *Under the Canopy: The Archaeology of Tropical Rain Forests*. Piscataway, NJ: Rutgers University Press, pp. 161–90.

Sémah, A.-M., Sémah, F., Moigne, A.-M., Ingicco, T., Purnomo, A., Simanjuntak, T., Widianto, H. 2016. The palaeoenvironmental context of the Palaeolithic of Java: A brief review. *Quaternary International* 416: 38–45.

Seneviratna, A. 1994. *Ancient Anuradhapura: The Monastic City*. Colombo: Archaeological Survey Department.

Senut, B., Pickford, M., Gommery, D., Mein, P., Cheboi, K., Coppens, Y. 2001. First hominid from the Miocene (Lukeino Formation, Kenya). *Comptes Rendus de l'Académie des Sciences, Paris* 332: 137–44.

Service, Elman. 1975. *Origins of the State and Civilization*. New York: Norton.

Shakespeare, W. 2008 (1609). *Troilus and Cressida: The Oxford Shakespeare*. Edited by K. Muir. Oxford: Oxford World's Classics.

Sharer, R. J., Traxler, L. P. 2006. *The Ancient Maya* (6th fully revised edition). Stanford, CA: Stanford University Press.

Sharp, Z. 2006. *Principles of Stable Isotope Geochemistry*. Upper Saddle River, NJ: Pearson Prentice Hall.

Shaw, A. 2016. *The Trees*. London: Bloomsbury.

Shea, J. J. 2011. *Homo sapiens* is as *Homo sapiens* was. *Current Anthropology* 52: 1–35.

Shea, J. J., Bar-Yosef, O. 2005. Who were the Skhul/Qafzeh people? An archaeological perspective on Eurasia's oldest modern humans. *Journal of the Israel Prehistoric Society* 35: 451–68.

Shea, J. J., Sisk, M. L. 2010. Complex projectile technology and *Homo sapiens* dispersal into Western Eurasia. *PaleoAnthropology* 2010: 100–22.

Sheil, D. 2003. Growth assessment in tropical trees: Large daily diameter fluctuations and their concealment by dendrometer bands. *Canadian Journal of Forest Research* 33: 2027–35.

Sheil, D. 2014. How plants water our planet: Advances and imperatives. *Trends in Plant Science* 19: 209–11.

Sheil, D. 2017. Exploring local perspectives and preferences in forest landscapes: Towards democratic conservation. In N. Sanz (ed.), *Tropical Forest Conservation: Long-Term Processesof Human Evolution, Cultural Adaptations and Consumption Patterns.* Mexico City: UNESCO, pp. 262–83.

Sheil, D., Basuki, I., German, L., Kuyper, T. W., Limberg, G., Puri, R. K., Sellato, B., van Noordwijk, M., Wollenberg, E. 2012. Do anthropogenic dark earths occur in the interior of Borneo? Some initial observations from East Kalimantan. *Forests* 3: 207–29.

Sheil, D., Boissière, M., Beaudoin, G. 2015. Unseen sentinels: Local monitoring and control in conservation's blind spots. *Ecology and Society* 20: 39.

Sheldon, F. H., Styring, A., Hosner, P. A. 2010. Bird species richness in a Bornean exotic tree plantation: A long-term perspective. *Biological Conservation* 143: 399–407.

Shen, G., Liu, J., Jin, L. 1997. Preliminary results on U-series dating of Panxian Dadong in Guizhou Province, S-W. China. *Acta Anthropologica Sinica* 16: 221–30.

Shen, G., Wang, W., Wang, Q., Zhao, J. X., Collerosn, K., Zhou, C. L., Tobias, P. V. 2002. U-series dating of Liujiang hominid site in Guangxi, southern China. *Journal of Human Evolution* 43: 817–29.

Shepard, G. H., Ramirez, H. 2011. 'Made in Brazil': Human dispersal of the Brazil nut (Bertholletia excels, Lecythidaceae) in Ancient Amazonia. *Economic Botany* 65: 44–65.

Shipton, C., Roberts, R., Archer, W., Armitage, S. J., Bita, C., Blinkhorn, J., Courtney-Mustaphi, C., Crowther, A., Curtis, R., d'Errico, F., Douka, K., Faulkner, P., Groucutt, H. S., Helm, R., Herries, A. I. R., Jembe, S., Kourampas, N., Lee-Thorp, J., Marchant, R., Mercader, J., Pitarch Marti, A., Prendergast, M. E., Rowson, B., Tengeza, A., Tibesasa, R., White, T. S., Petraglia, M. D., Boivin, N. 2018. 78,000-year-old record of Middle and Later Stone Age innovation in an East African tropical forest. *Nature Communications* 9: 1832. doi:10.1038/s41467-018-04057-3.

Shono, K., Cadaweng, E. A., Durst, P. B. 2007. Application of assisted natural regeneration to restore degraded tropical forestlands. *Restoration Ecology* 15: 620–6.

Shukla, A., Mehrotra, R. C., Guleria, J. S. 2013. Emergence and extinction of Dipterocarpaceae in western India with reference to climate change: Fossil wood evidences. *Journal of Earth Systems Science* 122: 1373–86.

Shukla, J., Nobre, C., Sellers, P. 1990. Amazon deforestation and climate change. *Science* 247: 1322–5.

Sidle, R. C., Ziegler, A. D., Negishi, J. N., Nik, A. R., Siew, R., Turkelboom, F. 2006. Erosion processes in steep terrain: Truths, myths, and uncertainties related to forest management in Southeast Asia. *Forest Ecology and Management* 224: 199–225.

Siegel, P. E., Jones, J. G., Pearsall, D. M., Dunning, N. P., Farrell, P., Duncan, N. A., Curtis, J. H., Singh, S. K. 2015. Paleoenvironmental evidence for first human colonization of the eastern Caribbean. *Quaternary Science Reviews* 129: 275–95.

Sierra, C. A., del Valle, J. I., Orrego, S. A., Moreno, F. H., Harmon, M. E., Zapata, M., Colorado, G. J., Herrera, M. A., Lara, W., Restrepo, D. E., Berrouet, L. M., Loaiza, L. M., Benjumea, J. F. 2007. Total carbon stocks in a tropical forest landscape of the Porce region, Colombia. *Forest Ecology and Management* 243: 299–309.

Simon, D., Adam-Bradford, A. 2016. Archaeology and contemporary dynamics for more sustainable, resilient cities in the peri-urban interface. *Water Science and Technology* 72: 57–83.

Simons, E. L., Ettel, P. C. 1970. *Gigantopithecus. Scientific American* 222: 77–85.

Singarayer, J. S., Valdes, P. J., Friedlingstein, P., Nelson, S. R., Beerling, D. J. 2011. Late Holocene methane rise caused by orbitally controlled increase in tropical sources. *Nature* 470: 82–5.

Singh, B. P. 1997. Neolithic and Chalcolithic pottery of the Middle Ganga Plains: A case study of Senuwar. In P. C. Pant, V. Jayaswal (eds.), *Ancient Ceramics: Historical Enquiries and Scientific Approaches*. Delhi: Agam Kala Prakashan, pp. 3–33.

Sinha, A., Cannariato, K. G., Stott, L. D., Cheng, H., Edwards, R. L., Yadava, M. G., Ramesh, R., Singh, I. B. 2007. A 900-year (600 to 1500 AD) record of the Indian summer monsoon precipitation from the core monsoon zone of India. *Geophysical Research Letters* 34: L16707.

Skoglund, P., Reich, D. 2016. A genomic view of the peopling of the Americas. *Current Opinion in Genetics and Development* 41: 27–35.

Skoglund, P., Mallick, S., Bortolini, M. C., Chennagiri, N., Hünemeier, T., Petzl-Erler, M. L., Salzano, F. M., Patterson, N., Reich, D. 2015. Genetic evidence for two founding populations of the Americas. *Nature* 525: 104–8.

Skoglund, P., Posth, C., Sirak, K., Spriggs, M., Valentin, F., Bedford, S., Clark, G. R., Reepmeyer, C., Petchey, F., Fernandes, D., Fu, Q., Harney, E., Lipson, M., Mallick, S., Novak, M., Rohland, N., Stewardson, K., Abdullah, S., Cox, M. P., Friedlaender, F. R., Friedlaender, J. S., Kivisild, T., Koki, G., Kusuma, P., Merriwether, D. A., Ricaut, F.-X., Wee, J. T. S., Patterson, N., Krause, J., Pinhasi, R., Reich, D. 2016. Genomic insights into the peopling of the southwest Pacific. *Nature* 538: 510–13.

Smith, A. 1975. Radiocarbon dates from Bosumpra Cave, Abetifi, Ghana. *Proceedings of the Prehistory Society* 41: 179–82.

Smith, B. 1999. *The Emergence of Agriculture*. New York: W. H. Freeman.

Smith, B. 2006. Eastern North America as an independent center of plant domestication. *Proceedings of the National Academy of Sciences of the United States of America* 103: 12223–8.

Smith, B. N., Epstein, S. 1971. Two categories of $^{13}C/^{12}C$ ratios for higher plants. *Plant Physiology* 47: 380–4.

Smith, C. B., Cohen, M. C. L., Pessenda, L. C. R., França, M. C., Guimarães, J. T. F., de Fátima Rossetti, D., Lara, R. J. 2011. Holocene coastal vegetation changes at the mouth of the Amazon River. *Review of Palaeobotany and Palynology* 168: 21–30.

Smith, E., Wishnie, M. 2000. Conservation and subsistence in small-scale societies. *Annual Review of Anthropology* 29: 493–524.

Smith, N. 2001. Are indigenous peoples conservationists? *Rationality and Society* 13: 229–61.

Snow, B. E., Shutler, R., Nelson, D., Vogel, J., Southon, J. 1986. Evidence of early rice cultivation in the Philippines. *Philippine Quarterly of Culture and Society* 14: 3–11.

Soares, P., Alshamali, F., Pereira, J. B., Fernandes, V., Silva, N. M., Afonso, C., Costa, M. D., Musilová, E., Macaulay, V., Richards, M. B., Cerny, V., Pereira, L. 2012. The expansion of mtDNA haplogroup L3 within and out of Africa. *Molecular Biology and Evolution* 29: 915–27.

Sodhi, N. S., Brooks, T. M., Koh, L. P., Acciaioli, G., Erb, M., Tan, A. K.-J., Curran, L. M., Brosius, P., Ming Lee, T., Patlis, J. M., Gumal, M., Lee, R. J. 2006. Biodiversity and human livelihood crises in the Malay Archipelago. *Conservation Biology* 20: 1811–13.

Sodhi, N. S., Posa, M. R. C., Ming Lee, T., Bickford, D., Pin Koh, L., Brook, B. W. 2010. The state and conservation of Southeast Asian biodiversity. *Biodiversity and Conservation* 19: 317–28.

Soekmono, R. 1976. *Chandi Borobudur: A Monument of Mankind*. Paris: UNESCO Press.

Solheim, W. G. 1981. Philippine prehistory. In G. Casal, R. Trota Jose Jr (eds.), *The People and Art of the Philippines*. Los Angeles: Museum of Cultural History, University of California, pp. 17–83.

Sorensen, K. W. 1993. Indonesian peat swamp forests and their role as a carbon sink. *Chemosphere* 27: 1065–82.

Southgate, D. 1998. *Tropical Forest Conservation: An Economic Assessment of the Alternatives in Latin America.* New York: Oxford University Press.

Southworth, F. 2005. *The Linguistic Archaeology of South Asia.* London: Routledge.

Spielberg, S. 1981. *Indiana Jones: Raiders of the Lost Ark.* San Francisco: Lucasfilm Ltd.

Spittel, R. L. 1924. *Wild Ceylon, describing in particular the lives of the present-day Veddas.* Colombo: Colombo Apothecaries.

Spittel, R. L. 1961. *Vanished Trails: The Last of the Veddas* (2nd edition). Colombo: Associated Newspapers of Ceylon.

Sponheimer, M., Lee-Thorp, J. A. 1999. Isotopic evidence for the diet of an early hominid, *Australopithecus africanus. Science* 283: 368–70.

Sponheimer, M., Lee-Thorp, J., de Ruiter, D., Codron, D., Codron, J., Baugh, A. T., Thackeray, F. 2005. Hominins, sedges, and termites: New carbon isotope data from the Sterkfontein Valley and Kruger National Park. *Journal of Human Evolution* 48: 301–12.

Sponheimer, M., Loudon, J. E., Codron, D., Howell, M. E., Pruetz, J. D., Codron, J., de Ruiter, D. J., Lee-Thorp, J. A. 2006a. Do 'savanna' chimpanzees consume C_4 resources? *Journal of Human Evolution* 51: 128–33.

Sponheimer, M., Passey, B. H., de Ruiter, D. J., Guatelli-Steinberg, D., Cerling, T. E., Lee-Thorp, J. A. 2006b. Isotopic evidence for dietary variability in the early hominin *Paranthropus robustus. Science* 314: 980–2.

Sponheimer, M., Alemseged, Z., Cerling, T. E., Grine, F. E., Kibel, W. H., Leakey, M. G., Lee-Thorp, J. A., Manthi, F. K., Reed, K. E., Wood, B. A., Wynn, J. G. 2013. Isotopic evidence of early hominin diets. *Proceedings of the National Academy of Sciences of the United States of America* 110: 10513–18.

Spriggs, M. 1997. *The Island Melanesians.* Oxford: Blackwell.

Spruce, R. 1908. *Notes of a Botanist on the Amazon and Andes,* 2 vols. (edited by A. R. Wallace). New York: Macmillan.

Stanford, C. B., Wallis, J., Matama, H., Goodall, J. 1994. Patterns of predation by chimpanzees on red colobus monkeys in Gombe National Park, 1982–1991. *American Journal of Physical Anthropology* 94: 213–28.

Stark, M. 2006. Early mainland Southeast Asian landscapes in the first millennium BC. *Annual Review of Anthropology* 35: 407–32.

Stark, M. 2015. Southeast Asian urbanism: From early city to classical state. In N. Yoffee (ed.), *The Cambridge World History, Volume III: Early Cities in Comparative Perspective, 4000 BCE–1200 CE.* Cambridge: Cambridge University Press, pp. 48–73.

Steadman, D. W. 2006. *Extinction and Biogeography of Tropical Pacific Birds.* Chicago: University of Chicago Press.

Stein, W. E., Mannolini, F., VanAller Hernick, L., Landing, E., Berry, C. M. 2007. Giant cladoxylopsid trees resolve the enigma of the Earth's earliest forest stumps at Gilboa. *Nature* 446: 904–7.

Stephens, M., Mattey, D., Gilbertson, D. D., Colin, V. M.-W. 2008. Shell gathering from mangroves and the seasonality of the Southeast Asian monsoon using high resolution stable isotope analysis of the tropical estuarine bivalve (*Gelona erosa*) from the Great Cave of Naih, Sarawak: Methods and reconnaissance of molluscs of early Holocene and modern times. *Journal of Archaeological Sciences* 35: 2689–97.

Stern, J. T. Jr, Susman, R. L. 1983. The locomotor anatomy of *Australopithecus afarensis. American Journal of Physical Anthropology* 60: 279–317.

Sternberg, L. S. L. 2001. Savanna-forest hysteresis in the tropics. *Global Ecology and Biogeography* 10: 369–78.

Sternberg, L. S. L., Mulkey, S. S., Wright, S. J. 1989. Oxygen isotope ratio stratification in a tropical moist forest. *Oecologia* 81: 51–6.

Stevens, S. (ed.). 1997. *Conservation through Cultural Survival: Indigenous Peoples and Protected Areas.* Washington, DC: Island Press.

Stewart, B. A., Dewar, G. I., Morley, M. W., Inglis, R. H., Wheeler, M., Jacobs, Z., Roberts, R. G. 2012. Afromontane foragers of the Late Pleistocene: Site formation, chronology and occupation pulsing at Melikane Rockshelter, Lesotho. *Quaternary International* 270: 40–60.

Stewart, B. A., Parker, A. G., Dewar, G. I., Morley, M., Allott, L. 2016. Follow the Senqu: Maloti-Drakensberg paleoenvironments and implications for early human dispersals into mountain systems. In S. Jones, B. A. Stewart (eds.), *Africa from MIS 6–2: Population Dynamics and Paleoenvironments*. Dordrecht: Springer, pp. 247–71.

Stewart, K. 1954. *Pygmies and Dream Giants*. New York: Harper & Row.

Stiles, D. 1992. The hunter-gatherer 'revisionist' debate. *Anthropology Today* 8: 13–17.

Stocker, B. D., Strassmann, K., Joos, F. 2011. Sensitivity of Holocene atmospheric CO_2 and the modern carbon budget to early human land use: Analyses with a process-based model. *Biogeosciences* 8: 69–88.

Stoner, K. E., Vulinec, K., Wright, S. J., Peres, C. A. 2007. Hunting and plant community dynamics in tropical forests: A synthesis and future directions. *Biotropica* 39: 385–92.

Storm, P., de Vos, J. 2006. Rediscovery of the late Pleistocene Punung hominin sites and the discovery of a new site Gunung Dawung in East Java. *Senckenbergiana lethaea* 86: 271–81.

Storm, P., Aziz, F., de Vos, J., Kosasih, D., Baskoro, S., Ngaliman, van den Hoek Ostende, L. W. 2005. Late Pleistocene *Homo sapiens* in a tropical rainforest fauna in East Java. *Journal of Human Evolution* 49: 536–45.

Street-Perrott, F. A., Huang, Y., Perrott, R. A., Eglinton, G., Barker, P., Khelifa, L. B., Harkness, D. D., Olago, D. 1997. Impact of lower atmospheric carbon dioxide on tropical mountain ecosystems. *Science* 278: 1422–6.

Strickland, S. S. 1986. Long-term development of Kejaman subsistence: An ecological study. *Sarawak Museum Journal* 36: 117–71.

Stringer C. 2016. The origin and evolution of *Homo sapiens*. *Philosophical Transactions of the Royal Society B: Biological Sciences* 371: 20150237.

Stringer, C., Galway-Witham, J. 2018. When did modern humans leave Africa? *Science* 359: 389–90.

Stringer, C. B., Grün, R., Schwarcz, H., Goldberg, P. 1989. ESR dates for hominid burial site of Es Skhul in Israel. *Nature* 338: 756–7.

Stronza, A. 2001. Anthropology of tourism: Forging new ground for ecotourism and other alternatives. *Annual Review of Anthropology* 30: 261–83.

Stump, D. 2013. On applied archaeology, indigenous knowledge, and the usable past. *Current Anthropology* 54: 268–98.

Su, D., Harrison, T. 2008. Ecological implications of the relative rarity of fossil hominins and Laetoli. *Journal of Human Evolution* 55: 672–81.

Subrahmanyam, S. 1993. *The Portuguese Empire in Asia, 1500–1700: A Political and Economic History*. London: Longmans.

Summerhayes, G. R. 2000. Far Western, Western and Eastern Lapita: A re-evaluation. *Asian Perspectives* 39: 109–38.

Summerhayes, G. R., Ford, A. 2014. Late Pleistocene colonisation and adaptation in New Guinea: Implications for modeling modern human behaviour. In R. Dennell, M. Poor (eds.), *Southern Asia, Australia and the Search for Human Origins*. Cambridge: Cambridge University Press, pp. 213–27.

Summerhayes, G. R., Leavesley, M., Fairbairn, A., Mandui, H., Field, J., Ford, A., Fullagar, R. 2010. Human adaptation and plant use in Highland New Guinea 49,000 to 44,000 years ago. *Science* 330: 78–81.

Summerhayes, G. R., Field, J. H., Shaw, B., Gaffney, D. 2017. The archaeology of forest exploitation and change in the tropics during the Pleistocene: The case of Northern Sahul (Pleistocene New Guinea). *Quaternary International* 448: 14–30.

Survival International. 2008. Uncontacted tribe photos no hoax. Survivalinternational.org. Retrieved 19 March 2018.

Susman, R. L., Susman, J. T. 1983. The locomotor anatomy of *Australopithecus afarensis*. *American Journal of Physical Anthropology* 60: 279–317.

Sutikna, T., Tocheri, M. W., Morwood, M. J., Saptomo, E. W., Jatmiko, Due Awe, R., Wasisto, S., Westaway, K. E., Aubert, M., Li, B., Zhao, J.-X., Storey, M., Alloway, B. V., Morley, M. W., Meijer, H. J. M., van den Bergh, G. D., Grün, R., Dosseto, A., Brumm, A., Jungers, W. L., Roberts, R. G. 2016. Revised stratigraphy and chronology for *Homo floresiensis* at Liang Bua in Indonesia. *Nature* 532: 366–9.

Swift, J. A., Miller, M. J., Kirch, P. V. 2017a. Stable isotope analysis of Pacific rat (*Rattus exulans*) from archaeological sites in Mangareva (French Polynesia): The use of commensal species for understanding human activity and ecosystem change. *Environmental Archaeology* 22: 283–97.

Swift, J. A., Molle, G., Conte, E. 2017b. Subsistence and coastal settlement at the Hane Dune Site, Ua Huka (Marquesas Islands): New insights from Pacific rat (*Rattus exulans*) stable isotope analysis. *Journal of Archaeological Science Reports* 15C: 161–8.

Swift, J. A., Roberts, P., Boivin, N., Kirch, P. 2018. Restructuring of nutrient flows in island ecosystems following human colonization evidenced by isotopic analysis of commensal rats. *Proceedings of the National Academy of Sciences of the United States of America*. doi: 10.1073/pnas.1805787115.

Syler, T. 2011. Why Sting and I set up the Rainforest Foundation Fund. Huffingtonpost.com. <https://www.huffingtonpost.com/trudie-styler/why-sting-and-i-set-up-th_b_98252.html>. Retrieved 19 March 2018.

Szefer, P., Carmona, C. P., Chmel, K., Konečná, M., Libra, M., Molem, K., Novotný, V., Segar, S. T., Švamberková, E., Topliceanu, T.-S., Lepš, J. 2017. Determinants of litter decomposition rates in a tropical forest: Functional traits, phylogeny and ecological succession. *Oikos* 126: 1101–11.

Tagami, T. 2013. High-resolution paleoclimate reconstruction of Tropical Asia using stable isotopic data from speleothems and trees. *Journal of Disaster Research* 8: 131–2.

Takemoto, H., Kawamoto, Y., Furuichi, T. 2015. How did bonobos come to range south of the Congo River? Reconsideration of the divergence of Pan Paniscus from other Pan populations. *Evolutionary Anthropology* 24: 170–84.

Takezawa, S., Cissé, M. 2004. Domestication des Céréales au Mema, Mali. In K. Sanogo, T. Togola (eds.), *XIth Congress of Panafrican Association Prehistory and Related Fields, Bamako, 07–12 Feb 2001*. Bamako: Soro Print Color, pp. 105–21.

Tang, Y., Yang, X., Cao, M., Baskin, C. C., Baskin, J. M. 2010. Buttress trees elevate soil heterogeneity and regulate seedling diversity in a tropical rainforest. *Plant Soil* 338: 301–9.

Tanner, E. V. J., Sheldrake, M. W. A., Turner, B. L. 2016. Changes in soil carbon and nutrients following 6 years of litter removal and addition in a tropical semi-evergreen rain forest. *Biogeosciences* 13: 6183–90.

Tappen, M. 1994. Bone weathering in the tropical rain forest. *Journal of Archaeological Science* 21: 667–73.

Tattersall, I., Schwartz, J. H. 2008. The morphological distinctiveness of *Homo sapiens and* its recognition in the fossil record: Clarifying the problem. *Evolutionary Anthropology* 17: 49–54.

Tattersfield, N. 1998. *The Forgotten Trade: Comprising the Log of the 'Daniel and Henry' of 1700 and Accounts of the Slave Trade from the Minor Ports of England 1698–1725*. London: Pimlico.

Taylor, D., Marchant, R., Hamilton, A. C. 2001. A reanalysis and interpretation of palynological data from the Kalambo Falls prehistoric site. In J. D. Clark (ed.), *Kalambo Falls Prehistoric Site*, vol. III. Cambridge: Cambridge University Press, pp. 66–81.

Taylor, N. 2009. The role of Lupemban core-axes in the hominin dispersal into the Congo basin, Central Africa: A functional assessment by use-wear analysis. PhD thesis, University of Liverpool, Liverpool.

Taylor, N. 2011. The origins of hunting and gathering in the Congo basin: A perspective on the Middle Stone Age Lupemban industry. *Before Farming* 1: 1–20.

Taylor, N. 2016. Across rainforests and woodlands: A systematic reappraisal of the Lupemban Middle Stone Age in Central Africa. In S. C. Jones, B. A. Stewart (eds.), *Africa from MIS 6–2: Population Dynamics and Paleoenvironments*. Dordrecht: Springer, pp. 273–99.

Terborgh, J. 1988. The big things that run the world: A sequel to E. O. Wilson. *Conservation Biology* 2: 402–3.

Terborgh, J. 1992. *Diversity and the Tropical Rainforest*. New York: Scientific American Library.

ter Steege, H. T., Pitman, N. C. A., Sabatier, D., Baraloto, C., Salomao, R. O., Guevara, J. E., Phillips, O. L., Castilho, C. V., Magnusson, W. E., Molinso, J.-F., Monteagudo, A., Vargas, P. N., Montero, J. C., Feldpausch, T. R., Coronado, E. N. H., Killeen, T. J., Mostacedo, B., Vasquez, R., Assis, R. L., Terborgh, J., Wittmann, F., Andrade, A., Laurance, W. F., Laurance, S. G. W., Marimon, B. S., Marimon Jr, B.-H., Vieira, I. C. G., Amaral, I. L., Brienen, R., Castellanos, H., López, D. C., Deuivenvoorden, J. F., Mogollón, H. F., Matos, F. D. A., Dávila, N., Garcia-Villacorta, R., Diaz, P. R. S., Costa, F., Emilio, T., Levis, C., Schietti, J., Souza, P., Alonso, A., Dallmeier, F., Montoya, A. J. D., Piedade, M. T. F., Araujo-Murakami, A., Arroyo-L., Gribel, R., Fine, P. V. A., Peres, C. A., Toledo, M., Aymard, C. G. A., Baker, T. R., Cerón, C., Engel, J., Henkel, T. W., Maas, P., Petronelli, P., Stropp, J., Zartman, C. E., Daly, D., Neill, D., Silveira, M., Paredes, M. R., Chave, J., Filho, D. A. L., Jorgensen, P. M., Fuentes, A., Schongart, J., Valverde, F. C., Di Fiore, A., Jimenez, E. M., Mora, M. C. P., Phillips, J. F., Rivas, G., van Andel, T. R., von Hildebrand, P., Hoffman, B., Zent, E. L., Malhi, Y., Prieto, A., Rudas, A., Ruschell, A. R., Silva, N., Vos, V., Zent, S., Oliveira, A. A., Schutz, A. C., Gonzales, T., Nascimento, M. T., Ramirez-Angulo, H., Sierra, R., Tirado, M., Medina, M. N. U., van der Heijden, G., Vela, C. I. A., Torre, E. V., Vriesendorp, C., Wang, O., Young, K. R., Baider, C., Balslev, H., Ferreira, C., Mesones, I., Torres-Lezama, A., Giraldo, L. E. U., Zagt, R., Alexiades, M. N., Hernandez, L., Huamantupa-Chuquimaco, I., Milliken, W., Cuenca, W. P., Pauletto, D., Sandoval, E. V., Gamarra, L. V., Dexter, K. G., Feeley, K., Lopez-Gonzalez, G., Silman, M. R. 2013. Hyperdominance in the Amazonian tree flora. *Science* 342: 325.

Tewari, R., Srivastava, R. K., Singh, K. K., Saraswat, K. S., Singh, I. B. 2003. Preliminary report of the excavation at Lahuradewa, District Sant Kabir Nagar, U.P. 2001–2002: Wider archaeological implications. *Pragdhara* 13: 37–68.

Tewari, R., Srivastava, R. K., Singh, K. K., Vinay, R., Trivedi, R. K., Singh, G. C. 2005. Recently excavated sites in the Ganga Plain and North Vindhyas: Some observations regarding the pre-urban context. *Pragdhara* 15: 39–49.

Thangaraj, K., Singh, L., Reddy, A. G., Rao, V. R., Sehgal, S. C., Underhill, P. A., Pierson, M., Frame, I. G., Hagelberg, E. 2003. Genetic affinities of the Andaman Islanders, a vanishing human population. *Current Biology* 13: 86–93.

Thangaraj, K., Chaubey, G., Kivisild, T., Reddy, A. G., Singh, V. K., Rasalkar, A. A., Singh, L. 2005. Reconstructing the origin of Andaman Islanders. *Science* 308: 996.

Thangaraj, K., Chaubey, G., Reddy, A. G., Singh, V. K., Singh, L. 2006. Unique origin of Andaman Islanders: Insight from autosomal loci. *Journal of Human Genetics* 51: 800–4.

The State of the Tropics Project. 2016. *State of the Tropics 2016 Report*. Queensland, Australia: James Cook University Press.

Thieret, J. W. 1958. Economic botany of the cycads. *Economic Botany* 12: 3–41.

Thomas, E., Alcazar Caicedo, C., McMichael, C. H., Corvera, R., Loo, J. 2015. Uncovering spatial patterns in the natural and human history of Brazil nut (*Bertholletia excelsa*) across the Amazon Basin. *Journal of Biogeography* 42: 1367–82.

Thomas, M. 2000. Late Quaternary environmental changes and the alluvial record in humid tropical environments. *Quaternary International* 72: 23–36.

Thompson, A. 2004. The future of Cambodia's past: A messianic Middle-Period Cambodian royal cult. In J. Marston, E. Guthrie (eds.), *History, Buddhism, and New Religious Movements in Cambodia*. Honolulu: University of Hawai'i Press, pp. 13–39.

Thompson, G. B. 1996. *The Excavation of Khok Phanom Di: A Prehistoric Site in Central Thailand: vol. IV. Subsistence and Environment: The Botanical Evidence (The Biological Remains, Part II)*. London: The Society of Antiquaries.

Thompson, L. G., Mosley-Thompson, E., Davis, M. E., Lin, P.-N., Henderson, K. A., Cole-Dai, J., Bolsan, J. F., Liu, K. G. 1995. Late glacial stage and Holocene tropical ice core records from Huascaran, Peru. *Science* 269: 46–50.

Thornton, J. 1992. *Africa and Africans in the Making of the Atlantic World*. Cambridge: Cambridge University Press.

Thorpe, S. K. S., Crompton, R. H. 2005. The locomotor ecology of wild orangutans (*Pongo pygmaeus abelii*) in the Gunung Leuser ecosystem, Sumatra, Indonesia: A multivariate analysis using log-linear modelling. *American Journal of Physical Anthropology* 127: 58–78.

Thorpe, S. K. S., Holder, R. L., Crompton, R. H. 2007. Origin of human bipedalism as an adaptation for locomotion on flexible branches. *Science* 316: 328–31.

Tian, H., Banger, K., Bo, T., Dadhwal, V. K. 2014. History of land use in India during 1880–2010: Large-scale land transformations reconstructed from satellite data and historical archives. *Global and Planetary Change* 121: 78–88.

TIES (The International Ecotourism Society). 2015. What is ecotourism? Ecotourism.org. <http://www.ecotourism.org/what-is-ecotourism>. Retrieved 25 March 2018.

Tieszen, L. L. 1991. Natural variations in the carbon isotopes of plants: Implications for archaeology, ecology and paleoecology. *Journal of Archaeological Science* 18: 227–48.

Timmermann, A., Friedrich, T. 2016. Late Pleistocene climate drivers of early human migration. *Nature* 538: 92–5.

Tishkoff, S. A., Reed, F. A, Friedlaender, F. R., Ehret, C., Ranciaro, A., Froment, A., Hirbo, J. B., Awomoyi, A. A., Bodo, J. M., Doumbo, O., Ibrahim, M., Juma, A. T., Kotze, M. J., Lema, G., Moore, J. H., Mortensen, H., Nyambo, T. B., Omar, S. A., Powell, K., Pretorius, G., Smith, M. W., Thera, M. A., Wambebe, C., Weber, J. L., Williams, S. M. 2009. The genetic structure and history of Africans and African Americans. *Science* 324: 1035–44.

Tobias, D., Mendelsohn, R. 1991. Valuing ecotourism in a tropical rainforest reserve. *Ambio* 20: 91–3.

Toledo, V., Ortiz-Espejel, B., Cortes, L., Moguel, P., Ordoñez, M. 2003. The multiple uses of tropical forests by indigenous peoples in Mexico: A case of adaptive management. *Conservation Ecology* 7: 9.

Tonkunaga, S., Oshima, H., Polhaupesy, A., Ito, Y. 1985. A palynological study of the Pucangan and Kabuh formations in the Sangiran area. In N. Watanabe, D. Kadar (eds.), *Quaternary Geology of the Hominid Fossil Bearing Formations in Java*. Bandung: Geological Research and Development Centre, pp. 199–217.

Toth, N., Schick, K. D., Savage-Rumbaugh, E. S., Sevcik, R. A., Rumbaugh, D. M. 1993. Pan the tool-maker: Investigations into the stone tool-making and tool-using capabilities of a Bonobo (*Pan paniscus*). *Journal of Archaeological Science* 20: 81–91.

Tougard, C., Chaimanee, Y., Suteethorn, V., Triamwichanon, S., Jaeger, J.-J. 1996. Extension of the geographic distribution of the giant panda (Ailuropoda) and search for the reasons for its progressive disappearance in Southeast Asia during the Latest Middle Pleistocene. *Comptes Rendus de l'Académie des Sciences, Paris* 323: 973–9.

Trauth, M. H., Maslin, M. A., Deino, A. L., Bergner, M. L., Strecker, M. R., Bergner, A. G. N., Dühnforth, M. 2007. High- and low-latitude controls and East African forcing of Plio-Pleistocene East African climate and early human evolution. *Journal of Human Evolution* 53: 475–86.

Trigger, B. 1989. *A History of Archaeological Thought.* Cambridge: Cambridge University Press.

Trigger, B. 2003. *Understanding Early Civilizations.* Cambridge: Cambridge University Press.

Turnbull, C. 1961. *The Forest People: A Study of the Pygmies of the Congo.* New York: Simon & Schuster.

Turnbull, C. 1965. *Wayward Servants: The Two Worlds of the African Pygmies.* Garden City, NY: American Museum of Natural History, Natural History Press.

Turner, B. L., Sabloff, J. A. 2012. Classic period collapse of the Central Maya Lowlands: Insights about human–environment relationships for sustainability. *Proceedings of the National Academy of Sciences of the United States of America* 109: 13908–14.

Turner, V. 1967. *The Forest of Symbols: Aspects of Ndembu Ritual.* Ithaca, NY: Cornell University Press.

Turney, C. S. M., Kershaw, A. P., Clemens, S. C., Branch, N., Moss, P. T., Fifield, L. K. 2004. Millennial and orbital variations of El Niño/Southern Oscillation and high-latitude climate in the last glacial period. *Nature* 428: 306–10.

Turton, S. M., Stork, N. 2006. Environmental impacts of tourism and recreation in the Wet Tropics. In N. Stork, S. M. Turton (eds.), *Living in a Dynamic Tropical Forest Landscape.* London: Wiley-Blackwell, pp. 47–58.

Tuttle, R. H. 2006. Seven decades of east African Miocene anthropoid studies. In H. Ishida, R. Tuttle, M. Pickford, N. Ogihara, M. Nakatsukasa (eds.), *Human Origins and Environmental Backgrounds.* Dordrecht: Springer, pp. 15–30.

Tuttle, R. H. 2014. *Apes and Human Evolution.* Cambridge, MA: Harvard University Press.

UNESCO. 2011. *World Heritage Forests.* Paris: World Heritage Vol. 61, UNESCO. <http://unesdoc.unesco.org/images/0021/002139/213912e.pdf>.

UNESCO. 2018. *Wet Tropics of Queensland.* <http://whc.unesco.org/en/list/486>. Retreieved 25 March 2018.

Ungar, P. S., Scott, R. S., Grine, F. E., Teaford, M. F. 2010. Molar microwear textures and the diets of *Australopithecus anamensis* and *Australopithecus afarensis. Philosophical Transactions of the Royal Society B: Biological Sciences* 365: 3345–54.

Uno, K. T., Cerling, T. E., Harris, J. M., Kunimatsu, Y., Leakey, M. G., Nakatsukasa, M., Nakaya, H. 2011. Late Miocene to Pliocene carbon isotope record of differential diet change among East African herbivores. *Proceedings of the National Academy of Sciences of the United States of America* 108: 6509–14.

Uno, K. T., Polissar, P. J., Jackson, K. E., de Menocal, P. B. 2016. Neogene biomarker record of vegetation change in eastern Africa. *Proceedings of the National Academy of Sciences of the United States of America* 113: 6355–63.

UN-REDD Programme Collaborative Online Workspace. 2018. UN-REDD Programme. <http://www.unredd.net/regions-and-countries/regions-and-countries-overview.html>. Retrieved 25 March 2018.

Upchurch, G. R., Jr, Wolfe, J. A. 1987. Mid-Cretaceous to Early Tertiary vegetation and climate: Evidence from fossil leaves and wood. In E. M. Friis, W. G. Chaloner, P. R. Crane (eds.), *The Origins of Angiosperms and Their Biological Consequences.* Cambridge: Cambridge University Press, pp. 75–105.

Usik, V. I., Rose, J. I., Hilbert, Y. H., Van Peer, P., Marks, A. E. 2013. Nubian complex reduction strategies in Dhofar, southern Oman. *Quaternary International* 300: 244–66.

Vadjunec, J. M., Schmink, M., Gomes, C. V. A. 2011. Rubber tapper citizens: Emerging places, policies, and shifting rural–urban identities in Acre, Brazil. *Journal of Cultural Geography* 28: 73–98.

Vaks, A., Bar-Matthews, M., Ayalon, A., Matthews, A., Halicz, L., Frumkin, A. 2007. Desert speleothems reveal climatic window for African exodus of early modern humans. *Geology* 35: 831–4.

Valdés, J. A., Fahsen, F. 2005. Disaster in sight: The Terminal Classic at Tikal and Uaxactun. In A. A. Demarest, P. M. Rice, D. S. Rice (eds.), *The Terminal Classic in the Maya Lowlands: Collapse, Transition, and Transformation.* Boulder, CO: University Press of Colorado, pp. 162–94.

Valentine, P. S., Hill, R. 2008. The establishment of a World Heritage Area. In N. Stork, S. M. Turton (eds.), *Living in a Dynamic Tropical Forest Landscape.* London: Wiley-Blackwell, pp. 81–93.

Valladas, H., Reyss, J. L., Joron, J. L., Valladas, G., Bar-Yosef, O., Vandermeersch, B. 1988. Thermoluminescence dating of Mousterian Troto-Cro-Magnon remains from Israel and the origin of modern man. *Nature* 331: 614–16.

Van den Bergh, G. D. 1999. The Late Neogene elephantoid-bearing faunas of Indonesia and their palaeozoogeographic implications: A study of the terrestrial faunal succession of Sulawesi, Flores and Java, including evidence for early hominid dispersal east of Wallace's Line. *Scripta Geologica* 117: 1–388.

Van den Bergh, G. D., Mubroto, B., Sondaar, P. Y., de Vos, J. 1996a. Did *Homo erectus* reach the island of Flores? *Bulletin of the Indo-Pacific Prehistory Association* 14: 27–36.

Van den Bergh, G. D., de Vos, J., Sondaar, P. Y., Aziz, F. 1996b. Pleistocene zoogeographic evolution of Java (Indonesia) and glacio-eustatic sea level fluctuations: A background for the presence of *Homo. Bulletin of the Indo-Pacific Prehistory Association* 14: 7–21.

Van den Bergh, G. D., Kaifu, Y., Kurniawan, I., Kono, R. T., Brumm, A., Setiyabudi, E., Aziz, F., Morwood, M. J. 2016. *Homo floresiensis*-like fossils from the early Middle Pleistocene of Flores. *Nature* 534: 245–8.

Van der Ent, R. J., Savenije, H. H. G., Schaefli, B., Steele-Dunne, S. C. 2010. Origin and fate of atmospheric moisture over continents. *Water Resource Research* 46: W09525.

Van der Heijden, M. G. A., Bardgett, R. D., Van Straalen, N. M. 2007. The unseen majority: Soil microbes as drivers of plant diversity and productivity in terrestrial ecosystems. *Ecology Letters* 11: 296–310.

Van der Kaars, S. 1998. Marine and terrestrial pollen records of the last glacial cycle from the Indonesian region: Bandung Basin and Banda Sea. *Paleoclimates—Data Modelling* 3: 209–19.

Van der Merwe, N. J., Medina, E. 1991. The canopy effect, carbon isotope ratios and foodwebs in Amazonia. *Journal of Archaeological Science* 18: 249–59.

Van der Merwe, N. J., Thackeray, J. F., Lee-Thorp, J. A., Luyt, J. 2003. The carbon isotope ecology and diet of *Australopithecus africanus* at Sterkfontein, South Africa. *Journal of Human Evolution* 44: 581–97.

Van der Merwe, N. J., Masao, F. T., Bamford, R. J. 2008. Isotopic evidence for contrasting diets of early hominins *Homo habilis* and *Australopithecus boisei* of Tanzania. *South African Journal of Science* 104: 153–5.

van Heerwaarden, J., Doebley, J., Briggs, W. H., Glaubitz, J. C., Goodman, M. M., Sánchez González, J. J., Ross-Ibarra, J., Gaut, B. S., Eguiarte, L. 2011. Genetic signals of origin, spread, and introgression in a large sample of maize landraces. *Proceedings of the National Academy of Sciences of the United States of America* 108: 1088–92.

van Hinsbergen, D. J. J., de Groot, L. V., van Schaik, S. J., Spakman, W., Bijl, P. K., Sluijs, A., Langereis, C. G., Brinkhuis, H. 2015. A paleolatitude calculator for paleoclimate studies. *PLoS ONE* 10: e0126946.

Van Noten, F. 1977. Excavations at Matupi Cave. *Antiquity* 51: 35–40.

Van Noten, F., Haesaerts, P., Cahen, D. 1972. Un habitat lupembien a Masango, Burundi. Rapport préliminaire. *Africa Tervuren* 18: 78–85.

Van Schaik, C. P., Ancrenaz, M., Borgen, G., Galdikas, B., Knott, C. D., Singleton, I., Suzuki, A., Utami, S. S., Merrill, M. 2003. Orangutan cultures and the evolution of material culture. *Science* 239: 102–5.

Van Straaten, O., Corre, M. D., Wolf, K., Tchienkoua, M., Cuellar, E., Matthews, R. B., Veldkamp, E. 2015. Conversion of lowland tropical forests to tree cash crop plantations loses up to one-half of stored soil organic carbon. *Proceedings of the National Academy of Sciences of the United States of America* 112: 9956–60.

Van Vliet, N., Milner-Gulland, E. J., Bousquet, F., Saqall, M., Nasi, R. 2009. Effect of small-scale heterogeneity of prey and hunter distributions on the sustainability of bushmeat hunting. *Conservation Biology* 24: 1327–37.

Vanhaeren, M., d'Errico, F., van Niekerk, K. L., Henshilwood, C. S., Erasmus, R. M. 2013. Thinking strings: Additional evidence for personal ornament use in the Middle Stone Age at Blombos Cave, South Africa. *Journal of Human Evolution* 64: 500–17.

Veiga, J. B., Tourrand, J. F., Piketty, M. G., Poccard-Chapuis, R., Alves, A. M., Thales, M. C. 2004. *Expansão e Trajetórias da Pecuária na Amazônia: Pará.* Brasília, Brazil: Editora Universidade de Brasília.

Venkataraman, V. V., Kraft, T. S., Dominy, N. J. 2013a. Tree climbing and human evolution. *Proceedings of the National Academy of Sciences of the United States of America* 110: 1237–42.

Venkataraman, V. V., Kraft, T. S., Desilva, J. M., Dominy, N. J. 2013b. Phenotypic plasticity of climbing-related traits in the ankle joint of great apes and rainforest hunter-gatherers. *Human Biology* 81: 309–28.

Venkataraman, V. V., Kraft, T. S., Dominy, N. J., Endicott, K. M. 2017. Hunter-gatherer residential mobility and the marginal value of rainforest patches. *Proceedings of the National Academy of Sciences of the United States of America* 114: 3097–102.

Verdu, P., Destro-Bisol, G. 2012. African pygmies: What's behind a name? *Human Biology* 84: 1–10.

Verdu, P., Austerlitz, F., Estoup, A., Vitalis, R., Georges, M., Théry, S., Froment, A., Le Bomin, S., Gessain, A., Hombert, J.-M., Van der Veen, L., Quintana-Murci, L., Bahuchet, S., Heyer, E. 2009. Origins and genetic diversity of pygmy hunter-gatherers from western Central Africa. *Current Biology* 19: 312–18.

Verdu, P., Becker, N. S. A., Froment, A., Georges, M., Grugni, V., Quintana-Murci, L., Hombert, J.-M., Van der Veen, L., Le Bomin, S., Bahuchet, S., Heyer, E., Austerlitz, F. 2013. Sociocultural behavior, sex-biased admixture, and effective population sizes in central African pygmies and non-pygmies. *Molecular Biology and Evolution* 30: 918–37.

Vernot, B., Tucci, S., Kelso, J., Schraiber, J. G., Wolf, A. B., Gittelman, R. M., Dannemann, M., Grote, S., McCoy, R. C., Norton, H., Scheinfeldt, L. B., Merriwether, D. A., Koki, G., Friedlaender, J. S., Wakefield, J., Pääbo, S., Akey, J. M. 2016. Excavating Neandertal and Denisovan DNA from the genomes of Melanesian individuals. *Science* 352. doi:10.1126/science.aad9416.

Vickers, W. 1995. From opportunism to nascent conservation: The case of the Siona-Secoya. *Human Nature* 5: 307–37.

Vieira, D. L. M., Holl, K. D., Peneireiro, F. M. 2009. Agro-successional restoration as a strategy to facilitate tropical forest recovery. *Restoration Ecology* 17: 451–9.

Villmoare, B., Kimbel, W. H., Seyoum, C., Campisano, C. J., DiMaggio, E. N., Rowan, J., Braun, D. R., Arrowsmith, J. R., Reed, K. E. 2015. Early *Homo* at 2.8 Ma from Ledi-Geraru, Afar, Ethiopia. *Science* 347: 1352–5.

Viveiros de Castro, E. 1998. Cosmological deixis and Amerindian perspectivism. *Journal of the Royal Anthropological Institute* 4: 469–88.

Voeks, R. A. 2004. Disturbance pharmacopoeias: Medicine and myth from the humid tropics. *Annals of the Association of American Geographers* 94: 868–88.

Vrba, E. 1995. The fossil record of African antelopes (Mammalia, Bovidae) in relation to human evolution and paleoclimate. In E. Vrba, G. Denton, L. Burckle, T. Partridge (eds.), *Paleoclimate and Evolution with Emphasis on Human Origins*. New Haven, CT: Yale University Press, pp. 385–424.

Walker, R., Moran, E., Anselin, L. 2000. Deforestation and cattle ranching in the Brazilian Amazon: External capital and household processes. *World Development* 28: 683–99.

Wallace, A. R. 1878. *Tropical Nature and Other Essays*. New York: Macmillan.

Wang, B. C., Sork, V. L., Leong, M. T., Smith, T. B. 2007. Hunting of mammals reduces seed removal and dispersal of the Afrotropical tree *Antrocaryon klaineanum* (anacardiaceae). *Biotropica* 39: 340–7.

Wang, P., Clemens, S., Beaufort, L., Braconnot, P., Ganssen, G., Jian, Z., Kershaw, P., Sarnthein, M. 2005. Evolution and variability of the Asian monsoon system: State of the art and outstanding issues. *Quaternary Science Reviews* 24: 595–629.

Ward, C. V., Leakey, M. G., Walker, A. 2001. Morphology of *Australopithecus anamensis* from Kanapoi and Allia Bay, Kenya. *Journal of Human Evolution* 41: 255–368.

Ward, S. C., Duren, D. L. 2002. *The Primate Fossil Record*. Cambridge: Cambridge University Press.

Waters, M. R., Stafford, T. W. Jr 2007. Redefining the age of Clovis: Implications for the peopling of the Americas. *Science* 315: 1122–6.

Watling, J., Iriarte, J., Mayle, F. E., Schaan, D., Pessenda, L. C. R., Loader, N. J., Street-Perrott, F. A., Dickau, R. E., Damasceno, A., Ranzi, A. 2017a. Impact of pre-Columbian 'geoglyph' builders on Amazonian forests. *Proceedings of the National Academy of Sciences of the United States of America* 114: 1868–73.

Watling, J., Iriarte, J., Mayle, F. E., Schaan, D., Pessenda, L. C. R., Loader, N.J., Street-Perrott, F.A., Dickau, R.E., Damasceno, A., Ranzi, A. 2017b. Reply to Piperno et al.: It is too soon to argue for localized, short-term human impacts in interfluvial Amazonia. *Proceedings of the National Academy of Sciences of the United States of America* 114: E4120–E4121.

Watson, V. D., Cole, J. D. 1977. *Prehistory of the Eastern Highlands of New Guinea*. Seattle: University of Washington Press.

Watts, S. 1997. *Epidemics and History: Disease, Power and Imperialism*. New Haven, CT: Yale University Press.

Webb, L. J. 1968. Environmental relationships of the structural types of Australian rain forest vegetation. *Ecology* 49: 296–311.

Webb, L. J., Tracey, J. G. 1981. Australian rainforest: Patterns and change. In A. Keast (ed.), *Ecological Biogeography of Australia*. The Hague: Dr W. Junk Publishers, pp. 605–94.

Weber, S., Lehman, H., Barela, T., Hawks, S., Harriman, D. 2010. Rice or millets: Early farming strategies in prehistoric central Thailand. *Archaeological and Anthropological Sciences* 2: 79–88.

Webster, D. 2002. *The Fall of the Ancient Maya*. London: Thames & Hudson.

Webster, D., Murtha, T., Straight, K. D., Silverstein, J., Martinez, H., Terry, R. E., Burnett, R. 2007. The Great Tikal Earthworks revisited. *Journal of Field Archaeology* 32: 41–64.

Welbergen, J. A., Klose, S. M., Markus, N., Eby, P. 2008. Climate change and the effects of temperature extreme on Australian flying-foxes. *Proceedings of the Royal Society B: Biological Sciences* 275. doi:10.1098/rspb.2007.1385.

Wengrow, D. 2006. *The Archaeology of Early Egypt: Social Transformations in North-East Africa, c. 10,000 to 2,650 BC*. Cambridge: Cambridge University Press.

Werth, D., Avissar, R. 2005. The local and global effects of African deforestation. *Geophysical Research Letters* 32: L12704.

Wesselman, H. B. 1985. Fossil micromammals as indicators of climatic change about 2.4 Myr ago in the Omo Valley, Ethiopia. *South African Journal of Science* 81: 260–1.

Westaway, K. E., Morwood, M. J., Roberts, R. G., Rokus, A. D., Zhao, J. X., Storm, P., Aziz, F., Van Den Bergh, G., Hadi, P., Jatmiko, De Vos, J. 2007. Age and biostratigraphic significance of the Punung rainforest fauna, East Java, Indonesia, and implications for *Pongo* and *Homo*. *Journal of Human Evolution* 53: 709–17.

Westaway, K. E., Morwood, M. J., Sutikna, T., Moore, M. W., Rokus, A. D., van den Bergh, G. D., Roberts, R. G., Saptomo, E. W. 2009. *Homo floresiensis* and the late Pleistocene environments of eastern Indonesia: Defining the nature of the relationship. *Quaternary Science Reviews* 28: 2897–912.

Westaway, K. E., Louys, J., Due Awe, R., Morwood, M. J., Price, G. J., Zhao, J.-X., Aubert, M., Joannes-Boyau, R., Smith, T. M., Skinner, M. M., Compton, T., Bailey, R. M., van den Bergh, G. D., de Vos, J., Pike, A. W. G., Stringer, C., Saptomo, E. W., Rizal, Y., Zaim, J., Santoso, W. D., Trihascaryo, A., Kinsley, L., Sulistyanto, B. 2017. An early modern human presence in Sumatra 73,000–63,000 years ago. *Nature* 548: 322–5.

Wheeler, P. E. 1984. The evolution of bipedality and loss of functional body hair in hominids. *Journal of Human Evolution* 13: 91–8.

Whistler, W. A., Elevitch, C. 2006. Broussonetia papyrifera (paper mulberry), ver. 2.1. In C. R. Elevitch (ed.), *Species Profiles for Pacific Island Agroforestry*. Hōlualoa, Hawai'i: Permanent Agriculture Resources (PAR).

White, F. 1983. *The Vegetation of Africa: A Descriptive Memoir to Accompany the UNESCO/AETFAT/UNSO Vegetation Map of Africa*. Paris: UNESCO.

White, J. P. 1972. *Ol Tumbuna: Archaeology Excavations in the Eastern Central Highlands, Papua New Guinea*. Terra Australis 2. Canberra: Department of Prehistory, Research School of Pacific Studies, The Australian National University.

White, L. A. 2007 (1949). *The Evolution of Culture: The Development of Civilization to the Fall of Rome*. London: Routledge.

White, T. D. 2014. Comment on 'Is the "savanna hypothesis" a dead concept for explaining the emergence of the earliest hominins?' *Current Anthropology* 55: 75–6.

White, T. D., Suwa, G. 1987. Hominid footprints at Laetoli: Facts and interpretations. *American Journal of Physical Anthropology* 72: 485–514.

White, T. D., Suwa, G., Asfaw, B. 1994. *Australopithecus ramidus*, a new species of early hominid from Aramis, Ethiopia. *Nature* 371: 306–12.

White, T. D., Asfaw, B., DeGusta, D., Gilbert, H., Richards, G. D., Suwa, G., Howell, F. C. 2003. Pleistocene *Homo sapiens* from Middle Awash, Ethiopia. *Nature* 423: 742–7.

White, T. D., WoldeGabriel, G., Asfaw, B., Ambrose, S., Beyene, Y., Bernor, R. L., Boisserie, J.-R., Currie, B., Gilbert, H., Haile-Selassie, Y., Hart, W. K., Hlusko, L. J., Howell, F. C., Kono, R. T., Lehmann, T., Louchart, A., Lovejoy, C. O., Renne, P. R., Saegusa, H., Vrba, E. S., Wesselman, H., Suwa, G. 2006. Asa Issie, Aramis and the origin of *Australopithecus*. *Nature* 440: 883–9.

White, T. D., Asfaw, B., Beyene, Y., Haile-Selassie, Y., Lovejoy, C. O., Suwa, G., WoldeGabriel, G. 2009. *Ardipithecus ramidus* and the paleobiology of early hominids. *Science* 326: 75–86.

Whiten, A., Goodall, J., McGrew, W. C., Nishida, T., Reynolds, V., Sugiyama, Y., Tutin, C. E. G., Wrangham, R. W., Boesch, C. 1999. Cultures in chimpanzees. *Nature* 399: 682–5.

Whitmore, T. C. 1977. A first look at *Agathis*. Tropical Forestry Papers 11. Oxford: University of Oxford Commonwealth Forest Institute.

Whitmore, T. C. 1984. *Tropical Rain Forests of the Far East* (2nd edition). Oxford: Clarendon Press.

Whitmore, T. C. 1989. Forty years of rain forest ecology. *Geojournal* 19: 347–60.

Whitmore, T. C. 1998. *An Introduction to Tropical Rainforests* (2nd edition). Oxford: Oxford University Press.

Whitmore, T. C., Peralta, R., Brown, K. 1985. Total species count in a Costa Rican tropical rain forest. *Journal of Tropical Ecology* 1: 375–8.

Whitten, T., Balmford, A. 2006. Who should pay for tropical forest conservation, and how could the costs be met? In C. Peres, W. Laurance (eds.), *Emerging Threats to Tropical Forests*. Chicago: University of Chicago Press, pp. 317–36.

Wichman, F. B. 2003. *Na Pua Alii O Kauai: Ruling Chiefs of Kauaì*. Honolulu: University of Hawai'i Press.

Wickham, C. 2009. *The Inheritance of Rome: Illuminating the Dark Ages 400–1000*. London: Penguin Books.

Wijeyapala, W. H. 1997. New light on the prehistory of Sri Lanka in the context of recent investigations of cave sites. PhD thesis, University of Peradeniya, Kandy.

Wikramanayake, E., Dinerstein, E., Loucks, C. J., Olson, D. M., Morrison, J., Lamoreux, J., McKnight, M., Hedao, P. 2002. *Terrestrial Ecoregions of the Indo-Pacific: A Conservation Assessment*. Washington, DC: Island Press.

Williams, M., Zalasiewicz, J., Haff, P., Schwägerl, C., Barnosky, A. D., Ellis, E. C. 2015. The Anthropocene biosphere. *Annual Review of Anthropology* 2: 196–219.

Willis, K. J., Gillison, L., Brncic, T. M. 2004. How 'virgin' is virgin rainforest? *Science* 304: 402–3.

Wilmsen, E. N., Denbow, J. R. 1990. Paradigmatic history of San-speaking peoples and current attempts at revision. *Current Anthropology* 31: 489–524.

Wilson, E. O. 1988. The current state of biological diversity. In E. O. Wilson (ed.), *Biodiversity*. Washington, DC: National Academy Press, pp. 3–18.

Wilson, H. B., Meijaard, E., Venter, O., Ancrenaz, M., Possingham, H. P. 2014. Conservation strategies for orangutans: Reintroduction versus habitat preservation and the benefits of sustainably logged forest. *PLoS ONE* 9: e102174.

Wilson, J. A. 1958. Egypt through the New Kingdom: Civilization without cities. In C. H. Kraeling, R. McC. Adams (eds.), *City Invincible*. Chicago: University of Chicago Press, pp. 124–36.

Wilson, S. M. 1985. Phytolith analysis at Kuk, an early agricultural site in Papua New Guinea. *Archaeology in Oceania* 20: 90–7.

Wilson, S. M. (ed.). 1999. *The Indigenous People of the Caribbean*. Florida Museum of Natural History: Ripley P. Bullen Series. Gainesville: University Press of Florida.

Wilson, S. M. 2007. *The Archaeology of the Caribbean*. Cambridge: Cambridge University Press.

Wing, S. L., Herrera, F., Jaramillo, C. A., Gómez-Navarro, C., Wilf, P., Labandeira, C. C. 2009. Late Paleocene fossils from the Cerrejón Formation, Colombia, are the earliest record of Neotropical rainforest. *Proceedings of the National Academy of Sciences of the United States of America* 106: 18627–32.

Winterhoff, E. 2007. The political economy of ancient Samoan basalt adze production and linkages to social status. PhD thesis, University of Oregon, Eugene.

Wittman, F., Junk, W. J., Piedade, M. T. F. 2004. The várzea forests in Amazonia: Flooding and the highly dynamic geomorphology interact with natural forest succession. *Forest Ecology and Management* 196: 199–212.

WoldeGabriel, G., White, T. D., Suwa, G., Renne, P., de Heinzelin, J., Hart, W. K., Heiken, G. 1994. Ecological and temporal placement of early Pliocene hominids at Aramis, Ethiopia. *Nature* 371: 330–3.

WoldeGabriel, G., Haile-Selassie, Y., Renne, P. R., Hart, W. K., Ambrose, S. H., Asfaw, B., Heiken, G., White, T. 2001. Geology and palaeontology of the late Miocene Middle Awash Valley, Afar Rift, Ethiopia. *Nature* 412: 175–8.

WoldeGabriel, G., Ambrose, S. H., Barboni, D., Bonnefille, R., Bremond, L., Currie, B., DeGusta, D., Hart, W. K., Murray, A. M., Renne, P. R., Jolly-Saad, M. C., Stewart, K. M., White, T. 2009. The geological, isotopic, botanical, invertebrate, and lower vertebrate surroundings of *Ardipithecus ramidus*. *Science* 326: 65–65e5.

Wolf, E. 1982. *Europe and the People Without History*. Berkeley: University of California Press.

Wolf, R. 2010. *The Paleo Solution: The Original Human Diet*. Las Vegas: Victor Belt Publishing.

Wolfe, L. M. 1938. *John of the Mountains: The Unpublished Journals of John Muir*. New York: A. A. Knopf.

Wolpoff, M. H., Spuhler, J. N., Smith, F. H., Radovcic, J., Pope, G., Frayer, D. W., Eckhardt, R., Clark, G. 1988. Modern human origins. *Science* 241: 772–4.

Wolpoff, M. H., Senut, B., Pickford, M., Hawks, J. 2002. Palaeoanthropology (communication arising): *Sahelanthropus* or 'Sahelpithecus'? *Nature* 419: 581–2.

Wolpoff, M. H., Hawks, J., Senut, B., Pickford, M., Ahern, J. 2006. An ape or the ape: Is the Toumaï Cranium TM 266 a hominid? *PaleoAnthropology* 2006: 36–50.

Wood, C. L., Lafferty, K. D., DeLeo, G., Young, H. S., Hudson, P. J., Kurs, A. M. 2014. Does biodiversity protect humans against infectious disease? *Ecology* 95: 817–32.

Wood, J. R., Wilmshurst, J. M., Worthy, T. H., Holzapfel, A. S., Cooper, A. 2012. A lost link between a flightless parrot and a parasitic plant and the potential role of coprolites in conservation paleobiology. *Conservation Biology* 26: 1091–9.

Wood, J. R., Perry, G. L. W., Wilmshurst, J. M. 2017. Using palaeoecology to determine baseline ecological requirements and interaction networks for de-extinction candidate species. *Functional Ecology* 31: 1012–20.

Woods, W. I., Teixeira, W. G., Lehmann, J., Steiner, C., WinklerPrins, A. M. G. A., Rebellato, L. 2009. *Amazonian Dark Earths: Wim Sombroek's Vision*. Dordrecht: Springer.

World Heritage Forests. 2011. *World Heritage Vol. 61*. UNESCO. <http://unesdoc.unesco.org/images/0021/002139/213912e.pdf>.

Wright, S. J. 2014. The El Niño southern oscillation influences tree performance in tropical rainforests. In E. Bermingham, C. W. Dick, C. Mortiz (eds.), *Tropical Rainforests: Past, Present and Future*. Chicago: University of Chicago Press, pp. 295–312.

Wurster, C. M., Bird, M. I., Bull, I. D., Creed, F., Bryant, C., Dungait, J. A. J., Paz, V. 2010. Forest contraction in north equatorial Southeast Asia during the Last Glacial Period. *Proceedings of the National Academy of Sciences of the United States of America* 107: 15508–11.

Wynn, J. G., Sponheimer, M., Kimbel, W. H., Alemseged, Z., Reed, K., Bedaso, Z. K., Wilson, J. N. 2013. Diet of *Australopithecus afarensis* from the Pliocene Hadar Formation, Ethiopia. *Proceedings of the National Academy of Sciences of the United States of America* 110: 10495–500.

Wynn, T. 1991. Tools, grammar, and the archaeology of cognition. *Cambridge Archaeological Journal* 1: 191–206.

Xhauflair, H., Pawlik, A., Forestier, H., Saos, T., Dizon, E., Gaillard, C. 2017. Use-related or contamination? Residue and use-wear mapping on stone tools used for experimental processing of pants from Southeast Asia. *Quaternary International* 427: 80–93.

Xiang, H., Gao, J., Yu, B., Zhou, H., Cai, D., Zhang, Y., Chen, X., Wang, X., Hofreiter, M., Zhao, X. 2014. Early Holocene chicken domestication in northern China. *Proceedings of the National Academy of Sciences of the United States of America* 111: 17564–9.

Xue, F., Sit, V. F. S. 2010. *Chinese City and Urbanism: Evolution and Development*. London: World Scientific.

Xue, X.-X., Delson, E. 1989. A new species of *Dryopithecus* from Gansu, China. *Chinese Science Bulletin* 34: 223–30.

Yanagisawa, H., Koshimura, S., Miyagi, T., Imamura, F. 2010. Tsunami damage reduction performance of a mangrove forest in Banda Aceh, Indonesia inferred from field data and a numerical model. *Journal of Geophysical Research: Oceans* 115: C06032.

Yang, Y. D., Liu, L., Chen, X., Speller, C. F. 2008. Wild or domesticated: DNA analysis of ancient water buffalo remains from north China. *Journal of Archaeological Science* 35: 2778–85.

Yasuoka, H. 2009. Concentrated distribution of wild yam patches: Historical ecology and the subsistence of African rainforest hunter-gatherers. *Human Ecology* 37: 577–87.

Yen, D. 1985. Wild plants and domestication in Pacific islands. In V. N. Misra, P. Bellwood (eds.), *Recent Advances in Indo-Pacific Prehistory*. New Delhi: Oxford University Press and IBH, pp. 315–26.

Yen, D. 1991. Domestication: The lessons from New Guinea. In A. Pawley (ed.), *Man and a Half: Essays in Pacific Anthropology and Ethnobiology in Honour of Ralph Bulmer*. Auckland: Polynesian Society, pp. 558–69.

Yerkes, R. W., Galili, E., Barkai, R. 2014. Activities at final Pre-Pottery Neolithic (PPNC) fishing village revealed through microwear analysis of bifacial flint tools from the submerged Atlit-Yam site, Israel. *Journal of Archaeological Science* 48: 120–8.

Yost, J. A., Kelly, P. M. 1983. Shotguns, blowguns, and spears: The analysis of technological efficiency. In R. B. Hames, W. T. Vickers (eds.), *Adaptive Responses of Native Amazonians*. New York: Academic Press, pp. 189–224.

Yule, C. M. 2010. Loss of biodiversity and ecosystem functioning in Indo-Malayan peat swamp forests. *Biodiversity Conservation* 19: 393–409.

Zachos, Z., Pagani, M., Sloan, L., Thomas, E., Billups, K. 2001. Trends, rhythms, and aberrations in global climate 65 Ma to Present. *Science* 292: 686–93.

Zagarell, A. 2002. Gender and social organization in the reliefs of the Nilgiri Hills. In K. Morrison, L. Junker (eds.), *Forager-Traders in South and Southeast Asia*. Cambridge: Cambridge University Press, pp. 77–104.

Zahawi, R. A., Holl, K. D. 2009. Comparing the performance of tree stakes and seedlings to restore abandoned tropical pastures. *Restoration Ecology* 17: 854–64.

Zaim, Y., Ciochon, R. L., Polanski, J. M., Grine, F. E., Bettis III, E. A., Rizal, Y., Franciscus, R. G., Larick, R. R., Heizler, M., Aswan, Eaves, L., Marsh, H. E. 2011. New 1.5 million-year-old *Homo erectus* maxilla from Sangiran (Central Java, Indonesia). *Journal of Human Evolution* 61: 363–76.

Zakharov, A. O. 2012. Epigraphy, political history and colective action in ancient Java. In M. L. Tjoa-Bonatz, A. Reinecke, D. Bonatz (eds.), *Connecting Empires and States: Selected Papers from the 13th International Conference of the European Association of Southeast Asian Archaeologists*. Singapore: NUS Press, pp. 82–90.

Zalasiewicz, J., Williams, M., Smith, A., Barry, T. L., Coe, A. L., Bown, P. R., Brenchley, P., Cantrill, D., Gale, A., Gibbard, P., Gregory, F. J., Hounslow, M. W., Kerr, A. C., Pearson, P., Knox, R., Powell, J., Waters, C., Marshall, J., Oates, M., Rawson, P., Stone, P. 2008. Are we now living in the Anthropocene? *GSA Today* 18: 4–8.

Zalasiewicz, J., Waters, C. N., Williams, M., Barnosky, A. D., Cearreta, A., Crutzen, P., Ellis, E., Ellis, M. A., Fairchild, I. J., Grinevald, J., Haff, P. K., Hajdas, I., Leinfelder, R., McNeill, J., Odada, E. O., Poirier, C., Richter, D., Steffen, W., Summerhayes, C., Syvitski, J. P. M., Vidas, D., Wagreich, M., Wing, S. L., Wolfe, A. P., An, Z., Oreskes, N. 2015. When did the Anthropocene begin? A mid-twentieth century boundary level is stratigraphically optimal. *Quaternary International* 383: 196–203.

Zarin, D. J., Schulze, M. D., Vidal, E. J., Lentini, M. 2007. Beyond reaping the first harvest: Management objectives for timber production in the Brazilian Amazon. *Conservation Biology* 21: 916–25.

Zavaleta, E. 1999. The emergence of waterfowl conservation among Yup'ik hunters in the Yukon-Kuskokwim Delta, Alaska. *Human Ecology* 27: 231–66.

Zegura, S. L., Karafet, T. M., Zhivotovsky, L. A., Hammer, M. F. 2004. High-resolution SNPs and microsatellite haplotypes point to a single, recent entry of Native American Y chromosomes into the Americas. *Molecular Biology and Evolution* 21: 164–75.

Zeppel, H. 2006. *Indigenous Ecotourism: Sustainable Development and Management*. Wallingford: Ecotourism Book Series.

Zhouri, A. 2010. 'Adverse forces' in the Brazilian Amazon: Developmentalism versus environmentalism and Indigenous rights. *Journal of Environment & Development* 19: 252–73.

Index

Note: Figures are indicated by *f* after the page number.